PROTEIN PURIFICATION

PROTEIN PURIFICATION

Principles, High-Resolution Methods, and Applications

SECOND EDITION

Edited by

Jan-Christer Janson
Biochemical Separation Centre
Uppsala University
Uppsala Biomedical Centre
S-751 23 Uppsala
Sweden

Lars Rydén
Department of Biochemistry
Uppsala University
Uppsala Biomedical Centre
S-751 23 Uppsala
Sweden

A JOHN WILEY & SONS, INC., PUBLICATION

New York • Chichester • Weinheim • Brisbane • Singapore • Toronto

This book is printed on acid-free paper. ⊗

Copyright © 1998 by John Wiley & Sons, Inc. All rights reserved.

Published simultaneously in Canada.

Library of Congress Cataloging-in-Publication Data:

Protein purification : principles, high resolution methods, and
    applications / edited by Jan-Christer Janson and Lars Rydén.—2nd
ed.
        p.   cm.
    Includes bibliographical references and index.
    ISBN 0-471-18626-0 (cloth : alk. paper)
    1. Proteins—Purification.   2. Chromatographic analysis.
3. Electrophoresis.   I. Janson, Jan-Christer.   II. Rydén, Lars.
QP551.P69754   1997
572'.6—dc21   97–13875

Printed in the United States of America.

10 9 8 7 6 5 4 3 2 1

CONTENTS

PREFACE

Since 1989, when the first edition of this book was launched, the development of biosciences has meant a revival of protein chemistry in the wake of the molecular biology revolution and the HUGO project. The total genome of baker's yeast is now sequenced, that of *E. coli* is not far behind, and within a not too distant future the feat of the total mapping of the human genome, which at the beginning seemed fictitious, is now within reach. This means that the attention of the world's bioscientific community will again, as in the 1960s and most of the 1970s, focus on the structure and function of the proteins. The PROTEOME era has thus begun, and with it follows the need of more efficient and more selective tools for the separation, isolation, and purification of the gene products, the proteins.

The development of new chromatographic separation media since 1989 has mainly been focused toward improvements demanded primarily by process development engineers in the biopharmaceutical industry. This has resulted in media with higher efficiencies, leading to shorter cycle times, primarily based on suspension polymerized styrene-divinylbenzene polymers with optimized internal pore size distributions, some allowing partial convective flow through the particles. This trend has received its ultimate solution in totally perfusive systems based on stacked membranes, or continuous "monolithic" columns made of cross-linked polymers, derivatized with various kinds of protein ad-sorptive groups. New composite media have been introduced primarily to increase the industrial applicability of size exclusion chromatography of pro-teins but also to increase binding capacity in, for example, ion exchange chromatography. The concept of "solid diffusion" in highly ionic group substi-tuted composite media is still awaiting its physicochemical explanation.

The demand for systems allowing direct capture of target proteins directly from whole cultures or cell homogenates, resulting in fewer process steps and concomitantly higher yields, has led to a revival of the fluidized bed concept. However, now optimized with regard to the design of both media and columns by the introduction of the more efficient one cycle technique called expanded bed adsorption.

As long as scientists have been engaged in the isolation and purification of proteins from crude extracts, there has been a demand for media with higher adsorptive selectivities. The extremely high variability in protein surface structure as well as their wide range of functional stabilities, makes it necessary for every protein chemist to have a stock of several different separation media, ion exchangers, hydrophobic interaction media, and a variety of general affinity media. Literature survey data presented in some of the chapters of this book reveal that on average somewhere between three and four steps are required to purify a protein to homogeneity. The hope for one-step purifications raised by the introduction of immobilized monoclonal antibodies has not yet been fulfilled. However, there is a renewed opportunity at hand to increase the selectivity of immobilized ligands in affinity chromatography and thus decrease the number of steps in the purification process. This opportunity has been raised by the recent rapid development in the design of a large variety of chemical and biological combinatorial libraries and high-speed screening technologies. It is easy to predict that over the next few years there will be an unprecedented number of new highly selective ligands, monospecific as well as group specific, introduced for the synthesis of new protein separation media.

Compared to the first edition of this book, there exists one additional chapter (Chapter 18) on large-scale electrophoretic processes. Three chapters (Chapters 15, 16, and 17) have been totally rewritten, Chapters 15 and 16 by new authors. Most other chapters have been thoroughly revised, and all have been updated regarding recent applications.

It is our hope that this new edition will receive the same overwhelmingly positive response as the first edition, and we would like to express our appreciation to Dr. Edmund H. Immergut and the staff of VCH Publishers, now John Wiley & Sons, Inc., for their patience and never-failing support of this project.

JAN-CHRISTER JANSON
LARS RYDÉN

Uppsala, Sweden

CONTRIBUTORS

FRANCISCO BATISTA-VIERA, C. de Bioquimica, F. de Quimica e, Instituto de Quimica, F. de Ciencias, Gral.Flores 2124. Casilla de Correo 1157, Montevideo, Uruguay

JOHN BREWER, BioLab Division, Pharmacia Biotech AB, S-751 82 Uppsala, Sweden

JAN CARLSSON, Biochemical Separation Centre, Uppsala University, Uppsala Biomedical Centre, Box 577, S-751 23 Uppsala, Sweden

KJELL-OVE ERIKSSON, J.T. Baker Inc., 222 Red School Lane, Phillipsburg, NJ 08865, USA

BO ERSSON, Biochemical Separation Centre, Uppsala University, Uppsala Biomedical Centre, Box 577, S-751 23 Uppsala, Sweden

ANGELIKA GÖRG, Lehrstuhl für Allgemeine Lebensmitteltechnologie, Technische Universität München, D-85350 Freising-Weihenstephan, Germany

LARS HAGEL, R&D Department, Pharmacia Biotech AB, S-751 82 Uppsala, Sweden

MILTON T.W. HEARN, Centre for Bioprocess Technology, Monash University, Clayton, Victoria, Australia

T. WILLIAM HUTCHENS, Department of Food Science and Technology, University of California, Davis, CA 95616–8598, USA

JAN-CHRISTER JANSON, Biochemical Separation Centre, Uppsala University, Uppsala Biomedical Centre, Box 577, S-751 23 Uppsala, Sweden

GÖTE JOHANSSON, Department of Biochemistry, University of Lund, Chemical Centre, Box 124, S-221 00 Lund, Sweden

JAN-ÅKE JÖNSSON, Department of Analytical Chemistry, University of Lund, Chemical Centre, Box 124, S-221 00 Lund, Sweden

LENNART KÅGEDAL, R&D Department, Pharmacia Biotech AB, S-751 82 Uppsala, Sweden

EVERT KARLSSON, Department of Biochemistry, Uppsala University, Uppsala Biomedical Centre, Box 576, S-751 23 Uppsala, Sweden

TORGNY LÅÅS, BioLab Division, Pharmacia Biotech AB, S-751 82 Uppsala, Sweden

JORGE A. LIZANA, BioLab Division, Pharmacia Biotech AB, S-751 82 Uppsala, Sweden

LARS RYDÉN, Department of Biochemistry, Uppsala University, Uppsala Biomedical Centre, Box 576, S-751 23 Uppsala, Sweden

MARIANNE SPARRMAN, BioIndustry Division, Pharmacia Biotech AB, S-751 82 Uppsala, Sweden

WOLFGANG THORMANN, Department of Clinical Pharmacology, University of Berne, Murtenstrasse 35, CH-3010 Berne, Switzerland

REINER WESTERMEIER, Pharmacia Biotech Europe GmbH, Münzinger Strasse 9, D-7800 Freiburg, Germany

PROTEIN PURIFICATION

PART I
Introduction

1 Introduction to Protein Purification

BO ERSSON
Biochemical Separation Centre
Uppsala University Biomedical Centre
Box 577, S-751 23 Uppsala, Sweden

LARS RYDÉN
Department of Biochemistry
Uppsala University Biomedical Centre
Box 576, S-751 23 Uppsala, Sweden

JAN-CHRISTER JANSON
Biochemical Separation Centre
Uppsala University Biomedical Centre
Box 577, S-751 23 Uppsala, Sweden

Protein Purification: Principles, High-Resolution Methods, and Applications, Second Edition.
Edited by Jan-Christer Janson and Lars Rydén.
ISBN 0-471-18626-0.   © 1998 Wiley-VCH, Inc.

1.1   INTRODUCTION

The development of techniques and methods for the separation and purification of biological macromolecules such as proteins has been an important prerequisite for many of the advancements made in bioscience and biotechnology over the past three decades. Improvements in materials and utilization of microprocessor-based instruments have made protein separations more predictable and controllable, although to many they are still more an art than a science. However, gone are the days when an investigator had to spend months in search of an efficient route to purify an enzyme or hormone from a cell extract. This is a consequence of the development of new generations of chromatographic media with increased efficiency and selectivity as well as of new automated chromatographic systems supplied with sophisticated interactive software packages and data bases. Also, new electrophoretic techniques and systems for fast analysis of protein composition and purity have contributed to increased efficiency in the evaluation phase of the purification process.

In the area of chromatography, the development of new porous resin supports, new cross-linked beaded agaroses, and new bonded porous silicas has enabled a rapid growth of high-resolution techniques [high-performance liquid chromatography (HPLC) and fast protein liquid chromatography (FPLC)] on an analytical and laboratory preparative scale as well as of industrial chromatography in columns with bed volumes of several hundred liters. The introduction of expanded bed adsorption has made it possible to rapidly isolate target proteins from whole-cell cultures or cell homogenates. Another field of increasing importance is micropreparative chromatography, a consequence of modern methods for amino acid and sequence analysis requiring submicrogram samples only. Data obtained are efficiently exploited by recombinant deoxyribonucleic acid (DNA) technology, and biological activities previously not amenable to proper biochemical study can now be ascribed to identifiable proteins and peptides.

A wide variety of chromatographic column packing materials such as gel-filtration media, ion exchangers, reversed-phase packings, hydrophobic interaction adsorbents, and affinity chromatography adsorbents are today commercially available. These are based on low pressure media (90- to 100-μm beads), medium pressure media (30–50 μm), and high pressure media (5–10 μm) to satisfy different requirements of efficiency, capacity, and cost.

However, not all problems in protein purification are solved by the acquisition of sophisticated laboratory equipment and column packings that give high selectivity and efficiency. Difficulties still remain in finding optimum conditions for protein extraction and sample pretreatment, as well as in choosing suitable methods for monitoring protein concentration and biological activity. These problems are discussed in this introductory chapter. There is also an overview of different protein separation techniques and their principles of operation. In the subsequent chapters, each individual technique is discussed in more detail. Finally, some basic equipment necessary for efficient protein purification work is described in this chapter.

Several useful books covering protein separation and purification from different points of view are available on the market or in libraries.[1-4] In *Methods in Enzymology,* especially Volumes 22, 34, 104, and 182,[5-8] a number of useful reviews and detailed application reports can be found. Also the booklets available from manufacturers of separation equipment and media can be helpful by providing detailed information regarding their products.

1.2 THE PROTEIN EXTRACT

1.2.1 Choice of Raw Material

In most cases, interest is focused on one particular biological activity, such as an enzyme, and the origin of this activity is often of little importance. Great care should then be taken in the selection of a suitable source for the enzyme. Among different sources there might be considerable variation with respect to the concentration of the enzyme, the availability and cost of the raw material, the stability of the enzyme, the presence of interfering activities and proteins, and difficulties in handling a particular raw material. Often it is compelling to choose a particular source because it has been described previously in the literature; however, sometimes it is advantageous to consider an alternative choice. One of the authors of this chapter had difficulty in obtaining guinea pig brains for the preparation of histamine *N*-methyltransferase. It turned out to be quite easy to prepare the same enzyme from pig kidney as an inexpensive and easily available source. In an attempt to study the subunit structure of human ceruloplasmin, a major difficulty was its sensitivity to proteolytic degradation. Here it was demonstrated that porcine ceruloplasmin is quite similar, is present in higher concentrations, and is not nearly as sensitive to proteases.

The traditional animal or microbial sources may today be replaced by genetically engineered microorganisms or cultured eukaryotic cells. Protein products of eukaryotic orgin, cloned and expressed in bacteria such as *Escherichia coli,* may either be located in the cytoplasm or be secreted through the cell membrane. In the latter case, they are either collected inside the periplasmic space or they are truly extracellular, secreted to the culture medium. Proteins that accumulate inside the periplasmic space may be selectively released either into the growth medium by changing the growth conditions,[9] or following cell harvesting and washing of resuspensed cell paste. Already at this stage, a considerable degree of purification is thus achieved by choosing a secreting strain as illustrated in Figure 1-1.

1.2.2 Extraction Methods

Some biological materials constitute themselves a clear or nearly clear protein solution suitable for direct application to chromatography columns after centrifugation or filtration. Some examples are blood serum, urine, milk, snake

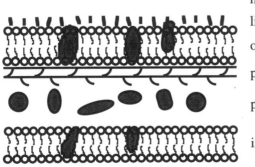

medium (<10 proteins)

lipopolysaccharides

outer membrane

peptidoglycan

periplasm (~100 proteins)

inner membrane

cytoplasm (~2000 proteins)

FIGURE 1-1. Location and approximate numbers of the proteins in *Escherichia coli.* (Courtesy of S. Ståhl, Royal Institute of Technology, Stockholm, Sweden.)

venoms, and—perhaps, most important—the extracellular medium after cultivation of microorganisms and mammalian cells as mentioned previously. It is normally an advantage to choose such a starting material because of the limited number of components and since extracellular proteins are comparatively stable. Some samples (e.g., urine or cell culture supernatants) are normally concentrated before purification begins.

In most cases, however, one has to extract the desired activity from a tissue or a cell paste. This means that a considerable number of contaminating molecular species are set free and that proteolytic activity will make the preparation work more difficult. The extraction of a particular protein from a solid source often involves a compromise between recovery and purity. Optimization of extraction conditions should favor the release of the desired protein and leave difficult-to-remove contaminants behind. Of particular concern is to find conditions under which the already extracted protein is not degraded or denatured while more is being released.

Various methods are available for the homogenization of cells or tissues. For further details and discussions, the reader is referred to the article by Kula and Schutte.[10] The extraction conditions are optimized by systematic variation of parameters such as the composition of the extraction medium (see below), time, temperature, and type of equipment used.

The proper design of an extraction method thus requires preliminary experiments in which aliquots are taken at various time intervals and analyzed for activity and protein content. Because the number of parameters can be very large, this part of the work has to be kept within limits by proper judgment. It is, however, not recommended to accept a single successful experiment. Further investigations of, in particular, the extraction time required often pay in the long run.

Major problems in preparing a protein are in general denaturation, proteolysis, and contamination with pyrogens, nucleic acids, bacteria and viruses. These can be limited by the proper choice of extraction medium as this chapter shows. However, many of the previous problems can be reduced by short preparation times and low temperature. It is thus good biochemical practice to carry out the first preparation steps as fast as possible and at the lowest possible temperature. However, low temperatures are not always necessary and are sometimes inconvenient. The working temperature is therefore one of the parameters that should be optimized carefully, especially if a preparation is to be done routinely in the laboratory.

The extract must be clarified by centrifugation before submission to column chromatography. A preparative laboratory centrifuge is normally sufficient for this step.

A common phenomenon when working with intracellularly expressed recombinant DNA proteins is their tendency to accumulate as insoluble aggregates (i.e., inclusion bodies, refractile bodies), which have to be solubilized and refolded to recover their native state. At first glance, the formation of insoluble aggregates in the cytoplasm might be considered a major problem.

However, as the inclusion bodies seem to be fairly well defined with regard to both particle size and density,[11] they should provide a unique means for rapid and efficient enrichment of the desired protein simply by providing low-speed fractional centrifugation and washing of resuspended sediment. The critical step is solubilization and refolding, often combined with chromatographic purification under denaturing conditions in the presence of high concentrations of urea or guanidine hydrochloride. This area has been reviewed by Marston.[12]

1.2.3 Extraction Medium

To arrive at a suitable composition for the extraction medium, the conditions in which the protein of interest is stable must first be studied. Second, the conditions at which the protein is most efficiently released from the cells or tissue needs to be considered. The final choice is usually a compromise between maximum recovery and maximum purity. The following factors have to be taken into consideration:

1. pH. Normally the pH value chosen is that of maximum activity of the protein. However, it should be noted that this is not always the pH that gives the most efficient extraction or is it necessarily the pH of maximum stability. For example, trypsin has an activity optimum at pH 8 to 9 but is much more stable at pH 3, where autolysis is avoided. The use of extreme pH values, (e.g., for the extraction of yeast enzymes in 0.5 *M* ammonia) is sometimes very efficient and is acceptable for some proteins without causing excessive denaturation.

2. Buffer salts. Most proteins are maximally soluble at moderate ionic strengths, 0.05 to 0.1, and these values are chosen if the buffer capacity is sufficient. Suitable buffer salts are given in Table 1-1. An acceptable buffer capacity is obtained within one pH unit from the pK_a values given. The proteins

TABLE 1-1. Buffer Salts Used in Protein Work

Buffer	pK_a-values	Properties
Sodium acetate	4.75	
Sodium bicarbonate	6.50; 10.25	
Sodium citrate	3.09; 4.75; 5.41	Binds Ca^{2+}
Ammonium acetate	4.75; 9.25	Volatile
Ammonium bicarbonate	6.50; 9.25; 10.25	Volatile
Tris-chloride	8.21	
Sodium phosphate	1.5; 7.5; 12.0	
Tris-phosphate	7.5; 8.21	

Buffer concentration refers to total concentration of buffering species. Buffer pH should be as close as possible to the pK_a value and not more than one pH unit from the pK_a.

as such also act as buffers, and the pH should be checked after the addition of large amounts of proteins to a weakly buffered solution. Some extractions do not give rise to acids and bases and thus do not need a high buffer capacity. In other cases, this might be necessary and occasional control of the pH value of an extract is recommended.

3. Detergents, chaotropic agents. In many extractions, the desired protein is bound to membranes or particles or is aggregated due to its hydrophobic character. In these cases, the hydrophobic interactions should be reduced by using either detergents or chaotropic agents (not both). Some of the commonly used detergents are listed in Table 1-2. Several of them do not denature globular proteins or interfere with their biological activity. Others, such as sodium dodecyl sulfate (SDS), will do that. Quite often it is not necessary to continue using a detergent in the buffer after the first step(s) in the purification, so its use is restricted to the extraction medium. In other cases, it might be necessary to use a detergent at all times. One is then actually purifying a protein–detergent complex. More information about detergents, including their chemical structures, can be found in Ref. 13.

Detergents are amphipatic molecules. When their concentration increases, they will eventually form micelles at the so-called critical micell concentration (CMC). Since micelles often complicate purification procedures, particularly column chromatography, concentrations below the CMC should be used.

Instead of detergents for dissolving aggregates, chaotropic agents such as urea or guanidine hydrochloride, or moderately hydrophobic organic compounds, such as ethylene glycol, could be used. Urea and guanidine hydrochloride have proven particularly useful for the extraction and solubilization of inclusion bodies.[12]

4. Reducing agents. The redox potential of the cytosol is lower than that of the surrounding medium where atmospheric oxygen is present. Intracellular

TABLE 1-2. Detergents Used for Solubilization of Proteins

Detergent	Ionic Character	Effect on Protein	Critical Micelle Concentration (% w/v)
Triton X-100	Nonionic	Mild, nondenaturing	0.02
Nonidet P-40	Nonionic		
Lubrol PX	Nonionic		0.006
Octyl glucoside	Nonionic		0.73
Tween 80	Nonionic		0.002
Sodium deoxycholate	Anionic		0.21
Sodium dodecyl sulfate	Anionic	Strongly denaturing	0.23
3-[(3-Cholamidopropyl) dimethyl amino] propanesulfonic acid	Zwitterionic		1.4

proteins often have exposed thiol groups, and these might become oxidized in the purification process. Thiol groups can be protected by reducing agents such as 1,4-dithioerythritol (DTE), 1,4-dithiothreitol (DTT), or mercaptoethanol (see Table 1-3). Normally 10 to 25 mM concentrations are sufficient to protect thiols without reducing internal disulfides. In other cases, a higher concentration might be needed.[14]

5. Chelators or metal ions. The presence of heavy metal ions can be detrimental to a biologically active protein mainly for two reasons. They can enhance the oxidation of thiols by molecular oxygen and they can form complexes with specific groups, which may cause problems. Heavy metals can be trapped by chelating agents. The most commonly used is ethylenediamine tetraacetic acid (EDTA) in the concentration range 10 to 25 mM. An alternative is ethyleneglycol-O,O'-bis(2-aminoethyl)-N,N,N',N' tetraacetic acid (EGTA), which is more specific for calcium. It should be noted that EDTA is a buffer. It is therefore best to add the disodium salt of EDTA before final pH adjustment. The chelating capacity of EDTA increases with increasing pH.

In other cases, stabilizing metal ions are needed. Many proteins are stabilized by calcium ions. The divalent ions calcium and magnesium are trapped by EDTA and cannot be used in combination with this chelator.

6. Proteolytic inhibitors. The most serious threats to protein stability are the omnipresent proteases. The simplest safeguard against proteolytic degradation is normally to work fast in the cold. An alternative or added precaution is to add protease inhibitors (Table 1-4), especially in connection with the extraction step. Often there is a need for a combination of inhibitors (e.g., for both serine proteases and metalloproteases). In general, protein inhibitors are expensive chemicals that may limit their use in large-scale applications. Proteolysis can also be reduced by rapid extraction of the fresh homogenate in an aqueous polymer two-phase system[15] or by adsorption of the proteases to hydrophobic interaction adsorbents.[16] Sometimes it is sufficient to adjust

TABLE 1-3. Reducing Agents

Agent	Structure
Mercaptoethanol	HS—CH$_2$—CH$_2$—OH
1,4–Dithioerythritol	CH$_2$SH H——OH H——OH CH$_2$SH
1,4–Dithiotreitol	CH$_2$SH H——OH HO——H CH$_2$SH

TABLE 1-4. Proteolytic Inhibitors

Inhibitor	Enzymes	Working Concentration
Diisopropyl fluorophosphate (DFP)	Serine proteases	Avoid DFP
Phenylmethylsulfonyl fluoride	Serine proteases	0.5–1 mM
EDTA	Metal-activated proteases	Around 5 mM
Cysteine reagents	Cysteine-dependent proteases	
Pepstatin A	Acid proteases	1 μM
Leupeptin	Serine proteases	1 μM

the pH to a value at which the proteases are inactive, but where the stability of the protein to be purified is maintained. The classic example is the purification of insulin from pancreas.

7. Bacteriostatics. It is wise to take precautions to avoid bacterial growth in the protein solutions. As mentioned in the previous paragraph on proteolytic inhibitors, the simplest remedy here is to use sterile-filtered buffer solutions as a routine in the laboratory. This will also reduce the risk of bacterial growth in columns. A common practice to avoid bacterial growth in chromatographic columns is to allow the column to flow at a reduced rate even when it is not in operation. Some buffers are more likely than others to support bacterial growth, (e.g., phosphate, acetate, and carbonate buffers at neutral pH values). Buffers at pH 3 and below or 9 and above usually prevent bacterial growth, but may occasionally allow growth of molds.

Whenever possible, it is recommended to add an antimicrobial agent to the buffer solutions. Often used are azide at 0.001 M or merthiolate at 0.005% and alcohols such as *n*-butanol at 1%. Azide has the drawback that it is a nucleophilic substance and binds metals. In cases where these substances may interfere with activity measurements or the chromatography itself, it is always possible to add the substances to solutions of the protein to be stored.

1.3 OVERVIEW OF FRACTIONATION TECHNIQUES

In early work, complex protein mixtures were fractionated mainly by extraction and precipitation methods. These methods are still used as preliminary steps for crude fractionation or to obtain more clear or more concentrated sample solutions. Preparative electrophoretic and chromatographic techniques developed during the 1950s and 1960s, made rational purification protocols possible and laid the foundation for the current situation. This section gives a short overview of the various techniques normally used in preparative biochemical work. Chapter 2 contains an introduction to chromatography, and

Chapter 12 gives an introduction to electrophoresis. Each individual chromatographic and electrophoretic separation technique is then treated in detail in subsequent chapters.

1.3.1   Precipitation, Extraction, and Centrifugation

Precipitation of a protein in an extract may be achieved by adding salts, organic solvents, or organic polymers, or by varying the pH or temperature of the solution. The most commonly used precipitation agents are listed in Table 1-5. The property of a particular salt as a precipitation agent is described by the so-called Hofmeister series.

Anions: PO_4^{3-}, SO_4^{2-}, CH_3COO^-, Cl^-, Br^-, NO_3^-, ClO_3^-, I^-, SCN^-

Cations: NH_4^+, K^+, Na^+, guanidine $C(NH_2)_3^+$

The *antichaotropic* salts on the left-hand side are the most efficient salting out agents. They increase the hydrophobic effect in the solution and promote protein aggregation by association of hydrophobic surfaces. The chaotropic salts on the right-hand side in the series decrease the hydrophobic effect, thus helping maintain the proteins in solution.

Organic solvents promote the precipitation of proteins due to the decrease in water activity in the solution as the water is replaced by organic solvent. They have been widely used as precipitation agents, especially in the fractionation of serum proteins. The following five variables are usually kept under control: concentration of organic solvent, protein concentration, pH, ionic strength, and temperature.[17] Low temperature during the precipitation operations is often necessary to avoid protein denaturation: The addition of an organic solvent decreases the freezing point of the solution and temperatures below 0°C can be used. In reversed-phase chromatography, proteins are chromatographed in solutions that contain up to about 50% organic solvent, often with retention of biological activity.

Organic polymers function in a way similar to that of organic solvents. The most widely used polymer is polyethylene glycol (PEG), with an average molecular weight of either 6,000 or 20,000. The main advantage of PEG over

TABLE 1-5. Precipitation Agents

Agent	Type	Properties
Ammonium sulfate	Salt	Easily soluble, stabilizing
Sodium sulfate	Salt	
Ethanol	Solvent	Flammable, risk of denaturation
Acetone	Solvent	Flammable, risk of denaturation
Polyethylene glycol	Polymer	Uncharged, unflammable

organic solvents is that it is more easily handled. It is inflammable, it is not poisonous, it is uncharged, and it is inexpensive. Rather low concentrations are required (often less than 25%) to precipitate most proteins. One disadvantage is that concentrated solutions of PEG are highly viscous. Polyethylene glycol can also be difficult to remove from protein solutions. However, after dilution with buffer the viscosity decreases, and since the substance is uncharged, the solution may be applied directly to an ion-exchange column to further separate the proteins and to remove the polymer.

pH adjustment has been used as a simple and inexpensive way to precipitate proteins. Proteins have their lowest solubility at their isoelectric point. This is sometimes used in serum fractionation and also in the purification of insulin.

In addition to pH, another parameter that influences precipitation of proteins in salt solutions is temperature (see as follows). Keeping the salt concentration constant and varying the temperature is another way of fractionating a protein solution.

The salting out of a protein can be described by the equation

$$\log S = B - K \cdot c$$

where S = the solubility of the protein in g/L of solution
B = intercept constant
K = salting out constant
c = salt concentration in mol/L.

The value of B depends on the salt used, the pH, the temperature, and the protein itself. K depends on the salt used and the protein. It should be stressed that the addition of a salt or another precipitating agent to a protein solution only decreases the solubility of the proteins. This is why a very dilute protein solution for precipitation may lead to low recovery, as a major part of the protein simply remains in solution. Reproducible results can only be achieved if all parameters mentioned previously, including the protein concentration, are kept constant.

Centrifugation is used routinely in the protein purification laboratory to recover precipitates. It can also be used to separate two immiscible liquid phases. Another application is density gradient centrifugation. Today this is used mostly for the fractionation of subcellular particles and nucleic acids. An alternative is the use of liquid–liquid phase extraction, which seems to offer several advantages over the more classical methods.

1.3.2 Electrophoresis

Electrophoresis in free solution or in macroporous gels such as 1 to 2% agarose separates proteins mainly according to their net electric charge. Electrophoresis in gels such as polyacrylamide separates mainly according to the molecular size of the proteins.

Analytical gel electrophoresis requiring microgram amounts of proteins is an important tool in bioscience and biotechnology (see Chapter 12). Convenient methods for the extraction of proteins after electrophoresis have been developed, especially protein blotting (see Chapter 16), making the technique micropreparative. There are also many instances where a very small amount of protein is sufficient for the analysis of size and composition, as well as the primary structure. Finally, there are cases where the starting material is extremely limited, for example, protein extracts from small amounts of tissue (biopsies, etc.). In these cases the protein "extract" might be just large enough for gel electrophoretic analysis.

Larger scale (milligrams to gram of protein) electrophoresis was an important method for the fractionation of protein extracts during the 1950s and early 1960s. It was carried out using columns packed with, for example, cellulose powder as the convection depressor as in the " Porath column."[18] An innate limitation of preparative column electrophoresis is the joule heat developed during the course of the experiment. This means that the column diameter, to allow sufficient cooling, should not exceed approximately 3 cm. Several hundred milligrams of protein can, however, be separated on such columns. Column zone electrophoresis has the advantage of allowing a precise discription of the separation parameters involved and is, in addition to gel filtration, the mildest separation technique available for proteins. It can be recommended for special situations, but practical aspects and the excessive time required preclude routine use. Methods for large- and medium-scale preparative electrophoresis have been developed (e.g., the flowing curtain electrophoresis of Hannig[19] and, more recently, the "Biostream" apparatus of Thomson[20]). For details, see Chapter 18.

Isoelectric focusing, the other main electrophoretic technique, separates proteins according to their isoelectric points. This technique gives very high resolution, but presents major difficulties as a preparative large- or medium-scale technique. Special equipment is required to allow cooling during the focusing. Proteins often precipitate at their isoelectric point, and this precipitate can contaminate the other bands when a vertical Sephadex bed or column with sucrose gradient is used as an anticonvection medium in the focusing experiment. Modern equipment for preparative isoelectric focusing[21] avoids these problems by dividing the separation chamber into smaller compartments. Another solution is to carry out the isoelectric focusing in a horizontal trough of sedimented gel particles such as Sephadex.[22] Here, precipitation in one zone will not disturb the other bands; however, recovery of proteins is more tedious.

For routine preparative protein fractionation, the electrophoretic techniques have become less important than chromatography. Ion-exchange chromatography depends on parameters similar to those for electrophoresis. Chromatofocusing fractionates proteins largely according to their isoelectric points and would therefore appear to be a more convenient alternative to preparative isoelectric focusing. However, recent developments have led to new preparative electrophoretic techniques and new types of intrumentation, as discussed in Chapter 18.

1.3.3 Chromatography

Separation by chromatography depends on the differential partition of proteins between a stationary phase (the chromatographic medium or the adsorbent) and a mobile phase (the buffer solution). Normally the stationary phase is packed into a vertical column of plastic, glass, or stainless steel, whereas the buffer is pumped through this column. An alternative is to stir the protein solution with the adsorbent batchwise and then pour the slurry onto an appropriate filter and make the washings and desorptions on the filter.

Column chromatography has proved to be an extremely efficient technique for the separation of proteins in biological extracts. Since the development of the first cellulose ion exchangers by Peterson and Sober[23] and of the first practical gel filtration media by Porath and Flodin,[24,25] a wide variety of adsorbents have been introduced that exploit various properties of the protein for the fractionation. The more important of these are listed below, together with the chromatographic method where they dominate the separation:

1. Size and shape—gel filtration
2. Net charge and distribution of charged groups—ion-exchange chromatography
3. Isoelectric point—chromatofocusing
4. Hydrophobicity—hydrophobic interaction chromatography and reversed-phase chromatography
5. Metal binding—immoblized metal ion affinity chromatography
6. Content of exposed thiol groups—covalent chromatography
7. Biospecific affinities for ligands, inhibitors, receptors, antibodies, etc.—affinity chromatography

The methods often have very different requirements with regard to chromatographic conditions. This applies to ionic strength, pH, and various additives such as detergents, reducing agents, and metals. By proper adjustment of the buffer composition, the conditions for adsorption and desorption of the desired protein can be optimized. It should be stressed that the result of a particular chromatographic separation often depends on more than one parameter. In ion-exchange chromatography, the charge interaction is the dominant parameter, but molecular weight and hydrophobic effects can also contribute to some degree, depending on the experimental conditions.

Highly specific methods, such as those based on bioaffinity, (e.g., antibody–antigen interaction), do in some cases give a highly pure protein in a single step. Normally, however, one has to combine several chromatographic methods to achieve complete purification of a protein from a crude biological extract. With the wide variety of chromatographic media available today, this can normally be done in a short period of time.

All of the chromatographic methods mentioned previously are treated in Part II, which begins with a general description of the concepts used in protein chromatography.

1.3.4 Expanded Bed Adsorption

The problem of removing cells and cell debris from large volumes of whole-cell cultures or cell homogenates has encouraged the development of technologies for the direct adsorption of target molecules from such feed stocks. In a fluidized bed the adsorbent particles are subject to an upward flow of liquid that keeps them separated from each other. The resulting increased voidage allows the unhindered passage of cells and cell debris. In a typical fluidized bed there is a total mixing of particles and sample in the reactor, leading to incomplete adsorption of the target molecules unless the feed stock is recycled. Expanded bed adsorption is a special case of fluidized bed adsorption[26] in which a specially designed agarose/crystalline quartz composite adsorbent, and ditto column sample distributor plate, stabilizes the fluidized bed to the extent that a plug flow situation prevails during the sample application cycle. This enables a high binding efficiency in a single sample application cycle. Stabilization is achieved by the formation of an adsorbent particle size and density gradient in the column (Fig. 1-2). The pressure drop required to keep a steady liquid flow through the expanded bed is accomplished by the design of the sample distributor plate at the column bottom inlet.

In a typical expanded bed adsorption experiment the adsorbent particles are poured into the column to a sedimented bed height of 15 cm. The particles had previously been equilibrated to the desired pH and salt concentration. The equilibration buffer is then pumped continuously upward through the bed at a flow rate of 300 cm/h, resulting in a bed expansion two- to threefold. After approximately 30 min the stabilization of the bed is complete, (i.e., the size and density gradient have been established). The sample is now applied at the same flow rate and, because of the higher viscosity, the bed expansion will increase a little. The upper adaptor of the column is kept a few centimeters above the bed level. After all the sample is applied, the column is washed with the equilibration buffer until all cells, cell debris, and nonadsorbed protein are removed. The upper adaptor is hydraulically forced downward in the column, concomitantly packing the bed. Elution of the adsorbed proteins takes place as in normal chromatography either stepwise or by applying a salt or pH gradient followed by regeneration and reequilibration. The adaptor is then hydraulically returned to its original upper position, and the bed is expanded before the next sample application cycle. The procedure is illustrated in Figure 1-3 and a photo of an experimental setup is shown in Figure 1-4.

Soon after its commercial introduction in 1993, expanded bed adsorption on ion-exchange media such as STREAMLINE® DEAE[27] became widely

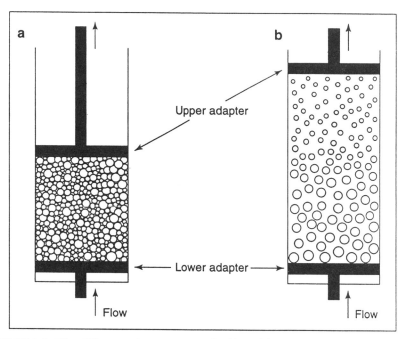

FIGURE 1-2. The difference between a packed bed (a) and an expanded bed (b) for protein adsorption. In the expanded bed, a homogeneous steady flow of liquid from the lower adaptor forces the adsorbent particles to rise to a position where their sedimentation rate is counterbalanced by the upward flow. Each size and density class of particles will find a stable equilibrium point at which they will only oscillate, neither rise nor sediment. A solid-phase gradient is thus formed, with the largest and most dense particles accumulating at the bottom and the smallest and less dense particles at the top of the expanded bed. The functional consequence of this is that a sample of whole-cell culture or crude cell homogenate will move upward by plug flow without distortion of the expanded bed, enabling efficient binding of target molecules to the gel particles in one cycle. (Reproduced by permission of author and publisher.[26])

accepted primarily by bioprocess engineers in the biopharmaceutical industry who saw the possibility of reducing the number of steps in the purification process, saving capital investment, and increasing product yield. Thus recombinant human serum albumin produced in the yeast *Pichia pastoris* was successfully isolated directly from a heat-treated, whole-cell culture on a 1-m-diameter STREAMLINE SP column.[28] Other published applications refer to the capturing of modified recombinant *Pseudomonas areuginosa* expressed in the periplasm of *E. coli,* using STREAMLINE DEAE,[29] and to the purification of recombinant human placental annexin V expressed intracellularly in *E. coli* with the same adsorbent.[30] In 1996 more than 100 pilot scale columns for expanded bed adsorption were in operation.

Another important application area for expanded bed adsorption is the capturing of monoclonal antibodies directly from hybridoma cell cultures. To

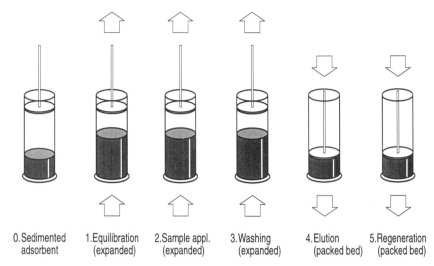

| 0. Sedimented adsorbent | 1. Equilibration (expanded) | 2. Sample appl. (expanded) | 3. Washing (expanded) | 4. Elution (packed bed) | 5. Regeneration (packed bed) |

FIGURE 1-3. Schematic illustration of an expanded bed adsorption cycle.

FIGURE 1-4. Setup of an expanded bed adsorption experiment using a 200-mm-diameter STREAMLINE column. The column is supplied with STREAMLINE rProtein A for the capturing of monoclonal antibodies directly from a whole hybridoma cell culture. (Courtesy of Pharmacia Biotech AB, Uppsala, Sweden.)

this end, a specific adsorbent has been developed based on an agarose/metal alloy composite gel to which a new type of recombinant protein A has been attached through an *in vivo*-fused N-terminal cysteine. This new adsorbent, STREAMLINE rProtein A, was used for the capturing of a humanized IgG$_4$ directly from a 100-L fermenter.[31]

1.3.5 Membrane Chromatography

The main argument for utilizing modified membranes as media for protein adsorption is to solve the problem of mass transport restriction in standard chromatography due to the slow diffusion of proteins in the pores of gel particles. In membranes, most pores allow convective flow, and the mass transport resistance is therefore minimized to film diffusion at the membrane matrix surface. The result is a more efficient adsorption–desorption cycle of target solutes, allowing considerably higher flow rates and thus considerably shorter separation times. The area has been reviewed by Thömmes and Kula.[32]

1.4 FRACTIONATION STRATEGIES

1.4.1 Introductory Comments

Before attempting to design a protein purification protocol, one should collect as much information as possible about the characteristics of the protein and preferably also about the properties of the most important impurities. Useful data involve approximate molecular weight and pI, degree of hydrophobicity, presence of carbohydrate (glycoprotein) or free sulfhydryl groups, and so on. Some of this information might be obtained already on a DNA level, if nucleotide sequence data are available, but is otherwise often collected easily by preliminary trials using crude extracts.

One should also establish criteria with regard to the stability of the protein to be purified. Important parameters affecting structure are temperature, pH, organic solvents, oxygen (air), heavy metals, and mechanical shear. Special concern should be addressed to the risk of proteolytic degradation.

According to a study of 100 published successful protein purification procedures,[33] the average number of steps in a purification process is four. Very seldom can a protein be obtained in pure form from a single chromatographic procedure, even when this is based on a unique biospecificity. In addition to the purification steps, there is often a need for concentrations and sometimes changes of buffers by dialysis or membrane filtrations.

The preparation scheme can be described as consisting of three stages:

1. The preliminary or initial fractionation stage (often called the capturing stage)

2. The intermediate purification stage
3. The final polishing stage

The purpose of the initial stage is to obtain a solution suitable for chromatography by clarification, coarse fractionation, and concentration of the protein extract. The purpose of the final stage is to remove aggregates and degradation products and to prepare a solution suitable for the final formulation of the purified protein.

Sometimes one or two of these stages coincide. An initial ion-exchange adsorption step can thus serve as a preliminary fractionation applied directly to the protein extract, or a gel filtration can give a product that is suitable as a final product; however, as the purposes of the three stages are different, it is useful to discuss them separately.

The design of the preparation scheme will be different depending on the material at hand and the purpose. If the starting material is very precious, one should favor high yield over speed and convenience. If one wants to extract several different proteins from a starting material, that will of course also influence the planning of the work. Finally, the final step is designed such that the product will be suitable for its purpose, which can vary. These aspects are discussed below.

1.4.2   Initial Fractionation

There are many methods for the clarification of protein solutions. Extracts of fungal or plant origin often contain phenolic substances or other pigments. These can be removed by adsorption to celite, either batchwise or on a short column.

Similarly, *lipid* material can be removed either by centrifugation, where the lipids will float and one thus needs to extract the protein solution from below, or by a chromatographic procedure. Lipids adsorb to a number of materials. Aerosil, a fused silica, has been used for the adsorption of lipids, but agarose is sometimes a simple choice.

Contamination with nucleic acids can in some cases, especially when preparing proteins from bacteria, constitute a problem due to their high viscosity. The classic way to solve this problem is to precipitate the nucleic acids. Streptomycin sulfate and polyethylenimine have been used as precipitants, as have protamine sulfate and manganese salts.[34] Another way to solve the problem is to add nucleases, which cut the nucleic acids into smaller pieces, thereby reducing the viscosity.

1.4.2.1   Clarification by Centrifugation and Microfiltration   The clarification of any cell homogenate is usually no problem on a laboratory scale, where refrigerated high-speed centrifuges covering operating speeds from 20,000 to 75,000 rpm generating from about $40,000 \times g$ to about $500,000 \times g$ can

be used. A useful review of centrifugation and centrifuges in preparative biochemistry is found in Ref.[35] As a complement to centrifugation, tangential or cross-flow microfiltration has received increased attention, especially for large-scale applications. For a review of the advantages of cross-flow microfiltration, see Ref.[36]

1.4.2.2 Ultrafiltration   Ultrafiltration has become a widely used technique in preparative biochemistry. Ultrafiltration membranes are available with different cutoff limits for separation of molecules from 1,000 Da up to 300,000 Da. The method is excellent for the separation of salts and other small molecules from a protein fraction with a higher molecular weight and at the same time can effect a concentration of the proteins. The process is gentle, fast, and comparatively inexpensive. For a review, see Ref.[37]

1.4.2.3 Precipitation   Crude extracts are seldom suitable for direct application to chromatographic columns. Preparative differential centrifugation seldom results in a sufficiently clear solution. This is one reason why it is often necessary to use other means for clarification that simultaneously concentrate the solution and, at the same time, remove most of the bulk proteins. Such an initial fractionation step should also result in the removal of proteases and membrane fragments that sometimes bind the protein of interest in the absence of detergents. The classic means is to make a fractional precipitation. First, bulk proteins in the solution are precipitated together with residual particulate matter and then the protein of interest can be precipitated from the resulting supernatant solution. Sometimes the protein of interest is allowed to remain in the mother liquor solution for direct application to chromatographic columns (e.g., hydrophobic interaction adsorption of proteins in ammonium sulfate solutions and ion-exchange chromatography of proteins in PEG mother liquors). The most commonly used precipitating agents are listed in Table 1-5, together with some of their properties. A typical precipitation curve is shown in Figure 1-5.

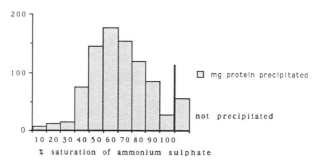

FIGURE 1-5. Example of a precipitation curve. Amount of protein precipitated by a stepwise increase of the ammonium sulfate concentration.

Of the various methods available for protein precipitation, ammonium sulfate has some disadvantages. The resulting protein solution often needs to be dialyzed to obtain an ionic strength that allows ion-exchange chromatography. This problem is avoided when using PEG. Organic solvents, particularly ethanol and acetone, often produce extremely fine, powder-like precipitates that are difficult to centrifuge and handle. They have also been shown to cause partial denaturation of proteins, which can, for example, prevent subsequent crystallization. This is why organic solvents are not recommended as first-choice precipitating agents.

1.4.2.4 Liquid–Liquid Phase Extraction A radically different way of making an initial fractionation is by partitioning in an aqueous polymer liquid–liquid two-phase system.[38] These systems often contain PEG as one phase constituent and another polymer, such as dextran or even salt, as the other. Under favorable conditions it is possible to obtain the protein of interest in the upper, normally the PEG, phase. The contaminating bulk protein as well as particles will become collected in the lower phase and can be removed by centrifugation. Particles sometimes stay at the interphase and are thus also removed in the centrifugation step. By covalent attachment of affinity ligands to PEG molecules these can be used for affinity partitioning. This technique is dealt with in detail in Chapter 11.

1.4.3 Chromatographic Steps

1.4.3.1 Choice of Adsorbent The first information of the chromatographic behavior of a protein is often obtained most simply by preliminary analytical scale experiments (e.g., by gel filtration and by ion-exchange chromatography using salt and pH gradients). In these runs, approximate values of molecular size and ionic properties such as isoelectric points are obtained, information that is fundamental to the further planning of the work. A more thorough survey of the behavior of the protein on various adsorbents can then be done using a panel of adsorbents. This can be carried out either in a panel of parallel columns or using tandem columns.

The parallel column approach has been developed by Scopes[39] for a panel of dye adsorbents. In this case he used up to 20 small columns containing various dye adsorbents. The columns were equilibrated with a predetermined application or starting buffer. A small volume of the protein extract was applied to each column and the protein content (280-nm absorption) and the activity in the effluent were measured. Then a predetermined terminating buffer was applied to each column and again the protein and enzyme activity in the effluent were determined. A column where the bulk of the proteins, but not the activity, was adsorbed was chosen as a "minus column" while an adsorbent where the reverse happened was chosen as a "plus column." These two columns in combination affected a considerable purification of the desired substance in the actual preparation. In a similar approach, a panel of parallel columns was used earlier by, for example, Shaltiel[40] for the evaluation of hydrophobic adsorbents. The technique can, however, be used for any setup of adsorbents, such as different ion exchangers, the same ion exchanger under different conditions, thiol gels, metal-chelating gels, and so on. The elution of the columns can also be performed with more than two elution buffers. The purpose, however, is to get a quick idea of the behavior of a previously unknown protein and thus the setup should not be enlarged beyond what can be handled easily in the laboratory.

If the adsorbents used have well-defined and continuously increasing adsorption capacities for proteins in general, the panel can also be arranged as tandem columns. This approach was used by Porath and co-workers for immobilized metal ion (IMAC) adsorbents[41]. Here, three columns (e.g., Zn, Fe, and Cu) were connected in series and a sample was pumped through all of them. After washing with starting buffer, the three columns were disconnected and eluted separately, mostly using gradients. The approach requires that the first column adsorbs few of the proteins present whereas the last adsorbs almost all of them. This technique is not as generally applicable as the use of parallel columns.

1.4.3.2 *Order of Chromatographic Steps*   A priori one would expect that the order in which the different chromatographic steps are applied in a protein purification protocol is of minor importance. The total purification factor should be constant and the product of the factors obtained in each individual step should be independent of the other steps of the protocol. In the ideal case, where each chromatographic technique is utilized optimally with regard to the resolution and recovery, that is, within the linear regions of the adsorption isotherms (see Chapter 2), with adequate sample volume to column volume ratios and with no adverse viscosity effects, this is probably true. However, real-life situations are always far from ideal or at least such that adaptation to ideality becomes highly impractical. For example, a fractionation gel filtration step can be optimized to give very high resolution (Chapter 3), but only at the cost of time and sample volume. To choose fractionation gel filtration as the first step when the sample volume might be much larger than the total volume of the column means repetitive injections and excessive and impractical total process times, which would probably also be deleterious to the proteins in the sample solution. Likewise, to choose affinity chromatography on immobilized monoclonal antibodies as the first step would probably result in an extraordinarily high purification factor. However, the high cost of such adsorbents prohibits the use of large columns, which makes repeated injections of sample in smaller columns almost mandatory. This leads to long process times and the risk of product losses and/or modifications due to proteolytic attack. Proteolytic activity can also threaten the stability and life length of the actual immunosorbent. Furthermore, protein-based adsorbents are difficult to maintain to a sufficiently high degree of hygiene. There are limitations with regard to means for regeneration (washing) and sterilization

(Chapter 10). This is why they should be saved for later steps of the purification protocol.

The consequence of these considerations is that there are a number of practical rather than theoretical reasons why one should choose certain chromatographic techniques[33] for the early steps and others for the final steps of a protein purification process. The choice is primarily governed by the following parameters:

1. The sample volume
2. The protein concentration and viscosity of the sample
3. The degree of purity of the protein product
4. The presence of nucleic acids, pyrogens, and proteolytic enzymes in the sample
5. The ease with which different types of adsorbents can be washed free from adsorbed contaminants and denatured protein

The last parameter governs the life length of the adsorbent and, together with its purchasing price, the material cost of the particular purification step.

In light of what has been said earlier, the logical sequence of chromatographic steps would be to start with more "robust" techniques that combine a concentration effect with high chemical and physical resistance and low material cost. The obvious candidates are ion-exchange chromatography and, to some extent, hydrophobic interaction chromatography. As the latter often requires the addition of salt for adequate protein binding, it is preferably applied after salt precipitation or after salt displacement from ion-exchange chromatography, thereby excluding the need for a desalting step. Thereafter, the protein fractions can preferably be applied to a more "specific" and more expensive adsorbent. The protocol is often finished with a gel-filtration step (Figure 1-6).

It is advisable to design the sequence of chromatographic steps in such a way that buffer changes and concentration steps are avoided. The peaks eluted from an ion exchanger can, regardless of the ionic strength, be applied to a gel filtration column. This step also functions as a desalting procedure, which means that the buffer used for the gel filtration should be chosen so as to allow direct application of the eluted peaks to the next chromatographic step. Different chromatographies have, in practice, widely different capacities, even though it is possible to adapt several of the methods to a larger scale. However, in the initial stages of a purification scheme, it is most convenient to start with the methods that allow the application of large volumes and that have the highest capacities. To this category belong, for example, ion exchange chromatography and hydrophobic interaction, but any adsorption chromatographic method can be used to concentrate larger volumes, especially in batchwise operations.

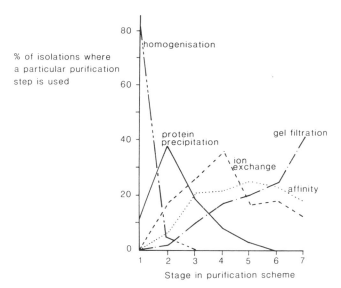

FIGURE 1-6. Analysis of the methods of purfication used at successive steps in purification schemes. The results are expressed as a percentage of the total number of steps at each stage. (Adapted from Ref. 33 by permission of the authors and publisher.)

1.4.4 The Final Step

The purpose of the final step is to remove possible aggregates or target protein molecules that have been posttranslationally modified and to condition the purified protein for its use or storage. The procedure will thus be different depending on the fate of the protein. Aggregates and low molecular weight degradation products are preferably removed by gel filtration and, if the protein is to be lyophilized, this step is also suitable for transfering the protein to a volatile buffer (see Table 1-1). This can sometimes be done by ion-exchange chromatography, but more seldom by other forms of chromatography. If the protein solution is intended to be frozen, stored as a solution, or used immediately, the requirements for specific buffer salts might be less stringent.

Several of the adsorption chromatographies might be adapted in such a way that they result in peaks of reasonably high protein concentration. This is an advantage when gel filtration is chosen as a final step. Gel filtration will always result in dilution of the sample and is often followed by a concentration step.

If the protein is to be used for physicochemical characterization, especially for molecular weight studies, gel filtration has the advantage of giving a protein solution of defined size and also in perfect equilibrium with a particular buffer. Biospecific methods, by definition, give a product that is homogeneous with respect to biological activity. This was taken advantage of for papain, where the enzyme eluted from a thiol column was twice as active as any previous preparation—most earlier enzyme batches apparently contained molecules in which the thiol necessary for activity was oxidized (see Chapter 9).

Proteins which after purification and formulation are intended for parenteral use in human beings should not contain endotoxins (lipopolysaccarides) or nucleic acids. The purification protocols must be designed such that these compounds are efficiently removed, and validation studies should be performed to prove this. To prepare sterile protein solutions, sterile filtration is used. For difficult to remove modifications of the target protein, more highly resolving techniques, such as reversed phase high performance liquid chromatography, are often applied (see Chapter 6).

1.5   MONITORING FRACTIONATION

Proper analysis is a prerequisite for successful protein purification. Most important is the establishment of a reliable assay of the biological activity. In addition, one needs to determine the protein content in order to be able to assess the efficiency of the different steps. It is beyond the scope of this chapter to go into details of the particular assay methods. This is covered by special literature dealing with the activity in question—hormone, enzyme, receptor, and so on.

Each preparation should be continually recorded in a purification table (Table 1–6). In combination with results from, for example, gel electrophoresis, this will serve as a guide for judging the reproducibility and the outcome of each preparation. In addition, each chromatography experiment should be accompanied by a suitable protocol such as the one examplified in Figure 1-7. However, the need for measurements of biological activity and protein concentration—especially the latter—should not be allowed to delay the preparation, and in many cases it is sufficient to save aliquots for analysis at one's convenience.

1.5.1   Measurements of Activity

In general, biochemical activities depend on the interaction between molecules. This can be measured in different ways. The classic method of enzyme

TABLE 1-6. Examples of a Purification Table

Material	Volume (ml)	Protein (mg/ml)	Total Protein (mg)	Activ. (U/ml)	Total Activ. (U)	Spec. Activ. (U/mg)	Yield (%)	Purific. (X)
Extraction	500	14	7,000	7	3,500	0.5	100	1
First purif. step	50	10	500	60	3,000	6	85	12

The activity (e.g., enzyme activity) is expressed as units (U) and specific activity as units per milligram of protein (U/mg).

Chromatography Data Sheet

Date..   Chrom. No...

Column
Length..................... cm V_T........... ml   Column filling:
Diameter................. cm V_0........... ml   Adsorbent...
C.s. area................... cm^2 V_{salt}...... ml   Gel...
Packing etc..   Lot No..
....   Preparation..
....

Equilibration of column
Buffer system or solvent..   Conc...................mol/l
pH.......................... Conductivity..................ohm^{-1}cm^{-1} Temp.................^{o}C
Detergent/Denaturant...   Conc............................
Bacteriostat...   Conc............................
Comments...

Sample
Description...   Designation................
Amount.....................mg Vol........... ml   Conc................ A_{280}...........A................
Buffer or solvent...
Specific acitivty etc...
Pretreatment...

Elution
Only buffer ❐   Gradient system...
....
....
Flowrate....................ml/h Prefraction.............. ml   Fraction size ....ml...........min
Comments...

Analyses
280 nm ❐.................nm ❐   pH ❐   Conductivity ❐ Volume ❐
Other reaction...   Aliquote volume........................
Procedure...
Comments...

Fate of fractions
Pools...
....
Comments...

FIGURE 1-7. Example of a chromatography protocol.

catalysis is only one of these. In addition, the monitoring of components can be done in several ways, such as spectrophotometry, measurements of radioactivity, and immunological methods. Some of them are:

1. Enzyme activity by direct spectroscopy
2. Enzyme activity by secondary measurements on aliquots
3. Binding of ligand
4. Binding of antibody

Immunological methods require that the protein studied has already been purified once to allow production of an antibody or an antiserum by immunization. The detection of the antigen–antibody precipitate can be done at almost any sensitivity down to the extreme sensitivity afforded by the use of sandwich techniques and radioactively or enzymatically labeled reagents (see, e.g., Chapter 14).

1.5.2 Determination of Protein Content

In general, a measure of protein content is obtained upon monitoring the effluent in chromatography by ultraviolet (UV) absorption. However, it is not always easy to relate these measurements to the protein content. In fact, the only certain measure of protein content is total amino acid analysis after hydrolysis. Strictly speaking, even this latter analysis suffers from some shortcomings, as tryptophan and cysteine normally have to be analyzed separately.

Large deviations from true protein values sometimes occur in the first steps in a purification scheme. The extract itself often contains substances that interfere with protein analyses. An overestimation might result, especially if measurements of absorption at 280 nm are used, as the solutions are often turbid and absorbing substances of nonprotein origin are present. This in turn will make the calculated values of specific activity erroneous.

Three main procedures for protein determination are used routinely:

1. Spectrophotometry at 280 nm
2. Colorimetry by Lowry–Folin–Ciocalteau reagent
3. Dye binding with Coomassie brilliant blue G-250

Each of these methods has its advantages and its disadvantages. UV-absorption measurements require knowledge of the extinction coefficient of the protein(s) to be measured. These vary widely. In the low end there is, for example, serum albumin with an OD of 0.6 for a 1-mg/ml solution and the extreme parvalbumin with no absorption at all in the 280-nm band. At the other extreme there is, for example, lysozyme with an OD of 2.7 for a 1-mg/ml solution. The values are due to a corresponding variation in the content of the aromatic amino acids tryptophan and, to a lesser extent, tyrosine. As a rule of thumb it is convenient to assume a mean extinction of 1.0 for a 1-mg/ml solution and this is often sufficient for practical purposes. When the protein is purified the extinction coefficient and the wavelength for maximum extinction should, however, be determined on a solution by spectral and amino acid analysis.

An alternative to measurements at 280 nm are low wavelength measurements at 225 nm or below. This absorption is due to the peptide bond, which has a maximum at 192 nm but still has considerable absorption at 205 nm (50% of maximum or an OD of 31 for a 1-mg/ml protein solution) and at 220 nm an OD of 11. These measurements are, of course, even more sensitive to contamination and they also require that buffers that are transparent in low UV regions are used. The use of sensitive, UV monitors at these wavelengths in chromatographic equipment thus allows an extreme sensitivity but their use is not possible unless great care is taken to avoid contaminants and impurities in the buffer salts.

The Lowry methods[42] are less problematic but also less sensitive. Aliquots for analysis should have protein concentrations of 0.1 mg/ml or more. It is often a good alternative in the beginning of a purification where direct UV might be impossible due to turbid solutions. The same applies to the use of Coomassie brilliant blue. This method is 5 to 10 times more sensitive than the colorimetric one but is more cumbersome to use.[43,44]

Both of the latter two methods are destructive, whereas the UV method allows the sample to be recovered.

1.5.3   Analytical Gel Electrophoresis

The gel electrophoresis techniques allow the investigator to get an idea of the complexity of the sample and, in particular, what the main contaminating species are. By using sieving electrophoresis, (e.g., in the presence of SDS) and isoelectric focusing, a considerable amount of information about the sample can be obtained in a couple of runs. The amount of each component, its molecular weight, its isoelectric point, and even its titration curve can be obtained (see Chapter 13). If an antiserum directed toward the complete mixture is available, it is also possible to see whether some of the components are immunologically related and thus also structurally related by means of crossed immunoelectrophoresis.

The gel electrophoresis techniques are introduced in Chapter 12.

1.6   THE FINAL PRODUCT

1.6.1   Buffer Change

The high resolution chromatographic steps for protein fractionation usually result in a product that is not directly suitable for the intended use, storage, or distribution. The salt content may be too high, the pH of the protein solution may be unsuitable for long-term storage, the concentration of the protein may be too low, or the solution may contain desorption agents from an affinity chromatography step that must be removed before the protein can be used.

The classic way of changing the buffer composition of a protein solution is dialysis. The protein solution is included in a dialysis bag consisting of cellophane or a similar semipermeable material. Salts and low molecular weight substances (<10,000) can diffuse through the membrane, whereas high molecular weight material remains within the dialysis bag. The bag is placed in a larger stirred vessel containing the desired buffer, which is changed several times.

A faster way of changing the buffer composition of protein solutions is by gel filtration on, for example, Sephadex® G25 or BioGel® P-6 equilibrated in the desired buffer. Proteins and other high molecular weight substances (>6,000) eluate at the void volume, whereas substances with lower molecular weights are retarded and are separated from the proteins. The method is fast, depending on the equipment and volume; the cycle time is often less than 1 h. As in every type of column chromatography the protein solution must not contain particles or colloidal material.

Ultrafiltration (diafiltration) is a third way of changing the buffer composition of a protein solution. With this technique the protein solution is diluted with the desired buffer, concentrated to the original volume, diluted again, and so on. After a number of cycles the original buffer has in practice changed to the dilution buffer. The last concentration cycle may be driven longer so that the protein solution after the buffer change is concentrated.

1.6.2   Concentration

Concentration is another operation often required after the final step in a protein purification procedure. Ultrafiltration is the most frequently applied technique for this purpose. Smaller volumes of protein solutions can alternatively be concentrated by inclusion in a dialysis bag that is covered with a high molecular weight substance that cannot penetrate the dialysis bag but creates an osmotic pressure that drives the liquid out through the dialysis membrane. Often used polymers for this purpose are polyethylene glycol and Ficoll.™ [45] All chromatographic techniques which adsorb protein can also be used for the concentration of protein solutions. Especially suitable is ion exchange chromatography due to its high capacity and easy handling of the ion exchange medium. Other concentration methods that have proved useful also in large scale applications are freeze concentration[46] and concentration using dry Sephadex.[47]

1.6.3   Drying

Most biological processes occur in water solution and one way to stop these is to freeze the protein sample. For minimum risk of inactivation or denaturation, a storage temperature of −70°C or below is required. If the protein under study cannot stand repeated freezing and thawing, storage in aliquots is recommended. Another way to stop biological processes is to remove the water. The method used most for biologically active proteins is freeze-drying or lyophilization. In this method the protein solution is frozen below the eutectic point of the solution to ensure that all liquid is frozen. The frozen solution is then placed in a chamber that can be set under high vacuum. In the chamber or connected to the chamber is a condensing surface, a cold trap, with a temperature of < −40°C. After the vacuum is applied the protein sample is gently heated such that it does not melt to speed up the sublimation of water to the condenser. Normally all proteins maintain their biological activity and are fully recovered upon adding water. A techique often used for commercially available biochemicals is to lyophilize aliquots of protein solution in ampoules.

1.7 LABORATORY EQUIPMENT

1.7.1 General Equipment

Laboratories for the preparation and separation of proteins may look very different, from the well-equipped special laboratory serving many research groups to the small laboratory with few people and limited resources. The large laboratory, in addition, often has dedicated service groups for special analyses and so on.

For the successful preparation and separation of proteins, certain basic equipment is needed. Standard laboratory glassware is not discussed. Basic equipment includes two or three balances, one spectrophotometer, one centrifuge, and one pH meter plus stirrers and micropipettes.

A good combination of balance equipment is two preparative balances: one double-range digital balance 0 to 1200 g or 0 to 3000 g with the possibility to weigh with 10-mg resolution and the other range 0 to 120 g or 0 to 300 g with 1-mg resolution; and one analytical balance for a 0- to 150-mg interval with 0.1-mg resolution. For measuring larger amounts of material, a simpler, and therefore inexpensive, balance suffices (e.g., a balance for use in the food industry). It is important that the balances are calibrated as well as that they are serviced regularly.

Spectrophotometers for protein work should cover the 190- to 800-nm wavelength interval. Absorption around 280 nm is used routinely for estimating protein concentration, whereas light in the visible region is often needed for measuring of different enzymatic activities. A double-beam UV-VIS spectrophotometer with a thermostated cuvette holder is a good choice. It is important that the wavelength setting is correctly calibrated and that the instrument shows low drift. Regular servicing of the instrument is recommended.

A refrigerated floor centrifuge is standard in a preparative protein laboratory. The largest rotor should take six flasks of 500 ml, whereas the smallest rotor should accommodate eight tubes of 50 ml. The maximum g force for those rotors are normally about 13,000 and 50,000 $\times$ g at the tube bottom, respectively. An additional rotor for six flasks of 250 ml will increase the flexibility of the centrifuge. Some centrifuges accommodate zonal rotors with accessories for continuous operation, allowing larger volumes containing relatively low contents of fine particles to be centrifuged at high g forces (up to 40,000 $\times$ g). Standard centrifuges normally need little or no service except changing motor brushes. If the rotors are carefully maintained, including thorough cleaning after each run when liquid is found inside the rotors, they will function for many years. The rotors just mentioned are angle rotors and the flasks are normally filled completely. The special, completely tight lids, which are available for most centrifuge flasks, are strongly recommended. For small, easily centrifuged samples, a simple desk-top centrifuge without cooling often is a good complement to the larger high-speed refrigerated centrifuge.

The maximum speed of the refrigerated floor centrifuge is normally 20,000 to 21,000 rpm. A new type of centrifuge allowing speeds up to 28,000 rpm has been introduced. At speeds over 20,000 rpm, the rotor compartment can be maintained under partial vacuum. Small samples can thus be centrifuged at 100,000 $\times$ g.

pH determinations are critical in the protein laboratory. Buffers should be checked and titrated routinely. For certain enzyme assays, equipment for monitoring the formation or consumption of protons is needed. The pH meter can then be equipped with an automatic burette that adds acid or base to hold the pH constant, and the consumption of acid or base is recorded. Combination electrodes are generally the most convenient. The linearity of the electrode changes with time, and for accurate measurements the electrode has to be calibrated on both sides of the interval within which the pH will be determined.

At least two types of stirrers are integral parts of the basic laboratory equipment. First, magnetic stirrers with variable speeds, at least one of which is equipped with a thermostat-controlled heating plate, are recommended. A variety of PTFE- and glass-coated magnets are required. For extractions, propeller stirrers with interchangeable propellers and with variable speed regulators are convenient. These should preferably be of the electronic type that compensate for variations in load. Stirrers where the speed is controlled only with a variable resistor should be avoided.

Pipetting small volumes, up to 1 ml, is a standard laboratory procedure. Micropipettes of different types, most commonly with adjustable volumes and with disposable tips, are used. Two or three pipettes of different sizes are normally sufficient. Check that the length of the pipette allows samples to be collected from the bottom of the longest type of test tube that is used in the laboratory.

For personnel safety, the laboratory must have a ventilated hood or fume cupboard where procedures involving handling or preparation of hazardous substances can be safely performed. Even such a simple and common procedure as dissolving sodium hydroxide in water creates aerosols that are very irritating; this procedure should be done in the hood.

1.7.2 Equipment for Homogenization

Starting materials for protein preparation can be obtained from plants, animals, and microorganisms. Plant and animal materials first have to be homogenized to individual cells but then the preparation procedures are in principle similar for the three different kinds of starting material.

Waring blenders or similar types of equipment and a "Turrax"-type apparatus are frequently used for the homogenization of plant and animal tissues. The material is first cut into pieces with a knife or a pair of scissors to a size that suits the type of blender used. At the same time the tissue or organ is trimmed and unwanted material such as fat, ligaments, and vessels are discarded before grinding the material in the blender. The blender can be used both dry and wet. Plant material like seeds, for example, can be ground dry if the time of grinding can be kept short. Dry ice can be added to keep the temperature low. Wet grinding prevents the formation of harmful dust particles and the heat formed in the grinding procedure is dissipated by the liquid. In most homogenization equipment, only about 10% of the energy input is normally used for the disintegration procedure, the rest of the energy is lost as heat. A common household meat grinder is also useful in many instances eventually in combination with a Waring blender. Further homogenization is provided by Elvehjem pestle-type homogenizers. There are different types (glass, Teflon®), sizes, and also motor-driven varieties of this homogenizer.

Microorganisms are much more difficult to disintegrate. The polysaccharide cell wall withstands most homogenization procedures used for plant and animal cells. The well-cited overview by Wimpenny[48] for breakage of microorganisms is still valid for small-scale preparations (Figure 1-8). For large-scale homogenization of microorganisms, which means kilogram quantities of cell paste, the most frequently used equipment is the Dyno Mill[49] and the Manton-Gaulin[50] homogenizer. Both are designed for continuous operation and are available in various sizes for the breakage of from 100 g up to several kilograms of microbial cell paste per hour.

1.7.3 Equipment for Chromatography

1.7.3.1 Column Design   To use packed chromatographic gel particles optimally, the column should be designed such that it does not significantly contribute to band spreading or peak distortion. Most modern columns are of the closed type with two fixed end pieces or with one fixed end piece and one adjustable adaptor. Sometimes it is convenient to use two adjustable adaptors,

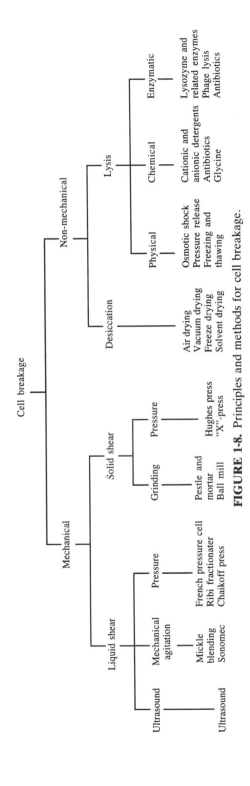

FIGURE 1-8. Principles and methods for cell breakage.

one at each end of the column. The end pieces or adaptors are equipped with porous frits made of sintered metal, glass, or plastic, pore diameters, and surface structures that should be optimized to avoid clogging and abrasion of the chromatographic particles. The frits should be easily exchangeable. For larger diameter columns the frits have to be combined with the flow distribution system. The prime requisites of such a system are that its volume should be negligible compared to the total column volume and that there should be no pressure drop in the system between the eluent inlet and the column wall. In many modern columns for low and medium pressure chromatography the frits have been replaced by a combination of a fine mesh (10 μm) polyamide fabric and a coarse mesh (e.g., 0.93 $\times$ 0.61 mm) polypropylene support net. The flow distribution layer will then be only 1 mm in depth and the support net will reduce the dead volume by as much as 32%. There is no indication that this uniform support net affects the sample application. This is the way many very large-diameter industrial columns are designed. For the largest columns ($>$40 cm diameter) the fine mesh nylon fabric is in most cases replaced by a high-quality stainless-steel mesh. In most large-diameter industrial columns the number of sample and eluent inlets and outlets has increased to four to six. The linear flow rate of the eluent in the column feed pipe is very high in large-diameter columns and it is recommended that this jet should not directly hit the fine mesh fabric (when using a frit this problem is less pronounced) which could cause flow inhomogeneity and band distortion. The effect is reduced by the application of a small disk covering the support net under each pipe inlet.

The materials used in the parts of the chromatographic system that come in contact with the eluent should preferably be noncorrosive and compatible with all normal protein samples and eluent buffers/solvents used. However, they should also preferably be resistant to the conditions applied in the cleaning and maintenance of the packed gel particles. This includes high concentrations of sodium hydroxide and solvents such as alcohols. The CIP concept (cleaning in place) has become an important part of process design in biotechnology. In many instances, sterilization by autoclaving is preferred. Most laboratory column tubes are made of borosilicate glass or plastics with end pieces and adaptors made of polypropylene, nylon, or fluorocarbon. Many HPLC columns are made of stainless steel to withstand the mechanical stress.

The column is connected to the pump and to the monitors by small-bore tubing made of stainless steel or titanium in HPLC systems or fluorocarbon, polypropylene, polyethene, or nylon in medium or low pressure systems. The inner diameters of the tubings should be optimized for each application. They should be as small as possible without too much contribution to the pressure drop of the system. To avoid zone mixing, the tube length should be reduced as much as possible. Ideally, the column should be attached directly to the monitor cell on top of the fraction collector. In normal laboratory systems, 1-mm-inner-diameter fluorocarbon tubing is recommended.

1.7.3.2 Pumps and Fraction Collectors The traditional laboratory-scale pumps for low-pressure systems (0.1 MPa) are of the peristaltic type, of which several brands are on the market. Care must be taken to keep the rollers and pumping tube clean. Otherwise, the tube life length is severely decreased. Peristaltic pumps are available with flow rates for a few milliliters per hour to cubic meters per hour. The pumping tube is available in several different materials, but for the protein work silicone tubing is recommended. Pumps with three or more channels can be used for gradient formation. The normal way to form a gradient is to use two connected vessels, with stirring in the output mixing vessel. In medium- and high-pressure chromatography, gradients are usually formed using two pumps connected to a programmable controller.

For HPLC, most pumps are of the piston type and are made of stainless steel or titanium, as are the sample injectors. In most cases they function according to the reciprocal displacement principle and are capable of pressues up to 35 MPa or more. Among medium-pressure (5 MPa) pumps is the positive displacement type in which a fluorocarbon-sealed piston made of titanium or stainless steel is allowed to move in a thick-walled borosilicate glass tube. When the piston reaches the end of the tube, a valve switches over to a second identical piston-equipped glass tube filled with eluent and which will continue feeding the column with eluent at the same pressure while the first is filled. Unless proper damping is provided, a pressure transient will occur, giving rise to spikes in the UV monitor recordings at each piston change.

Traditional fraction collectors consist of a large plastic or metallic ring with holes bored at the periphery moved by an electric motor controlled by a timer. This type of fraction collector requires a relatively large bench area. Alternatives are compact designs with the holes for the test tubes arranged in a spiral or with the test tubes placed in racks. These collectors are controlled by microprocessors, and different tube sizes and fraction collection times can be programmed in advance. In large-scale chromatography, when the results normally are predictable from pilot experiments on a small scale, only a few fractions are needed and the fractions are obtained by using magnetic valves connected to the controlling equipment.

1.7.3.3 Monitoring Equipment The classic way of monitoring a chromatographic experiment is to take fractions from the outlet of the column in a fraction collector and analyze each tube manually for the different substances one wishes to determine. This is very time-consuming work, and flow monitors connected directly to the column eluate have largely replaced this practice. The most commonly used parameter in protein work is the absorption at 280 nm and many detectors are available for this purpose. In new diode array monitors, even a spectrum can be obtained directly on the eluate. In ion-exchange chromatography the ionic strength is of interest, and conductometers with flow cells are available. The pH of the eluate can also be monitored in, for example, chromatofocusing by continuously working monitors. Even complicated enzyme assays can be made directly on line by the use of autoanalyzers. Here a small part of the process stream is shunted through the autoanalyzer. This part of the sample is normally destroyed during the analysis.

1.7.3.4 Chromatography Systems The introduction of the FPLC system[51] in 1982 meant a new way of thinking in the design of dedicated chromatography equipment for protein separation and purification. For the first time a microprocessor-controlled straightforward system approach was launched where the parts and components matched the performance of a new chromatography material. Parts in contact with sample and buffers are made of glass, plastic, or titanium to minimize corrosion and nonspecific adsorption of the sample components. In combination with prepacked columns, containing the 10-μm-diameter MonoBead ion exchangers, that give high-performance separations at moderate pressure drops, the FPLC system became extremely popular and has since been installed in several thousand biochemical laboratories.

Many followers to the FPLC system have been introduced by several manufacturers. One of the latest is ÄKTAexplorer[51] (Figure 1-9), which is a preassembled chromatography system configured for fast and easy development and optimization of a variety of purification methods using the "Adviser" included in its UNICORN™ software. This contains method development templates for media screening and method scouting, covering ion-exchange chromatography, hydrophobic interaction chromatography, reversed-phase chromatography, affinity chromatography, and gel filtration chromatography,

FIGURE 1-9. The ÄKTAexplorer chromatography system. (Courtesy of Pharmacia Biotech AB, Uppsala, Sweden.)

as well as purification protocols for recombinant proteins from *E. coli*, peptides, and oligonucleotides. Included is also a general-purpose method for column cleaning. When a prepacked column is selected from those listed in the data base, the system automatically sets the running parameters to those best suited to that particular column. This allows direct and optimal use of the column without any previous special knowledge or experience. An automatic buffer preparation function covering a broad pH range eliminates manual buffer preparation and titration needed for every pH change, particularly in ion exchange chromatography. The ÄKTAexplorer can also manage system control and data access via a PC network that gives a complete overview of operations. Results can be automatically saved on a server, and evaluation and generation of reports can be made locally or at remote PCs. The pumps have an exceptionally wide flow rate operation range (0.01 to 100 ml/min in isocratic mode and 0.5 to 100 ml/min in gradient mode) at pressures of up to 10 Mpa previously not covered by a single chromatography system. The whole system is mounted on a swivel platform and requires minimal bench space.

1.7.4 Equipment for Chromatographic and Electrophoretic Analyses

For the characterization of the product at various stages of a protein preparation in principle, the same separation techniques that are used for preparation of the protein can be used. The techniques are, however, scaled down to the micro- to milligram scale, and sometimes dedicated micropreparative chromatographic equipment is preferred[52]. Even more important are analyses by electrophoretic separation techniques. Equipment for the different electrophoretic techniques is described in connection with the techniques themselves in Chapters 12–18.

1.8 REFERENCES

1. J.A. Asenjo, ed., *Separation Processes in Biotechnology*, Marcel Dekker, New York, 1990.
2. S.M. Wheelwright, *Protein Purification, Design and Scale-up of Downstream Processing*, Hanser Verlag, Munich, 1991.
3. R. Scopes, *Protein Purification, Principles and Practice*, 3rd edition, Springer-Verlag, New York/Heidelberg/Berlin, 1993.
4. F. Franks, ed., *Protein Biotechnology*, The Humana Press, Totowa, NJ, 1993.
5. *Meth. Enzymol.*, *22*, W.B. Jakoby, ed., Academic Press, New York, San Fransisco, London (1971).
6. *Meth. Enzymol.*, *34*, W.B. Jakoby, and M. Wilchek, eds., Academic Press, New York and London (1974).
7. *Meth. Enzymol.*, *104*, W.B. Jacoby, ed., Academic Press, Orlando (1984).
8. *Meth. Enzymol.*, *182*, M.P. Deutscher, ed., Academic Press, San Diego (1990).

9. T. Moks, L. Abrahamsen, B. Österlöf, S. Josephson, M. Östling, S.-O. Enfors, I. Persson, B. Nilsson, M. Uhlén, *Biotechnology, 5*, 379–382 (1987).

10. M.-R. Kula, H. Schütte, *Biotechnology Progress, 3(1)*, 31–42 (1987).

11. G. Taylor, M. Hoare, D.R. Gray, F.A.O. Marston, *Biotechnology, 4*, 553–557 (1986).

12. F.A.O. Marston, *Biochem. J., 240*, 1–12 (1986).

13. L.M. Hjelmeland, *Meth. Enzymol., 124*, 135–164 (1986).

14. L. Rydén, H.F. Deutsch, *J. Biol. Chem., 253*, 519–524 (1978).

15. A. Veide, T. Lindback, S.-O. Enfors, *Enzyme Microb. Technol., 6*, 325–330 (1984).

16. P. Hedman, J.G. Gustavsson, *Dev. Biol. Standard., 59*, 31–39, S. Karger, Basel (1985).

17. E.J. Cohn, L.E. Strong, W.L. Hughes, Jr., D.J. Mulford, J.N. Ashworth, M. Melin, H.L. Taylor, *J. Am. Chem. Soc., 68*, 459 (1946).

18. J. Porath, *Science Tools, 11*, 21–29 (1964).

19. K. Hannig, *Hoppe-Seyler's Zeit. Physiol. Chem., 338*, 211–227 (1964).

20. P. Mattock, G.F. Aitchison, A.R. Thomson, *Separ. Purif. Meth., 9*, 1–68 (1980).

21. M. Bier, in *Recovery and Purification in Biotechnology*, ACS Symposium Separation Series 314, 185–192, J.A. Asenjo, and J. Hong, eds., American Chemical Society (1986).

22. B.J. Radola, in *Meth. Enzymol., 104*, 256–275, W.B. Jacoby, ed., Academic Press (1984).

23. E.A. Peterson, H.A. Sober, *J. Am. Chem. Soc., 78*, 751–755 (1956).

24. J. Porath, P. Flodin, *Nature, 183*, 1657–1658 (1959).

25. J.-C. Janson, *Chromatographia, 23*, 361–369 (1987).

26. H.A. Chase, *Trends in Biotechnol., 12*, 296–303 (1994).

27. Pharmacia Biotech AB, Uppsala, Sweden.

28. M. Noda, A. Sumi, T. Ohmura, K. Yokoyama, *European Patent Application*, EP0699687A2 (1996).

29. H.J. Johansson, C. Jägersten, J. Shiloach, *J. Biotechnol., 48*, 9–14, (1996).

30. A.-K. Barnfield Frej, R. Hjorth, Å. Hammarström, *Biotechnol. Bioeng., 44*, 922–929 (1994).

31. J. Bonnerjea, unpublished work 1996.

32. J. Thömmes, M.-R. Kula, *Biotechnol. Progr., 11*, 357–367 (1995).

33. J. Bonnerjea, S. Oh, M. Hoare, P. Dunnill, *Biotechnology, 4(11)*, 954–958 (1986).

34. J.J. Higgins, D.J. Lewis, W.H. Daly, F.G. Mosqueira, P. Dunnill, M.D. Lilly, *Biotechnol. Bioeng., 20*, 159–182 (1978).

35. C.M. Ambler, F.W. Keith, in *Separation and Purification* 3rd edition, E.S. Perry and A. Weissberger, eds., pp. 295–347, J. Wiley & Sons, New York, 1978.

36. D. Derise, V. Gekas, *Proc. Biochem., 23*, 105–116 (1988).

37. H. Strathman, *Trends Biotechnol., 3*, 112–118 (1985).

38. H. Walter, D.E. Brooks, D. Fisher, eds., *Partitioning in Aqueous Two-Phase Systems*, Academic Press, Orlando, 1985.

39. R. Scopes, *J. Chromatogr., 376*, 131–140 (1986)

40. S. Shaltiel, *Meth. Enzymol., 34,* 126–140 (1974).

41. J. Porath, B. Ohlin, *Biochemistry, 22,* 1621–1630 (1983).

42. O.H. Lowry, N.J. Rosebrough, A.L. Farr, R.J. Randall, *J. Biol. Chem., 193,* 265–275 (1951).

43. J. J. Sedmak, S. E. Grossberg, *Anal. Biochem.,* 79, 544–552 (1977).

44. T. Spector, *Anal. Biochem.,* 86, 142–146 (1978).

45. Ficoll™ is a polysucrose from Pharmacia Biotech AB, Uppsala, Sweden.

46. J.-C. Janson, B. Ersson, J. Porath, *Biotechnol. Bioeng.* XVl, 21–39 (1974).

47. P. Flodin, B. Gelotte, J. Porath, *Nature, 188,* 493 (1960).

48. J.W.T. Wimpenny, *Proc. Biochem., 2(7),* 41–44 (1967).

49. Dyno Mill; Willy A. Bachofen, Manufacturing Engineers, Basel, Switzerland.

50. Manton-Gaulin; APV GAULIN International SA, P.O. Box 58, 1200 AB Hilversum, Holland.

51. Pharmacia Biotech AB, Uppsala, Sweden.

52. SMART® System (Pharmacia Biotech AB, Uppsala, Sweden).

PART II
Chromatography

2 Introduction to Chromatography

JAN-CHRISTER JANSON

Biochemical Separation Centre
Uppsala University Biomedical Centre
Box 577, S-751 23 Uppsala, Sweden

JAN-ÅKE JÖNSSON

Department of Analytical Chemistry
University of Lund
Box 124, S-221 00 Lund, Sweden

Protein Purification: Principles, High-Resolution Methods, and Applications, Second Edition.
Edited by Jan-Christer Janson and Lars Rydén.
ISBN 0-471-18626-0.   © 1998 Wiley-VCH, Inc.

2.1   BASIC CONCEPTS AND VERSIONS OF CHROMATOGRAPHY

The term *chromatography* refers to a group of separation techniques that are characterized by a distribution of the molecules to be separated between two phases, one stationary and the other immobile. Molecules with a high tendency to stay in the stationary phase will move through the system at a lower velocity than those that favor the mobile phase.

The most common physical configuration is column chromatography in which the stationary phase is packed into a tube, a column, through which the mobile phase, the eluent, is pumped. The sample to be separated is introduced into one end of the column. The various sample components travel with different velocities through the column and are subsequently detected and collected at the other end. Other configurations, such as thin-layer chromatography and paper chromatography, are also used.

In the context of this book, the mobile phase is always a liquid, most often an aqueous buffer. Consequently, these techniques are versions of liquid chromatography. For the separation of more volatile compounds, gas chromatography, (i.e., with a gaseous mobile phase) is an extremely powerful and widely applied technique.

For protein separation, several versions of liquid chromatography are used, differing mainly in the types of stationary phase (Table 2-1). One of these,

TABLE 2-1. Versions of Protein Liquid Chromatography

Separation Principle	Type of Chromatography
Size and shape	Gel filtration
Net charge	Ion-exchange chromatography
Isoelectric point	Chromatofocusing
Hydrophobicity	Hydrobphobic interaction chromatography
	Reversed-phase chromatography
Biological function	Affinity chromatography
Antigenicity	Immunoadsorption
Carbohydrate content	Lectin affinity chromatography
Content of free sulfhydryl groups	Chemisorption ("covalent chromatography")
Metal binding	Immobilized metal ion affinity chromatography
Miscellaneous	Hydroxyapatite chromatography
	Dye affinity chromatography

gel filtration chromatography (also called size-exclusion chromatography), is based on quite different principles than are other versions of liquid chromatography. Therefore, much of the theoretical description of that technique must be made separately. In this book, the basic principles of gel filtration chromatography are described in Chapter 3, whereas other techniques are dealt with briefly here. A thorough treatment of the theory and principles of chromatography is outside the scope of this book, but can be found in specialized texts.[1-3]

2.2   THE STATIONARY PHASE

The stationary phase in a chromatographic experiment is composed of a porous matrix and imbibed immobile solvent. Typically, the solvent constitutes most of the stationary phase, often more than 90%, and such materials are generally referred to as gels. In protein chromatography the solvents are normally aqueous buffers and the gel-forming materials are usually composed of hydrophilic polymers. Modern stationary phases are almost exclusively bead shaped, with average particle diameters ranging from a few to approximately 100 μm.

In principle, one may distinguish between two types of gels, xerogels and aerogels. The xerogels are characterized by their ability to shrink and swell in the absence and presence of the solvent used, whereas the volume of aerogels is independent of the solvent. Typical xerogels are cross-linked dextran gels (Sephadex®) and cross-linked polyacrylamide gels. Typical aerogels are porous glass, silica, and most gels based on macroreticular organic polymer gels such as polystyrene gels and polymethacrylate gels.

2.2.1   Matrix Properties

In addition to being hydrophilic, an ideal general matrix for protein chromatography should not contain groups that spontaneously bind protein molecules. However, it should contain functional groups that allow the controlled synthesis of a wide variety of protein adsorbents. Furthermore, the matrix should be chemically and physically stable in order to withstand extreme conditions during derivatization and maintenance (regeneration, sterilization, etc.) and be rigid enough to allow high linear flow rates (5 cm/min or more) in columns packed with particles with diameters down to a few microns. Finally, the matrix substance should allow the production of gels with a broad range of controllable porosities.

A wide variety of material has been used for the design of protein chromatography matrices. These can be classified as being inorganic, synthetic organic polymers, or polysaccharides. Among all three groups we find traditional standard chromatography media as well as modern high-performance chromatography media. Examples of these are listed here:

Inorganic Materials

- Porous silica
- Controlled pore glass
- Hydroxyapatite

Synthetic Organic Polymers

- Polyacrylamide
- Polymethacrylate
- Polystyren

Polysaccharides

- Cellulose
- Dextran
- Agarose

None of these materials fulfills all criteria for an ideal general matrix for protein chromatography: they are all compromises. Since a combination of hydrophilicity with chemical and physical inertness is best achieved by use of alcohol hydroxyls or amido groups, the most widely used standard chromatographic media for proteins are based on neutral polysaccharides and polyacrylamide.

Among the polysaccharides, cellulose (Figure 2-1 and Figure 2-2) is still a widely used matrix for the synthesis of protein ion exchangers more than 30 years after its introduction by Peterson and Sober in 1956.[4] Modern cellulose gel media are marketed under the trade names Whatman™,[5] Cellulofine™,[6] and Sephacel®.[7] Cross-linked dextran (Figure 2-1 and Figure 2-3) was introduced in 1959 by Porath and Flodin[8] and is marketed under the trade name Sephadex.[7] This material is best known as a gel filtration medium[9] but is also widely used as a matrix for ion exchangers.[10] An account of the history of the development of Sephadex is given in Ref. 11. The neutral hydrophilicity of dextran gels makes them compatible with most proteins and quantitative recovery is thus the rule. Their most important application area today is in desalting and buffer exchange of protein solutions. Protein fractionation by gel filtration is nowadays performed largely with composite gel matrices (see below).

Agarose (Figure 2-1 and Figure 2-3) is a low-charge fraction of the seaweed polysaccharide agar. It was introduced as a medium for chromatography by Hjertén in 1962[12] and marketed under trade names such as Sepharose® and Superose®,[7] Ultrogel™ A,[13] and BioGel™ A.[14] Bead-shaped gels suitable for chromatography are formed by cooling 2 to 15 % aqueous solutions of agarose dispersed in an apolar organic solvent in the presence of suitable emulsifiers.[15] The agarose gel structure is an open three-dimensional network of fibers

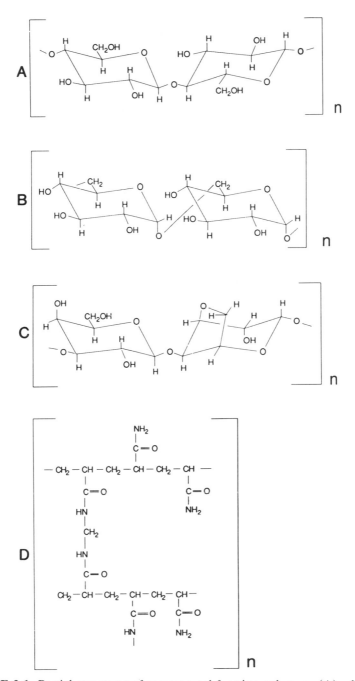

FIGURE 2-1. Partial structures of common gel-forming polymers: (A) cellulose (β-1,4-linked D-glucose), (B) dextran (α-1,6-linked D-glucose; the glycosidic bond here is stretched out for layout reasons), (C) agarose (alternating 1,3-linked β-D-galactose and 1,4-linked, 3,6-anhydro-α-L-galactose), and (D) polyacrylamide cross-linked with N,N-methylenebisacrylamide.

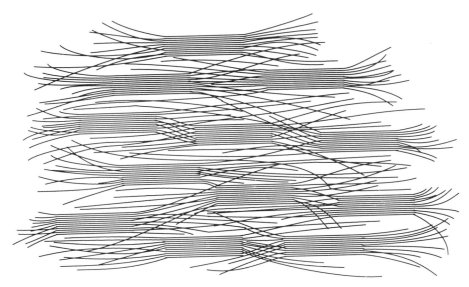

FIGURE 2-2. Schematic representation of part of a cellulose fiber composed of ordered (crystalline) and disordered (amorphous) regions.

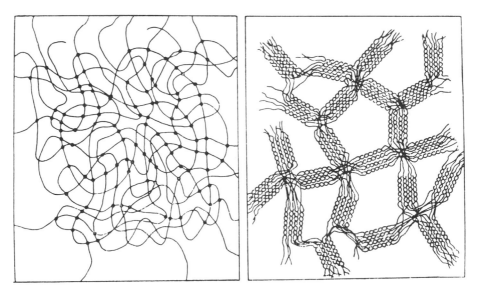

FIGURE 2-3. Schematic representation of the gel structures of agarose (right) and point cross-linked polymers such as polyacrylamide and dextran (Sephadex) (left). (Reproduced by permission of the authors[16] and publishers.)

composed of spontaneously aggregated galactan helices[16] (Figure 2-4). In a 90-μm average diameter 4% agarose gel the total surface area is approximately 5 m^2 ml^{-1} with an average pore diameter of approximately 30 nm.[17] By chemical cross-linking[18,19] of the spontaneously aggregated galactan polymers, the gel rigidity is improved considerably, allowing the manufacture and use of particles down to a 10-μm diameter for a 12% agarose gel.[20] The cross-linking does not change the gel pore structure, but the number of hydroxyl groups is reduced by around 50%, which, however, does not significantly affect the binding capacity for proteins of ion exchangers and affinity chromatography adsorbents prepared from cross-linked agarose gels.

From a functional point of view it is convenient to distinguish between *microporous* and *macroporous* (or macroreticular) gel matrics. The microporous gels are prepared by point cross-linking of linear polymers such as dextran and polyacrylamide (Figure 2-3). This type of gel lends itself ideally to molecular-sieving separations (size-exclusion chromatography, gel filtration). At porosities suitable for protein chromatography these gels become impractically soft and this is why they often participate in the design of composite gels (see below).

Macroporous gels are most often obtained from aggregated and physically cross-linked polymers. Agarose, macroreticular polyacrylamide, silica, and several synthetic organic polymers belong to this group. These gels are best suited for the design of stationary phases intended for ion-exchange chroma-

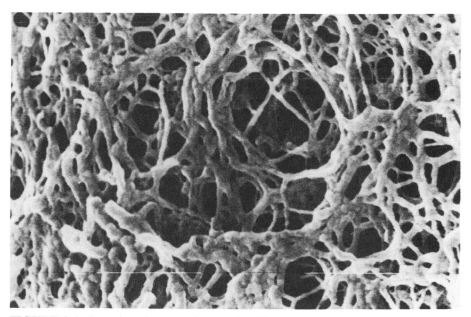

FIGURE 2-4. Scanning electron micrograph of 2% agarose gel. The white bar represents 500 nm. (Preparation and photo by A. Medin[29]).

tography, affinity chromatography, and other adsorption chromatographic techniques. By introducing microporous gel-forming polymers into the pores of macroreticular gels, composite gels are formed that combine the strength of and reduce the weakness of each separate gel moiety. Examples of such gels are Sephacryl® (poly-N,N'-bisacrylamide-dextran) (Figure 2-5),[7] Superdex® (agarose-dextran),[7] Ultrogel™ AcA (agarose-polyacrylamide),[13] and Hyper D (silica-polyacrylamide).[13]

The advent of protein HPLC increased the demand for matrix rigidity. This gave rise to new derivatization procedures for porous silica to make it compatible with proteins, (e.g., the introduction of diol silanols).[21] Despite attempts to stabilize the silica surface by using, for example, zirconium oxide,[22] the major weakness of this matrix is still its instability at alkaline pH. This has encouraged the development of rigid matrices based on porous synthetic organic polymers[23] and agarose.[24] Hydroxyapatite represents a widely used inorganic matrix with high selectivity for a wide variety of proteins and also nucleic acids.[25] Bead-shaped hydroxyapatite particles intended for HPLC with diameters of only a few microns are commercially available in prepacked columns from several sources.

2.2.1.1 *Pore Size Distribution*   An important parameter to consider in the design of media for protein chromatography is the pore size distribution of the gel matrix network.[26] This will affect the shape of the selectivity curve in size-exclusion chromatography (see Chapter 3), as well as the relative diffusion coefficients of proteins when separated by various kinds of adsorption chromatography. One way to estimate the apparent relative pore size distribution of chromatographic gel media is by correlation to the elution behavior of well-defined fractions of a neutral and inert linear homopolymer such as dextran. The resulting calculated distribution coefficients K_d and K_{av} are then plotted against the hydrodynamic (viscosity) radius calculated for each dextran fraction using the equation $R_\eta = 0.0271 \, M_p^{0.498}$, as suggested by Hagel.[27] Figure 2-6 shows the apparent pore size distributions, obtained in this way, of a variety of common chromatographic media for proteins.

The behavior of Hyper D is is remarkably deviant from the other gel media in this study. The pores are unexpectedly small, despite its proven high capacity for relatively high molecular weight proteins. This has been explained in terms of solid diffusion in this possibly polyacrylamide-coated silica composite ("gel-in-a-shell"). The phenomenon has been discussed in comparison to the behavior of the perfusion medium POROS 50.[28]

The apparent relative pore size distribution curve for 2% agarose (Sepharose 2B) is surprisingly similar to that of Poros MC, a perfusion medium. Data might be interpreted such that it should be possible to get perfusion effects also in this agarose gel. Figure 2-4 shows a scanning electron micrograph (SEM) of a 2% agarose gel. The SEM structure image indicates that an agarose gel might be characterized as built up of randomly oriented linear fiber cylinders. Using equations derived from work by Happel and Brenner[29]

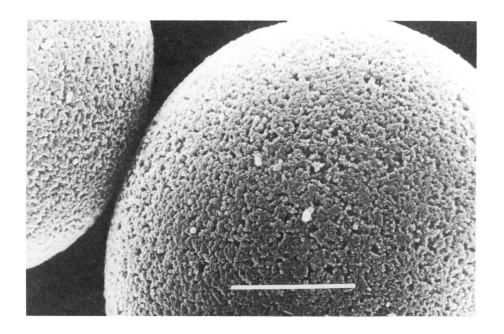

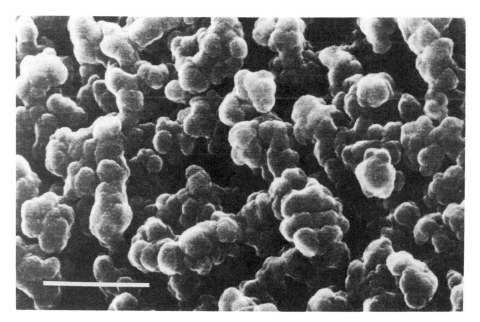

FIGURE 2-5. Scanning electron micrograph of Sephacryl S-500.[7] The white bars represent 10 μm (top) and 0.5 μm (bottom). (Preparation and photo by A. Medin, Uppsala, Sweden.)

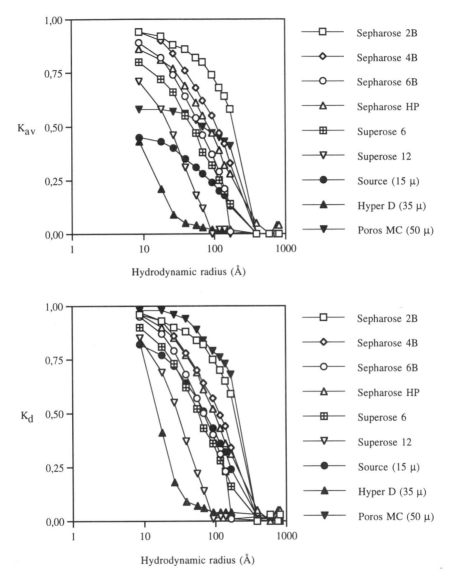

FIGURE 2-6. Distribution coefficients K_d and K_{av} plotted against the hydrodynamic (viscosity) radius for several dextran fractions for a variety of gel media frequently used in protein chromatography (J.-C. Janson, unpublished data). Data on the dextran fractions used can be found in Ref. 26.

for liquid flow through a network of randomly distributed, infinitely long, rigid, and cylindrically shaped fibers, a pressure drop of approximately 10 bar/cm is required to force a perfusive liquid flow of 0.16 ml/cm^2/min through a gel layer with a thickness of 1 cm (assuming a fiber diameter of 10 nm and a fiber volume of 2%), based on such a matrix structure. Figure 2-7 shows the equations used and the resulting graph obtained by plotting the flow rate as a function of the pressure drop through a 2-cm-diameter gel cylinder. Published data on agarose gel structure[30] report an average matrix fiber diameter of 10 nm for a 2% (w/v) agarose gel. This suggests that it should be possible to force a reasonable flow of water through a 2% agarose gel of appropriate thickness without applying excessive pressure drops, provided that the gel matrix is rigid enough. If not, the agarose gel rigidity is known to be significantly improved by adequate cross-linking.[19,20]

2.2.2 Stationary Phase Technology

The shape, rigidity, and particle size distribution profile of the gel matrix are important parameters governing the performance of the stationary phase.

$$Q_1 = \frac{\Delta p}{L}\frac{\pi r_c^2}{8\mu}(1-c)\,r^4\left(\frac{4}{c}-1-\frac{3}{c^2}(1+\ln\sqrt{c}\,)\right)$$

$$Q_2 = \frac{\Delta p}{L}\frac{\pi r_c^2}{4\mu}\frac{(1-c)}{c^2}\,r^4\left(\frac{c^2-1}{2(c^2+1)}-\ln\sqrt{c}\right)$$

$$Q = \frac{Q_1+2Q_2}{3}$$

Q_1 = Flow parallel to the fibre cylinders
Q_2 = Flow perpendicular to the fibre cylinders
Q = Total flow
μ = Viscosity
Δp= Pressure drop
L = Column length
c = Fibre concentration
r = Fibre radius
r_c = Column radius

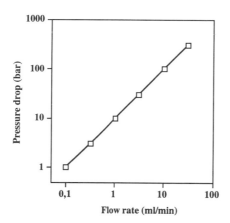

FIGURE 2-7. Equations for liquid flow through a network of randomly distributed, infinitely long, rigid, and cylindrically shaped fibers (derived from Happel and Brenner[29]). The graph shows the corresponding theoretical flow rate as a function of pressure drop for a 2-cm-diameter and 1-cm-long cylinder built up of a 2% (v/v) network of 10-nm diameter fibers.

The original batch polymerization procedures followed by grinding and sieve classification of the irregular particles have, with few exceptions (e.g., inexpensive silica variants), largely been replaced by bead polymerization technologies.

Cross-linked dextran gels, (Sephadex) are thus produced by allowing aqueous solutions of dextran containing an excess of alkali to be dispersed in an emulsifier-containing apolar solvent.[31] The average droplet diameter and size distribution are controlled by the emulsifier concentration and the type and speed of the stirrer. Cross-linking occurs upon addition of a predetermined quantity of epichlorohydrin. The pore diameter and size distribution of the formed gel are governed by the concentration and the molecular weight distribution of the dextran and by the concentration of epichlorohydrin.

Standard polyacrylamide gels (Figure 2-1 and Figure 2-3) for chromatography are formed by bead polymerization of aqueous droplets containing acrylamide (monomer) and methylenebisacrylamide (cross-linker) dispersed in an emulsifier-containing apolar solvent.[32] The gel pore dimensions depend on the total concentration of monomer (%T) and the relative concentration of cross-linker (%C), respectively.

One major step forward in stationary-phase technology was the bead polymerization technique developed by Ugelstad,[33] which is based on the controlled swelling of submicron particles in the presence of monomer and which gives rise to practically monodisperse beads (Figure 2-8) of any desired diameter up to approximately 30 μm. Such particles are commercially available with diameters of 10, 15 and 30 μm and with controlled porosities.[23]

As mentioned earlier, the development of new cross-linking procedures for agarose has enabled the use of this polysaccharide for synthesis of stationary phases that fulfill most of the requirements of a modern high performance matrix. Stationary phases based on 10-,[20,24] 30-,[34] 90-,[35] and 200-μm[36] average diameter beads of cross-linked agarose that cover the demands of protein chromatography from micrograms to several kilograms per cycle and with cycle times from a few minutes to less than 2 h in adsorption chromatography applications, such as ion-exchange chromatography and affinity chromatography.

A procedure for the preparation of a new type of superporous agarose gel media has been reported.[37] Agarose beads (300–500 μm diameter) containing perfusive pores of around 30 μm diameter could be prepared using a double emulsification process.

2.2.3   Introduction of Ligand Groups

The ideal base matrix for the synthesis of stationary phases for adsorption chromatography techniques is identical to the ideal gel-filtration medium in all respects (as indicated in the beginning of Section 2.2.1) but one: it should not give rise to excessive mass transport constraints but should have an open, nonsieving, and easily accessible pore structure, allowing unhindered diffusion

FIGURE 2-8. Scanning electron micrograph of MonoBeads.[7]

of the protein molecules to their adsorption sites on the gel matrix polymer surface.

The surface should be covered with hydrophilic functional groups that do not themselves participate in protein binding but should, however, be amenable to derivatization for the synthesis of ion exchangers and other adsorbents. However, all base matrices interact with at least some proteins under certain conditions. Cross-linking and derivatization procedures may also inadvertently introduce nonspecific adsorption sites that may complicate the interpretation of purification data.

By introducing new chemical structures on the surface of the matrix it is possible to design stationary phases that interact more or less specifically with a particular protein. Least specific are gels for ion-exchange chromatography (IEC), which separates proteins primarily on the basis of their content and distribution of charged groups, and hydrophobic interaction chromatography (HIC) adsorbents, which interact with hydrophobic patches and crevices on the protein surface. Low specificity should, however, not be confused with low selectivity. Both IEC and HIC can be very selective indeed when time can be spent on the optimization of the operating conditions. Stationary phases with the highest specificities are found among the affinity chromatography adsorbents, particularly among immunoadsorbents based on immobilized monoclonal antibodies. Medium specificities are obtained using so-called general ligands, (i.e., group-specific ligands such as coenzyme analogs).

The synthesis of an ion exchanger usually takes place by a one-step chemical reaction in which a charged molecule containing a reactive group (often a halogenide) is allowed to react with the alcohol hydroxyl group-containing matrix (such as a polysaccharide) under strongly alkaline conditions. Most DEAE and CM ion exchangers are produced in this way. The synthesis of an affinity chromatography adsorbent, however, is normally a three-step procedure. In the first step the matrix is activated by the introduction of reactive groups. In the second step the ligand is covalently attached to the matrix by reaction with the activated group. In the third and last step the excess reactive groups are inactivated by blocking with a low molecular weight substance that displays low adsorptivity when coupled. Different coupling methods are discussed in some detail in Chapter 10. Some of the more general approaches are mentioned below for hydroxyl group containing matrices. For synthetic, organic polymer-based matrices and for silica gels, special techniques are required (see Chapter 10).

As with ion exchangers, activation reactions normally take place at alkaline pH. The most frequently used reagents for hydroxyl group-containing matrices are CNBr, bisepoxides, or divinylsulfone, introducing reactive cyanoesters, epoxides, and vinylsulfones, respectively. These can then be reacted with nucleophilic groups such as thiols, amines, and alcohols. The blocking is often carried out with an excess of ethanolamine. Several alternatives to these coupling methods are available, also for the binding via carboxylates. Amino groups are the most commonly used functional groups for the coupling of low molecular ligands as well as proteins. Carbohydrates (sugars, oligosaccharides) can be attached to epoxides and vinylsulfones via their hydroxyls.

2.2.4 Ligand–Protein Interactions

The binding of a dissolved protein to an immobilized ligand is due to one or several of the following interactions:

1. Ion–ion or ion–dipole bonds
2. Hydrogen bonds
3. Dispersion or van der Waals forces

4. Aromatic or $\pi-\pi$ interactions

5. Hydrophobic effect or hydrophobic interaction

Sometimes covalent bonds are formed, as in covalent chromatography (see Chapter 9), or metal ligand bonding, as in immobilized metal ion affinity chromatography (see Chapter 8). The hydrophobic interaction differs from others in that it has a large entropic component.

The system solution–protein–stationary phase at each moment minimizes its free energy, giving rise to the binding of the protein to the ligand. This interaction is normally the result of many secondary bonds formed, and several of the previously mentioned factors might contribute. Each one of these might be weak, often around 1 kcal/mol, but several of them add up to a considerable binding constant. A K_d of 10^{-5}, which gives rise to retardation, corresponds to a binding energy of 7 kcal/mol in standard conditions according to

$$\Delta G = RT \ln K_d$$

and thereby four to eight weak interactions (see, e.g., Ref. 38). A general treatment of the subject is given in Ref. 39.

The surface of globular proteins typically has around 45% hydrophobic residues and 55% charged or hydrophilic uncharged residues accessible to surrounding water (Figure 2-9). In the latter groups the positive charges of lysine, arginine, and histidine side chains, as well as the α-amino group, change according to pH (Table 4-2). The negative charge of glutamic acid and aspartic acid side chains and the α-carboxyl group as well are pH dependent. Of these the histidine side chain and the α-amino groups titrate within the pH region that is commonly used in chromatography. The cysteine thiol side chain is not typically available in extracellular proteins, but might be so in intracellular proteins. It does not play a role as an anion, although it has a pK of about 8.5. Other hydrophilic groups such as asparagine, glutamine, serine, and threonine as well as peptide bonds are mostly involved in hydrogen-bonding interactions.

Although these interactions determine the strength of the binding of proteins, the specificity of binding is to a large extent explained by the fit between the ligand and the protein. This is particularly true in the case of affinity chromatography. But even if other types of chromatography might to a larger extent depend on one of the interactions discussed it is certainly not a sufficient explanation for protein selectivity. Ion exchange thus depends on ionic forces and hydrophobic interaction chromatography on hydrophobic interactions. Still, net charge or net hydrophobicity cannot alone explain the differences in retardation of different proteins. Here also the distribution of interacting groups on the protein surface and thereby the fit between the protein and the ligand is of major importance as well as secondary interactions, such as H-bonding.

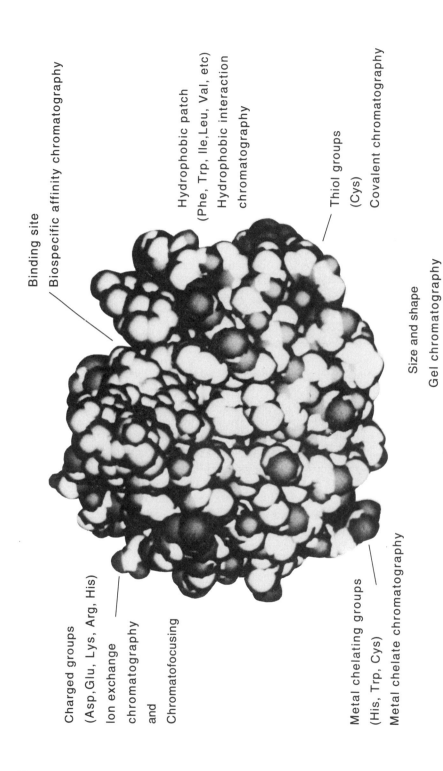

Binding site
Biospecific affinity chromatography

Hydrophobic patch
(Phe, Trp, Ile, Leu, Val, etc)
Hydrophobic interaction
chromatography

Thiol groups
(Cys)
Covalent chromatography

Size and shape
Gel chromatography

Charged groups
(Asp, Glu, Lys, Arg, His)
Ion exchange
chromatography
and
Chromatofocusing

Metal chelating groups
(His, Trp, Cys)
Metal chelate chromatography

The different interactions are each influenced by the solvent in some particular way. Increased ionic strength thus decreases ionic interactions, whereas hydrophobic interactions are favored. The different ways to elute adsorbed proteins are discussed in detail for each type of chromatography. It should, however, be kept in mind that since a single type of interaction is seldom responsible for binding, nonconventional approaches are sometimes worth trying.

2.3 CHROMATOGRAPHIC THEORY

2.3.1 Chromatographic Quantities

A large number of quantitative concepts are used to describe a chromatographic experiment. Some of the most significant of these are discussed here.

2.3.1.1 Retention Parameters   A chromatogram is a plot of concentration at the column exit versus time or eluent volume. The volume that has passed the column from the introduction of sample until the emergence of a certain component as a peak in the chromatogram is the *retention volume, V_R* (see Figure 2-10).

Instead of retention volume, *retention time, t_R,* is often used, which may be easier to measure. Obviously, $V_R = F_C t_R$, where F_C is the volumetric flow rate. The numerical value of the retention volume depends largely on the column used. To obtain a normalized retention quantity, the *retention factor k* is generally used. It is equal to the number of *void volumes, V_0,* needed to elute the compound of interest minus one. Thus

$$k = \frac{V_R}{V_0} - 1 \text{ or } V_R = V_0(1 + k) \qquad (2\text{-}1)$$

In Figure 2-10 an alternative x scale is graduated in units of k. Often the retention factor is termed *capacity factor* and is written k'. This is, however,

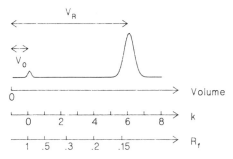

FIGURE 2-10. Relationship among various retention parameters.

in variance with official recommendations for the terminology of chromatography. In this chapter, the terminology recommendations by IUPAC[40] are followed where relevant.

Another normalized retention measure is the *retardation factor, R_F*, which is the ratio between the velocity of the actual component and that of the mobile phase. Consequently

$$R_F = \frac{V_0}{V_R} = \frac{1}{(1 + k)} \tag{2-2}$$

R_F is especially used in noncolumn versions of chromatography (e.g., thin-layer chromatography). When plotted in a chromatogram as in Figure 2-10, a scale in units of R_F is nonlinear.

The void volume or holdup volume, V_0, is usually seen as the volume of mobile phase in the column. With liquid chromatographic columns, packed with small particles or gels, this simple correspondence is unfortunately not exact, as small molecules penetrate the pores to a greater extent than large molecules and thereby distribute in a larger liquid volume. Even in the absence of physical or chemical interactions with the stationary phase itself, this effect leads to the separation of molecules of different sizes: molecules of a particular size will emerge at the column outlet after passage of a volume that is equal to the column volume that is available to that size. This is the principle of gel-filtration chromatography, where the porosity characteristics of the materials are chosen to maximize this effect. In other types of chromatography, the effect should be minimized by using pores that are wide enough to accommodate all molecules of interest. This is especially important for chromatography of proteins.

2.3.1.2   Peak Shape and Width; the Theoretical Plate Concept

A separation of two compounds is possible only when their velocities (R_F values) and thereby their k and V_R differ by some amount.

As is obvious from Figure 2-11, additional parameters related to the width of the peaks are needed to specify the resolution of two compounds. To describe the width of chromatographic peaks, the concept of *theoretical plates, N,* is commonly used.[40] This term is related to fractional distillation and is

FIGURE 2-11. (a) A chromatogram showing two resolved peaks. (b) Same retention volumes as in a, but wider peaks.

somewhat misleading as there are no "plates," either "theoretical" or otherwise in a chromatographic column. The proper definition is based on statistical theory:

$$N = \frac{\mu^2}{\sigma^2} \tag{2-3}$$

where μ^2 and σ^2 are the mean and variance of the chromatographic peak, respectively. In practice, one of the following formulae is used to calculate N from a chromatogram:

$$N = 5.55 \left(\frac{t_R}{w_{1/2}}\right)^2 \tag{2-4a}$$

$$N = 16 \left(\frac{t_R}{w_b}\right)^2 \tag{2-4b}$$

$$N = 6.28 \left(\frac{t_R}{A_p}\right)^2 \tag{2-4c}$$

Here, $w_{1/2}$ and w_b are the width of the peak at half the peak height, h_p, and at the base, respectively (see Figure 2-12). A_p is the area of the peak.

To apply these formulae, it is not important whether t_R or V_R is used. It is, however, crucial that the widths are measured in the same units at t_R. Typically they are measured in centimeters directly from the recorder paper using a ruler. Equation (2-4c) is intended to be used with an electronic integrator or a computer system, which usually supplies values for t_R, h_p, and A_p, but often no peak widths.

Equations (2-4a–c) are derived from the gaussian (normal) probability curve, which is a fair description of a chromatographic peak in simple cases.

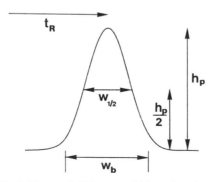

FIGURE 2-12. Definitions of different widths of a chromatographic peak.

It should be emphasized that they are only valid in isocratic chromatography and when the peak is reasonably symmetrical. For tailing peaks, which are often obtained, it is not meaningful to calculate the number of theoretical plates.

The theoretical plate concept is also used widely to characterize the performance of a chromatographic column. To a first approximation, all (symmetrical) peaks in a chromatogram show roughly the same plate number. Consequently, this number can be considered as a property of the colummn used. A large plate number means narrow peaks and thus a "good" column.

As the plate number is approximately proportional to the column length, L, column quality can also be expressed in terms of theoretical plates per meter column or height of a theoretical plate (HETP), H, which is defined by

$$H = \frac{L}{N} \tag{2-5}$$

Thus, for a "good" column the value of H is small.

2.3.1.3 The van Deemter Equation   According to the theory originally developed by J. J. van Deemter and co-workers[41] for gas chromatography, the plate height H is a sum of three independent contributions, which depend on the flow rate of the mobile phase according to the equation

$$H = A + \frac{B}{u} + C \cdot u \tag{2-6}$$

Here, u is the *linear flow rate* and A, B, and C are constants. A plot of the van Deemter equation will show a minimum at a certain flow rate (see Figure 2-13). This is the "optimal" flow rate at which a maximal plate number is obtained. In practice, especially for liquid chromatography of macromolecules, the "optimal" flow rate is often unpractically low and columns are operated

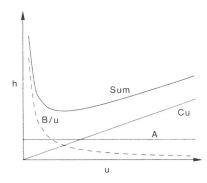

FIGURE 2-13. Plot of the three terms in the van Deemter equation and their sum.

at considerably higher flow rates. The third term in the van Deemter equation then becomes especially important.

A detailed interpretation of the terms is complex but rewarding. The classical work in this field is Giddings.[1] Other more recent and fairly complete treatments also exist.[3]

Briefly, the first term, A, the "eddy" term, describes peak broadening resulting from the complicated geometry in a packed column: Some molecules will find a relatively straight path through the column, whereas others may follow a longer, more tortuous path. The second term, B/u, results from the diffusion of sample molecules along the column. The third term, $C \cdot u$, originates from various kinetic parameters, such as slow transfer of molecules into and out of pores or within the stationary phase, noninstantaneous equilibrium, and so on. Both the A term and the C term depend on the diameter d_p of the column packing particles. A is directly proportional to d_p, whereas the relation for C is more complex: the contribution to the C term of slow mass transfer in pores and so on is proportional to d_p^2 whereas the nonequilibrium contribution is independent of d_p.[1] Using smaller particles thus leads to a lower plate height and a larger plate number. Observe that the effect might be largest for the C term, thereby moving the minimum in the van Deemter plot toward higher flow rates and decreasing the plate height, especially for high flow rates, thereby permitting efficient separations in shorter times. The price to be paid is the necessity of using high-pressure pumps to force the mobile phase at high-flow rates through columns with high resistance. This is the reason why the "P" in "HPLC" can mean "performance," "pressure," and "price."

To facilitate comparisons among columns of different types, the concepts of *reduced plate height, h,* and *reduced flow rate, v,* are often used. They are defined as

$$h = \frac{H}{d_p} \text{ and } v = \frac{u \cdot d_p}{D_M} \tag{2-7}$$

where D_M is the diffusion coefficient to the solute in question in the mobile phase. The adoption of reduced parameters removes the influence of d_p and D_M from the A and B terms in the van Deemter equation, whereas the C term still depends on these parameters.

2.3.1.4 Perfusion Chromatography

The concept of *perfusion chromatography* has been introduced by Afeyan and coworkers.[42] Column-packing particles with large pores, passing through the entire particle, are used. Similar materials were also termed "gigaporous."[43] A part of the mobile phase may flow through these pores. This significantly decreases the resistance to mass transfer caused by diffusion in pores as the molecules to be separated are brought to the inner of the particle by convection, not only by diffusion. The result is a smaller C term in Eq. (2-6) and the possibility for rapid separations without the need for especially small particles and the accompanying need for high pressure. These advantages are especially noted for chromatography of macromolecules due to the low diffusion coefficients of large molecules.

2.3.1.5 Resolution   In a chromatogram of a complex sample, a great number of peaks are to be separated, but it is only necessary to consider one pair of peaks at a time. The *peak resolution, R_S*, of this pair is defined by

$$R_s = \frac{t_{R2} - t_{R1}}{\frac{1}{2}(w_{b1} - w_{b2})} \tag{2-8}$$

that is, the distance between the peak maxima is divided by the mean peak width (see Figure 2-14). Just as for Eqs. (2-4a–c), retention volumes can be used instead of times. For $R_S < 1$, the peaks are incompletely separated, for $R_S = 1$, they just touch each other at the base, whereas for $R_S > 1$, there will be a stretch of baseline between the peaks. Optimal separation conditions are reached when $R_S \geq 1$ for all pairs of interest. If R_S is considerably greater than 1 for all pairs, the resolution is unnecessarily good and the separation can be made faster.

Upon combination with Eq. (2-4b), the resolution can be written:

$$R_s = \frac{\alpha - 1}{\alpha + 1} \cdot \frac{\overline{k}}{1 + \overline{k}} \cdot \frac{\sqrt{N}}{2} \tag{2-9}$$

Here, $\overline{k}$ is the mean of the retention factors for the two peaks, $(k_1 + k_2)/2$ and α is the relative retention or separation factor, k_1/k_2.

The resolution depends, according to Eq. (2-9), on three more or less independent factors: the selectivity factor, the retention factor, and the plate number factor. To optimize the resolution, each of these factors should be considered.

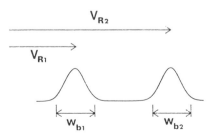

FIGURE 2-14. Peak widths and retention volumes for the definition of resolution.

The selectivity factor, $(\alpha - 1)/(\alpha + 1)$, is usually the most important. For example, an increase of α from 1.01 to 1.02 (i.e., only 1% change) will double the resolution. However, if $\alpha = 1$, no separation is possible. The retention factor, $k/(1 + k)$, increases with k, but when k is larger than 5 the increase is marginal and will mainly result in slower separation. If $k < 1$, the resolution is unnecessarily low. Finally, an increase in the plate number, N, (e.g., by using a longer column), will increase resolution. However, a fourfold increase in column length is necessary to double the resolution.

Facing an inadequate resolution, the most effective remedy would probably be to change the chromatographic conditions aiming at a higher relative retention. Changes in pH or in the composition of the eluent are the usual approaches tried to increase α. A mere increase of column length might be impractical and would considerably increase separation time (but not necessarily the total time spent on the entire task, especially if only a limited number of runs are to be made).

If the problem is the opposite and more pleasant one, (i.e., to speed up an unnecessarily effective separation), the use of a shorter column will be the best solution, offering a considerable time savings on each run. Also, a higher relative retention is beneficial here as it permits the use of still shorter and faster columns.

2.3.2 Retention in Adsorption Chromatography

Most versions of chromatography used for protein separation (except gel-filtration chromatography) may be more or less adequately treated together under the term adsorption chromatography. This implies that the sample molecules are adsorbed onto the surface (or in a thin surface layer) of the stationary phase. The precise nature of the adsorption forces varies among the techniques and can for the moment be disregarded.

2.3.2.1 *The Adsorption Isotherm*   The central concept in adsorption is the adsorption isotherm, which is a plot of the sample concentration on the adsorbing surface of the stationary phase versus the sample concentration in the mobile phase. These concentrations are written C_S and C_M, respectively. The concentration C_S may be expressed either in units of moles per surface area or moles per gram of adsorbent. C_M is expressed in moles per liter.

A general shape of such a plot is shown in Figure 2-15. The curved shape of the isotherm originates from competition among sample molecules for adsorption sites. The most simple description of this competition is due to Langmuir.[44]

Assume that the adsorbent has a fixed number of equal ligands or adsorption sites, S, to which the sample protein molecules, P, bind one to one in a reversible way. The following equilibrium then applies:

$$P + S \rightleftharpoons PS; \; K = \frac{[PS]}{[P] \cdot [S]} \tag{2-10}$$

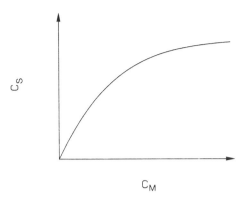

FIGURE 2-15. Typical adsorption isotherm.

where K is an association constant. Clearly $[PS] = C_S$ $[P] = C_M$ and $[S] +$ $[PS] = Q$, the *adsorbent* or *binding capacity,* that is, the number of sites per unit surface area (or weight). This has nothing to do with the "capacity factor," k, as defined in Eq. (2-1). We can easily solve the equilibrium equation for C_S, which leads to Langmuir's adsorption isotherm equation:

$$C_s = \frac{Q \cdot K \cdot C_M}{1 + K \cdot C_M} \tag{2-11}$$

For most adsorption equilibria we must consider the competition of two different counter ligands, one of which is the sample molecule, P, and one is a component, E, of the eluent.

$$P + ES \rightleftharpoons E + PS; \frac{[PS][E]}{[P][ES]} = K_{p/E} \tag{2-12}$$

Here, $[PS] = C_S$, $[P] = C_M$, and $[PS] + [ES] = Q$, and $K_{P/E}$ is a selectivity constant, the quotient of the relevant association constants.

From this, Eq. (2-11) is again obtained with $K = K_{P/E}/[E]$. For a fixed eluent concentration the Langmuir equation is thus also valid for competition equilibria. The model is also immediately applicable to an ion-exchange equilibrium, where P signifies the protein ions, E is an ion in the eluent, and Q is the ion-exchanger capacity.

In practical protein purification applications, there may be (and usually are) interfering molecules of several kinds in the sample. Also, in this case, the Langmuir model thus still applies, but with a conditional constant that incorporates the influences of the competing molecules.

As the natures and concentrations of interfering sample components are generally unknown or at least incompletely known, the value of measured $K_{P/E}$ values for the prediction of retention volumes is limited.

. The assumptions upon which the derivations just mentioned are based may not be entirely applicable in all cases: the adsorption sites may be unequal, there may be multisite binding (certainly relevant for macromolecules), and there may be interactions (e.g., repulsion) between adsorbed molecules. In these cases the Langmuir model is not quantitatively correct, but in practice it often has a reasonable semiquantitative validity.

2.3.2.2 Chromatographic Retention   In one of the first papers on chromatographic theory,[45] the relationship between the retention volume and the adsorption isotherm was derived:

$$V_R = V_0 + A_S \frac{dC_S}{dC_M} \tag{2-13}$$

Here A_S is the total area of the adsorbent in the column or, alternatively, the weight of the adsorbent, depending on the definition of C_S (see earlier discussion). Inserting Eq. (2-11), we obtain after differentiation:

$$V_R = V_0 + \frac{A_S \cdot O \cdot K}{(1 + KC_M)^2} \tag{2-14}$$

which is the proper expression of the retention volume, assuming a Langmuir adsorption isotherm.

If the concentration C_M of sample is small enough, the denominator in the second term of Eq. (2-14) is practically unity, leading to

$$V_R = V_0 + A_S \cdot Q \cdot K = V_0 + n_T \cdot K \tag{2-15}$$

where n_T is the total number of adsorption sites in the column. From Eq. (2-1), we find that

$$k = \frac{n_T}{V_0} \cdot K \tag{2-16}$$

The retention volume here is independent of sample concentration and is solely determined by column parameters and by the equilibrium constant, K. As mentioned in Section 2.3.1.1, V_0 may vary slightly with molecular size, which influences the retention.

This case is termed *linear chromatography*. In analytical applications of chromatography, this is the preferred case. It is characterized by a simple theoretical description of peak retention and dispersion (broadening). The concepts related to theoretical plates as described earlier are valid. The term "linear" refers to the fact that it is equivalent to the assumption of a linear adsorption isotherm. In reality the isotherm must be curved (see Figure 2-15), but at sufficiently low concentrations the curvature becomes negligible.

In many cases, especially when chromatography is used for preparative purposes (which is important in the context of this book), the assumption of linearity is not valid due to the relatively high concentrations of sample. If this happens, the retention volume varies with sample concentration, C_M, as described by Eq. (2-16). This leads to asymmetric "tailing" peaks, as parts of a peak with low concentrations are retarded more than the parts with high concentrations. Figure 2-16 shows tailing peaks of different sizes. The width of a tailing peak depends partly on the dispersion factors as described previously, but these are interrelated with the peak broadening due to the tailing itself in an untractable way. Consequently, theoretical plates and related quantities are not applicable to tailing peaks and should not be calculated. It is even impossible to treat this general problem of nonlinear chromatography in a mathematical way. Equation (2–15) describes fairly well the retention volume of the maximum of a tailing peak with C_M equal to the concentration at the maximum, but for the rest of the peak, no accurate mathematical description exists. A detailed review on this topic has been published.[46]

Observe that although the component of interest may not be present at a high concentration, other components may and often are. All components can then compete for the same adsorption (ion-exchange) sites and, consequently, influence each others retention.

The conditions prevailing when a mixture of proteins is chromatographed are very complex and attempts to calculate retention volumes, and so on from batch experiments or chromatography of pure compounds can be very inaccurate.

An erroneous interpretation of retention in nonlinear chromatography, usually implicitly expressed as a nonconstant capacity factor, is found in several texts. The capacity factor k can be written [see Eq. (2-16) and the definition of k]

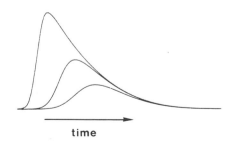

time

FIGURE 2-16. Tailing peaks of different sizes.

$$k = \frac{A_S}{V_0} \cdot \frac{C_S}{C_M} \tag{2-17}$$

This is correct in the case of linear chromatography and is often stated as an alternative definition of k. However, this is not generally true, and applying Eq. (2-14) to nonlinear chromatography is dangerous. It would imply that

$$V_R = V_0 + A_S \cdot \frac{C_S}{C_M} \tag{2-18}$$

instead of the correct Eq. (2-13). The difference between Eq. (2-13) and Eq. (2-18) is conceptually difficult and leads seemingly to a paradox. The matter has been clarified by Helfferich.[47]

2.4 CHROMATOGRAPHIC PROCEDURES

2.4.1 Sample Introduction

The usual way to perform a chromatographic experiment involves plug injection, (e.g., the introduction of a small volume of sample at the beginning of the column). The width of this plug obviously influences the width of the resulting peaks. In theoretical discussions, the width of the plug is usually assumed to be negligible. However, in preparative chromatography, wider plugs may be tolerated, thereby permitting the introduction of a larger volume of sample. See the example in Figure 2-17.

As the introduction of larger amounts of sample may cause overloading and nonlinear conditions (see Section 2.3.2.2), the effect is generally more complex than an increase in peak widths.

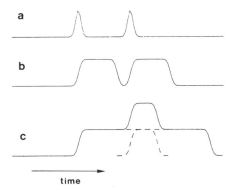

a

b

c

time

FIGURE 2-17. The influence of sample size in preparative chromatography. (a) Small (analytical) sample size. (b) Sample size optimized with regard to sample throughput. (c) Too large samples size (frontal chromatogram).

A special case is *frontal chromatography,* where the pure eluent is exchanged for a solution of sample that is pumped into the column. The sample will appear, after some time, as a more or less steep concentration step in the detector. The midpoint of the step corresponds to the retention time and the height of the fully developed step to the initial sample concentration. With several components, several steps will build up and only the fastest component will be (partly) separated from the other components. If the sample is again exchanged for pure eluent, negative steps will occur and the result is equivalent to a large plug. Figure 2-17c is an example of such a frontal chromatogram. The technique is used mostly for the determination of physicochemical parameters by chromatography, where it has some advantages, and for sampling and preconcentration of dilute samples. It may also unintentionally be encountered in preparative chromatography after the introduction of excessively wide plugs. In the latter case, it is important to realize that a chromatogram such as that in Figure 2-l7c contains two components, not three.

2.4.2 Chromatographic Development

2.4.2.1 Isocratic Elution   The most simple mode of chromatographic development is the isocratic elution mode, wherein all conditions are held constant throughout the experiment. Each component of the sample will thus travel through the column according to Eq. (2-13) to Eq. (2-15).

The retention times of the various components and thereby the degree of separation are determined by the corresponding values of K. As shown in Section 2.3.2.l, this constant may be interpreted in several ways, depending on the physical process on which the separation is based. Generally, a necessary condition for a successful isocratic elution is that all sample components elute in a reasonable time and that compounds of interest are not eluted too early [which would destroy resolution, cf. Eq. (2-9)]. This can only be accomplished for a narrow range of sample types, and the application of isocratic elution to a wide range of sample types leads to disturbing trade-offs regarding separation power and time. This is sometimes referred to as the *general elution problem.* To solve this problem, it is necessary to employ *gradient elution,* where the composition of the eluent is changed during the development. The change may be continuous, usually linear, or stepwise, and the object is to decrease K successively for each component with time.

2.4.2.2 Gradient Elution   This increase in "elution power" may be accomplished in various ways for different chromatographic techniques: in reversed-phase liquid chromatography the polarity of the eluent is decreased, thereby decreasing the partition coefficients; in ion-exchange chromatography the concentration of an eluent ion is increased, leading to a decrease in the apparent K [see Eq. (2-12)]; and in several other techniques the successive addition of competing compounds will affect the apparent K in the desired direction.

In the beginning of the chromatographic run, a low elution power is chosen. Then, the components that are most loosely bound to the adsorption (etc.) sites will elute under favorable conditions. Other more strongly held compounds will be successively eluted as the elution power is increased. The result is usually an increased resolution over a wide range of sample compounds.

The retention times in a gradient elution run cannot be calculated directly from Eq. (2-15) (and similar equations), as the parameter K varies with time. If the variation of K with time is known, the retention time can in principle be calculated by integration, a procedure that is rarely applied.

The plate number concept, as defined previously, is not relevant in the case of gradient elution, and the application of Eq. (2-4) to a gradient chromatogram leads to gross overestimates of n. Beware of excessive column performance claims produced in this way.

2.4.2.3 Displacement Chromatography An extreme case of gradient elution is displacement chromatography, which is based on competitive binding between the sample components themselves and an additional compound, the terminal displacer or developer, which is more strongly adsorbed than any of the proteins in the sample. During the course of the experiment the continuously added terminal displacer will push the sample components in front of itself, forcing them to displace each other, thereby forming a so-called displacement train in which the different molecules will arrange themselves in the order of their interaction strength with the adsorption sites of the column. In order to improve the resolution between the sample components, spacers, (i.e., molecules with intermediate adsorption strengths), are usually added. Examples of such compounds are the carrier ampholytes used in isoelectric focusing or carboxymethyl dextrans. A detailed description of displacement chromatography is given in Ref. 48. A review of displacement chromatography of proteins is given in Ref. 49 and of peptides in Ref. 50. A brief discussion can also be found in Chapter 4 of this book (see Section 4.7.2.4).

2.4.3 Determination of the Column Capacity and Association Constants

The chromatographic retention depends, as seen earlier [e.g., in Eq. (2-14)], on essentially two parameters apart from column dimensions: namely, the column binding capacity Q and the (apparent) association constant K.

The capacity Q for a particular protein is a complex function of several parameters: matrix composition and matrix pore structure, particle diameter and particle size distribution profile, protein molecular weight and solubility, forward and backward rate constants for the binding reaction, bulk-, film-, and gel pore-diffusion constants of the protein, and, finally, the event of possible competitive binding and displacement effects of other proteins present in the sample solution. It is convenient and useful to distinguish between the *nominal* binding capacity of a particular stationary phase such as an ion exchanger and its *dynamic* or *functional* binding capacity. The nominal binding capacity of an ion exchanger can, for example, be determined by an acid–base titration.

This value, however, does not tell how much protein will bind under normal operating conditions in a running column. This can only be measured by chromatographic methods and can, for a single column, vary considerably for different proteins. For a single protein the dynamic binding capacity varies significantly with the flow rate used during adsorption and washing.

The determination of the dynamic, or functional, binding capacity is usually performed by frontal chromatography (see Section 2.4.1) under isocratic conditions. Figure 2-18 shows typical fronts in nonlinear adsorption chromatography.

A positive step injection (i.e., the exchange of pure eluent for a solution of sample with concentration C_0) will produce a sharp front at the outlet as shown in Figure 2-18b, whereas a negative step injection (removal of the sample feed) produces a tailing, "diffuse" concentration step (Figure 2-18a). It is easiest to treat the second case, as the tailing slope follows Eq. (2-14). The area A in Figure 2-18 is found by integration:

$$A = \int_0^{C_0} V_R \cdot dC_M \tag{2-19}$$

If the experiment is arranged to make K very large, we obtain with Eq. (2-14):

$$A = V_0 C_0 + A_S Q \tag{2-20}$$

Area A in Figure 2-18b corresponds to the amount of sample that has been held on the column, whereas the analogous area in Figure 2-18a corresponds

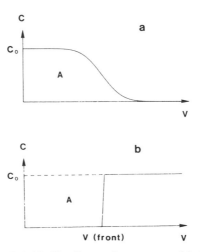

FIGURE 2-18. Nonlinear chromatographic fronts.

to the same amount that is released when sample feed is removed. Consequently, these areas are equal. The capacity Q can easily be calculated from a sharp frontal chromatogram, as the area A in such a case will be nearly rectangular,

$$Q = \frac{C_0}{A_S} (V_{R,front} - V_0) \qquad (2\text{-}21)$$

provided that K is large. $V_{R,front}$ is the retention volume measured from the start of sample introduction to the sharp front. In order to obtain a larger K, a suitable sample component (we intend to measure a property of the column, so there is a choice of sample) and an eluent with low eluting power, free from competing compounds, should be chosen. To obtain valid values of Q, it is advisable to repeat the measurement under different conditions. In some cases, the binding equilibrium might be slow. This is often the case in affinity chromatography. The result is that the shape of the front is not sharp but "rounded." The measurements described are still valid, provided that area A is calculated by integration of the recorder signal.

The determination of the association (partition) coefficients, K, can in some cases be performed by nonchromatographic methods. Hutchens and Yip[51] determined binding constant and capacities of several proteins to IMAC gels using Scatchard plots, (i.e., equilibrium binding). The values obtained agreed nicely with those derived from frontal chromatography. The chromatographic determination of K is simple: the retention volume of a peak obtained after the injection of a narrow sample plug of low concentration gives directly [Eq. (2-15)] $A_S \cdot Q \cdot K$. If $A_S \cdot Q$ is determined as described earlier, K is easily calculated.

2.5 CHROMATOGRAPHIC TECHNIQUES

2.5.1 General Comments

Modern column chromatography utilizes sophisticated equipment to obtain high-resolution separations. However, for some applications this might not be necessary. The most simple way to carry out an adsorption experiment is batchwise, by simply stirring the adsorbent with the protein sample, choosing proper conditions for adsorption and subsequent desorption. This has the advantage of being applicable with large volumes of protein solution, which do not necessarily need to be clear, and is sometimes of use at an early stage in a purification procedure or simply to test whether adsorption occurs under the conditions selected.

Column chromatography can be performed as low pressure (or standard), medium pressure, or high-pressure liquid chromatography. Medium-pressure techniques also include fast protein liquid chromatography, for example the

Pharmacia chromatography system FPLC®. For desalting experiments, adsorption tests, and so on, simple column chromatography equipment is often sufficient. High-resolution results can also be obtained with standard, low-pressure equipment, provided the selectivity of the column packing material is sufficiently high. High-resolution fast techniques require small diameter beads and equipment able to withstand the often high pressures necessary to force the buffer or solvent through the column.

In all three techniques the size, and thus the capacity, of the chromatographic column can vary considerably. Medium scale (i.e., from a milligram to a single gram of protein) is usually the easiest to handle. Microgram scale requires sensitive analytical techniques, high purity of buffers and solvents, although less volume is consumed, and often special care has to be taken to avoid adsorption to the walls of vessels. Equipment for large-scale chromatography runs into other difficulties, in particular the high cost of larger amounts of modern HPLC media. However, most industrial process chromatography separations of proteins today are based on 90-μm-diameter standard media such as Sepharose Fast Flow. Useful information on large-scale chromatography can be found in Ref. 52–57.

The most crucial point in column chromatography is to achieve a good column packing. This is therefore treated in some detail in the final section of this chapter. FPLC and HPLC columns are normally delivered prepacked from the manufacturer. However, most other gels are delivered in bulk and must be packed in the user's laboratory.

2.5.2 Packing of Columns

In any chromatographic experiment the result obtained can never be better than the quality of the column packing allows. This is why it always pays off in the long run to learn how to pack the most commonly used stationary phases. One should be particularly cautious when packing columns for gel filtration and other isocratic techniques. Detailed packing instructions are usually provided by the manufacturers of a particular gel material. Here only some general principles are discussed.

The packing techniques used differ depending on the type of stationary phase. The most important discriminating parameter is the rigidity of the gel matrix. It is thus convenient to distinguish among soft, semirigid, and rigid gel matrices. Particle shape, diameter, and size distribution are also important parameters to consider in column packing. The first step is to mount the column with its extension tube on a steady laboratory stand and to ensure that the column tube is perfectly vertical. The stationary phase slurry is degassed to remove all trapped air.

Rigid gel materials such as silica with particle diameters in the range of 5 to 15 μm are preferably slurry packed in dry acetone or chloroform. Slurry concentrations around 10% and packing pressures up to around 300 kg cm^{-2} usually give satisfactory results.

Semirigid gels such as Sephacryl HR, Sepharose FF, and Superose are preferably packed in two steps. In the first step, a homogeneous slurry (10 to 50%) containing all of the stationary phase intended for the column is poured into the column fitted with an extension tube. The extension tube is connected to a pump that is adjusted to a medium to high flow rate. For a 45-μm Sephacryl HR, this means around 30 cm h^{-1} for columns 40 to 100 cm in length. When the bed has settled, the second step in the column packing involves a doubling of the flow rate to, in this case, around 60 cm h^{-1}. The packing of the column should not be considered complete until four to five column volumes of packing buffer have been pumped through. The crucial feature of this procedure is that it prevents the formation of a plug of hard-packed gel at the bottom frit of the column, which will inevitably occur upon packing in one step at constant pressure with an initial high flow rate. This plug will block the flow and give rise to badly packed columns.

Every column should be tested for packing quality. A zone of acetone (0.5% V_t) at 30 cm h^{-1} is suitable for this purpose. The reduced plate height should fall in the range of 2 to 3, at the lower end for experienced column packers and at the upper end for beginners.

The packing buffer composition for semirigid gels does not seem to be critical. Similar results are obtained with distilled water and with various buffer salt solutions. The most convenient way is to use the slurry obtained by shaking the original bottle in which the gel is delivered and diluting with distilled water to the desired slurry concentration. The column bottom frit or filter mesh should be wetted and all air removed; 20% ethanol is recommended for this purpose. A few centimeters of this solution can be left in the column before the addition of the gel slurry.

A critical point in the column packing is the application of the adaptor on top of the packed bed. As a general recommendation, the adaptor should compress the upper part of the bed approximately 5 mm. Some workers prefer to pack the columns upside down toward the adaptor. In this way they get the best possible starting conditions for the sample zone, which is especially desirable in gel filtration. The bottom end piece is then allowed to compress the packed bed a few millimeters as just described.

For short columns (5 to 15 cm in length) normally used in various types of adsorption chromatography (ion-exchange chromatography, affinity chromatography, etc.), packing quality is less critical. Often one-step procedures based on either constant flow or contant pressure give equally satisfactory results. Larger diameter, semirigid stationary phases such as derivatives of Sepharose FF (90 μm) give too little flow resistance to be packed efficiently with normal laboratory pumps. This is why many workers pack this matrix using compressed air or nitrogen as a pressure source and regulate the flow with a needle valve at the column outlet. Sepharose FF is preferably packed in two steps in the way described earlier. The bed is thus allowed to settle at a linear flow rate of around 3 cm min^{-1}, and the final packing takes place at around 5 cm min^{-1}. One should be aware of the danger of using pressurized vessels and should never allow the pressure to exceed the ratings of the equipment used. For normal laboratory columns, a pressure drop of 2 kg cm^{-2} is sufficient to give efficient packing of Sepharose FF. As a general precaution, the column and accessories should always be placed behind a protective screen or cover during column packing. Some workers claim that the packing of Sepharose FF is facilitated by the presence of 0.05 to 0.1% Tween 20.

Soft gel matrices such as cellulose, Sephadex with higher G numbers, non-cross-linked Sepharose, and conventional polyacrylamide gels with low degrees of cross-linking are, in principle, packed in the same way as the semirigid gel materials with the exception that the flow rates used are considerably lower and are never allowed to approach the maximum flow rate obtainable in a particular column. Neither, of course, should the operating flow rate exceed the packing flow rate of the bed.

2.6 REFERENCES

1. J.C. Giddings, *Dynamics of Chromatography,* Marcel Dekker, New York, 1965.
2. E. Heftmann (ed.), *Chromatography,* 5th edition, Elsevier, Amsterdam, 1992.
3. J.Å. Jönsson (ed.), *Chromatographic Theory and Basic Principles,* Marcel Dekker, New York, 1987.
4. E.A. Peterson, H.A. Sober, *J. Am. Chem. Soc., 78,* 751–755 (1956).
5. Whatman BioSystems Ltd., Springfield Mill, Maidstone, Kent ME14 2LE, England.
6. Cellulofine® manufactured by Chisso Corp. Ltd., Kumamoto, Japan, and marketed by Amicon Div., W.R. Grace & Co., Danvers, MA 01923.
7. Pharmacia Biotech AB, Uppsala, Sweden.
8. J. Porath, P. Flodin, *Nature, 193,* 1657–1659 (1959).
9. *Gel Filtration, Theory and Practice.* Booklet published by Pharmacia Biotech AB, Uppsala, Sweden.
10. *Ion Exchange Chromatography. Principles and Methods.* Booklet published by Pharmacia Biotech AB, Uppsala, Sweden.
11. J.C. Janson, *Chromatographia, 23,* 361–369 (1987).
12. S. Hjertén, *Arch. Biochem. Biophys., 99,* 466–475 (1962).
13. BioSepra S.A., Villeneuve La Garenne, France.
14. BioRad Chemical Division, Richmond, CA 94804.
15. S. Hjertén, *Biochim. Biophys. Acta, 79,* 393–398 (1964).
16. S. Arnott, A. Fulmer, W.E. Scott, I.C.M. Dea, R. Moorhouse, D.A. Rees, *J. Mol. Biol. 90,* 269–284 (1974).
17. A. Amsterdam, Z. Er-el, S. Shaltiel, *Arch. Biochem. Biophys., 17,* 673–678 (1975).
18. J. Porath, J-C. Janson, T. Låås, *J. Chromatogr., 60,* 167–177 (1971).
19. J. Porath, T. Låås, J.-C. Janson, *J. Chromatogr., 103,* 49–62 (1975).
20. T. Andersson, M. Carlsson, L. Hagel, P.-Å. Pernemalm, J.-C. Janson, *J. Chromatogr., 326,* 33–44 (1984).

21. F.E. Reginer, R. Noel, *J. Chromatogr. Sci.*, *14*, 316–320 (1976).

22. R.W. Stout, J.J. DeStefano, *J. Chromatogr.*, *326*, 63–78 (1985).

23. MonoBead™ (10 μm) and Source™ (15 and 30 μm). Polystyrene/divinyl bensene based stationary phases. Pharmacia Biotech AB, Uppsala, Sweden.

24. Superose™. Agarose based, 10 and 13 μm diameter stationary phases. Pharmacia Biotech AB, Uppsala, Sweden. See also Ref. 20.

25. A. Tiselius, S. Hjertén, Ö. Levin, *Arch. Biochem. Biophys.*, *65*, 132–155 (1956).

26. L. Hagel, M. Östberg, T. Andersson, *J. Chromatogr. A*, *743*, 33–42 (1996).

27. L. Hagel, in *Aqueous Size Exclusion Chromatography*, P.L. Dubin, ed., pp. 119, Elsevier, Amsterdam, 1988.

28. L.E. Weaver, Jr., G. Carta, *Biotechnol. Prog.*, *12*, 342–355 (1996).

29. J. Happel, H. Brenner, *Low Reynold Number Hydrodynamics*, pp. 392–396, Martinus Nijhoff Publishers, Dordrecht, 1986.

30. A. Medin, *Studies on Structure and Properties of Agarose*, Ph.D. Thesis, University, Biochemical Separation Center, Uppsala, Sweden (1995).

31. P. Flodin, *Dextran gels and their applications in gel filtration*, Ph.D. Dissertation, Pharmacia AB, Uppsala, Sweden, pp. 1–85, 1962.

32. S. Hjertén, in *Methods of Immunology and Immunochemistry*, M.W. Chase, C.A. Williams, eds., Vol. 2, pp. 142–148, Academic Press, New York, 1968.

33. J. Ugelstad, L. Söderberg, A. Berge, J. Bergström, *Nature*, *303*, 95–96 (1983).

34. Sepharose High Performance.™ Agarose based, 34 μm diameter stationary phases, Pharmacia Biotech AB, Uppsala, Sweden.

35. Sepharose Fast Flow.™ Agarose based, 90 μm diameter stationary phases. Pharmacia Biotech AB, Uppsala, Sweden.

36. Sepharose Big Beads™ and STREAMLINE™ 200 μm diameter agarose based media for industrial scale chromatography and expanded bed adsorption, respectively. Pharmacia Biotech AB, Uppsala, Sweden.

37. P.-E. Gustavsson, P.-O. Larsson, *J. Chromatogr. A*, *734*, 231–240 (1996).

38. A. R. Fersht, J-P. Shi, J. Knill-Jones, D. M. Lowe, A. J. Wilkinson, D. M. Blow, P. Brick, P. Carter, M. M. Y. Waye, G. Winter, *Nature*, *314*, 235–238 (1985).

39. T. E. Creighton, *Proteins, Structures and Molecular Properties*, Freeman, New York, 1983.

40. Nomenclature for Chromatography, IUPAC, *Pure Appl. Chem.*, *65*, 819–872 (1993).

41. J.J. Van Deemter, F.J. Zuiderweg, A. Klinkenberg, *Chem. Eng. Sci. 5*, 271 (1956).

42. N.B. Afeyan, N.F. Gordon, I. Mazsaroff, L. Varady, S.P. Fulton, Y.B. Yang, F.E. Regnier, *J. Chromatogr.*, *519*, 1–29 (1990).

43. D.D. Frey, E. Schweinheim, C. Horváth, *Biotechnol. Prog.*, *9*, 273–284 (1993).

44. I. Langmuir, *J. Am. Chem. Soc.*, *40*, 1361 (1918).

45. D. de Vault, *J. Am. Chem. Soc.*, *65*, 632 (1943).

46. A. Katti and G. Guiochon, in *Advances in Chromatography*, J.C. Giddings E. Grushka, P.R. Brown, eds., Vol. 31, pp. 1–118, Marcel Dekker, New York, 1992.

47. F. Helfferich, *J. Chem. Educ.*, *41*, 410 (1964).

48. C. Horvath, A. Nahum, J.H. Frenz, *J. Chromatogr.*, *218*, 365–393 (1981).

49. E.A. Peterson, A.R. Torres, in *Methods Enzymol.* W.B. Jacoby, ed., Vol. 104, pp. 113–133, Academic Press, New York, 1984.

50. S.M. Cramer, C. Horvath, *Prep. Chromatogr. 1*, 29–49 (1988).

51. T.W. Hutchens, T.-T. Yip, J. Porath, *Anal. Biochem., 170*, 168–182 (1988).

52. J.-C. Janson, P. Hedman, *Advances in Biochemical Engineering, in A. Fiechter, ed., Vol. 25*, pp. 43–99, Springer-Verlag, Heidelberg, 1982.

53. M.R. Ladisch, R.C. Willson, C.C. Painton, S.E. Builder, eds., *Protein Purification; From Molecular Mechanisms to Large-Scale Processes,* ACS Symposium Series 427, American Chemical Society, Washington, DC, 1990.

54. G. Ganestos, P.E. Barker, eds., *Preparative and Production Chromatography,* Chromatographic Science Series, Vol. 61, Marcel Dekker, New York, 1992.

55. H.-J. Rehm, G. Reed, A. Pühler, P. Stadler, eds., *Biotechnology,* 2nd ed., Vol. 3, *Bioprocessing,* G. Stephanopolous, ed., VCH Verlagsgesellshaft mbH, Weinheim, 1993.

56. G. Subramanian, ed., *Process Scale Chromatography,* VCH Verlagsgesellshaft mbH, Weinheim, 1995.

57. G. Sofer, L. Hagel, eds., *Process Chromatography: Optimization, Scale-up and Validation,* Academic Press, London, 1997.

3 Gel Filtration

LARS HAGEL

Pharmacia Biotech AB
S-751 82 Uppsala, Sweden

Protein Purification: Principles, High-Resolution Methods, and Applications, Second Edition.
Edited by Jan-Christer Janson and Lars Rydén.
ISBN 0-471-18626-0. © 1998 Wiley-VCH, Inc.

3.1 INTRODUCTION

Molecular sizing properties of chromatography materials were noted already in the 1940s but the first attempts to separate biomolecules by size were described in 1955 by Lindqvist and Storgårds[1] and Lathe and Ruthven.[2] The separation material used was swollen maize starch which, due to the limited mechanical strength, only could be run at very low flow rates. In the late 1950s, Porath and Flodin noticed, by chance, that cross-linked dextran used as stabilizing medium in column electrophoresis experiments showed size-separating properties.[3] In 1959, this discovery led to the introduction of Sephadex®, composed of dextran cross-linked with epichlorohydrin to enhance the mechanical stability.[4] With the introduction of a commercial product designed for molecular size separations and having different purposely made separation ranges (due to the degree of cross-linking), the new technique, named gel filtration (GF) as suggested by Arne Tiselius, was soon applied to various tasks, such as desalting of protein solutions, purification of protein mixtures, and determination of molecular mass distributions of aqueous polymers, such as clinical dextran. Several types of beaded matrices, based on agarose, and polyacrylamide, to mention a few materials, were subsequently developed and marketed for gel filtration.[5] In 1964 the technique was extended to nonaqueous solutions, initially for the purpose of determining molecular mass distributions of organic polymers, by the introduction of a polystyrene-based matrix by Moore,[6] who, however, called the technique gel permeation chromatography (GPC).

Separation of solutes by their molecular size has, in addition to gel filtration and gel permeation, been given a variety of designations, and gel chromatography, exclusion chromatography, molecular sieving chromatography, steric exclusion chromatography, and size exclusion chromatography are synonymously used in older literature.[7] Today, there seems to be consensus to use size exclusion chromatography (SEC) as a general designation of the separation principle, as it is a formally correct descriptive term of the process, and to maintain the name gel filtration for the application of SEC in aqueous solvents and the name gel permeation chromatography to describe the application of SEC in organic solvents.

Many models have been proposed to explain the separation mechanisms in gel filtration and the validities of these have been thoroughly discussed.[8,9] The separation process is schematically illustrated in Figure 3-1. The separation may simply be regarded as due to the different amount of time different solutes stay within the liquid phase that is entrapped by the matrix. This time

INSTRUMENTATION

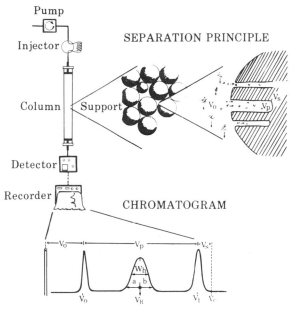

FIGURE 3–1. Fundamentals of gel filtration. Proteins injected into the column are separated according to decreasing size due to incompatibility between the solute dimensions and the pore size of the support. V_0, void volume between the support particles; V_p, pore volume (this is now labeled V_i); V_s, matrix volume of the support; V_R, elution volume of the protein; V_t, total liquid volume of the bed; and V_c, total geometric volume of the column. Column plate number $N = 5.54 \times (V_R/w_h)^2$, where w_h is the peak width at half peak height and $A_s = b/a$ is the asymmetry factor at 10% peak height.

is of course related to the fraction of the pores that is accessible to the solute. The interpretation of this fraction in terms of pore dimensions and gel structure, together with various expressions for solute size, results in slightly different equations for relating the distribution coefficient to the size of the solute. Interestingly enough, all these equations propose a linear relationship between the logarithms of the two parameters.[9] A more general approach was depicted by Casassa[10] who used a stochastic model to relate the nonavailable fraction of the pores and the dimension of the solute. It can be concluded that regions of inaccessibility of the pore volume result in a loss in entropy of the molecules.[11] This loss in entropy is due to the smaller number of possible conformations of the molecules within the pores as compared to an equal volume segment outside the pore. Also, in this case, a linear relationship between the logarithm of the distribution coefficient and the logarithm of solute size is expected.

Even though gel filtration is an uncomplicated and straightforward technique, there are some points worth considering before starting experimental work. The actual sample may require a special pH, solvent, additives, or pretreatment to yield a true solution. The next step is to select a gel that will cope with the chosen solvent and pH and that has a suitable separation range. Possible adsorption properties of the gel must also be considered. The nature of the separation and the sample may put demands on such parameters as resolution, separation time, and sample load, which in turn are partly dependent on the selected gel. These parameters are, however, also affected by the choice of column dimensions and the packing efficiency of the column. Obviously, for different separation problems, such as desalting, preparative fractionations, or analytical separations, different requirements should be stressed. Economic factors and the possibilities of scaling up may also be important.

A thorough description of laboratory techniques for gel filtration has been given by Fischer.[7] An extensive description of size exclusion chromatography has been presented by Yau and co-workers.[12] A review of modern SEC has been given by Hagel and Janson.[5] Reviews of articles dealing with size exclusion chromatography have been regularly published by Barth and co-workers and also provide excellent overviews of the latest accomplishments in gel filtration.[13]

Solutes may be separated according to size by other techniques than size exclusion chromatography. One example is hydrodynamic chromatography, where solutes are separated in the interstitial volume between porous or nonporous beads.[14] A related technique, termed slalom chromatography, is applied for the separation of elongated solutes such as DNA, which is trapped in the interstices between gel beads in a packed column and the solutes are eluted in order of increasing size.[15] Field flow fractionation is very suitable for the separation of larger solutes and is therefore a complementary technique to SEC.[16] The use of these techniques for size separation of biomacromolecules as compared to gel filtration is briefly reviewed in the applications section.

The aim of this chapter is to describe some fundamental practical and applied theoretical aspects of experimental work with gel filtration chromatography of biomacromolecules with special reference to separation of proteins. However, due to interest in the separation and characterization of larger molecules such as DNA as well as smaller molecules such as peptides and carbohydrates, the applicability of gel filtration for these solutes will be discussed when appropriate.

3.2   SELECTION OF THE SUPPORT

The first constraint on the choice of gel medium is due to the solvent and pH required to provide a good solvent for the sample. Most media withstand the common solvents and pH's used (i.e., pH 2 to 12) to dissolve proteins, with the exception of silica-based materials. The importance of selecting a gel with a suitable separation range increases with the complexity of the separation problem (i.e., desalting < preparative purification < analytical applications). For difficult separation problems, the resolution of the gel may be critical and, in these instances, properties such as bead size, slope of the selectivity curve, and separation volume become important. A third factor to take into account is the effect of the sorption properties of the matrix under running conditions. Even though virtually no matrix can be expected to be completely free from sorption properties, the nature and degree of these properties vary with the nature of the matrix.[17-19] Sometimes these properties have been used to achieve increased separation of the sample components.[19-21] However, it is important to realize that these mixed-mode separations may lack reproducibility and that they should not be classified as gel filtration!

3.2.1 Characteristics of Available Media

Since the advent of the cross-linked dextran, Sephadex in 1959, gels based on agarose, polyacrylamide, and combinations of these media have been commercially exploited. These traditional media are characterized by large bead sizes (i.e., typically 100 to 250 μm), low matrix volumes (e.g., less than 5%), and high deformability. Because of the large bead size, porous glass may also be included in this category. The chemistry and properties of these gels have been thoroughly described[7,19] and compared.[17] Data for some of these supports are compiled in Table 3-1.

Gels presently used for high-resolution gel filtration of proteins are based on silica, hydrophilized vinyl polymers, or highly cross-linked agarose with bead sizes typically between 5 and 50 μm. The smaller particle size yields more efficient columns, resulting in narrower peaks, and may be employed for achieving higher resolution and/or faster separations. However, these supports are more expensive than traditional media and are often only available as prepacked columns. The chromatographic properties of columns prepacked with silica-based media have been reviewed by Pfannkoch and co-workers.[18] The gel properties and application of the theoretical aspects of chromatography will be discussed with special reference to media designed for high-resolution gel filtration. Characteristics of some available media are given in Table 3-2. The porous structure of one support as visualized by scanning electron microscopy is shown in Figure 3-2. The actual structure is much more complex than implied by Figure 3-1.

3.2.2 Chemical and Sorptive Properties

The pH resistance of standard gels covers the approximate range pH 2 to 10. High-resolution gels have a somewhat higher pH stability (e.g., pH 1 to 14), with the exception of silica-based materials[23] (see Tables 3-1 and 3-2). To reduce the dissolution of silica at high pH, a precolumn packed with silica may be placed before the injector. To improve the stability of silica, treatment

TABLE 3-1. Some Traditional Media for Gel Filtration of Proteins and Peptides

Media	Type	Supplier[a]	Particle Size of Hydrated Beads (μm)	Fractionation Range for Globular Solutes ($M_r \cdot 10^{-3}$)	pH Stability
Sephadex	Dextran	1			2–10
G-50			20–100, 100–300[b]	1.5–30	
G-75			20–100, 100–300[b]	3–80	
G-100			20–100, 100–300[b]	4–100	
G-150			20–100, 100–300[b]	5–300	
G-200			20–100, 100–300[b]	5–600	
Sepharose	Agarose	1			4–9
6B			45–165	10–4,000	
4B			45–165	60–20,000	
2B			60–200	70–40,000	
Sepharose CL	Agarose	1			3–14
6B			45–165	10–4,000	
4B			45–165	60–20,000	
2B			60–200	70–40,000	
Ultrogel	Agarose	2			Not stated
A6			60–140	25–2,400	
A4			60–140	55–9,000	
A2			60–140	120–23,000	
Bio-Gel	Agarose	3			4–13
A-0.5m			40–80, 80–150, 150–300	1–500	
A-1.5m			40–80, 80–150, 150–300	1–1,500	
A-5m			40–80, 80–150, 150–300	10–5,000	
A-15m			40–80, 80–150, 150–300	40–15,000	
A-50m			40–80, 80–150, 150–300	100–50,000	

Matrix	Composition	Supplier	Bead diameter (μm)	Fractionation range (kDa)	pH stability
Bio-Gel	Polyacrylamide	3			2–10
P-10			<40, 40–80, 80–150	1.5–20	
P-30			<80, 80–150, 150–300	2.5–40	
P-60			<80, 80–150, 150–300	3–60	
P-100			<80, 80–150, 150–300	5–100	
P-150			<80, 80–150, 150–300	15–150	
P-200			<80, 80–150, 150–300	30–200	
P-300			<80, 80–150, 150–300	60–400	
Sephacryl	Dextran/bisacrylamide	1			2–11
S-100 HR			25–75	1–100	
S-200 HR			25–75	5–250	
S-300 HR			25–75	10–1,500	
S-400 HR			25–75	20–8,000	
Ultrogel	Agarose/acrylamide	2			3–10
AcA 202			60–140	1–15	
AcA 54			60–140	5–70	
AcA 44			60–140	10–130	
AcA 34			60–140	20–350	
AcA 22			60–140	100–1,200	
Glycophase CPG	Silica	4			<8
100			37–74	1–30	
200			37–74	2.5–100	
460			37–74	11–320	

[a] Data given as stated by the suppliers: 1, Pharmacia Biotech (Uppsala, Sweden); 2, Reáctifs IBF (Villeneuve la Garenne, France); 3, Bio-Rad Laboratories (Richmond, CA); and 4, Pierce Eurochemie BV (Rotterdam, Holland).

[b] Calculated from the stated dry bead diameter from $d_{p,\,wet} = d_{p,\,dry}(1 + Wr \cdot \rho)^{1/3}$, where Wr is water regain of bead and ρ is density of matrix.

TABLE 3-2. Characteristics of Some Media for High-resolution Gel Filtration of Proteins and Peptides[a]

Media	Type	Supplier	Particle Size (μm)	Fractionation Range of Globular Solutes (M_r × 10^{-3})	Operation Range[b] (M_r × 10^{-3})	Molecular Mass Selectivity (−ΔK_D/Δ log M_r)	Particle Porosity [V_i/(V_i + V_s)]	Permeability (V_t/V_0)	Void Fraction (V_0/V_c)	pH Stability	Buffer Composition
Prepacked columns											
Superdex Peptide	Agarose/dextran	1	13	0.1–7				1.2		1–14	
Superdex 75	Agarose/dextran	1	13	3–70				1.6		1–14	
Superdex 200	Agarose/dextran	1	13	10–600				1.7		1–14	
Superose 12	Agarose	1	10	1–2,000	0.5–600	0.28	0.84	1.84	0.30	1–14	0.05 M phosphate +
Superose 6	Agarose	1	13	5–40,000	5–10,000	0.23	0.93	1.87	0.33	1–14	0.15 M NaCl, pH 7.0[24]
TSK SW 2000	Silica	2	10	0.5–60	0.7–100	0.52	0.66	0.95	0.39	2.5–7.5	0.05 M Tris–HCl +
TSK SW 3000	Silica	2	10	1–300	2–400	0.43	0.81	1.33	0.38	2.5–7.5	0.5 M Na_2ClO_4, pH 7.5[25]
TSK SW 4000	Silica	2	13	5–1,000	6–10,000	0.28	0.89	1.40	0.39	2.5–7.5	
Synchropak[c] GPC 100	Silica	3	10	3–630		0.44	0.79	1.23	0.39		0.1–0.6 M phosphate pH 7[18]
Waters I-125[c]	Silica	4	10	0.8–450		0.36	0.55	0.92	0.38		0.08 M phosphate + 0.32 M NaCl in 20% ethanol[26]
Bio-Sil SEC 125	Silica	5	5, 10	5–100							
Bio-Sil SEC 250	Silica	5	5, 10	10–300							

Name	Matrix	Supplier	Particle size	Fract. range	Fract. range (globular)					pH	Buffer
Bio-Sil SEC 400	Silica	5	5, 10	20–1,000	14–100		0.52	0.78	0.40		
DuPont GF-250[d]	Silica	6	4	10–250						3–8.5	
DuPont GF-450	Silica	6		25–900						3–8.5	
PL-GFC 300 Å	Polymer	7	8, 10	1–700							
Bulk media											
Superdex 30	Agarose/ dextran	1	34	<10				1.8		1–14	
Superdex 75 prep grade	Agarose/ dextran	1	34	3–70				2.1		1–14	
Superdex 200 prep grade	Agarose	1	34	100–600				2.2		1–14	
Fractogel TSK HW-55S[e]	Vinyl polymer	2	25–40	1–1,000	0.2–100	0.18	0.96	2.17	0.31	1–14	25 mM Tris–HCl + 0.3 M NaCl, pH 7.5[27]
Superose 12 prep grade	Agarose	1	20–40	1–2,000	1–600	0.32	0.92	1.86	0.33	1–14	0.05 M phosphate + 0.15 M NaCl, pH 7.0[24]
Superose 6 prep grade	Agarose	1	20–40	5–40,000	5–10,000	0.24	0.97	2.23	0.30	1–14	

[a] Data as stated by the supplier: 1, Pharmacia Biotech (Uppsala, Sweden); 2, Toyo Soda (Tokyo, Japan); 3, SynChrom (Linden, IN); 4, Waters (Milford, MA); 5, Bio-Rad (Hercules, CA); 6, DuPont (Wilmington, DE); 7, Polymer Laboratories (Shropshire, UK).

[b] For globular proteins, calculated from $0.1 \leq K_D \leq 0.9$.

[c] Data from Ref. 18.

[d] Data calculated from Ref. 22.

[e] Toyopearl HW type.

FIGURE 3–2. Surface pore structure of an agarose matrix (Superose® 6) as visualized by scanning electron microscopy after coating the surface with a layer of gold. Magnification factor is 100,000×.

with zirconium has been suggested. However, deterioration of even this material was noted with prolonged use at pH 7.[22]

The surface of traditional supports prepared from natural polymers (e.g., agarose or dextran gels) contains predominantly hydroxyl groups and provides a good environment for hydrophilic proteins. Unfortunately, the hydrophilicity is reduced somewhat by the introduction of cross-linking reagents (e.g., epichlorohydrin). Matrices composed of the styrene-divinylbenzene copolymer must be chemically treated to increase the hydrophilicity of the surface before the material is suitable for gel filtration. The surface of silica and porous glass must also be derivatized or coated to prevent excessive adsorption of proteins.

The presence of charged groups may be due to small amounts of natural-occurring acidic groups in the raw materials (e.g., sulfate groups in agarose or carboxylic acid groups in dextran). Ionic sites may be introduced by acid or alkaline hydrolyses of the matrix by prolonged exposure to an extreme pH.

The structure of the matrix may yield specific interactions such as aromatic adsorption of, for example, tryptophan-rich peptides, on Sephadex G-25, G-15, and G-10.[29] The effect is assumed to be caused by an interaction between aromatic amino acids and ether bridges introduced by the cross-linker.[30] Biospecific interactions where the matrix resembles an enzyme substrate or an affinity site may also be noted (e.g., interaction between lectins and glucosidic sites of gels).

Adsorption properties of standard gel filtration media have been utilized advantageously to improve the separation of substances, and the basic principles of the phenomena have been fully described (e.g., see Refs. 30–38).

Nonspecific interactions between sample and high-resolution gels are mostly due to an ionic interaction with residual silanol groups or carboxylic acid groups and hydrophobic interactions with coating or cross-linking sites. Pfannkoch and co-workers[18] tested various interactions by measuring the distribution coefficients of citrate, arginine and phenylethanol at ionic strengths from 0.026 to 2.40 in a phosphate buffer at pH 7. Deviations of the distribution coefficient from unity indicated anionic exclusion, cationic adsorption, and hydrophobic effects. Large differences between different coated silica materials were noted.

The mechanism underlying the interaction between protein and gel filtration media was discussed from the theory of potential barrier chromatography, where the total effect is a sum of contributions from the electric double-layer interaction, van der Waals forces, and repulsive short-range interactions.[37,38] It was concluded that true size separations was, in some cases, only achieved under conditions where different forces, (e.g. hydrophobic and ionic) were balancing each other.[38]

Changes in the mobile-phase pH or ionic strength may induce conformational changes of proteins, leading to the exposure of hydrophobic groups which may interact with the matrix. Also, lipophilic interactions are often encountered for small macromolecules, such as peptides, where no tertiary structure aids in shielding these groups from a hydrophilic environment.

A general recommendation for all types of materials is to use a buffer with intermediate ionic strength (e.g., 0.15 M) to suppress ionic effects but still not promote hydrophobic interactions. This will also effectively mask the Donnan effect, (i.e., the increase of the electrolyte concentration in the pores caused by the presence of macroions in the void volume).[39] Buffers commonly used for different materials are listed in Table 3-2, and more detailed information on buffer preparation and additives is found in Section 3.5.

3.2.3 Selectivity Curve and Separation Range

In contrast to other types of media (e.g., media for ion exchange or reversed-phase chromatography), the selectivity of a gel filtration medium is not adjustable by changing the composition of the mobile phase (as long as this change does not influence the solute shape or pore structure). The selectivity of the gel is thus an inherent property of the material, and the separation volume of the column is limited by the total pore volume of the packed bed.

The selectivity curve of the separation material is obtained by plotting the elution volume, or some function thereof, versus an expression of the solute size. Very often, the distribution coefficient, K_D, is related to the logarithm of the molecular mass, M_r, of the solute. K_D is a column-independent variable and is calculated from the elution, or retention, volume, V_R (also denoted V_e), the void volume, V_0, and the pore volume, V_i according to

$$K_D = (V_R - V_0)/V_i = (V_R - V_0)/(V_t - V_0) \qquad (3\text{-}1)$$

where V_t is the total liquid volume of the bed (see Figure 3-1). The plot of K_D versus $\log M_r$ will yield a sigmoidal selectivity curve that in the middle range, may be approximated by

$$K_D \approx a - b \log M_r \qquad (3\text{-}2)$$

The determination of these parameters is discussed in Section 3.7.3. The slope of the selectivity curve (i.e., the value of b) depends on the width of the pore size distribution of the material, and a/b is related to the mean pore size (see Section 3.7.4). A narrow pore size distribution will result in high selectivity (i.e., a large value of b) but a small separation range for the gel, as $0 < K_D < 1$ for ideal gel filtration. The sigmoidal nature of the selectivity curve reduces the practical working range of the gel to approximately $0.1 < K_D < 0.9$. It is also important to realize that the exclusion limit quoted by many manufacturers is often an extrapolated value of the selectivity curve and will therefore only give an indication of the size of molecules that will be excluded from the matrix and not the ultimate separation range. Data on selectivity and separation range for some materials are found in Table 3-2. The separation range may be extended by the combination of materials with different pore sizes. If the bead sizes (and densities) of the materials are similar, they may be mixed before packing, but it is more usual to pack the two gels separately in the same or different columns. The gels are often placed in order of descending pore size. However, it has been shown that the column arrangement may be optional and is preferentially random.[40] It is very important to use materials with overlapping selectivity curves or else artifacts may be created.[41,42]

3.2.4 Support Pore Volume

The pore volume of a material, V_i, can be calculated by subtracting the void volume, V_0, from the total liquid volume, V_t. The matrix or support volume, V_S, is obtained by subtracting the total liquid volume from the geometric volume of the column, V_c (cf. Figure 3-1). The relative pore volume of the gel is expressed by

$$V_{i,rel} = V_i/(V_s + V_i) = (V_t - V_0)/(V_c - V_0) \qquad (3\text{-}3)$$

A low pore volume will result in a low separation volume of the column, as solutes are eluted between V_0 and $V_0 + V_i$.

3.3 THEORETICAL CONSIDERATIONS

3.3.1 Estimation of Molecular Size by Gel Filtration

In ideal gel filtration, the retention of solutes is governed solely by the differences between the solute dimensions and the pore dimensions. The relationship between size and molecular weight of solutes is strongly dependent on solute shape, which may be illustrated by the relationship between the radius of gyration, R_g, of a solute and its relative molecular mass, M_r

$$R_g \propto M_r^a \qquad (3\text{-}4)$$

where $a = 1$ for rods, $a \approx \frac{1}{2}$ for flexible coils, and $a = \frac{1}{3}$ for spheres.[43] The influence of the solute shape on the retention in gel filtration is illustrated in Figure 3-3. It is readily seen that calibration versus molecular mass is only meaningful for solutes of similar shape. The influence of shape on the slope of the selectivity curve may be inferred from Eq. (3-2) by $-b = dK_D/d \log M_r = dK_D/d \log R \cdot d \log R/d \log M_r$, where the first term is the "true" selectivity of the gel (i.e., the change of distribution coefficient with a change in solute size) and the second term gives the change in solute size with a change in molecular mass [i.e., the exponent a in Eq. (3-4)].

The shapes of solvated proteins vary considerably (i.e., there are spherical proteins, slightly asymmetrical globular proteins, rod-shaped fibrous proteins,

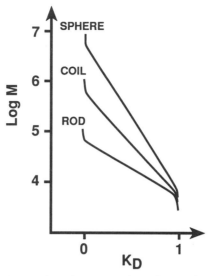

FIGURE 3–3. Theoretical calibration curve for solutes of various shapes. (Reproduced from Ref. 43, with kind permission of the authors and the publisher.)

and denatured flexible coil structures). The frictional coefficient, f, of a solvated protein, obtained from sedimentation or diffusion experiments, may be compared to the theoretical frictional coefficient, f_0, of a nonsolvated sphere of equal volume. It is tempting to interpret deviations of the frictional ratio, f/f_0, from unity as deviations of the protein shape from a sphere. This is facilitated by the theoretical relationship between the frictional ratio of an ellipsoid of rotation and the geometry of the particle (e.g., $f/f_0 = 1.25$ corresponds to a prolate ellipsoid with an axial ratio of $1:5$).[44] However, the frictional ratio is a function of solute solvation as well as asymmetry (and surface roughness), and therefore interpretation of the shape of, for example, globular proteins as ellipsoidal hydrodynamic particles may not be realistic. It has been concluded that globular proteins (e.g., with $f/f_0 < 1.25$) are neither highly solvated nor highly asymmetric.[44] The situation may be illustrated by reference to apoferritin. This molecule has a perfect spherical shape but yields a frictional ratio of 1.29 due to the large degree of solvation of this hollow sphere shaped molecule.[45]

Information on shape may be obtained by comparison of data from hydrodynamic measurements (e.g., intrinsic viscosity and sedimentation or diffusion coefficients) with the radius of gyration from light scattering. A gel filtration column calibrated with such reference substances may be utilized to provide structural information about solutes of known molecular weight (cf. Figure 3-3). Conversely, the column may be used for the estimation of molecular weights of compounds with shapes similar to the shapes of the reference substances.

Tanford and co-workers[46] proposed calibration of a Sephadex column by plotting K_D versus protein Stokes radius for the accurate (i.e., 5%) estimation of the molecular mass of proteins in a detergent solution. The Stokes radius, R_{St}, is defined as the radius of a sphere that would have the same frictional coefficient as the protein.[44] The molecular mass of a compact globular protein may be inferred from the Stokes radius through

$$\log M_r \approx 0.140 \pm 0.041 + \log(R_{St}^3/v) \qquad (3\text{-}5)$$

where the partial specific volume, v, if unknown, may be assigned a value of 0.74.[45]

The size of solutes may also be estimated from viscosity data yielding an expression of the hydrodynamic volume.

$$V_h = |\eta| M_r / v N_a \qquad (3\text{-}6)$$

where η is the intrinsic viscosity (i.e., volume of molecules per unit mass), N_a is Avogadro's number, and v is Simhas factor. The value of this factor depends on the shape of the molecule and is 2.5 for spheres and >2.5 for ellipsoids (e.g., an axial ratio of 1:5 corresponds to $v \approx 5$). For flexible polymers and spherical solutes, a viscosity radius, R_η or R_h, may be calculated from the hydrodynamic volume by $R_h = (V_h/\pi \cdot 3/4)^{1/3}$. This estimate is often called the hydrodynamic radius. However, R_{St} is also obtained from hydrodynamic measurements and therefore the term hydrodynamic viscosity radius is preferred for R_h or R_η. The following relationship between the molecular weight of compact globular proteins and hydrodynamic volume may be calculated from Tanford[44]:

$$V_h \approx 1.544 \times M_r \qquad (3\text{-}7)$$

The geometric radius calculated from this formula, assuming a spherical shape of the protein, will yield an underestimate of the size (since the formula was derived for globular proteins). Assuming a spherical shape of proteins (i.e., by assuming a value of 2.5 for Simhas factor in the calculation of hydrodynamic volume) yields $R_h \approx (0.82 \pm 0.02) \cdot M_r^{1/3}$. A similar relationship was calculated by Squire for solvated proteins assuming a spherical shape (i.e., $R_h \approx 0.794 \cdot M_r^{1/3}$).[47] Solving Eq. (3-5) for R_{St} yields $R_{St} \approx (0.808 \pm 0.025) \cdot M_r^{1/3}$. These relationships, derived from different assumptions, indicate that a reasonably accurate estimate of the radius of compact globular proteins is given by

$$R \approx 0.81 \times M_r^{1/3} \qquad (3\text{-}8)$$

As pointed out in the work of Tanford and co-workers,[46] it was earlier not known whether partition in the gel was responsive to Stokes radius or viscosity radius. However, observed differences between the two estimates were smaller than 10% for globular proteins, approximately 15% for random coils, and up to 100% for fibrous proteins. This is in qualitative agreement with Figure 3-3. In another study, denatured proteins of quite different conformations eluted according to identical calibration curves when the solute size was inferred from viscosity data.[48] Benoit and co-workers[49] proposed that the hydrodynamic volume is the retention decisive parameter for the solute in SEC. This approach has been shown to be applicable to a wide variety of solute shapes and solute–solvent systems.[47–55] Calibration of a column using hydrodynamic volume (or viscosity radius) has therefore been called universal calibration. This type of calibration curve is illustrated in Figure 3-4, which shows that globular proteins and rod-shaped virus particles elute according to the same curve. A similar result would not have been obtained if retention had been plotted versus R_{St}.[46] It has been shown that universal calibration is only applicable to compact solutes (i.e., globular proteins) and flexible polymers and that rigid rods (i.e., DNA of intermediate length) will deviate from a universal calibration curve.[56] As the length of DNA increases, the flexibility increases, which explains the behavior observed in Figure 3-4.[56]

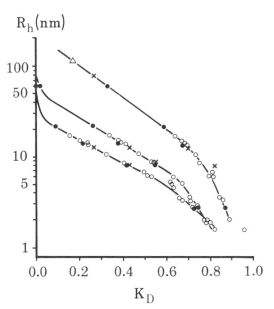

FIGURE 3–4. Universal calibration curve for evaluation of sample viscosity radius. Columns: TSK 6000 PW (upper curve), TSK 5000 PW (middle curve), and Superose 6 (lower curve). Sample: Δ, tobacco mosaic virus (TMV) dimers; ●, TMV, spectrin, tropomysin, and ovomucoid; x, DNA and α-actinin; and ○, proteins. (Reproduced from Ref. 28, with kind permission of the author and the publisher.)

Thus, the general applicability of size estimates increases in the order of molecular mass (or molecular weight) ≪ Stokes radius < hydrodynamic volume (or viscosity radius). The radius of gyration is proportional to the viscosity radius of spherical solutes and flexible polymers, but not to that of rigid macromolecules. The use of a well-defined polymer, such as dextran, to calibrate the column according to the hydrodynamic volume seems to be the most appropriate procedure for the characterization of an unknown solute with the aid of gel filtration. However, calibration using a polymer sample requires that the size related to the peak retention volume is known or else the column may be calibrated using an iterative procedure (see Section 3.7.3).

The solute size is often related to the distribution coefficient in a log-linear plot. Various relationships between size and K_D that have been found useful are given in Table 3-3. Many of the relationships are claimed to yield linear calibration curves, which probably reflect that variations in solute properties are sufficiently large to disguise the sigmoidal nature of the curve. The largest linear portion of the calibration curve is obtained from a single-pore-size support, but significant deviations from the linear portion are noticed for $K_D > 0.9$ and $K_D < 0.1$.[62] When selecting the mode of calibration it is important to examine critically the validity of the data for the calibration probes. Parameters such as shape and density of the molecule and the experimental conditions for obtaining data on Stokes radius have been used to explain disparities between different sets of experimental results.[49,50,63]

The conformation of proteins may be normalized to random coils to avoid the ambiguities sometimes involved with less well-defined solute shapes. Denaturating media promote the disruption of noncovalent bonds, and the native protein reverts to the state of a random polymer coil. Because urea is a poor denaturing agent for many proteins, the use of 6 to 8 M guanidine hydrochloride (GuHCl) is preferred.[48,64] The disulfide bonds are broken by reductive cleavage with mercaptoethanol or dithiothreitol and subsequent re-oxidation prevented by carboxylmethylation.[64] Proteins may also be denatura-

TABLE 3-3. Relationships Used for Calibration of Analytical Gel Filtration Columns

Relationship	Shape	Reference
$K_D^{1/3} = a - b \times M_r^{1/2}$	Linear	57
$K_{av} = \exp[-a(R_{St} + b)^2]$	Nonlinear	58
$K_D = a + b \times \log M_r$	Partly linear	59
$\mathrm{erfc}(1/K_D) = a + b \times R_{St}$	Linear	60
$K_D = a + b \times \log ([\eta]M_r)$	Partly linear	49
$\log(1 - K_D) = a + b \times \log M_r$	Linear	11
$1/V_R = a + b \times R_{St}$	Linear	61

In the formulas, a and b are operational constants, erfc is the error function, R_{St} is the Stokes radius of the solute, and $[\eta]$ is the limiting viscosity number of the solute.

ted with the aid of low concentrations of detergents, such as sodium dodecyl sulfate (SDS).[48,64,65] It should be realized that this treatment will yield a protein–detergent complex of a size that is considerably larger than that of the native protein and that charged groups are introduced when using ionic detergents.[66] The hydrodynamic properties of the protein–SDS complex indicate that the complex is a rod-like particle.[67] The shape of the complex will grow more spherical as the molecular weight of the protein decreases and the contribution of SDS to the size of the complex may be dominating. This sets a lower limit, of about 15,000, to molecular weights of proteins that may be estimated by gel filtration in SDS.[48] Procedures for fast analytical gel filtration in GuHCl and SDS (0.1%) have been presented and were shown to offer attractive alternatives to SDS gel electrophoresis.[68,69]

3.3.2 Column Efficiency and Peak Zone Broadening

A separation strategy should, in general, focus on obtaining a high selectivity (i.e., peak-to-peak distance) than a high efficiency (i.e., narrow peaks), unless peak dilution is a critical factor. However, as mentioned earlier, the maximum selectivity of gel filtration is inherently determined by the pore size distribution of the gel, and the efficiency is the only variable that is affected by the running conditions. Hence much effort has been spent to find an adequate description of the various phenomena that control peak broadening in order to optimize the experimental conditions.

The peak width of a solute can be related to the plate height by[70]

$$H = L/N = L/(V_R/\sigma)^2 = L(w_b/4V_R)^2 \qquad (3\text{-}9)$$

where H is the height equivalent to a theoretical plate (sometimes denoted HETP) for the solute, N is the number of plates per column length L, and w_b is the base width (i.e., 4σ) of a gaussian elution profile. The plate height in Eq. (3-9) is the sum of contributions to peak broadening from different parts of the chromatographic system. The extra-column dispersion can, with the aid of a proper experimental setup, (see Section 3.6.5), be neglected compared to column effects.

The plate height equation for a chromatography column was described by van Deemter[71] and later adapted to gel filtration by Giddings and Mallik.[72] Several other plate height equations have been derived starting from slightly different assumptions (e.g., the empirical Knox equation has gained wide popularity), but a review of the various equations shows that the van Deemter equation is presently the most accurate one for describing zone broadening in SEC.[73] This equation is generally written as

$$H = A + B/u + Cu \qquad (3\text{-}10)$$

The first term arises from multiple path dispersion of the solute, the second term describes the effect of axial diffusion of the solute, and the third term is due to nonequilibrium conditions in the separation process at the interstitial fluid velocity u. The terms are given by[70,74]

$$A = \Sigma_i \, (1/2 \, \lambda_i \, d_p + D_m/\omega_i \, d_p^2 \, u)^{-1}$$
$$B = 2[\gamma_m \, D_m + \gamma_s \, D_m \, (V_R/V_0 - 1)]$$
$$C = V_0/V_R(1 - V_0/V_R)d_p^2 \, /30\gamma_s \, D_m$$

where λ_i and ω_i are geometrical factors of order unity, d_p is the average particle diameter of the support, D_m is the diffusion coefficient of the solute in the mobile phase, γ_m and γ_s are obstruction factors to diffusion in the extra-particle space and the pores, respectively, and V_0/V_R is the ratio of zone velocity to mobile-phase velocity. Fortunately, several simplifications of the expressions can be made. Thus, due to the slow diffusion of proteins, the second term of A can be neglected compared to the first term. The value of γ_m was found to be close to 0.6.[70] The plate height equation for macromolecules can thus be approximated by

$$H = 2 \lambda d_p + 2[0.6 \, D_m + \gamma_s \, D_m \, (V_R/V_0 - 1)]/u + \qquad (3\text{-}11)$$
$$V_0/V_R(1 - V_0/V_R)d_p^2 \, u/30\gamma_s \, D_m$$

The different shapes of the van Deemter plot for small solutes and large macromolecules at the flow rates commonly used in gel filtration are illustrated in Figure 3-5. The contribution from the B term is seen to be very small for macromolecules, and this term is negligible compared to the C term at high reduced velocities (e.g., $d_p u/\gamma_s \, D_m \gg 5$). This condition is fulfilled for all the materials listed in Table 3-2 under the experimental conditions commonly used, and the plate height equation for macromolecules in this case may be reduced to

$$H = 2 \lambda d_p + V_0/V_R(1 - V_0/V_R) \, d_p^2 \, u/30 \, \gamma_s \, D_m \qquad (3\text{-}12)$$

This equation may be used to predict the zone broadening of a protein, of known molecular weight, under various experimental conditions. An identical conclusion regarding zone broadening of polymers in SEC was made by Dawkins.[75] Figure 3-5 also illustrates that the minimum value of the plate height is independent of the solute diffusivity. It can be shown that the flow rate giving this minimum is proportional to D_m/d_p (see Section 3.6.3).

The value of λ can be estimated from the minimum reduced plate height of *any* solute. We have found λ to be close to 1 for experimental results on many gel filtration columns (e.g., see Figure 3-5), although it may be inferred

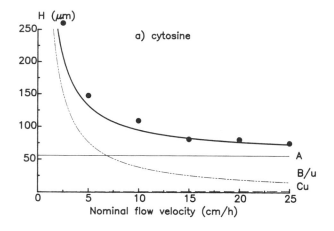

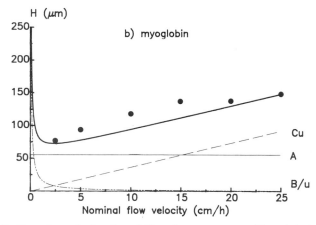

FIGURE 3–5. Zone broadening in gel filtration as measured by plate height, H, for small and large solutes as a function of fluid velocity. A, B, and C are the terms in the van Deemter equation as calculated from Eq. (3-11) with $d_p = 33$ μm and for (a) cytosine and (b) myoglobin. Dots represent experimental data for Superose 6 prep grade.[24]

that $\lambda = 0.5$ will be in accordance with a minimum of $H = 2d_p$. Also, Yamamoto and co-workers[76] reported that a value of $\lambda = 1$ was in accordance with experimentally found axial dispersion in columns packed with Sephadex as well as ion exchange gels.

The value of γ_s expresses the hindered diffusion of solutes within the porous network, D_s, as compared to free diffusion in solution, D. The effective pore diffusivity, D_{eff}, describing the flux of material through pores having uniform cylindrical shape, is given by[77]

$$D_{eff} = D\, K_D \varepsilon_p [1 - 2.104\,(R/r) + 2.09\,(R/r)^3 - 0.95\,(R/r)^5]/\tau \quad (3\text{-}13)$$

where R/r is the ratio of solute to pore radius (this is equal to $1-\sqrt{K_D}$ for a cylindrical pore model), ε_p is the particle porosity and τ is the tortuosity factor used to compensate for variations in effective pore length. The value of τ may be obtained from batch experiments as suggested by Liapis and Arve.[78] However, when no data is available τ is arbitrarily set to $1/\varepsilon_p$.[79] The restricted diffusion coefficient is given by $D_S = D_{eff}/K_D\varepsilon_p$. This relationship was, with slight modifications, used by Ackers and Steere[80] to calculate apparent pore radii of membranes. Another version of the expression for D_S, where the tortuosity factor was ignored, was proposed for predictions of zone broadening in SEC experiments.[81] Unfortunately, none of the equations yield data in sufficient agreement with experimentally found obstruction factors.[82–84] Diffusivities of macromolecules in pores were found to be 5 to 20% of the free diffusion. The difficulties in obtaining a general expression for hindered diffusion may be elucidated by work where a simple relationship was derived from the gel model of Ogston, that is, $D_s/D = \exp[-(\ln K_{av})^{1/2}]$, was fitted to experimentally determined diffusion coefficients of proteins in Sepharose®.[85] The authors obtained a good fit to experimental data only in the high K_{av} region (i.e., $K_{av} > 0.8$) and, in fact, the simple relationship $3*K_{av}/4$ seemed to provide a better general fit to their data in the investigated range $0.3 < K_{av} < 0.9$. The model applied gave geometric estimates of fiber thickness and radius of cavities of half-expected values as compared to other investigations. Also, the authors noted that the fit to the theoretical curve was not as good as those obtained for other types of gels, which indicates inherent differences between the gels. Thus, the lack of applicable models for pore structure of gel filtration media prevents accurate relationships for hindered pore diffusion to be established from theoretical ground. More research is needed to find appropriate expressions to describe the effective diffusivities of solutes in porous networks; unfortunately, as restrictions to free diffusion of solutes may be anticipated to be regulated by local properties of solutes and matrix, it is questionable if a general formula for hindered diffusion applicable to all solutes and types of matrices can be found. Meanwhile, the relationships reported must be regarded as empirical and only applicable to the type of matrix and solute tested. From one work, describing the hindered diffusion of proteins, the simple expression $\gamma_s \approx K_D/4$ for $0.2 < K_D < 0.8$ may be calculated. It is worth noting that even the pessimistic estimation of $\gamma_s \approx K_D/4$ probably yields too high of estimates of the hindered diffusion of high molecular weight solutes, as the experimental broadening is often larger than anticipated from the van Deemter equation (e.g., for thyroglobulin chromatographed on Superose® 6 prep grade).[86]

The diffusivity (cm²/s) of globular proteins may be derived from a formula given by Tanford to[44]

$$D_{25,H_2O} \approx 2.6 \times 10^{-5}\, M^{-1/3} \tag{3-14}$$

which is in good agreement (i.e., better than 6%) with experimentally found data.[82,83] The influence of temperature, T, and viscosity, η, may be estimated from $D = 8.89\,(T/\eta) \cdot M_r^{-1/3}$.[5]

Thus, by using these approximations, the zone broadening of any solute of known molecular weight may be estimated with the aid of Eqs. (3-14), (3-12), (3-2), and (3-1). The approach is illustrated in Figure 3-5, which shows that experimentally found data are in reasonable good agreement with the predicted plate height. Furthermore, by also using Eq. (3-9), a rough theoretical simulation of the chromatogram may be obtained.

3.3.3   Parameters Affecting Resolution

The ultimate goal for any separation is the acceptable resolution of a set of sample components. Resolution is greatly affected by the experimental conditions (see Section 3.6). More important, however, is the selection of a gel with optimal properties (e.g., selectivity) for the particular separation problem.

The resolution between two solutes is determined by their elution volumes and peak widths. In the linear part of the selectivity curve, the difference in elution volumes may be expressed in terms of molecular mass from Eqs. (3-1) and (3-2) as

$$\Delta V_R = -V_i \times b \times \Delta \log M_r \tag{3-15}$$

The resolution of two adjacent components, with molecular masses M_{r1} and M_{r2}, is, with the aid of Eqs. (3-9) and (3-15), expressed in a fundamental equation for resolution, R_s, in gel filtration

$$R_s = \frac{2(V_{R_2} - V_{R_1})}{w_{b_2} + w_{b_1}} = \frac{1}{4} \log M_{r1}/M_{r2} \left(\frac{b}{V_0/V_i + K_D} \right) \sqrt{L}/\sqrt{H} \tag{3-16}$$

The resolution is thus affected by the differences in the sample molecular masses $\log M_{r1}/M_{r2}$, the porosity-dependent quotient $b/(V_0/V_i + K_D)$, the column length L, and how well the column is packed, H. Equation (3-16) involves the simplification $(V_0/V_i + K_{D1})H_1^{1/2} + (V_0/V_i + K_{D2})H_2^{1/2} \approx 2\,(V_0/V_i + K_D)H^{1/2}$, where $K_D = \frac{1}{2}(K_{D1} + K_{D2})$ and $H = \frac{1}{2}(H_1 + H_2)$. By letting $(V_R - V_t)/V_t = k$, the equation is analogous with the resolution equation used in, for example, reversed-phase chromatography (see Chapter 6).

Equations (3-12) and (3-16) show that increased resolution is favored by increasing the slope of the calibration curve, the column length, and the permeability (i.e., V_i/V_0) and by decreasing the particle size and, in most cases, the flow rate. Operating at a low K_D value will theoretically also increase the resolution (the gain in selectivity is larger than the loss in efficiency). [Another conclusion was made by Hjertén who states that "the resolution decreases with increasing R" (and thus decreasing V_R and K_D, R is equal to the relative zone mobility, V_0/V_R).[87] Unfortunately, he did not take into account that ΔR varies with variations in R. Substitution of $\Delta V_R \cdot R^2/V_0$ for ΔR in Eq. (17) in

Ref. 87 shows that the resolution increases with increasing R. The effects of these parameters are illustrated in Figure 3-6, which shows that a large value of the slope of the selectivity curve is an important factor for achieving resolution (Figure 3-6A, a and d) and that a low value may be compensated for by a large value of the permeability (Figure 3-6A, b and c). The positive effect of operating at a low K_D value (i.e., at $K_D \approx 0.2$) is illustrated in Figure 3-6B. However, the effect is small compared to the positive effect of a large pore volume.

Since the slope of the calibration curve, b, is directly related to the shape of the solute (expressed as $d \log R / d \log M_r$, see Section 3.3.1), the resolution of rod-shaped molecules is higher than the resolution of globular solutes of similar molecular masses (see also Fig. 3-3). In a study of the theoretical limits for the resolving capacity of size-exclusion chromatography, it was concluded that separation of dimer and monomer forms of a globular protein should readily be obtained on a 30-cm column packed with a 10-μm bead, yielding 10,000 plates.[88] However, separation of proteins differing in molecular mass by less than 20% would be difficult also with columns of extreme plate counts (i.e., having plate counts of 100,000). However, the separation of rod-shaped molecules, such as DNA of intermediate size, differing 10% in molecular mass might be possible with a high-porous medium. These figures seems to set the limits for the resolution capacity of gel filtration.

The resolution increases significantly and the separation time decreases (cf. Section 3.6.3) using very small particle sizes (i.e., below 5 μm). However, the trade-off is a substantial increase in column back pressure, which puts special requirements on the instrumentation used, if the objective is very fast separations. Under such conditions the risk of temperature effects due to frictional heat or shear degradation of elongated solutes, caused by the high flow velocity in the narrow void channels, must be considered. Furthermore, an increase in the pressure resistance of a material is mostly achieved by increasing the matrix content and this is likely to have an adverse effect on sorptive properties and resolution (i.e., due to low pore volume). The inherent limitation of particle size for SEC is set by the large pore size necessary for separating macromolecules and the requirement for a large ratio of void channel radius (i.e., particle radius) to pore radius to avoid size exclusion effects in the extra-particle space. Guiochon and Martin[89] calculated the optimal particle size for SEC of macromolecules to 1–2 μm at the modest separation time of 1 h. However, as shown by Verzele and co-workers,[90] it is very difficult to pack such small particles efficiently, which seems to suggest an optimal particle size around 5 μm with present packing technology. Also, it may be noted that columns of small particle size media are predominantly supplied prepacked, that these columns are shorter than standard laboratory columns (since the material is expensive), and that the increased matrix volume results in a low separating volume. The total effect is that the peak capacity of columns for size exclusion chromatography expressed by

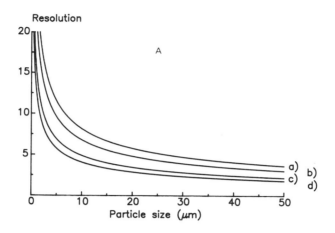

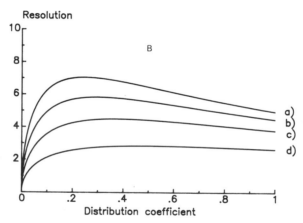

FIGURE 3–6. Factors affecting the resolution in gel filtration. (A) Resolution as a function of particle size (i.e., efficiency) of the support at various slopes of the selectivity curve, b, and permeabilities, V_i/V_0, of the support.

Curve	(a)	(b)	(c)	(d)
b	0.4	0.2	0.4	0.2
V_i/V_0	1.0	2.5	0.5	1.0

Calculated from Eq. (3-16) with $K_D = 0.5$, $L = 30$ cm, and $H = 2 \times d_p$. (B) Resolution as a function of distribution coefficient at various permeabilities of the support. $V_i/V_0 =$ (a) 2.0; (b) 1.5; (c) 1.0; and (d) 0.5. Calculated from Eqs. (3-11) and (3-16) with $d_p = 10$ μm, $D_m = 7 \times 10^{-7}$ cm²/s, $L = 30$ cm, $u = 1$ cm/min, $M_{r1}/M_{r2} = 10$ and $b = 0.3$.

$$n_{SEC} = 1 + (V_i/V_t) \cdot (N)^{1/2}/(4R_s)$$ (3-17)

is not exceeding 13 for as well microparticulate media (e.g., 4 μm Zorbax GF-250) as for laboratory packing media (e.g., 33 μm Superose 6 prep grade) and, furthermore, that traditional media such as Sepharose® CL 6B yield competitive peak capacities.[86] The gain in using media of small particle size is the reduction of separation time (see Section 3.6.3).

The sample volume will contribute considerably to the width of the sample peak unless the volume is small as compared to the volumetric dispersion caused by the column itself. Whereas this fact is a matter of keeping the concentration and volume as low as detector sensitivity permits in analytical gel filtration (see Section 3.7.3), the balance in preparative gel filtration is made with respect to volume, viscosity, and flow rate to yield the desired throughput (see Section 3.7.2).

3.4 PACKING THE COLUMN

3.4.1 Column Materials and Accessories

Whereas most column materials are suitable for aqueous buffer solutions commonly used in protein chromatography, there are often limitations in the use of organic solvents or extremes of pH with different materials. The resistance increases in the order acrylic plastic < polypropylene < glass. However, it is important to realize that the plungers, seals, and other components may have a lower solvent resistance than the column tube. When using steel columns, special precautions should be taken to reduce problems with corrosion (e.g., extremes of pH and the use of halide ions should be avoided). Recommendations for use should be provided by the manufacturer. Columns suitable for large-scale gel filtration where the adaptor is adjusted by hydraulic assistance have been introduced. A selection of columns suitable for gel filtration is compiled in Table 3-4.

The column of choice should preferably be transparent, pressure and solvent resistant, and equipped with an adjustable end adaptor. The dimensions may be chosen according to the application at hand, but most laboratory columns have a 4- to 16-mm inner diameter (i.d.) and a length of 25 to 70 cm. The choice of inner diameter may be based on the desired sample load or on considerations of the extra dispersion effects from the wall region, which extends 30 particle diameters away from the wall.[91] It is thus safe to use an 8-mm i.d. column for 10-μm beads, but not for 100-μm material. The length of the column is primarily chosen according to the resolution that is required [cf. Eq. (3-16)]. Before packing or repacking the column, all parts should be thoroughly cleaned. The bed support filters should be new and have a mesh size compatible with the particle size of the gel. Avoid touching the surface of nets and filters as a fingerprint may cause uneven sample application

TABLE 3-4. Empty Columns Suitable for Self-packing of Gel Filtration Columns

Column Type	Supplier[a]	Column Dimensions (i.d. × L, mm)	Pressure Specification	Tube Material
Laboratory columns				
C column system	A	10 × 100–26 × 1,000	Up to 0.1 MPa	Borosilicate glass
Econo column	B	5 × 50–50 × 700	Up to 0.1 MPa	Borosilicate glass
XK columns	A	16 × 200–50 × 1,000	Up to 0.5 MPa	Borosilicate glass
HR columns	A	5 × 200–16 × 500	3–5 MPa	Borosilicate glass
G columns	C	1 × 250–44 × 1,000	3–7 MPa	
Large-scale columns				
INdEX[TM]	A	100 × 500–200 × 950	0.3 MPa	Borosilicate glass
Vantage	C	60 × 500–250 × 1,000	0.7 MPa	TPX (polymer)
Superformance	D	50 × 500–450 × 1,000	0.2–1.4 MPa	Glass
BPG series	A	100 × 500–300 × 450	0.4–0.8 MPa	Borosilicate glass

[a] A, Pharmacia Biotech (Uppsala, Sweden); B, Bio-Rad Laboratories (Hercules, CA); C, W.R. Grace & Co., Amicon Division (Danvers, MA); and D, Merck (Darmstadt, Germany).

and poor separation. It is also wise to inspect the bed support parts for burrs and so on to ensure the best possible sample application after the column is packed. The use of screens instead of frits is advantageous in avoiding clogging and extra sample dispersion.[92]

3.4.2 Packing Procedure

The gel should be pretreated according to the manufacturer's instructions. However, it is often advantageous to decant the diluted slurry to remove fines and to use an excessive amount of gel to let nondispersed lumps settle before pouring the gel into the packing reservoir.

An efficiently packed column will have a low, nonseparating, void volume and a homogeneous bed that prevents channeling and disturbed flow paths. The ways to achieve an efficient packing seem in general to be derived empirically. Thus, small rigid beads are often packed at high pressures (>4 MPa) in solvents that are chosen to prevent interparticle interactions. Various techniques have been used to pack soft and semirigid materials.[24,93,94] This situation probably reflects the many parameters influencing the packing density and bed stability, parameters that are unique for each type of support.[90] A guideline on how to pack these materials may be given by the pressure–flow curve that is obtained from running a flow gradient through the column.[24] The pressure drop over the bed is proportional to the flow velocity (see Section 3.4.4). At sufficiently high-flow velocities the beads of semirigid materials are compressed (and particles of rigid materials may be partly crushed) by the large cumulative drag force acting on the particles. In the compression region (most likely found at the column outlet) the flow resistance of the packed bed increases drastically as a result of a decrease in void fraction due to local compression of nonrigid beads. It seems to be most appropriate to pack semirigid materials at a constant flow rate close to, but before, the compression region. Packing at a constant flow rate will generate a column with a uniform packing density due to the constant friction force acting on the particles.[95] It has been suggested that columns should be packed in the compression region to produce a bed with a low, nonseparating void volume. However, the effect on the average void fraction is small, and the increase in back pressure and irregular flow paths caused by partial blockage of the bed by collapsed particles may be disadvantageous. Furthermore, an impaired stability of the compressed bed, resulting in a reduced column lifetime, may also be expected.

It is important to stabilize the bed, after the end adaptor has been inserted, by running the column at a somewhat higher flow rate.[24,95] This second step probably predominantly influences the upper, more loosely packed part of the bed and may be performed using a constant flow rate or constant pressure since the bed is already formed.

To obtain a perfect sample application, it is essential to obtain a homogeneous and well-packed zone at the inlet of the column. This is easily achieved by packing the column toward the inlet adaptor (using two adaptors).[96] Running a column in the reverse direction to the packing direction is generally not recommended for rigid materials (i.e., silica), as this may cause rearrangement of particles in the bed. This effect is normally not seen with the elastic semirigid materials.

3.4.3   Evaluation of Column Packing

Many of the supports for high-resolution gel filtration are only available as prepacked columns, and there is an increasing demand for columns prepacked with more traditional materials. As shown by Potschka,[28] the efficiencies of similar prepacked materials vary substantially in the hands of a user, and the evaluation of prepacked columns may be as necessary as it is for user-packed columns to ensure optimal performance.

The column packing may be evaluated by running a low molecular weight noninteracting solute such as acetone, cytidine, sodium nitrate, glucose, or D_2O at a high flow rate (cf. Figure 3-5) and calculating the reduced plate height from

$$h = L/[5.54(V_R/w_h)^2 d_p] \tag{3-18}$$

where w_h is the peak width at half peak height. For an efficiently packed column, the reduced plate height should be close to 2.[97] Reduced plate heights of 1.5 have been reported for very efficiently packed columns.[82,83] The quality of the peak shape may be expressed by the asymmetry factor, A_s (see Figure 3-1), which should be close to unity (i.e., 0.9 to 1.1). The sample application can be evaluated visually by running a colored sample as depicted in Figure 3-7.

When the column packing is acceptable, it is often useful to run some well-defined proteins to measure the actual column performance in the desired separation range. For high-resolution columns this is done easily by running a few samples containing mixtures of proteins. Some mixtures that have proved to be useful in the high and middle molecular weight ranges are given in Table 3-5.

The homogeneity of the bed may be checked by comparing the pressure drop over the bed at a certain flow rate to that calculated from Eq. (3-19) (see Table 3-6). A large deviation (e.g., larger than 50%) indicates a nonuniform packing with a compressed region of beads in the column.

3.4.4   Pressure Drop over Packed Beds

The back pressure generated when a liquid is flowing through a packed bed can be calculated from the formula given by Darcy or the more explicit expression derived by Hagen and Poiseuille.[98] Originally, the relationship was utilized to obtain information on particle dimensions from the observed

FIGURE 3–7. Visual evaluation of column packing.

pressure drop.[98] This procedure has also been used in liquid chromatography to estimate an apparent particle size.[99] Many workers, including Blake, Kozeny, and Carman, have contributed to refining the equation given by Hagen and Poiseuille to the present formulation[98]

$$\Delta P = u_{nom} \times \frac{L \cdot \eta}{d_p^{\,2}} \times \frac{(1 - \varepsilon)^2}{\varepsilon^3} \, k \times 36 \qquad (3\text{-}19)$$

where ΔP is the pressure drop in Pa, u_{nom} is the nominal liquid velocity in centimeters per minute, η is the viscosity of the solvent (1.49×10^{-5} min $\times$ N/m, H_2O, 25°C), L is the bed height in cm, d_p is the harmonic mean value of the particle size distribution in centimeters, ε is the void fraction (i.e., void volume over column volume), and k is the aspect factor. For spherical beads the aspect factor is close to 5,[98] and the last term in Eq. (3-19) is then reduced to $k \times 36 \times (1 - \varepsilon)^2/\varepsilon^3 \approx 180 \, (1 - \varepsilon)^2/\varepsilon^3$. This term is called the column resistance factor and is typically 500 to 700 for spherical silica materials.[100] This corresponds to a void fraction of 0.42 to 0.45. Semirigid materials having a void fraction of 30% (see Table 3-2) will have a column resistance factor of approximately 2,250. It is important to notice the influence of the temperature dependence of the viscosity on the generated pressure drop. The relative viscosity of water is 1.6 at 4°C, which yields almost a twofold increase in the pressure drop in the cold room as compared to ordinary room temperature.

TABLE 3-5. Solute Mixtures for Calibration of Gel Filtration Columns

Solute	Supplier[a]	Molecular Mass	Radius (Å)[b]	Concentration[c] (mg/ml)
Mixture 1				
Cytidine	A	243		0.1
Bacitracin[d]	B	1,450		10
Cytochrome c	C	12,400	18.8	1
β-Lactoglobulin	C	35,000	26.5	2
Bovine serum albumin	E	67,000	32.9	6
Immunoglobulin G	D	160,000	44.0	2
Thyroglobulin	E	669,000	70.8	4
Mixture 2				
Acetone	A			5
Myoglobulin	C	17,600	21.1	1
Ovalbumin	E	43,000	28.4	2.5
Transferrin	C	81,000	35.0	1.5
Aldolase	E	158,000	43.8	6
Ferritin	E	440,000	61.6	1
Blue Dextran 2000	E			2

[a] A, Merck (Darmstadt, Germany); B, Serva (Heidelberg, Germany); C, Sigma (St. Louis, MO); D, Pharmacia Biopharmaceuticals (Stockholm, Sweden); and E, Pharmacia Biotech (Uppsala, Sweden).

[b] Calculated from Eq. (3-8).

[c] Suggested to be detected at 0.2 AUFS, 280 nm. Adjustment according to instrumentation used and zone broadening will be necessary. Figures include weight of additives.

[d] Freshly prepared.

The theoretical pressure drops of columns packed with materials of different particle sizes and void fractions can be inferred from Table 3-6. A large void fraction will lead to a low back pressure but also to poor resolution due to the decrease in permeability [cf. Eq. (3-16)].

The pressure drop over beds packed with nonsolid supports always deviates from the linear relationship predicted by Eq. (3-19) at sufficiently high flow velocities. The steep increase in pressure noticed with the softer materials is due to compression of the beads and a subsequent reduction of the void fraction in a small zone near the column outlet.[24,101] The compression is due to the cumulative drag force caused by the flowing liquid. This force increases with increasing flow velocity, viscosity of the liquid, and bed height.[102] Thus, dividing a bed into shorter segments (i.e., stacked columns) will enable the use of relative higher flow velocities before the beads are compressed.[103] The drag force is counteracted by the supporting force of the column walls (i.e., of narrow columns). Elastic materials are not destroyed by the compression, and the original performance of a column may be restored by repacking it with the material.[24]

TABLE 3-6. Theoretical Pressure Drop, in KPa, Over Packed Beds of Various Void Fractions and Particle Sizes[a]

Void Fraction	Particle Size (μm)									
(V_0/V_c)	2	5	10	15	20	30	35	50	75	100
0.50	134	21	5	2.4	1.3	0.6	0.4	0.2	0.10	0.05
0.45	223	36	9	4.0	2.2	1.0	0.7	0.4	0.16	0.09
0.40	377	60	15	6.7	3.8	1.7	1.2	0.6	0.27	0.15
0.35	661	106	26	11.7	6.6	2.9	2.2	1.1	0.47	0.26
0.30	1,217	195	49	21.6	12.2	5.4	4.0	1.9	0.87	0.49
0.25	2,414	386	97	42.9	24.1	10.7	7.9	3.9	1.72	0.97
0.20	5,364	858	215	95.4	53.6	23.8	17.5	8.6	3.81	2.15

[a] Calculated, per cm bed height, at a nominal fluid velocity of 1 cm/min from Eq. (3-19). Conversion factors: 100 KPa = 1 bar = 14.5 psi. The theoretical pressure drop is obtained by multiplying the figure in the table with the column length (cm) and the nominal fluid velocity (cm/min).

Comparison of compressibility of different materials may be obtained by normalizing the pressure drop with respect to column length and particle size, as in Figure 3-8. It is then important that the column diameter is sufficiently large to avoid supportive wall effects (e.g., a ratio of column to particle diameter of 1×10^3 seems necessary).

The matrix rigidity may be improved by an increased cross-linking of the polymer[104,105] or by allowing a larger matrix volume of the support.[18] Increased matrix rigidity is thus achieved at the expense of pore volume and resolution.

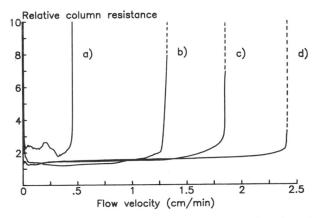

FIGURE 3-8. Variation in the column resistance factor as a function of fluid velocity for materials of different mechanical strenght (i.e., degree of cross-linking). Experimental data are normalized with respect to particle size, bed height, and initial void fraction of the beds. Supports: (a) Sepharose 6B, (b) Sepharose CL 6B, (c) Superose 6 prep grade, and (d) Superose 6.

Therefore, high matrix rigidity is not an objective per se, but should be related to the back pressures that are generated by the modest flow velocities that are applicable to the separation of high molecular weight solutes (cf. Section 3.6.3).

A gradual buildup of column pressure drop may be caused by the migration of fines toward the column outlet. Fines may be present due to incomplete decantation (e.g., up to 10 decantations may be necessary to remove the major part of fines), to bead fragments from the shear of fragile beads, or to mechanical crushing (e.g., due to excessive axial compression of silica materials).

3.5   SAMPLE AND BUFFER PREPARATION

3.5.1   Buffers and Additives

The sample is preferably dissolved in the same medium as that used for the gel filtration step, unless the procedure is a desalting step or buffer exchange. It is generally advisable to use aqueous buffer systems with buffering capacities in the pH range of 6 to 8 as this will produce a good environment for many proteins, as well as cope with most gel filtration matrices. The acid dissociation constant, pK_a, of the buffering substance should be close (i.e., 0.5 pH) to the desired pH to yield the optimal buffering capacity of the solution. Traditionally, phosphate ($pK_{a2} = 7.2$) and tris(hydroxymethyl)aminomethane ($pK_a = 8.1$) have been used at neutral pH. In cases where the buffering capacity of these substances is too low or the properties of the substance are incompatible with the sample (e.g., phosphate is known to inhibit certain enzymes[106]), the biological buffers proposed by Good and co-workers may be more suitable.[107] The universal citrate–phosphate–borate buffer described by Teorell and Stenhagen can be used to study the effect of pH on the separation in the pH range of 2 to 12 with only a slight variation in ionic strength.[106] However, borate buffer may interact with glycopeptides.[108] When running a preparative purification or a desalting step, volatile buffer salts such as ammonium acetate or ammonium bicarbonate may be preferred as these are readily removed by freeze drying. Suggestions for other volatile buffer systems covering the pH range of 2 to 12 are found in Perrin and Dempsey.[106]

To avoid ionic interactions between the solute and the matrix, the ionic strength of the buffer is often increased to 0.05–0.50 M by the addition of a salt. Sodium chloride is used most frequently, but because the halides are very corrosive, these salts should be replaced with, for instance, sulfates whenever stainless steel is in contact with the liquid. In some situations the salting-out effect of sulfate may create hydrophobic interactions with the matrix. In those cases, chaotropic ions, such as perchlorate, may be used to increase the ionic strength of the buffer.[25] Interpretation of elution volume in terms of molecular weight from analysis at a high ionic strength must be made with care, as the conformation of proteins and the expansion of polymers may vary with variations in ionic strength.

Nonionic interactions have been reported for some types of matrices. These effects may be reduced by increasing the pH, by decreasing the ionic strength, or by adding small amounts of detergents, ethyleneglycol, or an organic modifier, such as 1-propanol (1%) or acetonitrile, to the buffer. Gel filtration of lipophilic solutes has been carried out in 45% acetonitrile.[109] A successful separation of membrane proteins on Sepharose CL-4B using a chloroform/methanol mixture as the eluent has been reported.[110]

The addition of detergents to the buffer is sometimes required to solubilize certain proteins (e.g., membrane proteins). This approach was used for the solubilization of integral membrane proteins in 0.1 M sodium dodecyl sulfate (SDS) followed by high-performance gel filtration in 5 to 50 mM SDS.[111]

Denaturing solvent such as 6 M guanidine hydrochloride is often used to break the hydrogen bonds that stabilize the tertiary structure of proteins and to transform them into random coils for subsequent molecular weight analysis (see Section 3.3.1).

The use of 0.02% sodium azide in the buffer has proven to be useful in preventing microbial growth (sodium azide must be handled with care as it may form explosive insoluble salts with heavy metals and is believed to be a mutagen). Other preservatives that have been tested include 0.05% trichlorobutanol and 0.01% thiomersal, but they are less effective.[7] Chlorhexidine (0.01%) is claimed to be an efficient bacteriostatic agent. Cationic preservatives are not recommended because they may adsorb to residual negatively charged groups on the matrix at a low ionic strength. If a bacteriostatic agent is unsuitable to use, then the problem of bacterial growth may be reduced by using freshly prepared and sterile-filtered (i.e., 0.22 μm) buffer to continuously flush the column.

The choice of buffer, pH, ionic strength, and additives must be made with respect for solvent–matrix interactions, the solubility and biochemical properties of the sample, and the limitations of the detection system used. Undesired effects may be caused by chelating or solubility properties of the buffer substance (e.g., borate–carbohydrate or phosphate–Ca^{2+} interactions),[106,111] self-association of proteins,[112] or high absorbency of buffer or additives at the wavelength used for the detection of the sample and interferences of these substances with chemical detection steps (cf. Section 3.6.4).[106]

Some buffer systems commonly used in high-performance gel filtration are given in Table 3-2. An extensive compilation of other buffer systems was given by Perrin and Dempsey.[106] It is advisable to prepare the buffer at the running temperature, as the pK_a value of the substance may vary considerably with temperature (e.g., $-$ 0.03 pH/°C). Finally, filtering the buffer solution through a 0.45-μm filter prior to use will save the column from being blocked by undissolved substances.

3.5.2 Sample Load

The sample load, expressed as the amount of sample applied, depends of course on sample concentration and sample volume. Whereas the constraint on the former is mostly due to the high viscosity of concentrated samples, the effect of the latter is governed by the dimensions of the chromatographic bed, the pore volume of the support, and the zone broadening of the column. The sample load is in general minimized in analytical gel filtration to assure the highest attainable resolution but is maximized in preparative gel filtration to yield maximal throughput of material.

As a general rule, the viscosity of the sample relative to the viscosity of the eluent should be less than 1.5. This corresponds to a maximal concentration of 70 mg/ml of a globular protein such as human serum albumin. This was experimentally verified in the authors laboratory on a column packed with Superose 6 prep grade, where a sample load of 100 mg/ml gave abnormally broad peaks, whereas 50 mg/ml gave normal sharp peaks. Fernandez and co-workers were able to study viscosity effects in a packed bed with the aid of magnetic resonance imaging (MRI). Viscous fingering effects were clearly noticed on the chromatogram of bovine serum albumin (as a slight shoulder on the rear part of the peak) at a concentration of 50 mg/ml. However, viscosity effects were noticed by MRI at concentrations as low as 10 mg/ml, but broadening of the zone due to column dispersion was found to reduce the effects.[113]

Viscous fingering effects are caused by invasion of a less viscous solvent into the rear part of the more viscous sample zone; such effects are more likely to occur for concentrated solutions of high molecular weight solutes. The effect is readily indicated by varying the concentration of the sample, as demonstrated in Figure 3-9. It may be expected that viscosity effects are more pronounced in beds of small particle size, as this results in narrow void channels and less sample dilution (as experimentally found in the MRI experiments cited earlier). If the viscosity of the sample is very high, then this may be compensated for by increasing the viscosity of the eluent by adding sucrose or dextran.[7] However, this requires the use of a lower flow rate to maintain low zone broadening and a low back pressure.

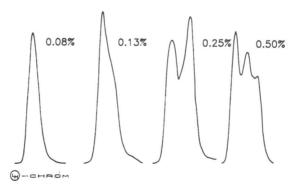

FIGURE 3–9. Disturbed elution profiles due to viscous fingering effects. Sample: 0.6 ml of native dextran.

The influence of sample volume on zone broadening can be estimated by comparing the width of the injected sample plug of volume, V_{inj},[114] and the final width of the peak as calculated by the plate height equation.

$$\sigma_{total}^2 = \sigma_{column}^2 + \sigma_{injection}^2 \approx V_R^2 (H/L) + V_{inj}^2/K_{inj} \qquad (3\text{-}20)$$

The theoretical value of K_{inj} (i.e., 12 for a square-wave pulse) has been achieved with carefully designed injection devices,[115] but values of 1 to 5 have been reported for ordinary valves.[114,116] A value of 5 has been found to be in accordance with experimental results.[117]

The relative contribution to the total variance, $\sigma_{rel}^2 = \sigma_{injection}^2/\sigma_{total}^2$, will then theoretically result from a sample volume of

$$V_{inj} = \sqrt{\sigma_{rel}^2/(1 - \sigma_{rel}^2)} \times A_c \times \sqrt{L} \times (1 + K_D V_i/V_0) \times$$
$$(V_0/V_c) \times \sqrt{h \times d_p} \times \sqrt{K_{inj}} \qquad (3\text{-}21)$$

This formula confirms the general expectations that high values of the cross-section area, A_c, the column length, the pore volume, and K_D will support a high sample volume. However, it is not always realized that a material of small particle size will only allow applications of small sample volumes, if optimal performance is to be retained. Equation (3-21) shows that the injection volume needs to be adjusted to a level proportional to the square root of the particle size if σ_{rel} is to be kept constant. Thus, while sample volumes of 1 to 2% of the bed volume do not impair the performance of traditional gel filtration media with a particle size of 100 μm,[7] this is certainly not true for 10-μm materials, where 0.3% is a more realistic figure.[117] The variation of peak width, expressed as reduced plate height, with a sample volume is shown in Figure 3-10. The most important implication of this figure is the limited use of the inherent efficiency of small, particle-sized materials when large sample volumes are applied.

The figures for sample load are of course only tentative, and due to the nature of the sample, the gel matrix, relative pore volume, and the desired separation, it may be experimentally found that other sample loads are applicable. Thus, small effects on the peak elution volume from the sample concentration have been noted in high-precision, analytical gel filtration.[118] Special precautions are required for gel filtration of highly charged solutes. For instance, the retention volume of hydroxyethylcellulosa increased at sample concentrations exceeding 1 mg/ml.[119] However, the effect on retention of proteins can be expected to be much more insensitive to concentration and ionic strength than for polymers. Nevertheless, it is a good practice to study the separation pattern at different sample loads to ascertain that the conditions used are appropriate.

The sample load in preparative gel filtration will be a compromise between peak width caused by a large injection volume and zone broadening caused

Reduced plate height

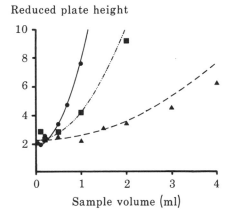

Sample volume (ml)

FIGURE 3–10. Effect of sample volume on the peak width (e.g., reduced plate height) of cytidine. Lines represent calculated relationships and dots represent experimental data. Columns: (●) Superose 6, 500 × 10 mm i.d.; (■) Superose 6 prep grade, 530 × 10 mm i.d.; and (◆) Superose 6 prep grade, 510 × 16 mm i.d. (Reproduced from Ref. 117 with kind permission from the publisher.)

by an excessive flow rate in order to process the required amount of sample in the available time at a predetermined resolution. The optimal injection volume for preparative gel filtration of macromolecules, $V_{inj,opt}$, may be inferred from Eq. (3-12) and Eq. (3-20) to[96]

$$V_{inj,opt} \approx (V_{feed} \times K_{inj} \times V_c \times V_i \times d_p^2/15\ D_m)^{1/3} \qquad (3\text{-}22)$$

where V_{feed} is the volume of sample that is to be processed per hour. The nominal solvent velocity is given by $u_{nom} = L \cdot (V_{feed}/V_{inj,opt})$. It was found that the maximal resolution was obtained at injection volumes of 2 and 6% of the column volume when a total sample volume of, respectively, 0.5 and 5% of the bed volume was to be processed per hour.[96] The maximal resolution decreased substantially when the processing rate was increased. The exact optimum is of course dependent on the initial resolution of the solutes and the requirements on final purity and recovery. In a study of the productivity of purification of IgG from transferrin on Superdex® 200, it was found that the maximum load was between 2 and 3% of the bed volume for the recovery of, respectively, 99.5 and 95.0% at a purity of 99.99%.[5] Equation (3-22) indicates that a good injection device is as important as a large column volume (although for large injection volumes the injected zone will approach a square-wave plug) and that a change in temperature (e.g., transferring the separation method from room temperature to cold room) will affect the optimal injection volume, as the diffusivity will be reduced due to the increase in solvent viscosity.

3.5.3 Calibration Substances

The choice of sample substances used for the determination of void volume, V_0, and total liquid volume, V_t, requires special attention. The probe for the void volume must be large enough to be totally excluded from the gel matrix but not so large as to be subjected to secondary exclusion in the void volume or separation due to hydrodynamic chromatography.[120] This will lead to the appearance of a peak before the actual void volume. To avoid these effects the radius of the probe should be less than 0.01 times the particle radius (i.e., ca. 500 Å). The solute must not be charged at the pH used as this can cause ionic adsorption or extra exclusion effects. The most common substance, Blue Dextran 2000, has an assigned molar mass of approximately 2×10^6, which would correspond to a hydrodynamic radius of 350 Å.[121] This substance is not monodisperse, and the low molecular weight part of the distribution may therefore permeate the matrix of a sufficiently porous gel. However, the high molecular weight region is useful for the determination of void volumes of matrices, with pore sizes of up to at least 1,000 Å. In cases where the blue dyestuff of this probe shows adsorption to the gel, it may be replaced by calf thymus DNA[18] or by lyophilized *Escherichia coli*.[9] The interstitial volume has been estimated from the elution volume of human trombocytes and fixed with formaldehyde, as Blue Dextran 2000 permeated the gel under study, Sephacryl S-1000.[122] The calibration curve of this gel was extended to 10^7 using intramolecularly cross-linked hemocyanin from *Helix pomatia*.[122]

Sodium azide, sodium nitrate, cytidine, or glycyltyrosine has been used as a marker of total liquid volume on highly porous gels with detection by UV absorption. Glucose, sodium chloride, or deuterium oxide may be used with detection by refractive index. However, special problems arise when measuring the pore volume of a gel of small pore size when the ionic strength of the medium has to be kept low. Because of the high surface area and low shielding effect, partitioning of solutes into the hydrated gel layer, adsorption, or ionic interactions may be more pronounced. In such cases the total volume is calculated from the geometric dimensions of the bed, and the distribution coefficient thus achieved is designed K_{av}. The difficulty of obtaining a correct estimate of the total liquid volume of some types of materials has been addressed.[123]

If the column is to be calibrated for estimating the molecular size of proteins, then the calibration substances should be well-defined proteins with similar characteristics (e.g., shape and density) as the sample (see Section 3.3.1). If suitable references are not available, then the column may be calibrated by a secondary reference, such as dextran, pullulan, ethyleneglycol, or similar well-characterized polymer, and data expressed in, for example, dextran radius (see Section 3.7.3).

Finally, before applying the sample it is a good experimental practice to filter the sample solution through a 0.45-μm filter unless the sample solution is perfectly clear. This will save the inlet filter from being partly blocked with subsequent impaired sample application as a result. The use of filtered buffer and sample solutions may preserve the column lifetime to more than 1,000 injections.[124]

3.6 THE CHROMATOGRAPHIC SYSTEM

A general view of the instrumentation commonly used in a gel filtration experiment is given in Figure 3-1. The instrumentation may, with respect to pressure resistance, be divided into three categories: high-pressure systems capable of overcoming pressure drops of several hundred bars, medium-pressure systems working up to 50 bars, and low-pressure systems used up to 1 bar. The main difference between the systems is the pumping principle (i.e., reciprocating pump with heads of stainless steel or titanium, syringe pump with wetted parts of glass, and tubing pump with inert tubing, respectively).

The choice of column dimensions is closely related to the application: traditionally, columns for desalting purposes are "short and fat" (e.g., 1 to 10 cm long with 1 to 50 cm i.d.), columns for fractionation purposes are 20 to 70 cm long with 3 to 5 cm i.d., and columns for analytical separations are "small and long" (e.g., 50 to 100 cm long with 0.5 to 1 cm i.d.) (see Table 3-4). These figures will change with miniaturization of support sizes and sample volumes, but the relative column sizes for the different application areas are likely to remain the same.

The choice of detector (i.e., flow cell) is not as critical as the choice of column dimensions. However, narrow-bore flow cells are generally not applicable in large-scale fractionations as they seldom tolerate the excessive back pressure generated over the cell at high flow rates. The recent development of on-line, mass-sensitive detectors provides powerful detection possibilities for analytical gel filtration.

3.6.1 Sample Application

There are several ways of applying the sample to the gel bed. On columns that are not equipped with an adaptor (e.g., small columns for desalting purposes), the sample may simply be layered on the bed surface with the aid of a Pasteur pipette or a similar device. More eluent is then carefully added and the elution of the sample started. This technique requires high operator skill but will, if properly performed, result in a very good sample application. A more convenient way is to use an injection device that is coupled to an adaptor through a valve. If the gravity flow of the column is high, the injection device may simply consist of a syringe attached to the valve. However, a more accurate sample application is achieved by using a loop connected to the valve. This is the only technique that is applicable to medium- and high-pressure systems. When running columns packed with beads of very small particle size (e.g., 3 to 5 μm), the construction of the injection device becomes critical (see Section 3.5.2). If very large sample volumes are to be applied, the use of an ordinary loop, consisting of a long, narrow-bore tubing, is not recommended. The sample may in these cases be applied through a Super-loop[125] or simply by using a peristaltic pump. The valve may be operated manually, electromagnetically, or pneumatically. It is important to flush the injection device and the valve extensively between sample applications to avoid carryover effects from the previous sample. In high-resolution gel filtra-tion the tubing lengths between the injection device and the adaptor should be kept as short as possible to minimize zone broadening. In some cases the addition of sucrose to increase the sample density and to reduce the broadening of the sample zone between the injector and the column has been used, but this is seldom necessary with modern injection devices as demonstrated by Figure 3-7.

3.6.2 Eluent Delivery

The eluent is contained in a reservoir that may be slightly heated (e.g., with the aid of a heating tape or an ordinary light bulb) to degas the eluent and avoid the creation of air bubbles in the gel bed when using the column for extensive periods of time. Using a degasser placed before the injection device is both convenient and effective.[126] This is not necessary when using a column packing that generates back pressures above 10 bars as introduced air bubbles will be dissolved in the eluent at this pressure.

   The eluent is delivered from the reservoir either directly to the column or through a pump. In the first case the hydrostatic pressure from the reservoir to the column outlet must be sufficient to create a flow through the column that exceeds the flow rate used. The actual flow rate is then preferably controlled by a pump placed at the column outlet. This procedure may only be used when the flow rate is low and the bead size is large. A Mariotte bottle is used as a reservoir to maintain a constant operating pressure over the bed (continuous degassing of the eluent is necessary when using a Mariotte bottle). A more convenient, and recommended, way is to place the pump before the injection device. This will also facilitate the collection of fractions after the column. Whatever system is used, measurement of the actual flow rate is of utmost importance due to the possibility of leakage in pump valves and so on and the large effect of an erroneous flow rate on the calculated molecular mass in analytical gel filtration.[127] The most common way to measure flow rate is by gravimetry, but it is also possible to use flow meters for continuous con-trol.[128] Even though high-precision pumps are preferable for most work, the long-term flow constancy of peristaltic pumps equipped with Tygon tubing is satisfactory in many cases. An indirect way of controlling the reliability of the pump is to include an internal standard (e.g., a low molecular weight solute) with each run. This may also provide a way to compensate for any long-term drift of the pump.

3.6.3 Optimal Flow Rate

The maximal flow rate that is applicable is restricted by both physical and physicochemical considerations.

The flow of liquid through the column generates a back pressure as described by Eq. (3-19). Significant deviations between the theoretical value of the pressure drop and the experimental one may indicate a nonuniform packing. Abnormally high pressure drops over filters, frits, or connection tubings may be caused by clogging from a precipitated sample or particulate materials. The column construction and materials, fittings, connections, pump type, and the mechanical rigidity of packing material limit the maximal flow rate that can be used.

The useful flow rate is in practice restricted by the low diffusion coefficients of macromolecules that may lead to excessive zone broadening when too high solvent velocities are used.[129] This is eliminated by assuring that equilibrium between the mobile and trapped phases is achieved. This can be done by decreasing the transfer distance of the solute (i.e., reducing the particle size of the support), increasing the available transfer time (i.e., reducing the flow rate of the solvent), and/or increasing the transfer speed (i.e., increasing the diffusivity, e.g., by increasing the temperature).

The zone broadening due to nonequilibrium is expressed by the C term in the van Deemter equation [Eq. (3-10)] and is illustrated in Figure 3-5. The optimal interstitial fluid velocity, yielding the minimum zone broadening, is found by differentiation of Eq. (3-11) with respect to u, which results in Eq. (3-23)

$$u_{\text{opt}} = \frac{D_m \times \gamma_s}{d_p} \times \sqrt{60 \times \left(\frac{2}{3 \times \gamma_s \times \frac{V_0}{V_R} \times \left(1 - \frac{V_0}{V_R}\right)} + \left(\frac{V_R}{V_0}\right)^2 \right)} \qquad (3\text{-}23)$$

showing that the optimal flow rate is proportional to the hindered diffusion of the solute and inversely proportional to the particle size of the support (and proportional to the column cross-sectional area since the flow rate is equal to the velocity times the cross-sectional area).

At u_{opt}, the major contribution to the zone broadening is the A term in Eq. (3-10). It seems practical to allow an interstitial velocity when the A term is not solely dominating (e.g., allowing an equally large contribution from the C term). That "practical" velocity, u_p, is given by

$$u_p = \frac{D_m \times \gamma_s}{d_p} \times \frac{60 \times \lambda}{\frac{V_0}{V_R} \times \left(1 - \frac{V_0}{V_R}\right)} \qquad (3\text{-}24)$$

With $\lambda \approx 1$, $\gamma_s = K_D/4$ and $V_0/V_R(1 - V_0/V_R) \approx 0.23$, Eq. (3-24) reduces to $u_p \approx 65\, K_D \times D_m/d_p$. Practical nominal velocities for various bead sizes and solutes are indicated in Table 3-7. It can be noted that no single flow rate is optimal for the entire separation range. Thus, at low flow rates, high molecular weight solutes will yield sharp peaks due to the large transfer time (low C term), but small molecular weight solutes will yield broad peaks due to excessive axial zone broadening (large B term) (cf. Figure 3-5). At high flow rates the conditions are opposite. However, by starting the elution at a low flow rate and increasing the flow rate during the run, a more optimal separation is achieved. This approach is illustrated in Figure 3-11. For this 10-μm support, the effect of solvent velocity on peak broadening is not as large as noticed earlier for 30-μm beads.[24] However, the increased resolution in the high molecular weight region is evident. The separation time can be decreased by a factor of two by increasing the flow rate during the run, as expected from Table 3-7. The flow gradient to apply may be calculated from Eq. (3-24).

3.6.4 Sample Detection

The most popular way to monitor the effluent for proteins is continuous measurement of the absorption in the ultraviolet range. Variable wavelength detectors are often expensive, and the use of a single- or dual-wavelength detector is in most cases adequate. Detectors where the wavelength may be selected at 546, 280, 254, 214, or 206 nm or other wavelengths are available. It is thus possible to select the proper wavelength for the specific detection problem, and for proteins, absorption at 280 nm is often utilized. For the detection of peptides that do not contain aromatic amino acids, lower wavelengths (i.e., below 220 nm) must be used. Care should be taken to avoid

TABLE 3-7. Practical Fluid Velocities for Optimal Zone Widths in Gel Filtration

Solute Molecular Mass (g/mol)	Diffusivity[a] $(cm^2/s) \cdot 10^7$	Nominal Fluid Velocity[b] (cm/h) at a Support Particle Size (μm) of								
		2	5	10	15	20	35	50	75	100
500	32.8	639	256	128	85	64	37	26	17	13
1,000	26.0	507	203	101	68	51	29	20	14	10
5,000	15.2	296	119	59	40	30	17	12	8	6
10,000	12.1	235	94	47	31	24	13	9	6	5
15,000	10.5	206	82	41	27	21	12	8	5	4
50,000	7.06	138	55	28	18	14	8	6	4	3
75,000	6.17	120	48	24	16	12	7	5	3	2
100,000	5.60	109	44	22	15	11	6	4	3	2
200,000	4.45	87	35	17	12	9	5	3	2	2
300,000	3.88	76	30	15	10	8	4	3	2	2
400,000	3.53	69	28	14	9	7	4	3	2	1
600,000	3.08	60	24	12	8	6	3	2	2	1

[a] Calculated from Eq. (3-14) assuming solutes being globular proteins.
[b] Calculated from $u_{nom} = u_p/3$ and $u_p \approx 65 \, K_D D_m/d_p$ with $K_D = 0.5$ and $D_m \approx D$.

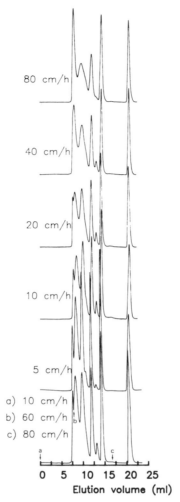

FIGURE 3–11. Separation of a protein mixture at various constant fluid velocities compared to stepwise increasing velocity. Column: Superose 12. Sample: thyroglobulin, ferritin, bovine serum albumin, myoglobin, and cytidine.

buffer solutions and additives that show appreciable absorption at the wavelength used for monitoring. Thus 0.03 M acetate or 0.2 M phosphate buffers, azide, merthiolate, and hibitane are not compatible with detection at 214 or 206 nm.[130] Proteins also show different molar absorptivities at different wavelengths, and chromatograms at 280 and 254 nm cannot be expected to coincide. This can be used to increase the detection selectivity with the response ratio method.[131] DNA absorbs at 360 nm, and the different absorptivity from proteins is frequently used for assaying DNA contamination in protein preparations. Differences in spectroscopic properties of solutes may be fully utilized with the aid of a fast scanning detector [e.g., a diode array detector (DAD)], where total spectrum is recorded in parts of a second.[132] When running unknown samples, the risk of losing peaks may be reduced by feeding the signal to a two-channel recorder set at two different sensitivity ranges.

The use of fluorescence monitoring is mainly applicable to compounds containing tryptophan or after chemical derivatization but, when applicable, the sensitivity and selectivity are substantially higher than with UV detection.

Refractive index (RI) detectors are frequently used for the detection of non-UV-absorbing species such as carbohydrates and polymers (e.g., dextran, PEG). This detector is quite general and will not only respond to variations in the solute concentration, but also to changes in the composition of the mobile phase (e.g., salt peaks or salt depletion) and to temperature and pressure fluctuations. The sensitivity of the RI detector is in most cases inferior to that obtained by UV detection.

The combination of a low-angle, laser light-scattering (LALLS) detector and an RI detector can be used to determine the molecular weight of the eluting solutes, provided that the refractive index increment of the solute is known.[133] The technique, as applied to proteins, yields data in good agreement with results from sequence determinations.[134] Improvements in detectors for the on-line measurement of viscosity and light scattering provide robust detectors suitable for laboratory use. Viscosity detectors are more suitable for detecting low molecular mass solutes, whereas light-scattering detectors are useful for detecting high molecular mass solutes. A multiple capillary detector yields information about the intrinsic viscosity of the sample.[135] Molecular dimensions (i.e., the viscosity average molar mass, M_v) may be calculated from the Mark–Houwink relation $|\eta| = KM_v^a$, where K and a are characteristic constants for the solute–solvent system used. Solute conformation may be inferred from the value of a (i.e., for rod-shaped molecules, a is roughly 1.8, whereas for flexible linear polymers, a is between 0.5 and 0.8).[44] The use of light-scattering detectors, either at low angle (LALLS) or multiple angle (MALLS), provides means for determining the weight average molecular mass of solutes, provided that the refractive index increment of the solute is known.[136] MALLS have been used for the size detection of proteins of molecular masses exceeding 6,000.[136]

Absolute size determination may be achieved by mass spectrometry (MS) and the recent development of on-line LC-MS detectors, where the flow rates commonly used in liquid chromatography (e.g., up to 1 ml/min) may be directly interfaced to the MS detector, providing an interesting, although expensive, option for analytical gel filtration. One example, given by Nylander and co-workers,[137] who identified cleavage products of neuropeptides by electrospray ionization mass spectrometry connected to a high-performance SEC column, Superdex® Peptide (3.2 × 300 mm).

One advantage of using a size detector for analytical gel filtration is that the column is "merely" acting as a separating device. Thus, solute–matrix interactions, variability in eluent velocity, or temperature will not influence the result as long as the sample components are resolved. Furthermore, calibration of the column is not necessary.

Electrochemical detection has been applied to column effluents, and determination of thiol-containing proteins after gel filtration has been described.[138] The use of a bromine detector, where bromine is allowed to react with the protein and the amount of bromine consumed is determined, may provide an interesting general technique for protein determination.[139]

In desalting procedures the registration of ionic strength is often of interest, and in these cases a conductivity detector equipped with a flow cell can be used. It is of course also possible to use an ion-selective electrode equipped with a flow cell for this purpose.

The use of flow-through detectors is mostly straightforward, although certain applications may reveal limitations in instrumental performance, including a slow electronic response to very fast analysis, refractive index effects in UV detection when changing mobile-phase composition, and unbalanced photomultipliers resulting from large differences in absorption of buffers in sample and reference cells. To avoid the formation of air bubbles in the cell, the pressure inside the cell may be increased with the aid of a narrow-bore tubing (e.g., 3 m $\times$ 0.02 mm i.d.) or a constant back-pressure device operating at 50 psi connected to the outlet of the cell.[131,140]

In cases where a more specific assay is required, the effluent may be collected in fractions and a chemical reaction carried out subsequently. In most cases it is also possible to carry out this reaction on line after the column to obtain a continuous assay.[141] The large potential of such methods may be exemplified by a method for the rapid diagnosis of myocardial infarction. The main components of serum were separated by fast gel filtration, and the content of myoglobin (down to 10 μg/ml) was selectively assayed by a postcolumn chemiluminescence reaction with luminol.[142]

3.6.5   System Dispersion

The theoretical contribution to extra-column zone broadening due to the chromatography system may be calculated from[114,116,140]

$$\sigma^2_{\text{system}} = V^2_{\text{inj}}/K_{\text{inj}} + \frac{d^4\,F\,l}{122 D_m} + 2V^2_{\text{cell}} \tag{3-25}$$

where the first term is related to the injection of the sample volume; the second term is the contribution from tubings or connections with length l and inner diameter d at the flow rate F, D_m being the diffusion coefficient of the solute; and the last term expresses the contribution from the flow cell volume, V_{cell}.

If the system is allowed to contribute 10% of the total zone broadening, a maximum of 40 μl of sample, 15 cm of tubing with 0.5 mm i.d., and a cell volume of 15 μl can be tolerated if a reduced plate height of 2 is maintained for a sample eluting in the void volume of a 300 × 10-mm i.d. column packed with a 10-μm material. For a 30-μm material packed in a 500 × 16-mm i.d. column, the corresponding figures are 300 μl sample, 30 cm tubing, and 50 μl cell volume. Thus, it is very important to minimize extra-column broadening if the separation power of smaller particle media is to be fully utilized. It should also be noted that a minimum of connections should be used and tubings with different diameters should be avoided in order to prevent turbulent mixing of the sample zones. The effect of sample volume on peak width in high-resolution gel filtration is treated more extensively in Section 3.5.2.

The risk of shear degradation of macromolecules when reducing the diameters of tubings or particle size while maintaining a high flow rate needs to be considered.[143,144] To avoid shear degradation, using tubings with an inner diameter of 0.5 mm or larger and keeping the flow velocity below 8 cm/min for 10-μm packing materials is recommended.[144]

3.7 APPLICATIONS

Applications of gel filtration may be classified into three principal categories: (1) desalting, buffer exchange, and group separation,where the protein and contaminating solutes differ substantially (e.g., more than a decade) in molecular size, and the requirements on the system and running parameters are relatively modest; (2) (protein) fractionation where the size difference is smaller and the requirements are higher, but related to the difficulty of the separation; and (3) determination of molecular size where high resolution is needed to assure integrity of the characterized solute.

Gel filtration has also been used in attempts to attain information about matrix properties through a method called inverse size exclusion. This may provide a valuable tool for obtaining apparent size exclusion dimensions of soft media for which standard methods are inapplicable. Other types of applications that are not traditional gel filtration include studies of kinetics of protein aggregation (e.g., see Ref. 145) and, quite recently, as a general approach for refolding proteins (e.g., recovery of recombinant proteins produced as insoluble incusion bodies).[146]

The interaction of certain gel filtration media with solutes may be used to separate substances. Of course this type of separation should not be classified as gel filtration. Other types of size-based separations are hydrodynamic chromatography, slalom chromatography, and field flow fractionation; in some cases, these may be a complement to ordinary gel filtration for the size separation of solutes.

Due to the recent interest in the biochemistry laboratories for the separation of peptides, carbohydrates, and nucleotides, references to these classes of molecules are made when appropriate.

3.7.1 Desalting

Gel filtration has been used for the desalting of protein solutions since the early 1960s.[93] Because of the difference in molecular size between the protein and contaminating solutes, the requirements on the gel and the chromatographic system are modest. The porosity of the gel is selected to exclude the solute to be desalted, and the particle size is chosen to give low surface area (to minimize adsorptive properties) and low back pressure instead of yielding columns of high efficiency. Since the protein is eluted in the void volume and the contaminating solutes have low molecular mass, the flow-sensitive dispersive effects of the column are minimal, and desalting may be carried out at high flow rates without impaired resolution. The buffer salts and additives should be volatile or the desalting should be carried out in distilled water. Desalting may be carried out at small scale as for the desalting of radioactive-labeled DNA in 1 ml Sephadex G-100 in a Pasteur pipette or at large scale as for desalting of large volumes of whey using a 2,500-L column of coarse Sephadex G-25.[5,147] Since the total pore volume is employed for the separation, very large volumes of sample may be desalted in one step, as illustrated in Figure 3-12. In this application, 400 ml of a hemoglobin solution, corresponding to a sample load of 37 % of the total bed volume, was desalted in one step on a column packed with Sephadex G-25. Figure 3-12 also illustrates the square-wave shape of the peak when applying large sample volumes and the small zone broadening (i.e., dilution factor) of the protein peak (in Figure 3-12A, the protein is diluted approximately nine times whereas in Figure 3-12B the protein is diluted only 10%). It may also be noticed that the mean effluent volume for the 400-ml sample is equal to the effluent volume of the 10-ml sample adjusted for the sample volumes (i.e., $V_R = 380 - 10/2 + 400/2 = 575$ ml).

Desalting in a micropreparative scale (i.e., desalting of 50 μl reduction/alkylation mixture) prior to sequence analysis using Sephadex G-25 prepacked into a 100 $\times$ 3.2-mm column was presented by Hellman and co-workers.[148]

In another application, McClung and Gonzales[149] separated plasmid DNA from RNA and protein using Superose 6 prep. grade. Plasmid DNA was eluted in the void volume, which permitted them to purify 0.5 mg plasmid DNA in 20 min, and the procedure was found to be advantageous compared to cesium chloride gradient centrifugation.

The procedure used for desalting may of course also be used for transferring the solute into another buffer system.

3.7.2 Fractionation of Protein Mixtures

The most favorable situation is when it is possible to use a gel that will exclude the protein of interest and include the contaminants or vice versa (i.e., as in desalting). In the normal case the difference in molecular size is perhaps only a factor of two, and a successful separation may require using a gel with optimal properties (i.e., a high pore volume, a narrow pore size distribution,

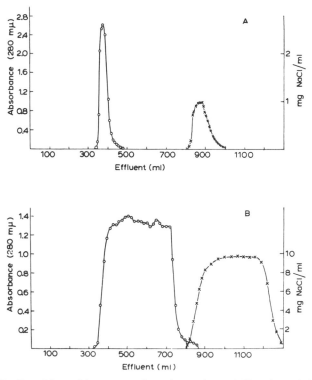

FIGURE 3–12. Desalting of large sample volumes by gel filtration. Column: 850 × 40 mm i.d. column packed with Sephadex G-25. Sample: O, hemoglobin; x, NaCl. Sample volume: A = 10 mL, B = 400 mL. Note that the retention volume is not corrected for the sample volume. (Reproduced from Ref. 93 with kind permission of the author and the publisher.)

and a suitable pore size to elute the protein of interest at $0.2 < K_D < 0.4$) (cf. Figure 3-6). From theoretical calculations, it was found that complete separation (i.e., with a resolution factor of 1.5) of molecules of spherical shape differing less than 30% in molecular mass may not be expected.[86] Since it is often desired to avoid additives, the adsorption properties of the gel should be small. Contaminants that show strong affinity for the gel matrix may be removed by filtering the sample solution through a small bed of the gel before applying the sample on the column.

To favor high sample loading, the pore volume and selectivity of the matrix should be high and the cross section of the column large enough to cope with the desired sample volume. It is also advantageous to use gels designed to facilitate the scale-up procedure from analytical to preparative runs. This requires that the medium is available in bulk quantities and that the packing procedure used to obtain efficient columns is simple. An example is given by the preparative purification of the conjugate of IgE–β-galactosidase in analytical and small-scale production shown in Figure 3-13 yielding amounts sufficient for diagnostic purposes.[150] Due to the complex sample, only partial purification of the conjugate was obtained. Preparative gel filtration was also employed for the intermediate purification of insulin-like growth factor-1 (M_r 7,600) from a fusion peptide (M_r 14,500) and uncleaved material using Superdex® 75, as shown by Hartmanis and co-workers.[151]

An example of small-scale fractionation of proteins was given by Hennes and co-workers[152] in a study of lipoproteins in hamsters. By employing a micropreparative system (SMART®, Pharmacia Biotech, Uppsala, Sweden), they were able to fractionate the proteins from only 20 μl of serum and thus the hamsters did not have to be sacrificed in the study and samples could be continuously withdrawn during the entire study.

The throughput (i.e., amount processed material per unit time) may be increased by applying a new sample at each time interval corresponding to $V_t - V_0$ (e.g., after 35 h in Figure 3-13b) since no material from the new sample is eluted during a time period corresponding to V_0. The gain in the production rate of consecutive runs equals the void fraction which, for many materials, is

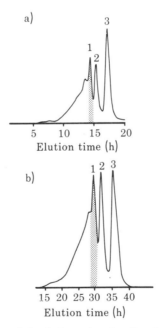

FIGURE 3–13. Purification of the 1–1 conjugate of anti-igE–β-galactosidase by gel filtration. (a) Laboratory scale. Column 950 × 16 mm i.d. packed with Superose 6 prep grade. Sample volume: 0.5 mL. (b) Pilot scale. Column 1000 × 26 mm i.d. packed with Superose 6 prep grade. Sample volume: 6 mL. Peaks are 1, anti-IgE–β-galactosidase; 2, β-galactosidase; and 3, anti-IgE. (Reproduced from Ref. 150 with kind permission of the publisher.)

roughly 35% (see Tables 3-1 and 3-2). The optimum injection volume and flow rate for a certain volume to be processed are given by Eq. (3-22).

With the use of high precision pumps and controllers to govern motorized valves, it is possible to use an automatic half-scale system (i.e., 700 × 50-mm i.d. column) with repetitive runs for the preparation of substances for laboratory use. Such a system was used for repetitive fractionation of glucose oligomers where 50 ml containing 10 g of a dextran mixture, having a weight average molecular mass of 1,100, was processed per cycle on two 900 × 65-mm i.d. columns of Sephadex G-25.[5]

3.7.3 Analytical Gel Filtration

In analytical applications the resolution is of utmost interest. This is affected by the selectivity (i.e., peak-to-peak distance) and the zone broadening, as discussed in Section 3.3.3. Equation (3-16) may be used to estimate the impact of various parameters on the resolution. It is advisable to check the column for efficiency and symmetry of a well-defined solute, and these values should be close to $2 \cdot d_p$ for the "plate height," at optimal flow rate, and 1.0 ± 0.1 for the peak symmetry at 10% peak height.[97]

The risk of sample overload must be considered when high resolution is desired. Some general aspects of sample volume and concentrations were given in Section 3.5.2. The applicability of these may be experimentally verified by first decreasing the concentration and then decreasing the sample volume and noting the width of the eluted zone. If this also decreases, then the parameter(s) should be adjusted to the region where no effect on the zone is noted. Overload effects are most often noticed for samples eluting in or near the void volume, and viscous fingering may result in double or triple peaks, as illustrated in Figure 3-9. This effect is readily precluded by decreasing the sample concentration. To eliminate the influence of sample concentration on peak elution volume, a procedure has been suggested where the sample is chromatographed at several concentrations and the elution volume is taken as the intercept of a plot of concentration versus elution volume.[118] However, this method is tedious and not necessary in most applications. As a general guide, a sample volume of 0.2% of the column volume and a concentration of up to 30 mg/ml for proteins, 6 mg/ml for tRNA, and 5 mg/ml for noncharged dextran of high molecular mass should be applicable.

Gel filtration may be used in the analytical mode to monitor a purification process selectively as illustrated in Figure 3-14. The purification of staphylococcal enterotoxin B was controlled by fast gel filtration using Superose 12.[153]

Interactions between the matrix and the solute should be absent in analytical gel filtration. Such interactions may be influenced by parameters that do not ordinarily affect the gel filtration process and thus cannot be expected to have been controlled by the manufacturer. Interactions may also vary unpredictably from lot to lot. Interactions may be indicated by variations in elution volume or peak shape with concentration, temperature, ionic strength, organic mod-

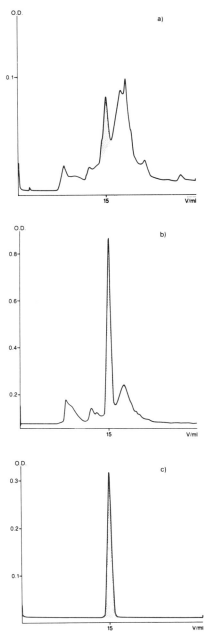

FIGURE 3–14. Monitoring of a large-scale purification process by analytical gel filtration. Column: Superose 12 HR 10/30. Sample: staphylococcal enterotoxin B in various stages of the purification process.[153] (a) After initial purification of 400-L cell supernatant on S Sepharose FF, (b), after intermediate purification by gradient elution on S Sepharose FF, and (c) after final gel filtration on Sephacryl S-200 HR. (Courtesy of H.O.J. Johansson, N.T. Pettersson, and J.H. Berglöf, Pharmacia Biotech.)

ifier, or chaotropic salts. However, the same effects may be caused by solute–solvent interactions (cf. Section 3.2.2). A general procedure for testing matrix properties has been given by Pfannkoch and co-workers.[18] One type of analytical gel filtration where interactions must be absent is in the measurement of equilibrium of protein association. A recent application of large zone analytical gel filtration showed that accurate data for the association constant of hemoglobin were achieved with Sephacryl® S-200 HR, but Toyopearl HW-50 S was not suitable, as indicated by tailing peaks.[145]

Analytical gel filtration is frequently used for the assay of molecular mass or molecular weight (i.e., mass) distributions (MWD) of hydrophilic macromolecules or polymers. Granath and Flodin[154] were the first to describe the use of gel filtration for the analysis of MWDs of clinical dextrans. This application requires the calibration of the column with well-characterized solutes of shapes identical to the sample (e.g., narrow dextran fractions), unless a size detector is used. A similar approach may be used for the estimation of protein molecular weight as exemplified by the characterization of chick interferon by Phillips and Wood in 1964. They were able to assign a molecular weight of approximately 40,000 to the interferon by eluting the sample on a column of Sephadex G-10, calibrated with standard proteins.[155] The obtained molecular weight was confirmed using an analytical ultracentrifuge. However, calibrating the column with standard proteins is only valid as long as the geometric properties of the sample and standards are similar. In many cases, different shapes of proteins must be normalized by the denaturation of the protein with SDS or quanidine hydrochloride.[48] The size of an intact protein may be inferred from the hydrodynamic volume obtained using the concept of universal calibration of the column. These procedures are discussed in Section 3.3.1. It must be noted that calibration of the column using the elution volume corresponding to the peak apex of a polymer requires that the molecular mass or molecular size corresponding to the peak apex is known. Another way is to assign the value $(M_w \times M_n)^{1/2}$ to the peak and then adjusting the calibration so that proper values of M_w, the weight average molecular mass, and M_n, the number average molecular mass, are obtained. Such iterative procedures have been successfully applied to the calibration of columns for analytical gel filtration.[59,156]

Calibration of the column using the column-independent distribution coefficient, K_D, is recommended if the column is going to be used over a long period of time or the performance is to be compared with other materials. Calculation of K_D requires the determination of the void volume, V_0, and the total liquid volume of the bed, V_t, as well as the sample elution volume. If a proper probe for V_0 is used, then the correct value for the void volume is obtained from the injection point to the peak apex. (Earlier a practice of calculating the void volume from the commencement of the sample application to the inflection point of the ascending part of the peak was recommended.[59,156] However, this results in a value of K_D for the peak apex being close to but not equal to zero and is therefore not suitable.) If the probe is too large, then a too low value of V_0 may be expected due to secondary exclusion effects, and if the probe partially penetrates the gel, a too high value of V_0 can be obtained. Suitable probes for the determination of void volume and total liquid volume are discussed in Section 3.5.3. The total liquid volume may be determined from the peak apex of a small molecule anticipated to penetrate the entire pore volume. For gels with very small pores, it may be difficult to find a suitable substance that will not interact with the liquid gel phase. In such cases the total volume is calculated from the geometric column volume, V_c, and the distribution coefficient is then designated K_{av}. The two distribution coefficients can be related to each other with the formula

$$K_D = K_{av} (1 + V_s/V_i) \tag{3-26}$$

where V_s is the volume of the support matrix and V_i is the pore volume. K_{av} of a totally permeating solute is thus a measure of the relative pore volume of the support.

If there is no need to calibrate the column with the aid of K_D or K_{av} (e.g., for the purpose of comparisons), then the elution volume, expressed by the first statistical moment of the peak, may serve as the dependent variable.

The injection point should be set when half the sample volume has been applied. However, when the sample volume is constant, the injection point may arbitrarily be set at the start of the sample application.

Thermostating the column is preferred to ensure minimal variation in the calibration curve if the column is to be used over a long time period. It is also a good practice to check the efficiency and symmetry by elution of a test probe at regular intervals. When the column is not in use, the flow rate may be reduced without impairing the calibration curve, but if the flow is switched off, then the column must generally be recalibrated. In this case, precautions to avoid bacterial growth will be necessary. If it is properly handled, an analytical column can be expected to have a lifetime of 1 to 5 years.

It is also possible to use the gel filtration column as a separation device only and to analyze the effluent for a size-related property. Thus, collected fractions may be characterized by chemical end-group analysis, light scattering, osmometry, viscosimetry, mass spectrometry, and so on. The requirement on the resolvability of the column is high (i.e., collected fractions need to be as pure as possible) but the requirements on other parameters (e.g., flow rate and temperature stability) are low. Furthermore, the column does not have to be calibrated and solute–matrix interactions will not influence the result, as long as components are resolved. Developments in on-line detectors suitable for the characterization of solute size in column effluents have provided the user with many options (see Section 3.6.4).

3.7.4 Determination of Pore Size

In gel filtration the elution volume of a solute is affected by the relationship between the solute size and the pore size of the matrix. Thus, by eluting molecules of well-known sizes and applying the established theory for gel filtration, approximations of the pore size can be readily achieved.[9]

By establishing the entire calibration curve with well-characterized standards, estimations of the pore size, pore volume, and surface area may be obtained.[157,158] In the method of Freeman and co-workers,[157] data are obtained from the inflection point of the plot of K_{av} versus the hydrodynamic radius of the solute. The method of Halász and co-workers[158] included an initial empirical calibration of the procedure to yield data in accordance with established methods. It has been reported that both methods give data in good agreement with absolute methods.[159] However, the method of Halász was found to yield a broader pore size distribution than that obtained by classical methods.[160] The shortcomings of the method are due to an incorrect assumption of the relationship between available pore volume and solute size (Halász assumed that a solute will either be excluded from the pore or enter the pore and then have access to the entire pore volume, which is not true, e.g., see Figure 3-1) as reviewed by Knox and Scott.[161] These authors mathematically derived the relationship between the pore size distribution curve and the SEC calibration curve to

$$F(r) = \frac{1}{2} \left[\frac{d^2 K_D}{d(\ln R)^2} \right]_{R=r} - \frac{3}{2} \left[\frac{dK_D}{d(\ln R)} \right]_{R=r} + K_D \qquad (3\text{-}27)$$

where $F(r)$ is the cumulative pore size distribution from cylindrical pores of radius 0 to r and K_D is the distribution coefficient for a solute of radius R. The equation was found to yield data in excellent agreement with mercury porosimetry of rigid materials.[161] However, the numerical calculation requires a very smooth calibration curve and may show instability. Thus, the authors recommended the calculation of pore size distribution from a simulation procedure.[162] An alternative direct calculation procedure based on pores being build up by the space between random solid spheres was used by Schou and co-workers[163] for characterizing silica materials. However, the pore models used may be more or less realistic, e.g. materials having cylindrical pores are probably very rare (the model is perhaps applicable to controlled porous glass), due to dissolution and redeposition of silica, beads composed of perfect spheres are probably also rare, and the pore model of polymer gels being composed of intersecting rods seems to yield inconsistent data.[85] The shortcomings of the existing models for pore structure may be exemplified by the difficulty to assign an equation for hindered pore diffusion (see Section 3.3.2).

However, there seems to be a need for a simple method for characterizing apparent size exclusion dimensions of gel filtration media for the purpose of comparisons of media and for predicting functional properties. A rough estimate of the pore size may be calculated from twice the solute size at the inflection point of the calibration curve.[62] (Another interesting observation that can be made from the theoretical calibration curves in Ref. 62 is that the linear part of the calibration curve intersects with log R at a value close to the average pore radius of the cylindrical pore model used. However, the variability in experimental data may yield a high uncertainty to the estimate obtained.) Hagel and co-workers showed that the procedure used by Jerabek[164] and supported by Knox and Ritchie[162] where the apparent pore size distribution is inferred by an iterative procedure to yield K_D values in agreement with the experimental data may be a feasible approach.[165] Data obtained may be used for functional predictions of size exclusion behavior, and it was stressed that only an apparent pore size is obtained with inverse SEC (or chromatographic porosimetry) as different estimates are obtained from the same experimental data, depending on the pore model applied.[165] The pore model generally applied is the one of cylindrical pore shape. This model is used for converting experimental data from "classical" methods such as nitrogen sorption and mercury porosimetry to pore size. It was shown that the selectivity curve of Superose 6 corresponded to a hypothetical support having cylindrical pores of a mean radius of 250 to 290 Å.[165] This is slightly larger than the value found by applying Eq. (3-27) due to the different shape of the apparent PSD curves (i.e., the latter method gave a very skewed PSD whereas a gaussian distribution was assumed in the former method).

These methods are of course only applicable to aqueous media for obtaining operational information on the pore size distribution, provided that reference data for the probe molecules are carefully selected to reduce disparities, as revealed for data on proteins and dextrans.[50] It must also be emphasized that an incorrect application of theory (e.g., by neglecting the increasing exclusion from part of the pore volume for permeating solutes of increasing size), as recently presented by a number of authors, will lead to erroneous results.[166–168] In one work the mean pore diameter for Superose 6 and TSK SW 3000 was reported to be substantially less than 142 and 120 Å,[168] which is unrealistic low (i.e., roughly half expected value) as compared with other reports.[62,169] This again illustrates the importance of interpreting data obtained and comparing these data with information from other sources to validate the approach used. A review of methods for the characterization of pore size of chromatography media has been given by Hagel.[62] A discussion of the various methods used for inverse SEC and data for apparent pore radius of some common media for gel filtration has been recently given by Hagel and co-workers.[165]

3.7.5 Mixed Mode Separations

In mixed mode separations the elution of a solute is not due to size parameters only, but is affected by adsorption or affinity to or exclusion from the matrix or layers of solvents or solutes adsorbed to the surface of the matrix. These types of solute–matrix interactions provide the basis for many of the separation principles dealt with in detail in subsequent chapters. Since mixed mode separations are sometimes noticed on gels designed for gel filtration, the subject will be discussed briefly here.

Because of its chemical nature and origin, the matrix may contain residual amounts of anionic groups, as in dextran and agarose gels,[7,19] or exposed anionic groups as for silica-based materials.[18] The cross-linking agent may introduce hydrophobic sites as will the bounded organic layer of silica materials. In rare cases, the matrix may also provide sites for biospecific interactions.[9] These adsorptive properties of the matrices are highly undesirable in gel filtration and may be reduced using a low pH to suppress ionization or a high ionic strength to decrease ionic interaction, adding detergents or modifiers to prevent hydrophobic interactions, or simply saturating the active sites with, for example, ovalbumin,[9,17] phospholipids,[65] or basic peptides.[170]

Interactions between the solute and the matrix may be utilized advantageously as shown for the initial purification of yeast cytochrome oxidase.[171] More than 500 ml of an extract of submitochondrial particles from yeast was applied at high ionic strength [i.e., 30% (w/w) of ammonium sulfate] to a column packed with Toyoperl (also called Fractogel). A highly purified and concentrated fraction of active enzyme was eluted by a gradient of decreasing salt content. The authors concluded that the very hydrophobic protein was probably adsorbed due to hydrogen bonding and that the interaction provides a favorable alternative to salt fractionation for the initial purification of yeast cytochrome oxidase. This separation would be classified as hydrophobic interaction chromatography (see Chapter 7).

Separations that are influenced by solute–matrix interactions should of course not be referred to as *gel filtration*. The retention volumes of solutes may in a nongel filtration mode vary substantially with changes in the mobile-phase composition, as shown in Figure 3-15. The general trend in this figure is that proteins having an isoelectric point above the pH of the buffer are retained due to ionic interaction with the silanol groups of the sorbent at low ionic strength whereas acidic proteins (being negatively charged) are subjected to ion exclusion at these conditions. High ionic strength reduces these ionic effects. However, as many solutes and matrix surfaces posess both ionic and hydrophobic sites, the optimum composition of the mobile phase will balance the different effects (e.g., at high ionic strength the retention volume may start to increase due to hydrophobic interaction).

Mixed mode separations may lead to unexpected results, such as the concentration of a solute zone upon passage through the column[19] or the separation of proteins according to their isoelectric points.[172] A separation of ribonuclease A and cytochrome c was accomplished utilizing cationic effects at low ionic strength.[173] Purification of cytochrome c was achieved by adsorption of cytochrome c at high ionic strength with subsequent elution with a decreasing salt gradient.[174] Several publications where mixed mode separation was utilized or may be suspected to influence the separation have been published.[172–177] It has also been pointed out that such effects may result in poor resolution compared to gel filtration alone.[178]

Naturally, this information is very valuable when selecting materials and conditions for which separations based on gel filtration alone are desired.

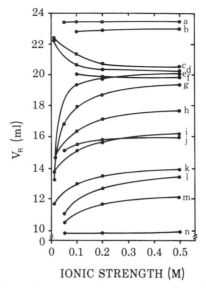

FIGURE 3–15. The effect of ionic strength on the retention volumes of native proteins of various isoelectric points. Column: TSK-G3000 SW. Buffer: 0.01 *M* sodium phosphate at pH 7.0 containing 0 to 0.5 *M* ammonium acetate. Proteins: (a) L-tyrosine, (b) vitamin B_{12}, (c) cytochrome c (isoelectric point 10.3), (d) ribonuclease (9.3), (e) α-lactalbumin, (f) chymotrypsinogen (9.2–9.6), (g) soybean trypsin inhibitor (4.2–4.5), (h) ovalbumin (4.7), (i) serum albumin (4.7–4.9), (j) transferrin (5.5), (k) serum albumin dimer (4.7–4.9), (l) apoferritin (4.6–5.0), (m) β-galactosidase, and (n) thyroglobulin (4.5). (Reproduced from Ref. 50 with kind permission of the authors and the publisher.)

Gel filtration is, by definition, a method for separating molecules of different size due to their different permeation of a porous matrix. This distinguishes gel filtration from some other size-separating techniques, such as hydrodynamic chromatography, slalom chromatography, and field-flow fractionation (FFF).

Hydrodynamic chromatography may be carried out in a column packed with nonporous beads.[14,120] The separation takes place in the void volume, and solutes are eluted according to decreasing size due to the fact that large solutes will, statistically, spend more time in the faster flowing center portion of the parabolic flow profile of the solvent. The advantage is that fast separations may be carried out but disadvantages such as shearing of large solutes and low peak capacity seem to limit the applicability of hydrodynamic chromatography.

Slalom chromatography was introduced quite recently as a result of the observation that elongated stiff molecules, such as DNA, could be trapped in the void volume of a column and eluted at a low flow rate.[15] The phenomena is probably related to the relaxation time of the solutes. Results published for separation of DNA show remarkable results since the resolution increases with increased flow rate.[179] Because the separation takes place in the void volume, the separating volume is small and the method is not suitable for preparative applications. Also, the technique is mainly applicable to stiff, rod-shaped solutes of appreciable size. Hydrodynamic chromatography effects will start to influence the result when the size of solutes decreases.

The third technique, field-flow fractionation explored by Giddings and co-workers,[16] shows some resemblance to hydrodynamic chromatography in that the solutes are carried forward due to the distribution of solutes across a parabolic flow profile in a thin separation channel. The differential distribution of sample components is caused by a force field applied perpendicular to the flow. An adoption of this technique, called asymmetrical flow field-flow fractionation, where solutes in the size range of 20 to 5,000 Å are separated according to their diffusion coefficient, has been employed for the analytical separation of monomers from multimers of monoclonal antibodies.[180] The authors found that FFF was advantageous as compared to size exclusion chromatography with respect to peak capacity, selectivity, separation time, and dynamic separation range for analytical applications. However, the peak capacity of FFF for completely resolved peaks may be estimated to 10 (the reported value was 15 at the resolution factor of 1), which is less than that expected for a gel filtration column operating at optimal conditions (i.e., 16) (see Section 3.3.3). The maximum selectivity as functionally expressed by the complete separation of monomer and dimer of a globular protein is of the same order as that expected for gel filtration columns, and the separation time, in the cited case 6.3 min, may be matched using gel filtration media of small particle size (e.g., a 10-cm-long column packed with 2 μm media operated at 120 cm/h, which still will yield high resolution in the molecular mass range investigated, i.e., 156,000, see Table 3-7). Thus, although analytical size exclusion chromatography, as its best, still provides a good alternative for the fast separation of proteins, FFF may be seen as a complementary technique, especially for separating very large solutes (i.e., exceeding 100 Å). However, FFF is not suitable for separating low molecular weight solutes (e.g., peptides) or for preparative size separations.

3.8 SYMBOLS USED IN GEL FILTRATION

As a result of the many different approaches taken to develop liquid chromatography as a separation technique and also to characterize chromatography supports, the nomenclature in liquid chromatography has, for some parameters, turned out to be inconsistent (e.g., in reversed-phase chromatography, the total liquid volume is not denoted V_t but V_0). This is unfortunate, and to avoid ambiguities the designation V_m for mobile-phase volume has been suggested instead of V_0.[181] A review including a detailed discussion of the sometimes contradictory proposals for nomenclature in liquid chromatography issued by IUPAC and ASTM has been given by Ettre.[181] The proposal, "Steric Exclusion Chromatography" for SEC,

TABLE 3-8. Symbols Used in Gel Filtration

Symbol	Parameter	Reference[a]	Alternative Symbol
V_R	Retention (elution volume)	IUPAC-93	V_e
V_0	Void volume	IUPAC-93	
V_i	Pore volume[b]	IUPAC-93	V_p
V_s	Matrix volume[c]		
V_t	Total liquid volume	IUPAC-93	
V_c	Geometric column volume	IUPAC-93	
V_0/V_c	Void fraction		
V_i/V_0	Permeability		V_p/V_0
K_D	Distribution coefficient	IUPAC-74	
K_{av}	Gel-phase distribution coefficient		
ε	Void fraction, V_0/V_c	IUPAC-93	
L	Column length, bed height	ASTM	
A_c	Column cross section area	ASTM	
d_p	Average particle diameter of support	IUPAC-93	
Δp	Pressure drop over the column	IUPAC-93	
η	Viscosity of solvent	ASTM	
F	Volumetric flow rate[d]	IUPAC-93	
u	Interstitial fluid velocity, $F/A_c \cdot 1/\varepsilon$	IUPAC-93	
u_{nom}	Nominal fluid velocity, F/A_c		
N	Number of theoretical plates in column	IUPAC-93	
H	Height equivalent to a theoretical plate, L/N	IUPAC-93	
h	Reduced plate height, H/d_p	IUPAC-93	
w_b	Peak width at base (4σ for a gaussian peak)	IUPAC-93	
w_h	Peak width at half peak height	IUPAC-93	
R_S	Resolution factor	IUPAC-93	
A_S	Asymmetry factor, b/a		
n_{RS}	Peak capacity factor at a resolution of R_S		
M_r	Relative molecular mass of solute	SIS 016174	
D	Diffusion coefficient of solute in free solution	IUPAC-93	
D_m	Diffusion coefficient of solute in mobile phase	IUPAC-93	
D_S	Diffusion coefficient of solute in gel pores	IUPAC-93	
γ_m	Obstruction factor in extra particle space, D_m/D		
γ_s	Obstruction factor in gel pores, D_S/D		
V_h	Hydrodynamic volume of solute	ASTM	
R_h	Hydrodynamic viscosity radius of solute, R_η		
R_{St}	Stokes radius of solute		

[a] References: IUPAC-93, symbols proposed by the International Union of Pure and Applied Chemistry (IUPAC) in 1993.[183] IUPAC-74, symbols proposed by the International Union of Pure and Applied Chemistry (IUPAC) in 1974.[184] ASTM symbols proposed by the American Society for Testing and Materials (ASTM) in 1979.[182]

[b] The symbol used in the earlier edition, V_p, has been changed.

[c] The symbol for the stationary phase as defined by IUPAC-93 does not include the solid matrix. However, in gel filtration, no stationary phase, as defined for sorption techniques, exists and it is convenient to let V_S denote the solid gel conceptually being responsible for the separation. The matrix volume has also been referred to by V_x.

[d] The symbol has been changed to comply with the standard proposed by IUPAC-93.

made by one ASTM Committee[182] has not yet found broad acceptance by workers in this field of chromatography, as judged from the vast literature on the subject. The most recent suggestion for nomenclature in liquid chromatography seems to offer a reasonable compilation of designations that may be accepted by the workers in the field.[183] This nomenclature has been adopted throughout this revision, affecting the designation for flow velocity (from the designation, F_c, suggested earlier by IUPAC[184] to F) and for pore volume, V_i, instead of V_p. The symbol for relative molecular mass has been altered to M_r from M (M being used for molarity). However, it may be noted that molecular weight is sometimes referred to as MW, which may be regarded as an abbreviation for molecular weight and not a symbol. Unfortunately, the nomenclature suggestion from IUPAC is not complete and some of the symbols used still lack formal acceptance. Thus, the designation V_s for solid matrix volume is not yet supported, but until a formal proposal is made, the use of V_s as earlier suggested will remain, especially since this practice is not ambiguous in gel filtration. The symbols given in Table 3-8 are believed to represent an acceptable compilation of designations used in size exclusion chromatography.

3.9 REFERENCES

1. B. Lindqvist, T. Storgårds, *Nature (London), 175,* 511 (1955).
2. G.H. Lathe, C.R. Ruthven, *J. Biochem. J., 60,* xxxiv (1955).
3. J.-C. Janson, *Chromatographia, 23,* 361 (1987).
4. J. Porath, P. Flodin, *Nature, 183,* 1657 (1959).
5. L. Hagel, J.-C. Janson, Size-exclusion chromatography, in E. Heftmann, ed., *Chromatography, 5th edition,* pp. A267–AA307. Elsevier, Amsterdam, 1992.
6. J.C. Moore, *Polym. Sci.,* Part A, *2,* 835 (1964).
7. L. Fischer, *Gel Filtration Chromatography,* Elsevier, Amsterdam, 1980.
8. H. Determann, in J.C. Giddings, R.A. Keller, eds., *Advances in Chromatography,* Vol. 8, pp. 35–42. Marcel Dekker, New York, 1969.
9. P. Andrews, in D. Glick ed., *Methods of Biochemical Analysis,* Vol. 18, pp. 2–53. Wiley-Interscience, New York, 1970.
10. E.J. Casassa, *J. Polym. Sci., Polym. Lett. 5,* 773 (1967).
11. A.M. Basedow, K.H. Ebert, J.H. Ederer, E. Fosshag, *J. Chromatogr., 192,* 259 (1980).
12. W.W. Yau, J.J. Kirkland, D.D. Bly, *Modern Size-Exclusion Liquid Chromatography Practice of Gel Permeation and Gel Filtration Chromatography,* Wiley-Interscience, New York, 1979, p. 3.
13. H.G. Barth, B.E. Boyes, C. Jackson, *Anal. Chem., 66,* 595R–620R (1994).
14. H. Small, *J. Coll. Interface Sci., 48,* 147 (1974).

15. J. Harabayashi, N. Ito, K. Noguchi, K.-I. Kasai, *Biochemistry, 29,* 9515 (1990).

16. J.C. Giddings, *J. Chromatogr., 125,* 3 (1976).

17. H.G. Barth, *J. Chromatogr. Sci., 18,* 409 (1980).

18. E. Pfannkoch, K.C. Lu, F.E. Regnier, H.G. Barth, *J. Chromatogr. Sci., 18,* 430 (1980).

19. P. Flodin, *Dextran Gels and Their Applications in Gel Filtration,* Meijels Bokindustri, Halmstad, Sweden, 1963, p. 74.

20. S. Aoyagi, K. Hirayanagi, T. Yoshimura, T. Ishikawa, T. *J. Chromatogr., 253,* 133 (1982).

21. Y. Kato, T. Hashimoto, *J. High Res. Chromatogr., Cromatogr. Comm., 6,* 45 (1983).

22. R.W. Stout, J.J. De Stefano, *J. Chromatogr., 326,* 63 (1985).

23. T.W. Hearn, F.E. Regnier, C.T. Wehr, *Am. Lab., 14,* 18–39 (1982).

24. L. Hagel, T. Andersson, *J. Chromatogr., 285,* 295 (1984).

25. Y. Kato, T. Hashimoto, *J. High Res. Chromatogr., Cromatogr. Comm., 8,* 79 (1985).

26. F. Hefti, *Anal. Biochem., 121,* 378 (1982).

27. Fractogel TSK, E. Merck, Darmstadt, 1984, p. 9.

28. M. Potschka, *Anal. Biochem., 162,* 47 (1987).

29. J.-C. Janson, *J. Chromatogr., 28,* 12 (1967).

30. H. Determan, I. Walter, *Nature, 219,* 604 (1968).

31. B.J. Gelotte, *J. Chromatogr., 3,* 330 (1960).

32. R.P. Bywater, N.V.B. Marsden, in E. Heftmann ed., *Chromatography, Fundamentals and Applications of Chromatographic and Electrophoresis Methods, Part A,* pp. A297–A305. Elsevier, Amsterdam, 1983.

33. M. Belew, J. Porath, J. Fohlman, J.-C. Janson, *J. Chromatogr., 147,* 205 (1978).

34. L. Politi, M. Moriggi, R. Nicoletti, R. Scandurra, *J. Chromatogr., 267,* 403 (1983).

35. C.L. de Ligny, W.J. Gelsema, A.M.P. Roozen, *J. Chromatogr. Sci., 21,* 174 (1983).

36. H. Schlueter, W. Zideg, *J. Chromatogr., 639,* 17 (1993).

37. E. Ruckenstein, V. Lesins, *Biotechnol. Bioeng., 28,* 432 (1986).

38. N.P. Golovchenko, I.A. Kataeva, V.K. Akimenko, *J. Chromatogr., 591,* 121 (1992).

39. L.W. Nichol, W.H. Sawyer, D.J. Winzor, *Biochem J., 112,* 259 (1969).

40. P.M. James, A.C. Ouano, *J. Appl. Polym. Sci., 17,* 1455 (1973).

41. P.C. Christopher, *J. Appl. Polym. Sci., 20,* 2989 (1976).

42. M.R. Ambler, L.J. Fetters, Y. Kesten, *J. Appl. Polym. Sci., 21,* 2439 (1977).

43. W.W. Yau, D.D. Bly, in T. Provder, ed., *Size Exclusion Chromatography (GPC),* p. 197. ACS Symposium Series, American Chemical Society, Washington, DC. 1980.

44. C. Tanford, *Physical Chemistry of Macromolecules,* Wiley, New York, 1961, Chapter 6.

45. C. de Haen, *Anal. Biochem., 166,* 235 (1987).

46. C. Tanford, Y. Nozaki, J.A. Reynolds, S. Mikano, *Biochemistry, 13,* 2369 (1974).

47. P.G. Squire, *J. Chromatogr., 210,* 443 (1981).

48. W.W. Fish, J.A. Reynolds, C. Tanford, *J. Biol. Chem., 245,* 5166 (1970).

49. H. Benoit, Z. Gzubisic, P. Rempp, D. Decker, J.G. Zilliox, *J. Chim. Phys., 63,* 1507 (1966).

50. R.P. Frigon, J.K. Leypoldt, S. Uyejl, L.W. Hendersson, *Anal. Chem., 55,* 1349 (1983).

51. Z. Grubisic, P. Rempp, H. Benoit, *Polym. Lett., 5,* 753 (1967).

52. D.J. Harmon, Elution. in R. Epton, ed., *Chromatography of Synthetic and Biological Polymers, Vol. 1, Column Packings, GPC, GF and Gradient,* pp. 122–145. Ellis Horwood, Chichester, 1978.

53. D. Berek, I Novák, Z. Grubisic-Gallot, H. Benoit, *J. Chromatogr., 53,* 55–61 (1970).

54. A.M. Basedow, K.H. Ebert, H. Ederer, H. Hunger, *Macromol. Chem., 177,* 1501–1524 (1976).

55. M.G. Styring, C. Price, C. Booth, *J. Chromatogr., 319,* 115 (1985).

56. P.L. Dubin, J.M. Principi, *Macromolecules, 22,* 1891 (1989).

57. J. Porath, *Pure Appl. Chem., 6,* 233 (1963).

58. T.C. Laurent, J. Killander, *J. Chromatogr., 14,* 317 (1964).

59. K.A. Granath, B.E. Kvist, *J. Chromatogr., 28,* 69 (1967).

60. G.K. Ackers, *J. Biol. Chem., 242,* 3237 (1967).

61. L.C. Davis, *J. Chromatogr. Sci., 21,* 214 (1983).

62. L. Hagel, in P. Dubin ed., *Aqueous Size-Exclusion Chromatography,* pp. 119–155. Elsevier, Amsterdam, 1988.

63. S.C. Meredith, G.R. Nathans, *Anal. Biochem., 121,* 234 (1982).

64. C. Tanford, in C.B. Anfinsen, Jr., M.L. Anson, J.T. Edsall, F.M. Richards, eds., *Advances in Protein Chemistry,* Vol. 23, p. 211. Academic Press, New York, 1968.

65. E. Mascher, P. Lundahl, *Biochim. Biophys. Acta, 856,* 505 (1986).

66. C. Tanford, J.A. Reynolds, *Biochim. Biophys. Acta, 457,* 133 (1976).

67. J.A. Reynolds, C. Tanford, *J. Biol. Chem., 245,* 5161 (1970).

68. N. Ui, *Anal. Biochem., 97,* 65 (1979).

69. T. Takagi, *J. Chromatogr., 219,* 123 (1981).

70. J.C. Giddings, in E. Heftmann, ed., *Chromatography: A Laboratory Handbook of Chromatographic and Elecrophoretic Methods,* pp. 27–45. Van Nostrand Reinhold Company, New York, 1975.

71. J.J. van Deemter, F.J. Zuiderweg, A. Klinkenberg, *Chem. Eng. Sci., 5,* 271 (1956).

72. J.C. Giddings, K.L. Mallik, *Anal. Chem., 38,* 997 (1966).

73. E. Katz, K.L. Ogan, P.W. Scott, *J. Chromatogr., 270,* 51 (1983).

74. J.H. Knox, P.H. Scott, *J. Chromatogr., 282,* 297 (1983).

75. J.V. Dawkins, in C. Booth and C. Price eds., *Comprehensive Polymer Science, Vol. 1, Polymer Characterisation,* pp. 231–258. Pergamon Press, Oxford, 1989.

76. S. Yamamoto, K. Nakanishi, R. Matsuno, *Ion-Exchange Chromatography of Proteins,* Marcel Dekker, New York, 1988, p. 167.

77. E.M. Renkin, *J. Gen. Physiol., 38,* 225, (1954).

78. B.H. Arve, A.I. Liapis, *AICHE J., 33,* 179 (1987).

79. B.H. Arve, Phazmacia AB, Uppsala, Sweden, personal communication, 1987.

80. G.K. Ackers, R.L. Steere, *Biochim. Biophys. Acta, 59,* 137 (1962).

81. B.F.D. Ghrist, M.A. Stadalius, L.R. Snyder, *J. Chromatogr., 387,* 1 (1987).

82. J.H. Knox, F. McLennan, *J. Chromatogr., 185,* 289 (1979).

83. J.K. Leypoldt, R.P. Frigon, L.W. Henderson, *J. Appl. Polym. Sci., 29,* 3533 (1984).

84. M.E. van Kreveld, N. van den Hoed, *J. Chromatogr., 149,* 71 (1978).

85. M. Moussaoui, M. Benlyas, P. Wahl, *J. Chromatogr,. 591,* 115 (1992).

86. L. Hagel, *J. Chromatogr., 591,* 47 (1992).

87. S. Hjertén, in A. Niederwiesser, G. Pataki, eds., *New Techniques in Amino Acid, Peptide and Protein Analyses,* pp. 227–247. Ann Arbor Science Publishers, Ann Arbor, 1971.

88. L. Hagel, *J. Chromatogr., 648,* 19 (1993).

89. G. Guiochon, M. Martin, *J. Chromatogr., 326,* 3 (1985).

90. M. Verzele, C. Dewaele, D. Duguet, *J. Chromatogr., 391,* 111 (1987).

91. J.H. Knox, G.E. Laird, P.A. Raven, *J. Chromatogr., 122,* 129 (1976).

92. J.L. Glajch, D.C. Warren, M.A. Kaiser, L.B. Rogers, *Anal. Chem., 50,* 1962 (1978).

93. P. Flodin, *J. Chromatogr., 5,* 103 (1961).

94. Y. Kato, K. Komiyta, T. Iwaeda, H. Sasaki, T. Hashimoto, *J. Chromatogr., 208,* 71 (1981).

95. S.A. Karapetyan, L.M. Yakushina, G.G. Vasijarow, V.V. Brazhinkov, *J. High Res. Chromatogr. Comm., 8,* 148 (1985).

96. L. Hagel, H. Lundström, T. Andersson, H. Lindblom, *J. Chromatogr. 476,* 329 (1989).

97. P.A. Bristow, J.H. Knox, *Chromatographia, 10,* 279 (1977).

98. T. Allen, *Particle Size Measurement,* Chapman and Hall, London, 1981, pp. 432–436.

99. P.A. Bristow, *J. Chromatogr., 149,* 13 (1978).

100. J.H. Knox, in C.F. Simpson, ed., *Practical High Performance Liquid Chromatography,* p. 25. Heyden and Son Ltd., London, 1976.

101. M.R. Ladisch, G.T. Tsao, *J. Chromatogr., 166,* 85 (1978).

102. S. Katoh, *TIBTECH, 5,* 328 (1987).

103. J.-C. Janson, *J. Agric. Food Chem., 19,* 581 (1971).

104. J. Porath, T. Låås, J.-C. Janson, *J. Chromatogr., 103,* 49 (1975).

105. T. Andersson, M. Carlsson, L. Hagel, P.-Å. Pernemalm, J.-C. Janson, *J. Chromatogr., 326,* 33 (1985).

106. D.D. Perrin, B. Dempsey, *Buffers for pH and Metal Ion Control,* Chapman and Hall, London, 1979, pp. 24–61.

107. N.E. Good, G.D. Winget, W. Winter, T.N. Connoly, S. Izawa, R.M.M. Singh, *Biochemistry, 5,* 467 (1966).

108. R.J. Rothman, L. Warren, *Biochim. Biophys. Acta, 955,* 143 (1988).

109. G.D. Swergold, C.S. Rubin, *Anal. Biochem., 131,* 295 (1983).

110. S.C. Meredith, *J. Biol. Chem., 259,* 11682 (1984).

111. E.A. Lance, C.W. Rhodes III, R. Nakon, *Anal. Biochem., 133,* 492 (1983).

112. E.T. Adams, Jr., L.-H. Tang, J.L. Sarquis, G.H. Barlow, W.M. Nonnan, in N. Catsimpoolas, ed., *Physical Aspects of Protein Interactions,* pp. 29–47. Elsevier, New York, 1978.

113. L.D. Plante, P.M. Romano, E.J. Fernandez, *Chem. Eng. Sci., 49,* 2229 (1994).

114. J.F.K. Huber, *Instrumentation for High-Performance Liquid Chromatography,* Elsevier, Amsterdam, 1978, pp. 1–9.

115. R.W. Stout, J.J. De Stefano, L.R. Snyder, *J. Chromatogr., 261,* 189 (1983).

116. J.L. Di Cesare, M.W. Dong, J.G. Atwood, in A. Zlatkis, ed., *Advances in Chromatography,* p. 366. Elsevier, Amsterdam, 1981.

117. L. Hagel, L. *J. Chromatogr., 324,* 422 (1985).

118. S. Moori, *J. Appl. Polym. Sci., 20,* 2157 (1976).

119. J.F. Kennedy, Z.S. Rivera, L.L. Lloyd, F.P. Warner, M.P.C. da Silva, *Carbohydr. Polym., 26,* 31 (1995).

120. P.G. Squire, A. Magnus, M.E. Hiimmel, *J. Chromatogr., 242,* 255 (1982).

121. K. Granath, *J. Colloid Sci., 13,* 308 (1958).

122. W.R.K. Barnikol, H. Pötzschke, *J. Chromatogr. A, 685,* 221 (1994).

123. H. Engelhardt, H. Müller, B. Dreyer, *Chromatographia, 19,* 240 (1985).

124. B.-L. Johansson, C. Ellström, *J. Chromatogr., 330,* 360 (1985).

125. Superloop 10 ml, Pharmacia Biotech AB, Uppsala, Sweden.

126. Degasser ERC-3510, Erma Optical Works Ltd., Tokyo, Japan.

127. D.D. Bly, H.J. Stoklosa, J.J. Kirkland, W.W. Yau, *Anal. Chem., 47,* 1810 (1975).

128. K. Asai, Y.-I. Kanno, A. Nakamoto, T. Hara, *J. Chromatogr., 126,* 369 (1976).

129. J.J. Hermans, *J. Appl. Polym. Sci., Part A-2, 6,* 1217 (1968).

130. R. Bishop, H. Lundin, *Application Note 315,* LKB, Bromma, 1978.

131. R.L. Stevenson, in T.M. Vickrey, ed., *Liquid Chromatography Detectors,* pp. 44, 65–72. Marcel Dekker, New York, 1983.

132. H. Elgass, A. Maute, R. Martin, S. George, *Int. Lab., 13,* 72 (1983).

133. A.C. Ouano, W. Kay, *J. Polym. Sci. Part A-I, 12,* 1151 (1974).

134. T. Takagi, *J. Biochem., 89,* 363 (1981).

135. W.W. Yau, *Chemtr.-Macromol. Chem., 1,* 1 (1990).

136. C. Jackson, L.M. Nilsson, P.J. Wyatt, *J. Appl. Polym. Sci., 43,* 99 (1989).

137. I. Nylander, K. Tan-No, A. Winter, J. Silberring, *Life Sci., 57,* 123 (1995).

138. M.L. Hitchman, F.W.M. Nyasulu, A. Aziz, D.D.K. Chingakule, *Anal. Chim. Acta, 155,* 219 (1983).

139. K. Isaksson, K.J. Lindquist, K. Lundström, *J. Chromatogr., 324,* 333 (1985).

140. P.A. Bristow, *Liquid Chromatography in Practice;* hetp: Wilmslow, England, 1976.

141. R.S. Deelder, M.G.F. Kroll, A.J.B. Beeren, J.H.M. Van den Bezg, *J. Chromatogr., 149,* 669 (1978).

142. V.G. Maltsev, T.M. Zimina, A.B. Khvatov, B.G. Belenkii, *J.Chromatogr., 416,* 45 (1987).

143. C.N. Trumbore, R.D. Tremblay, J.T. Penrose, M. Mercer, F.M. Kelleher, *J. Chromatogr., 280,* 43 (1983).

144. H.G. Barth, F.J. Carlin, *J. Liquid Chromatogr., 7,* 1717 (1984).

145. E. Nenortas, D. Beckett, *Anal. Biochem., 222,* 366 (1994).

146. M.H. Werner, G.M. Clore, A.M. Gronenborn, A. Kondoh, R.J. Fisher, *FEBS Lett., 345,* 125 (1994).

147. T. Maniatis, E.F. Fritsch, J. Sambrook, in *Molecular Cloning, A Laboratory Manual,* Cold Spring Harbor Laboratory Press, Cold Spring Harbor, New York, 1982, p. 464.

148. U. Hellman, E. Wiksell, B.-M. Karlsson, *A new approach to micropreparative desalting exemplified by desalting a reduction/alkylation mixture,* Presented at Eight International Conference on Methods in Protein Sequence Analysis, Kiruna, Sweden, July 1–6, 1990.

149. J.K. McClung, R.A. Gonzales, *Anal. Biochem., 117,* 378 (1989).

150. T. Andersson, L. Hagel, *Anal. Biochem., 141,* 461 (1984).

151. H. Lundström, A. Karlström, G. Forsberg, M. Hartmanis, *Separation and purification of recombinant growth factors using Superdex™ 75 pg, a new 34 μm gel filtration medium,* Presented at First Conference on Advances in Purification of Recombinant Proteins, Interlaken, Switzerland, March 14–17, 1989.

152. U. Hennes, W. Gross, A. Edelman, *Sci. Tools, 36,* 10 (1992).

153. H.O.J. Johansson, N.T. Pettersson, J.H. Berglöf, Abstract of Papers of the 4th European Congress on Biotechnology, Amsterdam, June 1984.

154. K. Granath, P. Flodin, *Macromol. Chem., 48,* 160 (1961).

155. A.W. Phillips, R.D. Wood, *Nature, 201,* 819 (1964).

156. G. Nilsson, K. Nilsson, *J. Chromatogr., 101,* 137 (1974).

157. D.H. Freeman, I.C. Poinescu, *Anal. Chem., 49,* 1183 (1977).

158. I. Halász, K. Martin, *Angew. Chem. Int. Ed. Engl., 17,* 901 (1978).

159. O. Chiantore, M. Guaita, *J. Chromatogr., 260,* 41 (1983).

160. W. Werner, I. Hálasz, *J. Chromatogr. Sci., 18,* 277 (1980).

161. J.H. Knox, H.P. Scott, *J. Chromatogr., 316,* 311 (1984).

162. J.H. Knox, H. Ritchie, *J. Chromatogr., 387,* 65 (1987).

163. O. Schou, Novo, Bagsvaerd, Denmark, personal communication, 1987.

164. K. Jerábek, *Anal. Chem., 57,* 1595 (1985).

165. L. Hagel, M. Östberg, T. Andersson, *J. Chromatogr.,* A, 743, 33 (1996).

166. M. Le Maire, A. Ghazi, J.V. Møller, L.P. Aggerbeck, *Biochem. J.,* 243, 399 (1987).

167. M. Andersson, A. Axelsson, G. Zacchi, *Bioseparations, 5,* 65 (1995).

168. J.E. Harlan, D. Picot, P.J. Loll, R.M. Garavito, *Anal. Biochem., 224,* 557 (1995).

169. K. Makino, H. Hatano, in P. Dubin ed., *Aqueous Size-Exclusion Chromatography,* p. 238. Elsevier, Amsterdam, 1988.

170. G.W. Link, Jr., P.L. Keller, R.W. Stout, A.J. Banes, *J. Chromatogr., 331,* 253 (1985).

171. A.V. Galkín, I.E. Kovaleva, S.L. Kalnov, *Anal. Biochem., 142,* 252 (1984).

172. W. Kopaciewicz, F.E. Regnier, *Anal. Biochem., 126,* 8 (1982).

173. R.A. Jenik, J.W. Porter, *Anal. Biochem., 111,* 184 (1981).

174. M. Gurkin, V. Patel, *Am. Lab.,* January, 64 (1982).

175. B. Renck, R. Einarsson, *J. Chromatogr., 197,* 278 (1980).

176. T. Hashimoto, H. Sasaki, M. Aiura, M.Y. Kato, *J. Chromatogr., 160,* 301 (1978).

177. D.E. Schmidt, Jr., R.W. Giese, D. Conron, B.L. Karger, *Anal. Chem., 52,* 177 (1980).

178. P.L. Dubin, I.J. Levy, *J. Chromatogr., 235,* 377 (1982).

179. J. Hirabayashi, K.-I. Kasai, *Anal Biochem., 178,* 336 (1989).

180. A. Litzén, J.K. Walter, H. Krischollek, K.-G. Wahlund, *Anal. Biochem., 212,* 469 (1993).

181. L.S. Ettre, *J. Chromatogr., 220,* 29 (1981).

182. ASTM E 682-79; Annual Book of ASTM Standards; American Society for Testing and Materials, Philadelphia, 1979, pp. 541–549.

183. L.S. Ettre, "Nomenclature for Chromatography," IUPAC Recommendations 1993, *Pure and Appl. Chem., 65,* 819 (1993).

184. D. Amborse, E. Bayer, O. Samuelson, *Pure and Appl. Chem., 37,* 437 (1974).

4 Ion-Exchange Chromatography

EVERT KARLSSON and LARS RYDEN
Department of Biochemistry
Uppsala University
Uppsala Biomedical Center
Box 576, S-751 23 Uppsala, Sweden

JOHN BREWER
Pharmacia Biotech AB
S-751 84 Uppsala, Sweden

Protein Purification: Principles, High-Resolution Methods, and Applications, Second Edition.
Edited by Jan-Christer Janson and Lars Rydén.
ISBN 0-471-18626-0. © 1998 Wiley-VCH, Inc.

4.1   INTRODUCTION

Ionic interactions as the basis for separation and purification of proteins by ion-exchange chromatography (IEC) have been applied successfully since the late-1940s. Thus there is much accumulated experience of this type of chromatography documented in the literature. It is traditionally the most utilized protein separation technique, included in about 75% of purification protocols, followed by affinity chromatography (60%) and gel filtration (50%).[1] Data obtained in a recent literature survey showed that IEC is still the most popular method, accounting for about 40% of the chromatographic steps included in protein purification protocols (Table 4-1). Many different buffers were used for equilibration of ion exchangers, and the columns were usually eluted with a salt (very often NaCl) gradient, rarely with a pH gradient. A stepwise change of the salt concentration was used in only one case.

Reasons for the popularity of IEC include its (1) high resolving power, (2) high protein-binding capacity, (3) versatility (there are several types of ion exchangers, and the composition of the buffer and pH can be varied over a mile range), (4) straightforward separation principle (primarily according to differences in charge), and (5) ease of performance. However, with macromolecules such as proteins, separation mechanisms are often complex. Thus, differences in the distribution of charges on the protein surface can be important. Nonelectrostatic interactions, such as hydrophobic interactions and hydro-

TABLE 4-1. Chromatographic Methods Used for Purification of Nonrecombinant and Recombinant Proteins

Method	No. of Times Used (%)	
	Recombinant Proteins ($n = 40$)	Nonrecombinant Proteins ($n = 94$)
Ion exchange	37 (35)	130 (41)
Electrophoresis	0	4 (1)
Chromatofocusing	0	3 (1)
Isoelectric focusing	1 (1)	0
Gel filtration	15 (14)	63 (20)
Reversed-phase chromatography	3 (3)	19 (6)
Hydrophobic interaction chromatography	7 (7)	31 (9)
Affinity chromatography	24 (23)	33 (10)
Immobilized metal ion affinity chromatography	4 (4)	6 (2)
Dye ligands	6 (6)	14 (4)
Covalent chromatography	0	1 (<1)
Hydroxyapatite chromatography	7 (7)	16 (5)
Sum	104 (100)	320 (100)
No. of chromatographic steps per protein	2.7	3.4

Data were collected from articles published in 1995 and 1996 in *Arch. Biochem. Biophys., Biochemistry, Biochem. J., Biochim. Biophys Acta, J. Biol. Chem., Protein Express. Purif.,* and *Toxicon.* Homogeneity was analyzed by polyacrylamide gel elctrophoresis of the native protein or in SDS under reducing conditions 131 times, by chromatofocusing 6 times, by isoelectric focusing 2 times, by reversed-phase chromtography 8 times, and by gel filtration 9 times (E. Karlsson, unpublished review).

gen bonding, and the nature of the buffer ions can also influence the separation. Similarly, separation in chromatofocusing reflects not only the pI of a protein, but also the shape of the titration curve in the vicinity of the pI. Chromatofocusing, which in fact is a special form of IEC, is treated in Chapter 5.

Even though IEC is an established technique in most biochemical laboratories, little is known about the fundamental mechanism behind protein binding to charged surfaces. Discussions regarding different adsorption models can be found in Refs. 2, 3, and 4.

The first use of an ion exchanger in protein chemistry may have been for the removal of pectinmethylesterase from a preparation of pectinpolygalactonurase. The contaminant was adsorbed to a polystyrene cation exchanger (Amberlite IR-100, sulfonic acid type).[5] Some years later the styrenemethacrylic acid resin Amberlite IRC-50 was used for successful chromatographies of a number of basic proteins,[6] such as cytochrome c,[7] ribonuclease,[8,9] and lysozyme.[10,11] Methacrylic acid resins have a high pK_a of 6.5 (at ionic strength 0.1 M) and are therefore suited for basic proteins. Because of their very hydrophobic matrices and low capacity, polystyrene resins are not suited for protein chromatography (see Section 4.4).

The introduction of ion exchangers with hydrophilic and macroporous matrices in the middle of the mid-1950s by Sober and Peterson[12-14] extended ion-exchange chromatography to most proteins. They synthesized carboxymethyl (CM) and diethylaminoethyl (DEAE) derivatives of cellulose. Adsorption of proteins by forces other than elelctrostatic is low and they have a high capacity, as the macroporous structure renders ion-exchange groups in the interior of the particles accessible to proteins. Since then, a number of chromatographic media, in particular beaded ones, have been synthesized. These include gels based on cross-linked dextrans, cross-linked agarose, synthetic hydrophilic polymers, and small, rigid beads used in HPLC.

This chapter is devoted mainly to practical aspects of ion-exchange chromatography. Hydroxyapatite chromatography will also be briefly treated, as it may be regarded as a variant of ion-exchange chromatography. The chapter concludes with some examples of practical applications.

4.2   THE ION-EXCHANGE PROCESS

4.2.1   Fundamental Concepts

The basis for IEC is the competition between ions of interest and other ions for oppositely charged groups on an ion exchanger. The interaction between small molecules and an ion exchanger depends on the net charge, as illustrated in Fig. 4-1, and the ionic strength of the medium. When the concentration of competing ions is low, the ions of interest bind to the ion exchanger, whereas when it is high, they are desorbed. The interaction between a protein and an ion exchanger depends not only on the net charge and the ionic strength, but

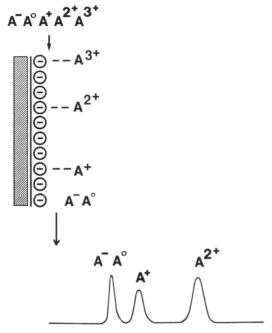

FIGURE 4-1. Principle of ion-exchange chromatography. Species with several positive charges (A^{3+}) are adsorbed to the column; those with few charges move slowly, whereas those with no net charge or a net charge of the same sign as the adsorbent pass through the column unretained. The resulting chromatogram is shown. In this case, the separation occurred according to differences in net charge. This is valid for small molecules. With proteins, many other factors besides the net charge can contribute to the separation (see text).

also on the surface charge distribution of the protein, pH, the nature of particular ions in the solvent, additives such as organic solvents, and properties of the ion exchanger.

The interaction F between two charges Z_a and Z_b separated by a distance r is given by Coulomb's law

$$F = Z_a Z_b / D r^2 \qquad (4\text{-}1)$$

where D is the dielectric constant of the medium. Water has a high dielectric constant, about 80, and most organic solvents have about 20. Thus, the electrostatic attraction between two oppositely charged groups is higher in a hydrophobic environment. For the value of D to be valid, there must be a sufficiently thick layer of solvent molecules between the two charged groups. If there is a direct interaction, as between say an active site residue of a protein and its inhibitor or substrate, the bond is in general stronger than would be predicted by Eq. (4-1). This type of interaction is exploited in affinity elution, where a charged substrate or inhibitor is used to elute a protein from an ion exchanger (see Section 4.7.2.3).

It is clear that the more highly charged a protein is, the more strongly it will bind to a given, oppositely charged ion exchanger. Similarly, more highly charged ion exchangers, (i.e., those with a higher degree of substitution with charged groups), usually bind proteins more effectively than weakly charged ones. Conditions, for example, pH, that alter the charge on either the protein or the ion exchanger will affect their interaction and may be used to influence the ion-exchange process.

4.2.2   The pH Parameter

The pH is one of the most important parameters in determining protein binding, as it determines the charge on both the protein and the ion exchanger. Although proteins have charges of both signs over a wide pH range, as a rule binding to an ion exchanger occurs only when there is a net charge on the protein of opposite sign to that on the ion exchanger. At pH values far away from the pI, proteins bind strongly and in practice do not desorb at low ionic strength. Near to its pI, the net charge of a protein is less and consequently it binds less strongly. Because the charges on the protein surface are distributed asymmetrically, binding can also occur close to the pI even when the overall charge is the same as that on the ion exchanger,[15,16] as exemplified by yeast phosphoglycerate kinase[17] that has a cluster of positive charges, which explains its binding to a cation exchanger at a pH when its net charge is slightly negative. The opposite has also been observed; rat adenylosuccinate synthetase (pI = 8.9) does not bind to a cation exchanger at neutral pH despite its positive net charge.[18]

An ion exchanger is normally used in conditions where its charges will not be significantly titrated by small shifts in pH. Strong ion exchangers (see Section 4.4.1) have pK_a values outside the pH range in which it is usual to work with proteins, and changes in the pH will therefore not change the charge of the ion exchanger.

Due to Donnan effects, protons are attracted and hydroxyl ions are expelled from the microenvironment of cation-exchange groups. The pH in the matrix is therefore usually about one unit lower than in the surrounding buffer. In anion exchangers the two ions move in opposite directions, resulting in an increase of pH by about one unit. If a protein is adsorbed at pH 5, it will be exposed to pH 4, and if it has poor stability at that pH, it may denature.[19] Low recoveries have been observed in IEC at pH 4 or below (see Section 4.4.1). At pH 3, many proteins have only positive charges and may therefore bind more strongly than at higher pH values when they have both positively and negatively charged groups. Exceptions are, for instance, phosphoproteins, which have negative charges at pH 3, and trypsin, which has an aspartic acid with a pK_a of 2.9 to 3.1 (see Section 4.3.2).

4.2.3 Influence of Ions

The proteins compete with other ions in the solvent for binding to the charged groups on the ion exchanger. At a low enough concentration of competing ions, proteins bind through interactions between several charged groups on the proteins, the Z value, and oppositely charged groups on the ion exchanger. At higher concentrations of competing ions, the proteins will start to be displaced from the ion exchanger; the least strongly bound are displaced and eluted from the column first. There is no general rule as to what salt concentration is needed to displace a protein with a certain net charge from an ion exchanger. Most proteins are eluted at a salt concentration lower than 1 M. The type of ion is also an important factor, as some ions displace proteins more efficiently than others, and ions that interact specifically with charged groups on the proteins are more efficient displacers (see Section 4.7.2.3). Ions differ not only in their elution power, but can also affect the resolution and the order in which proteins are eluted, or the selectivity (see Section 4.5.2).

4.2.4 Other Factors: Hydrophobic Interaction and Hydrogen Bonding

Although Coulombic forces are normally responsible for the binding between the stationary phase and proteins, other interactions may also occur, for example, hydrophobic interactions between, in particular, resin-based ion exchangers, including some modern ion exchangers used in HPLC, and hydrophobic residues in proteins (see Section 4.9.5). Similarly, in some cases, hydrogen bonds are apparently formed between a protein and an ion exchanger. These additional effects are sometimes crucial for obtaining a separation of two similar proteins, but the effects are difficult to predict and thus difficult to exploit rationally and might of course just as well constitute a disadvantage.

4.3 CHARGE PROPERTIES OF PROTEINS

4.3.1 Charged Groups in Proteins

Seven out of the 20 amino acid side chains of proteins contain groups that are weak acids or bases (Table 4-2).[20,21] Several of them belong to the most common amino acid residues of proteins. Thus nearly all proteins have both positive and negative charges at pH values used in IEC. It should be mentioned that the amide groups of asparagine and glutamine for all practical purposes are neither basic nor acidic (although they do accept a proton at about pH −0.5).

The charged groups nearly always reside on the protein surface. The main exceptions are when internal metal ions in metalloproteins use histidines and cysteines (e.g., in Cu, Zn, and Fe proteins) or glutamates and aspartates (mainly in Ca and Fe proteins) as ligands to the metal. In these cases, the site

TABLE 4-2. Charged Amino Acid Side Chains and Some Other Groups in Proteins[a]

Group	Structure	pK_a Proteins	pK_a Free Amino Acids	Occurrence in Proteins (%)[b]
Arginine	Guanidino	12.0	12.5	5.7
Aspartic acid	Carboxylate	3.9–4.0	3.7	5.3
Cysteine	Thiol	9.0–9.5	10.6	1.73
Glutamic acid	Carboxylate	4.3–4.5	4.3	6.2
Histidine	Imidazole	6.0–7.0	6.1	2.2
Lysine	ε-Amino	10.4–11.1	10.7	5.7
Tyrosine	Phenol	10.0–10.3	10.4	3.2
α-Amino	Amino	6.8–8.0		
α-Carboxyl	Carboxylate	3.5–4.3		
Sialic acid	Carboxylate			Sialoglycoproteins
γ-Carboxyglutamate	Carboxylate			Blood coagulation factors
Phosphoserine, etc.	Phosphate			Phosphoproteins

[a] Data from Creighton.[20]

[b] In sequences of 1021 unrelated proteins.

of the proton is occupied by the metal, which can be displaced at a low enough pH. It should be added that metal proteins do not form a special case. About one-third of all proteins described belong to this category. Only rarely have internal salt bridges been observed. Other prosthetic groups may also bind to the peptide chain via a charged amino acid side chain (e.g., pyridoxal phosphate forms covalent bonds with lysine residues).

Further charged groups (Table 4-2) are provided by the N-terminal α-amino and the C-terminal α-carboxyl groups. The various posttranslational modifications that many proteins undergo may also provide new titratable groups. These are most often acidic, such as phosphates in the phosphoproteins, γ-carboxylates in blood-clotting factors, or sialic acid residues on oligosaccharides in serum glycoproteins.[20] Basic groups are introduced less often, but methylation of lysine and histidine in, for example, histones should be mentioned as it turns these side chains into much stronger bases.[20] Finally a modification may also eliminate the acid–base properties of a group, as in the acylation of the N terminus by acetic or formic acid, the cyclization of an N-terminal glutamate to pyroglutamate, or the amidation or esterification of the C-terminal α-carboxylate.

The various posttranslational modifications are often incomplete, as reflected by a heterogeneity seen in ion-exchange chromatography. This is exemplified by pancreatic ribonucleases A, B, C, and D with different degrees of glycosylation[22] or ceruloplasmin forms I and II.[23] A further source of charge heterogeneity is covalent modification, which occurs during handling of the proteins and affects only part of the protein, such as nicking (hydrolysis of a single peptide bond) of the peptide chain, and loss of amide groups.

Experimental chemical modifications also give rise to heterogeneity in proteins. Even if the different isomers formed have the same net charge it may still be possible to separate them by IEC.[24,25] Thus when a cobra neuro-toxin with six amino groups (one α and five ε) was acetylated, the six monoace-tyl derivatives could be separated on the cation exchanger Bio-Rex 70 (see also Section 4.9.4), a reminder of the importance of protein charge distribution for binding to the ion exchanger.

4.3.2 Protein Titration Properties

Most proteins have their isoelectric points at pH below 7.[26,27] They thus have a surplus of acidic groups. Therefore, in many applications of IEC a first and often successful approach is to use an anion exchanger at slightly basic conditions, around pH 8.

Proteins that have an excess of basic groups and thus a high pI are called basic proteins. Pancreatic ribonuclease is a moderately basic protein, with a pI of 9.5 to 9.7.[28] More strongly basic proteins are not encountered to often; examples include mitochondrial cytochrome c (pI of 10.1 to 10.2[28]), chicken egg white lysozyme (pI of 11.3 to 11.6[28]), and the histones. Cytochrome c is attached to reductase and oxidase in the mitochondrial inner membrane by ionic forces, whereas histones bind to the DNA phosphates to form nucleo-somes in chromatin.

Extremely acidic proteins with a surplus of acidic groups and a pI below 5 are also found. Their acidity may be due to aspartates as in, for example, pepsin (pI of 2.9 to 3.1[27]), but often posttranslationally added groups explain the acidity of many sialoglycoproteins.

Titration curves of proteins not only provide information on the position of the point of zero net charge, the pI, but also on how net charge varies with pH. The weak acids and bases in proteins normally titrate over a very wide range. It is especially interesting to study the assembly of titration curves for all the proteins in a sample as a guide to possible separation strategies. Figure 4-2 shows an example of such a collection. It is clear from these curves that the order in which the proteins in a sample may be arranged according to net charge varies with pH. It can, therefore, be expected that the proteins will elute from an IEC column in different orders at different pH values. It is also clear that even if a less successful result is obtained at one pH, it is worth trying either a different pH with the same ion exchanger or an ion exchanger of the opposite charge.

4.3.3 Factors Influencing Protein Charge

Protein surface charge depends on the dissociation constants, pK, of the weak acids and bases on its surface. The pK of an acid is defined as the pH of 50%

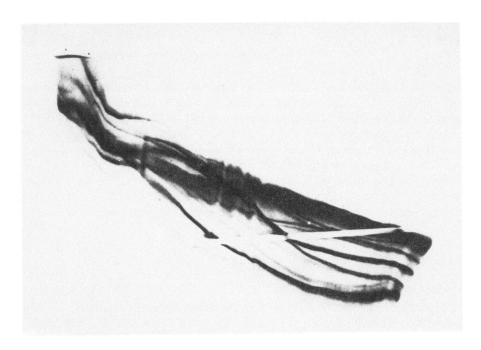

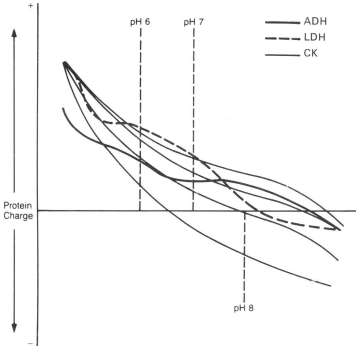

FIGURE 4-2. Titration curves of proteins showing how protein net charge varies with pH. (Top) Actual titration curves of proteins in a chicken skeletal muscle as obtained by the technique described in Chapter 12 and visualized by protein staining. (Bottom) The result of staining with the zymogram technique to identify alcohol dehydrogenase (ADH), lactate dehydrogenase (LDH), and creatine kinase (CK). The same sample was submitted to sequential ion-exchange chromatography on a cation and anion exchanger (see Section 4.9.6). (Reproduced with permission from Ref. 171.)

titration or protonation. An acidic group provides a net charge of -0.5 at pH $=$ pK and a basic group provides a net charge $+0.5$ at its pK value. Only so-called acidic dissociation constants, pK_a, are used. At one pH unit from the pK a charged group is titrated to 91% as calculated from the Henderson–Hasselbalch equation. Thus, acidic and basic groups provide 0.1 charges one unit above and below their pK values, respectively.

When the main chain amino and carboxyl groups are part of peptide bonds, this influences the pK of groups in the side chain. The pK_a values tend to move closer to neutrality when compared to the free amino acid (Table 4-2). However, in proteins, many more factors modify the pK_a of the side chains, such as the influence from the neighboring groups and the position in the tertiary structure. Deviations from standard pK_a values can be seen for side chains that fulfill special roles in proteins. In α-neurotoxins (bind to nicotinic acetylcholine receptors), Tyr 25, which participates in stabilization of the structure by hydrogen bonding and reacts only very slowly with tetranitromethane, has a pK_a of 11.3, whereas Tyr 35, which is fully exposed to the solvent and reactive, has a pK_a of 9.7.[29] Even larger deviations from standard pK_a values are known. For instance, pepsin has two aspartates in its active site that are connected via hydrogen bonds. The two carboxyls interact closely and, due to this interaction, one of the carboxyls (Asp 32) has the unusually low pK_a of 1.5. The other carboxyl (Asp 215) has a pK_a of 4.7. The situation resembles ionization of maleic acid, which has pK_a values of 1.9 and 6.2. A salt bridge between Asp 70 and His 31 in T4 lysozyme alters the pK_a to 0.5 and 9.8, respectively. In the unfolded state the residues have pK_a values of 3.5–4 and 6.8, respectively.[20]

From this last observation it is evident that protein surface charge properties change as the conformation changes. The changes can be brought about by the addition of structure-modifying agents, such as cosmotropic and chaotropic ions,[30,31] solvents, polyethylene glycol, urea, or detergents, or they can be part of the biological function as substrates or ligands. The latter is exploited in affinity elution (see Section 4.7.2.3).

4.3.4 The Chromatographic Contact Region and the Z Value

The chromatographic contact region is the region of the protein surface that determines its chromatographic behavior.[32] The contact region varies depending on the type of chromatography (e.g., in affinity chromatography the region is small and includes only amino acids in the active site that bind to the biospecific adsorbent). In IEC, the contact region consisting of several charged regions, ionotopes,[33] can include the major part of the protein surface.[32,33] If there is a heterogeneous distribution of charge groups on a protein surface, regions with a high concentration of charged groups can determine the chromatographic behavior. This is probably the explanation why proteins can bind to an ion exchanger at the pI when the net charge is zero or even when the net charge is the same as that of the ion exchanger; for instance, phosphoglycerate kinase has a cluster of positive charges that makes the protein bind to a cation exchanger at pH values when the net charge is slightly negative (see Section 4.2.2).

Changes in the chromatographic contact region alter the chromatographic behavior. Lysozyme variants differing by a few amino acid mutations have been separated by cation-exchange chromatography.[32,34] Chemical modifications can change the contact region. The six monoacetyl derivatives of a cobra neurotoxin have the same net charge, but they lose a positive charge from different locations in the molecule. They do not have the same surface charge distribution and they are separated by ion-exchange chromatography on Bio-Rex 70 (see Section 4.9.4). Similarly, mono derivatives of fasciculin 2, an acetylcholinesterase inhibitor from a mamba venom, obtained after neutralization of the charge of amino groups,[35] arginine residue,[36] and carboxylates[37] (see Section 4.9.5) were separated on a cation exchanger.

The chromatographic contact region can include amino acids from opposite sides of the protein molecule.[32,34] For instance, in fasciculin 2, amino groups are located on all sides of the molecule, and acetylation of each amino group changes the chromatographic behavior. Amino acids at different locations will be able to bind to an ion exchanger if a protein molecule orients in many different ways. Removal of a positive charge changes the contact region and the orientation in the vicinity of cation-exchange groups. A carboxylate will not bind to cation-exchange groups, but carboxyl-modified derivatives of fasciculin were separated on a cation exchanger (see Section 4.9.5). Removal of a negative charge changed the orientation of the various derivatives and therefore also their chromatographic behavior.

The Z value is the number of charged sites on the protein that interact with an ion exchanger. The retention is determined at isocratic elution (constant ionic strength). The Z value is obtained from the relation $\log k' = Z \log[D_o] + \log I$, as the slope of $\log k'$ versus $\log 1/[D_o]$ plots; k' is the capacity factor equal to $(t_c - t_o)/t_o$ (t_c is the retention time for the component and t_o is the retention time for a non-retarded component), $[D_o]$ is the molar concentration of the displacing ion and I is a constant.[32,38] As the Z value increases, a protein is bound more strongly and is consequently more difficult to elute. Boardman and Partridge[39] showed that the retardation of cytochrome c on Amberlite IRC-50 increased rapidly with increasing Z values.

The Z value depends on many factors: pH (changes the charge of the protein and the ion exchanger), the protein conformation, the presence of other ions, the amount of protein adsorbed to the adsorbent, and the properties of the ion exchanger.

The Z-values for conalbumin (pI of 6.0 to 6.6[27]), β-lactoglobulin (pI of 5.2[28]), ovalbumin (pI of 4.7[40]), and soybean trypsin inhibitor (pI of 4.6[28]) were determined in the pH interval of 4 to 8 and in the presence of Na^+ or Mg^{2+}.[41] A strong anion exchanger (Mono Q) with a constant charge between pH 4 to 8 was used. The Z values varied between 1 and 3–5; the increase was greatest between pH 4 and 5.5 when the carboxyls were titrated, reached a maximum at pH 7, and remained constant or decreased slightly when the pH was increased to 8. The Z values were higher in Mg^{2+} than in Na^+, which may depend on conformational changes. The proteins had Z values of 1 to 3.5 at pH = pI when the net charge was zero, indicating that there is little relation between the net charge and the Z value. Binding to an ion exchanger at pI may depend on a cluster of anionic groups (see Section 4.2.2).

Since the amino acids in the chromatographic contact region are distributed over a large part of the protein surface, and even situated at opposite sides of the molecule, it is evident that all of them cannot bind simultaneously to an ion exchanger. For instance, soybean trypsin inhibitor has 31 anionic groups (one α-carboxyl, 17 Asp, 13 Glu) and 22 cationic groups (one α-amino, 2 His, 10 Lys, 9 Arg)[42] and its net charge at pH 8 is -11.5 (assuming no charge on the histidine residues and pK_a about 8 of the α-amino group). The Z values at pH 8 were 3.3 in the presence of Na^+ and 4.2 in the presence of Mg^{2+}. Thus, only 3 to 4 of the 31 anionic groups could bind simultaneously to the ion exchanger.

The Z values are average values resulting from several factors; a protein molecule may be oriented in several ways and present different numbers of sites to the ion exchanger,[43] all binding sites on the protein may not be fully ionized,[43] and, due to steric factors, the distance between the charged groups on the protein and ion exchanger is not always the same.

The Z value decreases with increasing load. The maximum value, 3.6, was at very low protein concentration and the minimum, 2.5, at saturating concentration.[44]

Multiple peaks of a protein can be obtained at high load. A gel filtration fraction of black mamba (*Dendroaspis polylepis*) venom was chromatographed on Bio-Rex 70. The load was about twice that of earlier chromatographies. Dendrotoxin I (a blocker of voltage-dependent potassium channels), which is the dominating component, accounting for more than 10% of the protein, eluted as a multiple peak whereas multiple peaks were not seen for other components. Rechromatography of a smaller aliquot of the pooled multiple peak gave only one peak.[45] A possible explanation is that at high load all dendrotoxin molecules were not able to bind to the same number of sites due to hindrance from neighboring molecules. Some of them were more easily eluted than others and multiple peaks were obtained. At low load, all molecules had enough space to bind to the same number of sites.

A protein can bind to more charged groups on a highly substituted ion exchanger than to an ion exchanger with a low degree of substitution. Symmetrical peaks (linear adsorption isotherms) in isocratic elution are obtained from an adsorbent with a low degree of substitution.[46,47]

The Z values for proteins in IEC are often rather small, as seen from the examples given here. A larger protein does not necessarily have more contact points with an ion exchanger. Soybean trypsin inhibitor [molecular weight (MW) 20,000] had a Z value of 3.3 whereas ovalbumin (43,500) had a Z value of 2.4 (pH 8, NaCl).[41] Z values at different pH of five proteins in the molecular weight range of 17,000 to 69,000 can be found in Ref. 40. However, the dimers of a cobra neurotoxin (MW 7820) provide an example where the size has a large effect on the retention. When chromatographed on Bio-Rex 70 at pH 6.5, the monomer eluted at 0.15 M ammonium acetate, whereas the two dimers eluted in a long extended peak between 0.5 and 0.7 M. On a cellulose exchanger, CM 52, at pH 5.2, the monomer eluted at 0.2 M ammonium acetate, whereas the dimers eluted in two resolved peaks at about 0.3 M (E. Karlsson, unpublished observation). Because chromatographic behavior depends on the protein conformation, structural changes can be detected by IEC.[48-50]

4.4 THE STATIONARY PHASE: ION EXCHANGERS

Ion exchangers consist of a matrix with either acidic or basic groups. The basic ion exchangers containing positive groups are called anion exchangers, whereas the acidic ones containing negative groups are called cation exchangers.

Ion exchangers are usually classified as weak and strong. The name refers to the pK_a values of their charged groups (by analogy with weak and strong acids or bases) and it does not say anything about the strength with which they bind proteins. At pH values far away from the pK_a, binding will be equally strong to either a weak or a strong ion exchanger. For instance, a carboxymethyl ion exchanger is a weak cation exchanger, as it has a weak acid -OCH_2COOH (pK_a 4 to 4.5) as the functional group. Similarly, a sulfopropyl ion exchanger with -$CH_2CH_2CH_2SO_3H$ (pK_a of 2) is a strong cation exchanger (Table 4-3).

Matrices are (1) hydrophobic polystyrene-based or partly hydrophobic polymethacrylate-based polymers, which will be called resins; (2) hydrophilic and macroporous synthetic or naturally occurring polymers, such as polyacrylamide, cellulose, dextran, and agarose; (3) various synthetic and macroporous hydrophilic polymers, which make hard or moderately hard beads for high-pressure applications; or (4) nonporous; molecules do not penetrate the interior of the ion exchange particles. For a discussion of ion-exchangers, see also Ref. 51.

4.4.1 Functional Groups and Acid–Base Properties

A number of different functional groups are used (Table 4-3). Most common are the amines in the anion exchangers and the carboxylic acids in the cation exchangers. Strong anion exchangers have quaternary amines whereas strong cation exchangers have sulfonates. Practically all ion exchangers have groups with a single positive or negative charge, except phosphate.

· Approximate pK_a values of the groups are given in Table 4-3. Precise values are, however, difficult to determine. Titration curves show that some ion exchangers do not have a single pK_a value (Figure 4-3). DEAE-ion exchangers

TABLE 4-3. Functional Groups Used in Ion Exchangers

Name	Designation	pK	Structure
Anion exchangers			
Diethylaminoethyl	DEAE	9.0 to 9.5	$-OCH_2N^+H(C_2H_5)_2$
Trimethylaminoethyl	TMAE	—	$-OCH_2CH_2N^+(CH_3)_3$
Dimethylaminoethyl	DMAE	ca. 10	$-OCH_2CH_2N^+H\,(CH_3)_2$
Trimethylhydroxypropyl	QA		$-OCH_2CH(OH)N^+H(C_2H_5)_2$
Quaternary amino ethyl	QAE		$-OCH_2CH_2N^+(C_2H_5)_2CH_2$
			$CH(OH)CH_3$
Quatenary amine	Q		$-OCH_2N^+(CH_3)_3$
Triethyl amine	TEAE	9.5^a	$-OCH_2N^+(C_2H_5)_3$
Cation exchangers			
Methacrylate		6.5	$CH_2=CH(CH_3)COOH$
Carboxymethyl	CM	3.5–4	$-OCH_2COOH$
Orthophosphate	P	3 and 6	$-OPO_3H_2$
Sulfoxyethyl	SE	2	$-OCH_2CH_2SO_3H$
Sulfopropyl	SP	2–2.5	$-OCH_2CH_2CH_2SO_3H$
Sulfonate	S	2	$-OCH_2SO_3H$

Data have been compiled from manufacturers' booklets. The pK values given mostly refer to an ionic strength of 0.1. See also text.

^{a} The pK value apparently does not refer to quatenary groups.

with a high degree of substitution contain a significant amount, up to 30%, of so-called tandem groups formed by further derivatization of an already coupled DEAE group. This leads to the formation of two additional kinds of charged groups: a quaternary amine and a DEAE group whose pK_a has been lowered to about pH 6 under the influence of the nearby quaternary nitrogen. Note that pK_a depends on the salt concentration[52] and could be 0.2 to 0.5 units higher for CM-cellulose and 1.5 units higher for Bio-Rex 70 in water as compared to 1 M NaCl. The pK_a values of DEAE-cellulose ion exchangers varied between 6.9 (in water) and 9.8 (in saturated NaCl, 6.24 M) and were 9.1 in 0.1 M NaCl.[53]

All cation exchangers have a limiting pH below which they cannot be used. They are normally not used at pH values where a significant part of the groups have lost their charge. As a rule of thumb, the pK_a is suggested as the lower limit. Strong cation exchangers such as sulfopropyl can be used at a lower pH than carboxymethyl ion exchangers. In the same way, weak anion exchangers have an upper pH limit for their use. For the quaternary amines, there is no upper limit as they will not lose the charge whatever the pH.

The pH interval in which an ion-exchange chromatography is carried out can also be restricted by reasons other than the charge properties of the ion exchanger. One is obvious: the pH range for the stability of the protein. But even at pH values where the proteins are stable, unwanted effects can occur. At pH 3.5, phospholipase A_2 from cobra venom adsorbed irreversibly to CM-

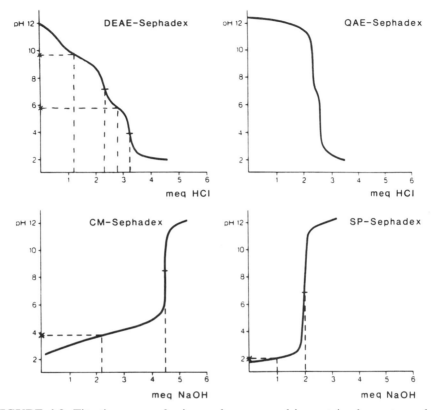

FIGURE 4-3. Titration curves for ion exchangers used in protein chromatography. The four curves show the titration of DEAE Sephadex A-50 (pK_a 9.5 and 5.8), CM Sephadex C-50 (pK_a 4.0), QAE Sephadex A-50 (the irregularity indicates that the ion exchanger is not fully quaternerized), and SP Sephadex C-50 (pK_a 2.0).

cellulose (unpublished observation) and cellulase to CM Sephadex C-50 (G. Pettersson, personal communication). Poor recovery of proteins has also been observed on a strong cation exchanger (Mono S) at pH 4 or below[54] (see Section 4.2.2, Donnan effect).

4.4.2   Capacity and Porosity

In general, ion exchangers are more densely substituted than other adsorbents used in protein chromatography. Common degrees of substitution as determined by titration (total ionic capacity), are 100 to 500 μmol/ml bed. This figure corresponds to a concentration of ion exchanging groups of 0.1 to 0.5 M.

The available or total capacity is the amount of protein that can be bound, and it depends on the size of the protein and the porosity of the ion exchanger. This capacity is also called static capacity. The dynamic capacity depends on the flow rate, and is equal to the static capacity at low flow rates when there is sufficient time for saturation of the ion exchanger. At higher flow rates, when there is not enough time, the dynamic capacity is lower than the static capacity. Tables 4-4 and 4-5 give the available capacity of ion exchangers for some standard proteins. A low ionic strength (0.01 to 0.1 M) was used to ensure strong binding of the protein and pH 5.0 for cation exchangers and pH 8.0 for anion exchangers. For some DEAE ion exchangers the protein-binding capacity has been determined at an ionic strength of 0.01 M when the DEAE group has a pK_a of 8.35,[53] and at pH 8.0 only 70% of the groups are ionized (calculated from Henderson–Hasselbalch's equation). Only part of the ionic groups are available to protein. Mono Q with a total substitution of 300 μmol/ml binds 65 mg (1 μmol) of human serum albumin (HSA) per milliliter. The Z value of HSA at pH 8 on Mono Q is 4.8 at pH 7.5[16] (i.e., about five negative groups on each albumin molecule bind to at least the same number of positively charged groups on the ion exchanger). Thus, 1 μmol of HSA binds to 5 μmol of charged groups of Mono Q, and less than 2% of its charged groups are involved in protein binding. The cellulose ion-exchanger DEAE-Sephacel has a total capacity of 170 μmol/ml and binds 160 mg bovine serum albumin (BSA) (2.4 μmol) per milliliter. The Z value of BSA on DEAE-Sephacel is not known, but it binds 0.9 mg albumin/μmol DEAE and Mono Q only 0.2, which should mean that a larger fraction of the charged groups of the cellulose ion exchanger is available for the binding of albumin.

The capacity depends not only on the number of charged groups, but also on their availability, which in turn depends on the pore size and the molecular weight of the protein. Thus, DEAE-Sephadex A-50 binds 250 mg hemoglobin per milliliter, whereas the corresponding gel with smaller pores, DEAE-Sephadex A-25, only binds 70 mg of the same protein. This is why the more recently developed ion exchangers are based on matrices with large pores.

TABLE 4-4. Capacity of Microgranular CM-cellulose (Whatman CM-52) for Rabbit Muscle Glycolytic Enzymes and Other Proteins at pH 6.0, at Ionic Strength 0.01 M[19,55]

	Molecular Weight	Capacity	
		mg/ml	μmol/ml
Chicken egg white lysozyme	14,300	130	9.0
Phosphoglycerate kinase	45,000	70	1.55
Phosphoglycerate mutase	60,000	40	0.67
Creatine kinase	82,000	35	0.43
Enolase	88,000	48	0.55
Lactate dehydrogenase	140,000	21	0.15
Glyceraldehydephosphate dehydrogenase	145,000	26	0.18
Aldolase	160,000	22	0.14
Pyruvate kinase	228,000	12.5	0.055

The relationship between capacity and molecular weight is illustrated by the fact that DEAE-Sephacel binds 160 mg/ml of albumin (M_r 67,000) and only 10 mg/ml of thyroglobulin (M_r 670,000). Apparently thyroglobulin does not penetrate the matrix and binds only to sites near the surface of the particles. The variation of capacity with molecular weight is illustrated in Table 4-4 for a cellulose ion exchanger.[19,55]

4.4.3   Ion Exchangers

A summary of the properties of some of the better-known ion exchangers commercially available is given in Table 4-5.

4.4.3.1   Resin Ion Exchangers   A classical group of ion exchangers are made by polymerizing styrene with divinylbenzene (DVB) as a cross-linker in the presence of a catalyst such as benzoylperoxide (Figure 4-4). Ion-exchange groups are then substituted into the matrix. Among the best known are the strongly acidic sulfonated cation exchangers such as Dowex 50-X8 or AG-X8. X8 denotes the percentage of the cross-linker in the polymerization mixture, which is not necessarily the same as that in the final product. The corresponding strong anion exchangers containing quaternary amines are Dowex 1 X-4 and AG 1-X4.

Polystyrene ion exchangers have a low capacity for proteins (one or a few milligrams per gram of ion exchanger) due to their small pore size.[52,56] They may also bind proteins almost irreversibly by hydrophobic interactions (see Section 4.5.3). They are consequently not used for protein chromatography, but have for many years been important in the chromatography of small molecules, for example, for desalting of water and for separation of peptides and amino acids (e.g., in the Moore and Stein amino acid analysis).

A resin obtained by copolymerization of methacrylic acid with divinylbenzene is available as Bio-Rex 70 (5% DVB) (Bio-Rad Laboratories) (a practically identical resin was earlier available as Amberlite IRC-50). Polymerization of methacrylic acid alone gives 11.6 mmol of carboxylic groups per gram. Since the actual capacity is 10 mmol/g, the aromatic constituent accounts for about 14% of the resin. In the packed chromatographic bed, this capacity corresponds to 3.5 mmol/ml or 3.5 M concentration of carboxylates. The pK_a is about 6.0 in 1 M NaCl, 6.5 in 0.1 M, and increases to 7.5 in 0.01 M. When the pH is lowered to 5.7, the particles aggregate. At this pH, proteins can be adsorbed irreversibly to the resin, probably because of increased hydrogen bonding. Bio-Rex 70 is used at a pH of 6.5 or above and can therefore act as a strong (3.5 M) buffer; the pH of the effluent is mainly determined by the equilibrium conditions of the resin and the ionic strength of the buffer. The acrylic acids resins are well suited for small basic proteins. They were used for the first purifications of proteins by IEC[6-11] (see Section 4.1). They have a resolving power for basic proteins that has not been surpassed by any other ion exchanger. They have proved of value for the fraction of the collection of basic proteins that make up the snake neurotoxins, as is illustrated later. Hydrogen bonding to the carboxylic acid groups and hydrophobic interaction with the matrix may contribute to the separation.

4.4.3.2 *Cellulose Ion Exchangers*

The cellulose ion exchangers for proteins were introduced in 1954 by Sober and Peterson.[12-14] Their work was preceded by a number of reports on the use of derivatized cellulose as ion exchangers,[57-59] but at the time it constituted a major breakthrough in protein chromatography. Cellulose ion exchangers are still in wide use, particularly for large-scale preparations of neutral and acidic proteins, the group to which most proteins belong.

Cellulose has a macroporous structure and is thus accessible to large macromolecules. Microfibrous cellulose binds four times its own weight of water. Its capacity for protein binding is high and its interaction with proteins due to hydrophobic or other nonionic interactions is low.

The cellulose ion exchangers are prepared from strong alkaline (mercerized) cellulose by derivatization. For instance, carboxymethyl groups are introduced by reaction with chloroacetic acid and diethylaminoethyl groups by reaction with 2-chlorotriethylamine. A typical degree of substitution is 1 mmol/g cellulose, which corresponds to 200 μmol/ml bed volume. It is mostly amorphous regions in the otherwise fibrous structure that are derivatized by this treatment. In alkaline conditions, however, the bundles of polyglucan chains (Figure 2-2) open up.

The cellulose ion exchangers exist in several different physical forms. The fibrous celluloses are those that are obtained directly from the derivatization process. They contain fibers of polyglucan chains mixed with amorphous regions. The microgranular ion exchangers are obtained by partial acid hydrolysis of cellulose, a treatment that removes a large part of the amorphous regions and creates large void spaces. The ordered structures are stabilized by cross-linking and derivatization. Microgranular adsorbents have a more uniform distribution of charges. Cellulose ion exchangers in beaded form have been prepared as described by Determan and co-workers.[60] They are available from Pharmacia Biotech by the name of Sephacel and from Serva as Servacel.

4.4.3.3 *Dextran and Agarose Ion Exchangers*

The beaded forms of dextran and agarose gels originally prepared for gel filtration have been derivatized to produce ion exchangers for protein chromatography.

The dextran-based ion exchangers are derived from Sephadex. The designations A and C denote anion and cation exchangers, respectively, whereas the numbers 25 and 50 refer to the starting gel, which is Sephadex G-25 or G-50. Both of these have a considerably enlarged pore size as compared to the underivatized gels, as repulsion of the charged groups expands the gel. This swelling depends on both the pH and the concentration of counterions, as well as on the degree of cross-linking. It is more important for the less cross-linked A-50 and C-50; it decreases when the ion-exchanging groups are neutral-

TABLE 4-5. Ion Exchangers for Protein Chromatography

Name	Matrix	Functional Group	Degree of Substitution (μmol/ml)	Available Capacity (mg/ml)	Company
Standard media					
DE 23	Fibrous cellulose	DEAE	150	60 BSA	Whatman
CM 23	Fibrous cellulose	CM	80	85 Lys	Whatman
DE 52	Microgranular cellulose	DEAE	190	130 BSA	Whatman
CM 52	Microgranular cellulose	CM	190	210 Lys	Whatman
DE 53	Microgranular cellulose	DEAE	400	150 BSA	Whatman
CM 32	Microgranular cellulose	CM	180	200 Lys	Whatman
DEAE Sephacel	Beaded cellulose	DEAE	170	160 BSA	Pharmacia
DEAE Sephadex A-25	Dextran, Sephadex G-25	DEAE	500	70 Hb	Pharmacia
QAE Sephadex A-25	Dextran, Sephadex G-25	QAE	500	50 Hb	Pharmacia
CM Sephadex C-25	Dextran, Sephadex G-25	CM	560	50 Hb	Pharmacia
SP Sephadex C-25	Dextran, Sephadex G-25	SP	300	30 Hb	Pharmacia
DEAE Sephadex A-50	Dextran, Sephadex G-50	DEAE	175	250 Hb	Pharmacia
QAE Sephadex A-50	Dextran, Sephadex G-50	QAE	100	200 Hb	Pharmacia
CM Sephadex C-50	Dextran, Sephadex G-50	CM	170	350 Hb	Pharmacia
SP Sephadex C-50	Dextran, Sephadex G-50	SP	90	270 Hb	Pharmacia
DEAE Sepharose CL-6B	Agarose, 6% cross-linked	DEAE	150	100 Hb	Pharmacia
CM Sepharose CL-6B	Agarose, 6% cross-linked	CM	120	100 Hb	Pharmacia
DEAE Bio-Gel A	Agarose	DEAE	20	45 Hb	Bio-Rad
CM Bio-Gel A	Agarose	CM	20	45 Hb	Bio-Rad
DEAE-Trisacryl M	Trisacrylate polymer	DEAE	300	90 Hb	Sepracor

CM-Trisacryl M	Trisacrylate polymer	CM	200	100 Hb	Sepracor
SP-Trisacryl M	Trisacrylate polymer	SP	230	150 BSA	Sepracor
Fractogel TMAE 650 (S)	Synthetic organic polymer	TMAE	—	100 BSA	Merck
Fractogel DEAE 650 (S)	Synthetic organic polymer	DEAE	—	100 BSA	Merck
Fractogel DMAE 650 (S)	Synthetic organic polymer	DMAE	—	100 BSA	Merck
Fractogel SO$_3$ 650 (S)	Synthetic organic polymer	S	—	100 Lys	Merck
Fractogel COO 650 (S)	Synthetic organic polymer	COO$^-$	—	100 Hb	Merck
High-performance media					
Mono Q	Synthetic organic polymer	Q	270–370	65 HSA	Pharmacia
Mono S	Synthetic organic polymer	S	140–180	75 RNase	Pharmacia
SOURCE 15Q	Synthetic organic polymer	Q	—	25	Pharmacia
SOURCE 15S	Synthetic organic polymer	S	—	25	Pharmacia
SOURCE 30Q	Synthetic organic polymer	Q	—	40 BSA	Pharmacia
SOURCE 30S	Synthetic organic polymer	S	—	80 Lys	Pharmacia
Q Sepharose High Performance	Cross-linked agarose	Q	200	70 BSA	Pharmacia
SP Sepharose High Performance	Cross-linked agarose	SP	200	70 RNase	Pharmacia
DEAE-5-PW	Synthetic organic polymer	DEAE	—	1.5–3	Bio-Rad
SP-5-PW	Synthetic organic polymer	SP	—	1.5–3	Bio-Rad
Mini S	Nonporous synthetic polymer	S	20	4–6	Pharmacia
Mini Q	Nonporous synthetic polymer	Q	75	4–6	Pharmacia
HRLC MA7P	Nonporous synthetic polymer	PEI	—	0.6–2	Bio-Rad
HRLC MA7C	Nonporous synthetic polymer	CM	—	0.6–2	Bio-Rad
DEAE-3SW	Silica	DEAE	—	120 BSA	TosoHaas
CM-3SW	Silica	CM	—	45 Hb	TosoHaas

Data have been compiled from the manufacturers' booklets. The proteins used in capacity measurements were bovine serum albumin (BSA), human serum albumin (HSA), lysozyme (Lys), bovine hemoglobin (Hb), and bovine ribonuclease (RNase).

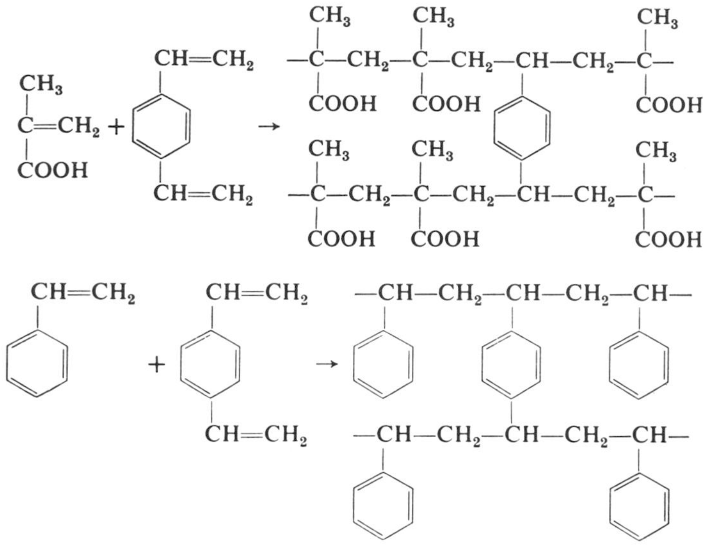

FIGURE 4-4. Covalent structures of the methacrylate ion exchanger Amberlite IRC-50 and Bio-Rex 70 (top) and polystyrene-based ion exchangers (bottom).

ized and is largest at a low ionic strength. A column of DEAE Sephadex A-50 or CM Sephadex C-50 will shrink to half its volume when the ionic strength is increased from 0.01 to 0.5 M. These effects do not seem to have any adverse consequences, at least not with continuous gradients.

Sepharose ion exchangers (Pharmacia Biotech) are derived from cross-linked 6% agarose (e.g., Sepharose CL-6B). These gels are more porous than the dextran gels, and volume changes with changes in ionic strength or pH are insignificant. Q Sepharose Fast Flow and S Sepharose Fast Flow are produced by a different cross-linking procedure. The beads are of the same size as conventional agarose, but are much more rigid and particularly suitable for large-scale work and other applications in which high flow rates are desirable. DEAE and CM Bio-Gel A ion exchangers (Bio-Rad Laboratories) also consist of cross-linked (4%) agarose beads. Agarose-based ion exchangers are useful for chromatography of larger proteins as they are more macroporous. Their degree of substitution is less than that of the dextran-based ion exchangers, but the available capacity for proteins might not be less if the pore size is the limiting factor.

4.4.3.4 Tentacle Ion Exchangers In Fractogel EMD tentacle ion exchangers, the charged groups are located on linear polymers grafted onto the matrix. Such an arrangement will decrease the probability of proteins interacting with the matrix by non-ionic forces,[61-63] but introduces the possibility for interaction with the polymer chain.

4.4.3.5 High-Performance Ion Exchangers In high- or medium-pressure chromatography, hard ion-exchange beads of small and uniform size are used. In general the particle size is 5–10 μm. These are based on silica matrices, for example, DEAE-3SW and CM-3SW (TosoHaas), coated with a hydrophilic layer of ion-exchange groups. The silica-based supports have the drawback that they are unstable at alkaline pH (see Chapter 2). A number of media are based on cross-linked noncarbohydrate polymers. Mono Q and Mono S (Pharmacia Biotech) have a charged hydrophilic surface layer on a macroporous polystyrene polymer.[64] The matrix is, practically speaking, monodisperse with a bead size of 10 μm and is stable to aqueous solutions in the pH range of 2 to 12. This stability enables them to be cleaned with NaOH (2 *M*) to remove lipids and aggregated proteins that otherwise shorten the life of high-performance columns. TSK ion exchangers, including DEAE-5PW and SP-5PW (particle size 5 μm), are produced by TosoHaas and are distributed by several suppliers such as Bio-Rad and Merck. They have ion-exchange groups on hydroxylated and therefore hydrophilic vinyl polymers. They arc stable between pH 2 to 12 and can be regenerated by washing with 0.1 to 0.5 *M* NaOH.[65] SOURCE Q and SOURCE S (Pharmacia Biotech) are also monodisperse with respect to bead size, but are somewhat larger (15 or 30 μm). The larger bead size allows them to be run at lower operating pressures, a factor that is important in designing large-scale processes.

Most high-performance gels are porous, with more than 95% of the surface inside the particles. Their large pore size (>30 nm) allows protein to enter. A pore diameter of 30 to 50 nm is recommended for proteins in the molecular weight range of 30,000 to 100,000. For even larger proteins the pore size should increase to 80 to 100 nm.[66,67]

The large number of HPLC ion exchangers available may be confusing when selecting the chromatographic support, but the choice may not be critical, as it has been shown that organic and silica-based stationary phases with the same ionic groups, porosity, and particle size have about the same chromatographic behavior.[67–69]

The high-performance gels allow considerably increased resolution as compared to conventional gels. Drawbacks include a higher cost and smaller scale. In practice, the HPLC gels are not used for crude extracts because of their limited capacity and the risk of fouling the column. Removal of bulk protein is therefore recommended before HPLC is introduced in the purification. Manufacturers suggest a practical loading up to 25 mg/ml bed volume for porous HPLC ion exchangers. However, at least in some cases, this value is considerably lower. Normal column volumes are in the range of 1 to 10 ml. Sample volumes can be large if the binding is sufficiently strong. Of course if only minor quantities of protein are at hand, the small scale is rather an advantage. The small scale is also sufficient for the analysis of proteins. Amino acid analysis requires 30 μg or less, and for determination of the N-terminal sequence, often the first 40 amino acids are determined directly with a gas-phase sequenator when 1 nmol or less is taken.

A large number of other ion exchangers for HPLC and conventional chromatography based on organic polymers are available, such as those based on polyacrylamide (Bio-Rad), Trisacryl, a polymer of N-tris(hydroxymethyl)-methylacrylamide (BioSepra), Separon, and hydroxymethylmethacrylate (Tessek).

For more information, see reviews by Boschetti[51] and Freiser and Gooding (review on silica-based HPLC ion exchangers).[70]

4.4.3.6 Nonporous Ion Exchangers

Nonporous ion exchangers, where all binding occurs at the surface of the particles, have been developed. The partioning process is speeded up due to the small diffusion distances. A higher resolution has been obtained five to six times faster than on porous ion exchangers. The recovery of small amounts of proteins (μg or less) is high. The surface area is small, although the particles are small (<3 μm). The maximum sample load is small, only 5 to 10% of that of porous ion exchangers. They are most suitable for fast analytical applications and micropreparative work. For more information, see Hashimoto.[70]

4.4.3.7 Perfusion Chromatography Ion Exchangers

Adsorption at the surface of a porous chromatographic support requires that solutes diffuse across an immobile film of liquid at the other surface of the adsorbent and then through a stagnant pool of liquid in the pores of the matrix. The diffusion into and out of the pores is a major cause of zone spreading and is more serious with proteins and other macromolecules. The rate at which the solute transport occurs determines both the flow rate at which a column can be operated and the zone spreading. The use of small particles diminishes the problem of mass transfer through the stagnant liquid pool.

One solution to this problem is to use adsorbents without pores. Analytical separations have been achieved on nonporous supports, but because of their small loading capacity, these columns are not suitable for preparative work.[71] Regnier and co-workers have used particles with 600- to 800-nm pores interconnected with smaller 50- to 150-nm pores. The solutes are transported by the flow through the large pores, and the diffusion distances through the small pores are short. The protein-binding capacity is less than that of conventional HPLC ion exchangers. However, they can be operated at 5 to 10 times higher flow rates than conventional HPLC ion exchangers without loss of resolution. The technique is called perfusion chromatography because the solute transport is partially by convection.[72,73] These ion exchangers are available from Perseptive Biosystems under the trade name POROS.

4.4.3.8 Continuous Bed Ion Exchangers

Ion exchangers in which the matrix is a continuous bed, rather than a bed packed with particulate material, have been prepared by copolymerization of N,N'-methylenebisacrylamide and acrylic acid (cation exchanger) or N-allyldimethylamine (anion exchanger) in the presence of a catalyst and high salt. The peak width was found to be constant, or even decreasing, with an increase in flow rate (i.e., resolution can increase with flow rate).[74-77]

4.4.3.9 Other Ion Exchangers

Ion exchangers with dipolar ligands, such as β-alanine, sulfanilic acid, and arginine, have been prepared by Porath and co-workers.[78,79] The electric field potential decreases faster around dipoles than around unit charges, which should facilitate desorption. Scopes[19] has suggested that dipolar ion exchangers should be useful for affinity elution.[19] Both anionic and cationic groups of the protein would simultaneously bind to the ion exchanger, and both anions and cations of the buffer would compete for the protein-binding sites of the adsorbent. A protein molecule in the vicinity of a charged dipole should orient itself differently as compared to one close to a single charge and therefore bind differently. A dipolar ion exchanger should not exploit differences in charge distribution in the same way as ordinary ion exchangers.

Liposomes immobilized in the pores of carrier gels offer a possibility for studying, for instance, biological transport processes. Immobilized liposomes containing anionic or cationic phospholipids have been used for ion-exchange chromatography. For instance, separation of ribonuclease A, cytochrome c, and lysozyme on phosphatidylserine liposomes was better than on Mono S, and a lower ionic strength was needed to elute them from the liposome ion exchanger.[80]

4.4.3.10 Nonionic Interactions on Ion Exchangers

Nonionic interactions of proteins with ion exchangers with hydrophilic matrices are normally low. However, under more extreme conditions of low pH and very low ionic strength, proteins can adsorb very strongly. Adsorption at pH 4 and below has already been mentioned (Section 4.4.1). Adsorption to cellulose ion exchangers from concentrated ammonium sulfate ($3\ M$)[81,82] and to a silica-based HPLC anion exchanger (Synchropak AX 300) at an ionic strength higher than $1\ M$ has been observed.[83] This might depend on both increased hydrogen bonding[82] and increased hydrophobic interactions.[83]

Even at normal conditions at neutral pH, separation can depend on nonionic interactions, such as hydrogen bonding to Bio-Rex 70 (see Section 4.5.3), and hydrophobic interactions with the HPLC ion-exchanger Bio-Gel SP (see Section 4.9.5).

4.5 THE MOBILE PHASE: BUFFERS AND SALTS

4.5.1 Buffer pH and Concentration

Normally the concentration of buffer salts during protein adsorption is low, around 0.01 to 0.05 M. Proteins adsorbed at a salt concentration far below the desorption concentration can be difficult to desorb and some denaturation can occur. A suitable adsorption concentration can be determined by simple test tube experiments (see Section 4.6.2). The residence time (i.e., the time a protein is bound to an adsorbent) can also be important. Proteins adsorbed to a column overnight can be more difficult to elute than proteins desorbed shortly after adsorption. During a prolonged stay on the column, proteins slowly change their conformation and bind more strongly to the ion exchanger (S. Hjertén, personal communication).

Because the concentration of the buffer during adsorption is low, Peterson and Sober[84] suggested that a buffering species having a pK in the vicinity of the starting pH is needed to provide sufficient buffer capacity to eliminate pH disturbances that might result from interaction with an incompletely titrated adsorbent or from the adsorption process itself. If not controlled, these may impair the resolution in the early part of the chromatogram or give rise to misleading peaks.

The maximum buffering capacity is at the pK_a, and at 1 unit away the capacity has decreased fivefold, as 90% of the protolytic system is in either the base or the acid form. Buffers are usually chosen after two criteria: (1) the buffer should have a high capacity, preferably with the pK_a of the buffering species less than 0.5 units from the working pH; and (2) the buffering species should not interact with the ion exchanger.

For an anion exchanger, a positive buffering ion, such as Tris (pK_a 8.2), is often used and usually with Cl⁻ as the counterion. For a cation exchanger, a negatively charged buffering ion is recommended, for example, phosphate, carbonate, acetate, or morpholinoethane sulfonate (MES), and the counterions are mostly Na⁺ or K⁺.[52,56] Examples of buffers for ion exchange selected according to these criteria are found in Refs. 19 and 101.

The pK_a varies with temperature,[19,101] for example, dpK_a/dT for Tris is −0.028 per degree. A Tris buffer pH adjusted at room temperature has about 0.05 units lower pH in a cold room (+4°C). The corresponding pH change for a phosphate buffer is 10 times smaller, $dpK_a/dT = -0.0028$. Sodium phosphates (Na_3PO_4 and Na_2HPO_4) have a lower solubility at 4°C than the corresponding potassium salts. Concentrated (>0.2 M) phosphate buffers (pH > 7) used in a cold room should be prepared as potassium salts to avoid crystallization. If the second buffer to be used for gradient elution is prepared by the addition of a neutral salt, such as NaCl, often to a concentration of 1 M, the pH will change because of the large increase in ionic strength.

It is often advantageous to use a volatile buffer (Table 4-6) that allows direct lyophilization of the pooled fractions after chromatography. In particular, small amounts of protein are likely to give bad recoveries in desalting or dialysis; these losses can be avoided by using volatile buffers. However, lyophilization often induces reversible changes in the protein structure.[85] Freeze-drying is therefore avoided before trying to crystallize a protein.

A further consideration when choosing buffers is the possibility of exploiting the special effects of some buffer salts (see below). The possibility that a particular buffer species will interfere with subsequent studies should also be

TABLE 4-6. Volatile Buffers[101]

Buffer System (Base–Acid)	pK_a Values	Special Properties
Formic acid	3.75	
Pyridine/formic acid	3.75; 5.25	Smell, $\varepsilon_{280} = 3.5$
Trimethylamine/formic acid	3.75; 9.25	
Pyridine/acetic acid	4.76; 5.25	Smell, $\varepsilon_{280} = 3.5$
Trimethylamine/HCl	9.26	
N-Ethylmorpholine/acetic acid	4.76; 7.8	
Ammonia/formic acid	3.75; 9.25	
Ammonia/acetic acid	4.76; 9.25	
Trimethylamine/carbonate	6.50; 9.25	
Ammonium bicarbonate	6.50; 9.26	
Ethanolamine/HCl	9.5	

taken into account. For example, low concentrations of Tris, which is nonvolatile, will esterify aspartic acid and glutamic acid during acid hydrolysis for amino acid analysis (D. Eaker, unpublished). From buffers containing histidine or glycine, trace amounts (nanomoles) of these amino acids may remain after desalting and interfere with subsequent amino acid analysis.

The above rules for selecting buffers are not mandatory. The question is to what extent do pH disturbances occur if the rules are not followed and to what extent do they deteriorate the resolution. Examples of minor pH changes in the initial part of the chromatogram and the apparently small effect on resolution are shown in Ref. 51. Ammonium bicarbonate at pH 7.9[86,87] and ammonium acetate at neutral pH have frequently been used for the isolation of neurotoxins and neurotoxin derivatives by ion-exchange chromatography on Bio-Rex 70 or Amberlite IRC-50 (Section 4.9.4). Both buffers have their minimum buffer capacity at respective pH values, but the resin Bio-Rex 70 is a strong buffer and the pH of the effluent is determined by the equilibrium conditions of the resin and by the ionic strength of the buffer. However, ammonium acetate at neutral pH has also been used frequently on ion exchangers, such as SP Sephadex C-25[88,89] and BioGel TSK SP,[89–94] that do not have high buffering capacity. Phosphate buffers have also been used for chromatography on anion exchangers, such as DEAE, although the buffering ions bind to the ion exchanger.

Some experiments were carried out to study the pH changes that occur during a linear gradient of various buffers. In each case, the column was eluted with 2 bed volumes of the concentrated buffer and then with the diluted buffer until the equilibrium pH was reached, which required more than 5 bed volumes.

After elution with the final buffer, ion exchangers require between 1 and 6 bed volumes of the staring buffer before they are reequilibrated.[51] After elution with 10 mM Tris-HCl and 1 M NaCl (pH 7.3), DEAE Sephacel had to be eluted with 2.3 bed volumes and DEAE Sepharose CL 6B with 4.8 bed volumes of 10 mM Tris-HCl (pH 7.3) for reequilibration. A more complete list is given by Boschetti.[51]

At the end of a gradient the column is usually eluted with the second buffer and then with the first buffer until equilibration is reached. Therefore, SP Sephadex C-25 or Bio-Gel TSK SP was eluted with 1.0 M ammonium acetate (pH 6.8) and then with 10 mM ammonium acetate (pH 6.8). At the break-through of the 10 mM buffer, the pH of the effluent was 7.5, and when it had dropped to 6.8, a gradient of 10 mM buffer versus 1.0 M buffer was started. In the first part of the gradient there was no or a small pH change, a slow drop of about 0.2 units, and a slow rise back to pH 6.8. Similarly, no or small pH changes were also observed on the two cation exchangers with a gradient of 10 mM Na-phosphate buffer (pH 6.8) versus 10 mM Na-phosphate and 1 M NaCl (pH 6.8). For isolation of muscarinic toxins (bind to muscarinic acetylcholine receptors), a column of SP Sephadex C-25 equilibrated with 10 mM ammonium acetate (pH 6.9) was used.[88,89] No pH change was observed when the sample was applied from the same buffer or during elution with a gradient of 0.50 M ammonium acetate (pH 6.9). The counterions ammonium and acetate were released at protein adsorption and the concentration was transiently increased without any pH change.

With a gradient of 10 mM Na-phosphate buffer (pH 7.6) versus 10 mM Na-phosphate and 1 M NaCl (pH 7.6) on a column of TSK-DEAE, similar small pH changes (an increase with about 0.2 units and a drop back to 7.6) were observed. However, if cation exchanger SP Sephadex C-25 was eluted with 10 mM Tris-HCl and 1 M NaCl (pH 7.3) and then with 10 mM Tris-HCl (pH 7.3), the pH of the effluent was 9.1 at the breakthrough and was still 8.8 after more than 10 bed volumes had passed through the column.

4.5.2   Effect of Ions on Chromatographic Behavior of Proteins

A nonbuffering salt, such as NaCl, is usually added to the buffer to elute proteins from an ion exchanger. The properties of these nonbuffering salts also influence the separation. It is possible to influence both resolution and selectivity (order of elution) by a change in the nonbuffering salts. A change in ionic composition is not likely to affect all proteins in the same way. The outcome is difficult to predict, except in affinity elution (see Section 4.7.3.2), and one is left to trial and error to arrive at a proper result.

The effect of ions on the chromatographic behavior of proteins has been studied extensively.[15,16,40,41,47,54,95–106]

Only a few examples are mentioned here. Polyvalent anions are better displacers (less retention) of small molecules from an anion exchanger than monovalent ions (compared at the same ionic strength). It is also often true for proteins, but there are also exceptions, for example, citrate was a better displacer at pH 8 of ovalbumin from an anion exchanger than chloride, but not of soybean trypsin inhibitor.[96,100] The retention of small molecules on a cation exchanger is in the order $Ba^{2+} < Ca^{2+} < Mg^{2+} < NH_4^+ < K^+ < Na^+ < Li^+$. The effect of these ions on retention of many proteins follows the same order, but there are also exceptions (e.g., NH_4^+ is a better displacer than Mg^{2+} of lysozyme from a cation exchanger).[99]

Not only displacing ions change the retention, but also ions that do not compete with the protein for the charged groups of the ion exchanger. A classical example is ribonuclease A, which elutes much earlier from a column of Amberlite IRC-50 in Na-phosphate buffer than in NaCl because of the binding of phosphate to a histidine residue in the active site.[107,108]

The effect of anions on the retention of lysozyme, chymotrypsinogen A, α-chymotrypsin, and cytochrome c on a cation exchanger was in the order MOPS < acetate < chloride < sulfate < phosphate, anion sodium.[99]

It can be advantageous to use a strong displacing salt for elution of a protein that binds strongly to an ion exchanger; ferritin was eluted less retarded and in a higher yield from an anion exchanger at pH 8 when $MgCl_2$ was used salt instead of NaCl.[54]

Both displacing and nondisplacing ions affect selectivity and resolution. Proteins were eluted from the anion exchanger Mono Q with gradients of NaCl, NaBr, NaI, and Na-acetate. For instance, α-lactalbumin eluted ahead of ovalbumin in NaCl and NaBr gradients, they eluted together with NaI, and in reversed order with Na-acetate.[95,101]

A mixture of five proteins were separated on a cation exchanger (SynChropak CM 300, pH 7) in $BaCl_2$, but not in NaCl or KCl.[99]

Yao and Hjertén used the anion exchanger QAE-agarose in Tris buffer pH 8 for the separation of various proteins. Carboxylate ions, particularly acetate, gave a better resolution than chloride.[47] Gooding and Schmuck[99] also observed that acetate gave a better resolution than chloride. They used a cation exchanger (SynChropak CM 300, pH 7).[99] Although neither chloride nor acetate bind to a cation exchanger, they affected the chromatography.

The choice of buffering ions can also influence the separation. In several instances, ammonium acetate buffers have increased the resolution. In addition, it has the advantage of being volatile and can be removed by lyophilization (see Section 4.7.3). A fresh solution of ammonium acetate has a pH of 7.0 but it loses ammonia and the pH drops to 6.7–6.8. Ammonium acetate has a very low buffering capacity at neutral pH, but when used with an ion exchanger that by itself has a high buffering capacity, such as Bio-Rex 70, ammonium acetate may very well be used (see Section 4.9.4). The pH will then be determined by the equilibrium conditions of the resin and the ionic strength of the eluent. However, ammonium acetate buffers at neutral pH have also been used successfully on cation exchangers with a low buffering capacity (see Section 4.5.1, Section 4.9.5, and Refs. 88–94). There are several examples of the successful use of ammonium acetate buffers for obtaining separations that otherwise may not have been possible by IEC. Two isotoxins from the sea snake *Enhydrina schistosa,* differing only by a Pro/Ile mutation, could be separated on the cation exchanger Bio-Rex 70 using ammonium acetate, but not using a Na-phosphate buffer.[109] The Ser mutant was more retarded, probably due to hydrogen bonding to the resin. Other examples of isotoxins being separated because of differences in hydrogen bonding are toxins differing only by an Ile/Ser mutation (toxins from the cobra *Naja naja*)[110] and a Tyr/Asn mutation (toxins from the green mamba *Dendroaspis angusticeps*).[88] They were separated on Bio-Rex 70 using ammonium acetate buffers. In each case the isotoxin having an amino acid with a side chain able to form more hydrogen bonds was retarded. The Ser mutant eluted after the Ile mutant and the Asn mutant (fasciculin 2)[111] after the Tyr mutant (fasciculin 1).[111]

The difference in water-structuring or, inversely, chaotropic properties of ions and their effect on the protein structure (see Chapter 7) may be one explanation for some of the effects. Cosmotropes (water-structure makers) bind water strongly, minimize the surface area, and stabilize proteins, whereas chaotropic ions (water-structure breakers) bind water weakly, maximize the surface area, and destabilize proteins.[30,31,112,113] The chaotropic effect of anions increases in the order[30,114] $SO_4^{2-} \sim PO_4^{3-} < HPO_4^{2-} < CH_3COO^- < HCO_3^- < Cl^- < Br^- < ClO^- < I^- < SCN^-$, whereas cations increase in the order $NH_4^+ < K^+ < Na^+ < Li^+ < Mg^{2+} < Ca^{2+} < Ba^{2+}$.

The effect of anions is greater than that of cations. Chloride has little effect on the water structure, in the range of 0.1 to 0.7 M, or on protein stability.[31] Anions before chloride are cosmotropes and those after chloride are chaotropes.

Ions that bind to proteins, for instance, phosphate to ribonuclease A, will change the chromatography. Divalent ions, such as Ca^{2+} and Mg^{2+}, may interact with proteins by complexing with amide moieties.[115,116] $CaCl_2$ also has a destabilizing effect on proteins.[112,117] The effect of different combinations of cosmotropic and chaotropic ions on Z values has been studied by Hearn and co-workers.[105]

4.5.3   Additives

Additional compounds are often included. The additives used are mostly those needed to increase solubility. Nonionic and zwitterionic detergents are used for the solubilization of membrane proteins[100,118] and organic solvents, such as chloroform and methanol, increase the solubility of hydrophobic proteins.[100] Organic solvents decrease the dielectric constant, which is about 80 for water and about 20 for many organic solvents. This increases the ionic forces, as is evident from Eq. (4-1). The strong adsorption of proteins to polystyrene resins might be partly due to a decreased dielectric constant in the hydrophobic microenvironment of the charged groups on the matrix. Zwitterions such as betaine and taurine decrease aggregate formation and strong binding to ion exchangers, thus increasing resolution.[101] Urea and chaotropic ions are used to increase solubility. A neutral polymer, such as polyethylene glycol (PEG), competes with protein for water molecules. This will lead to increased interaction with the ion exchanger and hence increased resolution. In some cases,

PEG improved the resolution.[119] Enzyme inhibitors are often added (see Section 4.9.8). Most common are inhibitors of proteolytic enzymes, for instance, when working with membrane proteins. Solubilization by detergents also releases large amounts of proteases. Examples of inhibitors include phenylmethylsulfonyl fluoride of serine proteases, iodoacetamide of thiol proteases, and metal chelators of metallo enzymes.

4.6   EXPERIMENTAL PLANNING AND PREPARATION

4.6.1   Choice of Mode

Ion-exchange chromatography may be carried out in a variety of ways according to the properties of the sample and the objective of the separation. Its high capacity and ability to adsorb and concentrate proteins from dilute solutions make it particularly useful in early stages of a separation scheme when both the sample volume and the protein mass are large. Batch adsorption in combination with stepwise elution is particularly appropriate as a preliminary step in order to reduce the sample volume and the amount of bulk protein to quantities, which are more easily handled in column chromatography.

Column chromatographic techniques are required if advantage is to be taken of the high-resolution capabilities of ion IEC. Special small particle size media, as used in high-performance chromatography, are necessary for the highest resolution, usually after a number of earlier fractionation steps. It is often inadvisable to apply crude extracts directly to high-performance columns as the risk of column blockage is then relatively high. Excellent results are obtained with modern standard media if proper thought is applied to experimental design, although in some situations, HPLC ion exchangers have definite advantages. These include analytical and other small-scale applications and cases where speed is essential, as when proteins are rapidly degraded.

4.6.2   Choice of Ion Exchanger and Buffer

Since the net charge of a protein, positive or negative, depends on the pH, it is clear that the choice of both the ion exchanger, cation or anion, and the buffer will be governed by the pH. The consequences of this choice can been seen from titration curves (e.g., Figure 4-2).

In general, it is best to choose conditions where the protein of interest is adsorbed to the ion exchanger. Electrophoretical titration curves (Figure 4-2) would provide an excellent guide for choosing both the ion exchanger and the pH. However, they are rather complicated and are not used often. The choice of ion exchanger and pH can be made by simple test tube experiments to see which ion exchanger adsorbs the protein of interest at the chosen pH, of course at a pH where the protein is stable. The ion exchanger (1 to 1.5 ml) equilibrated with the starting buffer and the sample in the same buffer are mixed for 1 to 2 min. After centrifugation, the supernatant is assayed for the protein.[120] If the protein is only partially adsorbed, the conditions can be used for isocratic elution, and the retention coeffcient can be calculated from the distribution coefficient (see Chapter 2). Desorption conditions can be examined in the same way.

Another simple way is to dissolve the sample in a dilute buffer with a slightly acid (alkaline) pH of about 5 (8) and apply it on a cation exchanger (anion exchanger) equilibrated with the same buffer. If the proteins of interest adsorb to one of the ion exchangers, they may be desorbed with a salt gradient (see Section 4.3.2).

Thus, it is possible to find conditions at which the protein of interest adsorbs. The column is then eluted with a buffer sufficiently concentrated to desorb the protein; 1 M will suffice in most cases. From this first experiment, it is possible to see if the slope of the gradient needs to be changed to improve resolution. This can be done by changing either the gradient volume or the salt concentration of the displacing buffer. pH gradients are also used (see Section 4.7.2).

With sensitive proteins the ionic strength to be used should be close to the ionic strength needed for desorption. Too strongly adsorbed proteins can be difficult to desorb or be inactivated. Several peaks arising from denaturation can be obtained. The recovery should be checked. A prolonged residence time on the column can also make proteins more difficult to elute (see Section 4.6.1).

Alternatively, by using starting conditions so that the protein of interest passes unretarded through the column while many other proteins are adsorbed, a high degree of purification can be obtained in the first ion-exchange step. This method has been used, for instance, in the purification of muscarinic toxins from mamba venoms.[88-90]

It is of course possible to use both anion and cation exchangers. The use of a mixture of the two, or two columns in series, is an established technique for deionizing water. This technique has found limited application in protein purification,[121-123] but, for example, when bovine serum was adsorbed to a mixed bed ion exchanger at 10 mM phosphate buffer, pH 7.5, IgG1 (pI of 7.5) was eluted 95% pure, whereas the other proteins followed in a linear salt gradient.[123] Normally, two consecutive chromatographies should be run instead.

There is generally little to be gained from choosing a weak as opposed to a strong ion exchanger. As mentioned earlier, the significant difference between them is the pH range over which they retain their full charge. Possible specific effects of choosing, for example, a DEAE-anion exchanger as opposed to a strong quaternary ion exchanger are serendipitous and can hardly be predicted beforehand, although a case where a strong ion exchanger was advantageous has been reported.[54,67] If extreme pH values are to be used, strong ion exchangers are essential. Thus at pH values below 5, cation exchangers to be fully ionized should be sulfonated or phosphorylated, whereas at pH values above 9, anion exchangers containing quaternary amines should be used.

The porosity of the matrix influences the choice of ion exchanger. Larger protein molecules require media with larger pore sizes. The largest pores are available in the cellulose and agarose (Sepharose) ion exchangers. Exclusion limits are normally given by the manufacturer. Ion exchangers based on Sephadex have smaller pore sizes and may be advantageous for smaller proteins. When fractionating low molecular weight proteins from snake venom, changing from SP Sephadex C-25 (exclusion limit 30,000) to SP Sephadex C-50 (exclusion limit 200,000) gave practically identical results. If the porosity is large enough, a further increase will not make a difference. The final choice among adsorbents with similar porosity, capacity, and hydrophilicity is probably often arbitrary, both with conventional and HPLC ion exchangers.[67,69]

In some cases, mechanisms other than ion exchange influence the separation. It has already been mentioned that methacrylate ion exchangers Amberlite IRC-50 and Bio-Rex 70 have a high resolving power for the separation of mixtures of small basic proteins, such as ribonuclease[8,107,108] and toxins.[25,88,89,109,110,124–126] Here hydrogen bonding has in some cases been shown to be important for the separation (see Section 4.5.2). The retardation on Bio-Gel TSK SP of a toxin derivative with a tyrosine residue modified with a hydrophobic reagent was most likely due to hydrophobic interactions (see Section 4.9.5). Similar interactions should also occur with other proteins, but snake toxins are small molecules (MW 7,000 to 8,000) and some of them differ only by a single amino acid mutation. Differences in their chromatographic behavior are therefore easier to explain than those observed with proteins differing more from each other.

Phosphocellulose shows specific interactions with enzymes binding phosphate-containing substrates. Purifications of aminoacyl-tRNA synthetases,[127] hexokinase,[128] phosphorylase, phosphoglucose isomerase, and creatine kinase[129] by adsorption to phosphocellulose and affinity elution have been reported.

If the sample size is large and fouling of the chromatographic column is expected at an early stage of the separation, a simple and inexpensive ion-exchange material that can be discarded after use is recommended. This will not allow the very fast and high-resolution separations that can be obtained with HPLC material, but these can be used later.

Buffers are discussed in Section 4.5. Some practical aspects are mentioned here. In HPLC, a blank gradient should be run before applying a sample to check for UV-absorbing material in the buffers. Phosphate can contain material that interferes with the monitoring in the lower UV range. The contaminants can be removed as described by Karkas et al.[130] Because halides can corrode stainless-steel pumps, systems with stainless-steel components should be washed thoroughly with water after use with salts that contain halides.

4.6.3 Column Dimensions

The amount of the ion exchanger can be calculated from the protein binding capacity. In practice, calculations are rarely made, as the amount of protein is often not well known. The samples often contain proteins of different molecular weights, which influence the protein-binding capacity. The columns are probably in general too large, and if the amount of the ion exchanger used is about 5 to 10 times higher than that calculated from the protein-binding capacity, this is hardly a serious disadvantage, except for economy. Thus phospholipase A2 from cobra venom adsorbed to CM-cellulose CM 52 and desorbed by gradient elution produced identical chromatograms in a 2- and 10-cm column. It has also been shown that column length has a relatively small effect on resolution. Columns 5 cm in length have approximately 75% of the resolving power of columns 5 times longer.[131,132]

If the ion exchange is carried out as a simple adsorption/desorption experiment, no more stationary phase is required than is needed to bind the protein(s) of interest. The sample will then not need to travel through a long column after desorption. Bed lengths that are not more than four to five times the bed diameter are recommended. Short columns are also preferred in gradient elution. Columns used for isocratic elution should be longer. Here several proteins migrate on the column simultaneously, and their resolution increases as the square root of the column length (Section 2.3.1.4).

4.6.4 Equilibration of Ion Exchanger

An ion exchanger will always have counterions at pH values where it is charged. The purpose of the equilibration procedure is to ensure that the ion exchanger is in equilibrium with the counterions.

If adsorbed material needs to be removed from the ion exchanger, follow the procedure for regeneration given by the manufacturer (see also Section 4.7.4). Note that DEAE-ion exchangers are unstable in the free base form and should be converted to the salt form if the adsorbent has been washed with NaOH, for instance, by washing with 1 M sodium acetate (pH 3, adjusted with HCl) (see Section 4.7.4).

If equilibration is done on a Büchner or glass filter, wash with the starting buffer, alternatively with the final buffer if it has a higher buffering capacity than the first buffer, for example, if the buffers are 10 mM and 0.5 M sodium phosphate buffers. Wash finally with the starting buffer until the pH and conductivity of the effluent are the same as that of the buffer. If equilibration is done in a beaker, add the starting buffer and adjust the pH of the slurry with acid or base. After settling and decantation, add a new portion of the buffer and adjust the pH if necessary.

The methacrylate ion exchangers should be equilibrated batchwise. If the resin is used with ammonium acetate, wash the resin with ammonium hydroxide and then neutralized with acetic acid. After settling and decantation, suspend the resin in ammonium acetate of the desired concentration and adjust the pH with acetic acid or ammonium hydroxide. Repeat the procedure, usually once or twice, until the resin is equilibrated (i.e., until the pH does not change on the addition of a new portion of ammonium acetate). For isolation of neurotoxins, the resin is usually equilibrated with 0.20 *M* ammonium acetate and not with the starting buffer.

4.6.5 Preparation of Sample

The sample is usually equilibrated with the starting buffer. This is often done by dissolving it in the starting buffer and checking the pH. However, proteins are ions and the ionic strength is therefore higher than that of the starting buffer. To ensure binding the sample can be diluted.

As in all chromatography, it is important that the sample is clarified by centrifugation or filtration (see Chapter 1).

4.6.6 Large-Scale Ion Exchange

Ion-exchange chromatography is well suited to purification of either large quantities of proteins, because of its high loading capacity, or purification from large volumes of sample, because of its ability to concentrate proteins from dilute solution.[133] Although process scale ion-exchange chromatography is routinely carried out with columns as large as 170 L, even a standard laboratory column with a volume of 500 ml can be sufficient to concentrate and purify 5 g of protein from as much as 10 L of a dilute sample. In practical terms, large-scale IEC differs mainly in the increased use of step, as opposed to continuous gradient, elution procedures. Otherwise exactly the same general procedures and principles are used as in a smaller scale.

For obvious reasons, it is clearly desirable to be able to plan a large-scale procedure at a smaller scale. This is done by carrying out a series of trial runs under carefully controlled and systematically varied conditions of sample loading and elution to find the conditions that give the desired result in terms of resolution, yield, and time. Once this is done at a small scale, the conditions can be accurately extrapolated to the larger scale by application of a few simple scaling factors.[134]

When the conditions of binding and elution, [i.e., sample composition, ion exchanger, eluent compositions, pH, loading factor (mg protein/ml ion exchanger)] are kept constant, then the column dimensions, eluent volumes, and flow rate are scaled up as follows.

Column volume is increased as the increase in sample volume. It is advisable to obtain this increase in volume by choosing a column with a larger diameter to avoid excessively long separation times.

Eluent volumes are increased as the increase in sample volume. This applies equally to stepwise elution schemes, to continuous gradient elution, and to all steps in the elution sequence.

The linear flow rate is kept constant. The volume flow rate will thus be increased by the same factor as the increase in the cross-sectional area of the column. If the large column is longer or shorter than the small one, then the elution times are increased or decreased accordingly.

It is important to note that these factors only apply if the sample loading, elution conditions, and the ion exchanger are the same in the two scales. If the sample loading as mg protein/ml ion exchanger is changed in the scaleup, then the results will be more difficult to predict, particularly at the high-loading capacities that are frequently used in large-scale work. Changes in the elution scheme will have the same effects on the separation as they would have had without the change of scale. The need to use the same ion exchanger means that it is usually inadvisable to optimize the separation using high-performance ion exchangers if subsequent large-scale separations will use standard ion exchangers.

The choice of ion exchanger for large-scale work follows the same principles as for small-scale separations, with the added proviso that care should be taken to choose a gel matrix that will withstand high linear flow rates in large-volume columns. Ion exchangers that have been produced for process scale ion-exchange chromatography, for example, Q Sepharose Fast Flow and S Sepharose Fast Flow, are also suitable for column chromatography in a more usual laboratory scale. However, when batch adsorption is used to concentrate and simplify the sample (see below), there is no need for high physical stability, and simpler media, for example, DEAE-cellulose, will suffice.

Columns for volumes up to ca. 1 L can be of the conventional design. Because it is generally desirable to use short, wide columns for ion exchange, very large columns should have end pieces that are specially designed to spread the incoming liquid flow evenly over a relatively large bed area. The multiple inlet design[135] is suitable.

Removal of particulate matter is specially important when large sample volumes are applied. Polypropylene depth filters (1 μm) that give a negligible pressure drop over the filter are available for this purpose. In the case of large sample volumes, it may be difficult or impossible to ensure that the sample is exactly equilibrated to the starting conditions for elution. In this case the sample ionic strength should be adjusted by dilution if necessary and the pH adjusted to a value that will give good binding. The column is then equilibrated with the starting buffer and eluted in the usual way. The choice of buffer systems for large-scale work is naturally influenced by cost considerations, but the needs for adequate buffering capacity remain and should not be sacrificed for small cost savings.

4.6.7 Batchwise Procedures

Although batchwise procedures are not strictly speaking chromatographic, they are nonetheless well worth considering as an alternative to column operations, particularly in connection with large-scale applications. Batchwise adsorption and desorption are also very useful for crude or viscous samples independent of scale.

The sample is mixed with gentle stirring with the ion exchanger in conditions that promote binding of the components of interest. The time taken for binding depends on the sample and on the ratio of the ion exchanger volume to the sample volume and should be determined by activity measurements. As a rule of thumb, an hour is sufficient for protein solutions that behave normally. Unbound components are washed away on a filter funnel under the same conditions, and the bound proteins are then desorbed by a stepwise increase in salt concentration or by a change in pH. This can be done either on the funnel or after the ion exchanger has been packed in a column. If the volume of the eluting buffer is kept small, a considerable increase in concentration of the protein can be achieved, particularly if a column is used.

After recovering the protein, the ion exchanger may be regenerated and washed with distilled water for further use, providing it is not heavily contaminated with particulate or lipid material. If this is the case, it is often best simply to discard it. However, if the ion exchanger needs to be recovered, procedures for regeneration are given by the manufacturers (see also Section 4.7.4).

The choice of cation or anion exchanger is made so as to achieve binding of the desired components, but it is not necessary to use the most sophisticated matrices. Indeed, inexpensive bulk ion exchangers that can be added in the dry form, for example, those based on cellulose or Sephadex, are often an excellent choice. The volume changes associated with ion exchangers based on Sephadex G-50 are even an advantage, as the shrinkage of the ion exchanger at high ionic strength improves recovery. Note that some ion exchangers will generate fines if they are stirred vigorously, thus increasing the time it takes for filtration.

4.7 CHROMATOGRAPHIC TECHNIQUES

4.7.1 Sample Application

In column procedures the sample solution is applied to the column by pumping. The proper flow rate for adsorption depends on several factors. In general it can be quite high if the protein of interest is far away from the desorption conditions in the starting buffer. However, for samples that are viscous, it is wise to reduce the flow rate during the application or to dilute the sample. For protein solutions more concentrated than about 30 mg/ml, this can be a concern, although only experience of the particular sample and experimental setup can give definite answers.

In general, flow rates should be expressed in the dimension length per time, most commonly centimeters per hour. Consequently, the maximal volume flow rate increases as the square of the column diameter, given that other factors are constant. Dextran ion exchangers can be run at about 15 cm/h, whereas high-performance media are typically run at about 300 cm/h. The rigid agarose and cellulose ion exchangers stand very high flow rates. For adsorption and elution, flow rates up to 50 cm/h are used. Much higher rates can be used for equilibration. Flow rates have to be decreased by about 50% when changing to cold room temperature due to increased viscosity. In general, flow rate is a parameter that should be checked if maximum resolution is of importance. Resolution is also lower in a cold room. Conversely, improved resolution will be obtained above room temperature.[100]

At this stage it is sometimes possible to see protein bands on the column with the naked eye. The banding is due to protein–protein displacement (see Chapter 2). Proteins that bind most strongly to the ion exchanger are found closest to the top of the column.

After application the column is washed with starting buffer. The proteins that do not adsorb are recovered in a breakthrough peak. Normally, washing continues until the UV absorption in the eluate decreases to a low value. However, no more than two column volumes should be necessary, and in the best case a single column volume suffices. The proteins that elute in the breakthrough peak have a retention of $k' = 1$. The retained proteins are then eluted by a buffer change.

4.7.2 Elution Techniques

The proteins that adsorb to the ion exchanger are eluted from the column by an increase in ionic strength, by inclusion of new ionic species, or by a change in pH. It is common practice to increase the concentration of a nonbuffering salt, such as NaCl, for elution. However, as already mentioned, exceptions occur, as when proteins are eluted from either anion or cation exchangers by an increase in the concentration of ammonium acetate or other volatile buffer or by increasing the concentration of a phosphate buffer.

If a pH change is used, anion exchangers should be eluted by a decrease in pH to make the adsorbed proteins less negative, whereas cation exchangers are eluted by an increase in pH to make the proteins less positively charged. pH gradients are less often used, but are not necessarily inferior to salt gradients. Mhatre and co-workers[136] used pH gradients to separate Fab fragments and other proteins with differences in isoelectric points as low as 0.1 (see also Section 4.9.3).

An alternative is to run two consecutive ion-exchange chromatographies at different pH values. Two proteins coeluting at one pH are not likely to coelute at another pH.

The changes can be brought about either step by step, by stepwise elution, or continuously, by gradient elution. If the chromatography is being run with constant solvent composition and the proteins leave the column more or less retarded, it is called isocratic elution. If a protein is being displaced due to its specific interaction with an added ionic species in the elution buffer, this is called affinity elution. Finally it is possible to add a general eluting agent to the buffer that displaces all proteins, which will move ahead of the displacer in the column. This is called displacement chromatography (see Section 4.7.2.4).

To make the best use of the resolving power of ion exchange, a protein should be submitted to chromatography in such a way that it elutes retarded from the column. This normally occurs in gradient elution. Only desorption/adsorption occurs if a protein elutes in a breakthrough peak in stepwise elution. In such a case, the protein did not interact with any ion-exchange groups during elution, but was only filtered through the column.

4.7.2.1 *Isocratic and Stepwise Elution*
Isocratic elution is used in cases where the sample and its properties are well known and the same kind of sample is run repeatedly, (i.e., used mostly for analytical separations). To achieve proper reproducibility, great care has to be taken to ensure that the column is properly equilibrated and that the sample ion composition does not vary. A well-known example of isocratic elution in an analytical context is the Moore and Stein method for amino acid analysis.

In isocratic elution, all components move simultaneously. Only sample volumes much smaller than the total bed volume can be applied. Separation power increases with the square root of the column length (see Chapter 2). The peaks widen later in the chromatogram and finally become so wide that they hardly can be detected. After several bed volumes of elution buffer—the practical upper limit is 3 to 10 bed volumes—the chromatography has to be interrupted.

Stepwise elution is a serial application of several isocratic elutions. At each step, only about a single total bed volume of eluent is passed through the column. Stepwise elution is often used for recovering a protein in a concentrated form in the breakthrough peak of the displacing buffer. Here it is best if the elution conditions for the desired component are known. The application conditions are then chosen so that the desired protein stays bound to the column while contaminants, often bulk proteins, pass through. After washing, the salt concentration or pH is changed so that the protein of interest is eluted in a breakthrough peak (see Section 4.9.7).

Stepwise elution can give a few disagreeable surprises, some of which are illustrated in Figure 4-5. A retarded peak is eluted with one buffer. The breakthrough peak of the following buffer may contain the same protein. In fact, if the capacity of the column was inadequate, the second component might also be found in the first breakthrough peak (also true for gradient elution).

For each step that produces a breakthrough peak, the resolving power might not be used optimally, with several components eluting together. Only peaks eluting after the breakthrough of the displacing buffer indicate chromatographic homogeneity.

Peaks eluting at the breakthrough of a displacing buffer have a typical shape: a sharp front and a tailing end (see 2 and 6 in Figure 4-5A). It is usually more difficult to choose the right conditions for step-wise elution. If the buffer is too weak, the protein might elute in a broad retarded peak, whereas if it is too strong it might elute together with several other components. In general,

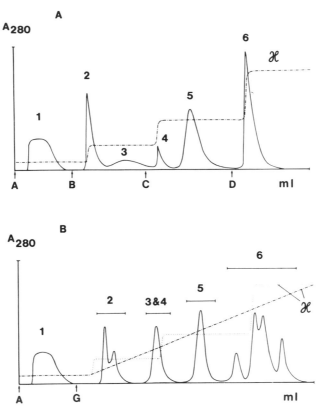

FIGURE 4-5. Elution of an ion-exchange column (A) stepwise with four buffers of increasing ionic strength (broken line) and (B) with a gradient. The gradient starts at G. The figure illustrates possible drawbacks of stepwise elution: poor resolution of several components (2 and 6) because too high an ionic strength was used and the false separation of one component into two peaks (3 and 4).

one should avoid stepwise elution with an unknown sample and instead use gradients.

4.7.2.2 Gradient Elution In gradient elution the concentration (more rarely pH) of the eluting buffer is changed continuously. At one concentration the least strongly adsorbed protein is desorbed, at somewhat higher concentration the second protein, and so on. The gradient elution has the separation obtained in isocratic elution, but because of the continuously increasing elution power the peaks do not become much broader as the gradient develops. Tailing due to nonlinear adsorption isotherms is also diminished. Gradients are the classical answer to the "general elution problem" (see Chapter 2). Normally, a single substance is not recovered artifactually in several peaks (for an exception, see Section 4.3.4) and a maximal use of the resolving capacity of the column is obtained.

Gradients are obtained by mixing starting and final buffer so that the percentage final buffer pumped into the column is gradually increased. Modern chromatographic pumps are equipped with a gradient device that can be programmed to give any conceivable gradient. Simpler gradient mixers can easily be constructed by connecting two cylinders by a tubing. One cylinder, equipped with a stirrer, contains the starting buffer of concentration C_s and the other cylinder contains the final buffer of concentration C_f. The eluting buffer is continuously withdrawn from the stirred chamber. The concentration of the eluent C_v after v ml from a total of V ml is described by the equation[137]

$$C_v = C_f - (C_f - C_s) \cdot (1\text{-}v/V)^{A_f/A_s} \tag{4-2}$$

Here A_f and A_s are the cross-sectional areas of the two cylinders. A linear gradient is obtained for $A_f/A_s = 1$, a concave for $A_f/A_s < 1$, and a convex for $A_f/A_s > 1$ (Figure 4-6). The total volume of a gradient might vary, but as a rule of thumb it can be made 10 times the column volume as a first try.

Linear gradients are recommended as a first choice. If the resolution is not satisfactory, the slope of the gradient is changed by changing either the gradient volume or the concentration of the final buffer. With more sophisticated gradient devices, as, for example, in HPLC, the gradient can be made more shallow in areas where the peaks are less well resolved. Similarly, concentration ranges without proteins eluting can be passed by a steep part of the gradient. However, one should not try to solve protein resolution problems only by gradient tinkering, but rather by carrying out the ion exchange at another pH (Figure 4-2 and Section 2.3).

4.7.2.3 *Affinity Elution*

In affinity elution,[16,19,55,127–129,138–141] adsorption of a protein is nonspecific but desorption is specific. In ion exchange the specific eluent is normally a substrate ion, and the name substrate elution is sometimes also used. The eluent has a charge of the same sign as the ion exchanger and binds specifically to the active site of the protein. An early example of affinity elution is the desorption of fructose 1,6-diphosphatase from a column of CM-cellulose with the negatively charged substrate 1,6-diphosphate.[16,42] The enzyme consisting of four subunits could bind four substrate molecules and thus decreased its net charge by 16 units, enough to elute it from the cation exchanger.

Scopes studied affinity elution of a number of enzymes.[16, 42, 104, 105] He found that on CM-cellulose the inclusion of a charged ligand had the same elution power as an increase in pH of 0.5 to 0.8 units at constant ionic strength. To avoid nonspecific elution, a "dummy ligand" with similar ionic properties was included in the buffer before the affinity ligand. Thus EDTA served as a dummy ligand for phosphoenolpyruvate. For a successful result, he found that

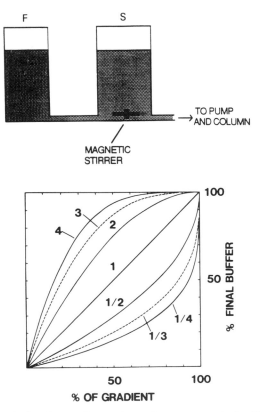

FIGURE 4-6. (Top) Simple gradient mixer of two connected cylindric vessels with cross-sectional areas A_S and A_f containing starting (S) and final (F) buffer, respectively. (Bottom) Shape of gradient obtained for different values of A_f/A_S. The concentration changes rapidly at the end of the concave gradients and becomes similar to stepwise elution.

the number of charges should be at least 4 per 100,000 of the molecular weight of the protein unless conformational changes facilitate desorption and that the ligand–protein dissociation constant should be below millimolar.

4.7.2.4 Displacement Chromatography In displacement chromatography the column is eluted with a substance, the displacer, with a high affinity for the adsorbent whereas other substances move at the same velocity through the column ahead of the displacer, forming rectangular zones that follow each other in a train. Displacement elution is not identical to isocratic elution, as the components do not displace each other, but move with different rates through the column. A difference between displacement chromatography and stepwise elution is that in the latter case one often searches for conditions where the protein of interest is eluted while other components stay behind.

Displacement analysis has rarely been used in ion exchange due to difficulties of finding suitable displacers. A few examples will, however, be mentioned.

In polyion displacement chromatography, a polyion with high affinity for the ion exchanger is used as displacer. With carboxymethyldextrans[142] and chondroitin sulfates[143] as displacers, a 100-mg sample of β-lactoglobulin was fractionated into two rectangular zones of the A and B forms after adsorption to a 3.5-ml column of TSK-DEAE equilibrated in sodium phosphate, pH 7.0. This method had a twofold higher capacity as compared to conventional ion exchange.

In ampholyte displacement chromatography, ampholytes intended for isoelectric focusing are used as displacers. Thus two forms of a β-N-acetyl-D-hexoseaminidase were separated by elution with a 4% solution of ampholytes, pH 8 to 10, from a column of CM-cellulose equilibrated in 0.02 M Tris-HCl, pH 7.6.[144] These forms of the enzymes were not resolved by either isoelectric focusing or ion-exchange chromatography. Because a pH gradient was obtained at the beginning of the chromatogram, it is perhaps likely that the separation depended on both chromatofocusing and displacement chromatography.[142]

A novel class of positively charged displacers based on pentaerythritol have been developed. They are dendritic polymers (i.e., the molecules branch out from a central atom with quaternary ammonium groups at the end of the branches). The outer previously mentioned displacers are linear polymers. Two of the dendritic polymers have molecular weights of about 500 and 1,600, whereas the linear polymers have molecular weights >2,000. Examples of the use of dendritic polymers as displacers from a cation exchanger and a discussion of displacement chromatography are found in Ref. 145.

4.7.3   Collection and Treatment of Sample

The pooling of fractions is best guided by measurement of the activity or another property of the protein of interest, but very often fractions are pooled after measuring the absorbance, usually at 280 nm, or after measuring the protein content by a colorimetric method, such as the Lowry method. Figure 4-7 illustrates the cross-contamination of two overlapping peaks of both equal and unequal size and shape. The underlying assumption is that absorbances reflect the amount of protein. This might be a reasonable approximation if absorption in the low UV region (<230 nm) is used, as the peptide bond is the main chromophore. At the 280-nm band, absorption coefficients can vary considerably. It is a good practice to run an absorption spectrum in the 200- to 400-nm region. Most proteins (those containing tryptophan and/or tyrosine) have an absorption maximum close to 280 nm and a minimum at 250 nm. If a "nontypical" protein spectrum is obtained, the sample is likely to contain nonproteinaceous substances.

Spectrophotometry is often used for estimating the protein content, but an accurate determination requires that the extinction coefficient is known, which is often not the case.

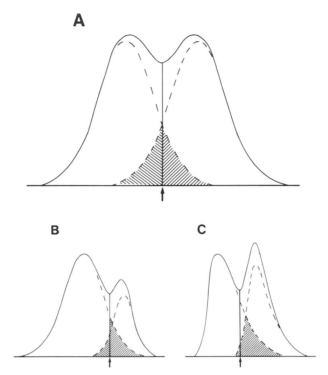

FIGURE 4-7. Estimation of cross-contamination in two overlapping peaks of equal or unequal size and shape. (A) Two equally large gaussian profiles. If pooled as indicated by the arrow, the cross-contamination is about 10%, much less than judged by the eye. The cross-contamination, C, can be found from measurement of the resolution R_s (see Chapter 2). Thus $R_s = 0.9$ gives $C = 4\%$; $R_S = 0.8$, $C = 6\%$; $R_S = 0.7$, $C = 9\%$; $R_S = 0.6$, $C = 13$; and $R_S = 0.5$, $C = 19\%$. (B) A smaller peak is more contaminated than a larger one. (C) A trailing peak (the first one) increases the contamination of the neighboring peak.

A protein solution of 1 mg/ml is often considered to have A_{280} (absorbance at 280 nm) = 1. Recalling that the absorbance in the vicinity of 280 nm depends on the content of tryptophan (molar absorptivity at neutral or acidic pH 5559 M^{-1} cm^{-1}), tyrosine (1260), and cystine (110), it is obvious that A_{280} (1 mg/ml) can vary much. These three amino acids account for about 90% of the total absorbance of a protein at 280 nm, the rest depends on the secondary and tertiary structure. However, this is not valid for proteins, such as hemoglobin, having a UV-absorbing cofactor.

Some empirical formulas give a good estimation of the protein content:

$$\text{Concentration (mg/ml)} = (A_{235} - A_{280})/2.51^{146}$$

$$\text{Concentration } (\mu\text{g/ml}) = (A_{215} - A_{225}) \times 144^{147,148}$$

$$A_{205} \text{ (1 mg/ml)} = 27.0 + 120 \times A_{280}/A_{205}^{19,149}$$

Buffers absorb differently at the two wavelengths used and this has to be corrected for. Most salts have a strong adsorption at 205 nm. Scopes suggested 5 mM phosphate buffer and 50 mM sodium sulfate (pH 7) to diminish the adsorption of proteins to cuvettes. A small aliquot is added, preferably to a 3-ml cuvette, and if the dilution is large (several hundredfold), the increase in absorbance because of the sample buffer can be neglected.[19,149]

The concentration of fractions is easily done by freeze-drying. This requires that a volatile buffer has been used or that a change is made to such a buffer (see Table 4-6). Ammonium acetate is suitable for lyophilization, but dilution to about 0.2 M is recommended. Ammonia is more volatile than acetic acid, and during freeze-drying the pH will drop. Some proteins can aggregate when freeze-dried from acetic acid and to avoid that the pH is adjusted to 7–8. It is often best to lyophilize twice, dissolving the sample in a small volume of 0.01 to 0.02 M ammonium acetate. Sonication is recommended to bring proteins adsorbed to the walls of vessels into solution. Lyophilization can produce reversible changes in structure (see Section 4.5.1).[85]

It is often worthwhile rerunning a protein that eluted in an asymmetric peak or close to another component. As illustrated in Figure 4-7 the cross-contamination of two not fully resolved peaks of equal size (A) is surprisingly low. A rechromatography of one of the peaks will remove most of the contaminating protein. If a contaminant is present in smaller amounts (B and C), it is much easier to remove it from the main component. A rerun also confirms that substance A is homogeneous with regard to the separation parameters used. This is easily demonstrated with radioactively labeled components. Sometimes earlier eluting peaks can be contaminated with material from those eluting later. For instance, the nonretarded peak 1 (Ref. 88) was contaminated with material from peaks 5 and 6 containing very potent inhibitors of acetylcholinesterase (fasciculins). When a protein starts eluting from an ion exchanger, it will displace small amounts of proteins still adsorbed to the column. If the contaminant is a very active substance, even small amounts will interfere with assays of the main component. Rechromatography will remove them.

It is more tricky when an originally homogeneous protein shows false heterogeneity in a chromatogram. This is not unusual in stepwise elution as described earlier in Section 4.7.2.1. However, it can also happen in well-conducted gradient chromatography for various reasons, for instance, when a protein complex dissociates on the column and components elute in different positions. This will very likely cause a loss of biological activity. Several presynaptic neurotoxins that inhibit the release of acetylcholine consist of subunits with very different charge properties held together by noncovalent forces. One of the subunits is a very basic protein, and if the toxin is adsorbed to a cation exchanger, this component will be bound strongly and will not elute together with the other subunits.[150–153]

Finally, some observations suggest that it is also possible to obtain multiple peaks of homogeneous proteins, especially when overloading the column. In such situations, rechromatography of a smaller aliquot should be performed (see Section 4.3.4). Experience also suggests that binding a protein to the column far below its desorption concentration may give rise to artifacts and several peaks arising from protein denaturation.

Freeze-drying frequently produces small amounts of aggregates which are detected by gel filtration,[110] and which may also be observed in IEC.

4.7.4   Regeneration, Cleaning, and Storage of Adsorbent

After each use, ion exchangers usually have some material bound, often denatured proteins. Depending on the samples, there may also be lipids. It is important to remove this material after the ion exchanger has been used for some time. Remaining material will block adsorption, impair performance, and may even contaminate the sample.

Regeneration procedures vary according to the stability of the matrix and the functional group. Manufacturers generally describe how their products should be handled, and only a few comments are given here. The agarose ion exchangers should be washed with high salt and then NaOH up to 0.5 M for a maximum of 12 h. The cellulose ion exchangers should be washed with high salt and 0.1 M NaOH. Exposure of dextran and agarose to pH below 3 should be avoided. They can be washed with 1 M sodium acetate (pH 3, adjusted HCl or other strong acid), 0.5 M NaOH, and 1 M sodium acetate (pH 3). Trisacryl adsorbents are, however, stable at pH 1 to 13. Lipoproteins and lipids can be removed by nonionic detergents or ethanol. HPLC media are regenerated in the column. Mono S and Mono Q can withstand high concentrations of alkali (2 M). Silica-based adsorbents are unstable to alkali, and pH above 8 should be avoided. Recommendations for the care of HPLC media have been published.[51,154,155]

Adsorbents that are to be stored for a long time are best transferred to 25% ethanol. HPLC media can likewise be stored in alcohol up to 100%.

4.8   HYDROXYAPATITE CHROMATOGRAPHY

Calcium phosphate had been used in protein adsorption applications with varying success for many decades when Tiselius and co-workers[156] showed that a much more useful adsorbent was obtained when the brushite, $Ca_3(PO_4)_2$, was changed into hydroxyapatite, $Ca_5(PO_4)_3OH$, by boiling in 1% NaOH for 1 hour. The hydroxyapatite crystals generated fines more easily, but these were removed by decantation before column packing.

In hydroxyapatite chromatography the column is normally equilibrated, and the sample is applied in a low concentration of phosphate buffer, pH 6.8, and the proteins are eluted in a concentration gradient of phosphate buffer. Sometimes quite shallow gradients (e.g., 0.02 to 0.05 M) give excellent results[157] whereas in other instances concentrations up to 0.35 M phosphate are used[158] (Section 4.9.8). The presence of other salts such as NaCl and $(NH_4)_2SO_4$ does not seem to affect the elution. This can be taken advantage of by applying a sample from an ion-exchange column eluted by a NaCl gradient directly to the apatite column. It has also been used for the purification of halophilic enzymes in the presence of 3.4 M NaCl.[159] Some other anions (e.g., citrate) can be used for the elution of apatite columns.

Hydroxyapatite seems to be particularly useful for the purification of medium and high molecular weight proteins and nucleic acids.[160] In general, low molecular weight solutes show very low affinity for hydroxyapatite.

The high selectivity of hydroxyapatite appears to be due to the competition for calcium ions on the crystal surface. This agrees with the finding that the content and distribution of acidic groups, particularly carboxylates, on the surface of proteins are major determinants of protein binding to hydroxyapatite[161]. In general, if proteins that chromatograph well on anion exchangers also bind to hydroxyapatite, the two methods do not have the same selectivity. Thus isoforms of human ceruloplasmin were readily separated on apatite but not by conventional ion-exchange chromatography (see Section 4.9.8). A special possibility is to use the technique for the purification of phosphoproteins.

A drawback of hydroxyapatite is its mechanical and chemical instability. This has been partly overcome by the introduction of ceramic fluoroapatite. This is available as 10-μm-average-diameter spherical particles in prepacked HPLC columns.[162,163] Apparently the fluoroapatite has the same chromatographic properties as the hydroxyapatite.

Hydroxyapatite is available from BioSepra (HA-Ultrogel, spherical particles of microcrystals in agarose beads), Bio-Rad (Bio-Gel-HTP, crystals and BioGel CHT, spherical, ceramic), and Merck (hydroxyapaptite, spheroidal particles).

Hydroxyapatite chromatography is discussed more thoroughly by Boschetti[51] and Kawasaki.[164]

4.9 APPLICATIONS

4.9.1 Isolation of a Fungal Cellulase by Chromatography on DEAE Sepharose[165]

The culture filtrate of the fungus *Trichoderma reesei* constitutes an excellent source for the preparation of cellulose-degrading enzymes. After buffer exchange on Sephadex G-25 equilibrated with 0.01 M ammonium acetate, pH 5.0. (the ion-exchange starting buffer), the filtrate was applied to a column (2.6 × 30 cm) of DEAE Sepharose CL-6B, equilibrated with 0.01 M ammonium acetate, pH 5.0. A 1,000-ml linear gradient of ammonium acetate at pH 5.0 from 0.01 to 0.5 M and a flow rate of 40 ml/h were used to develop the column. Three different activities, called A, B, and C, were separated (Figure 4-8).

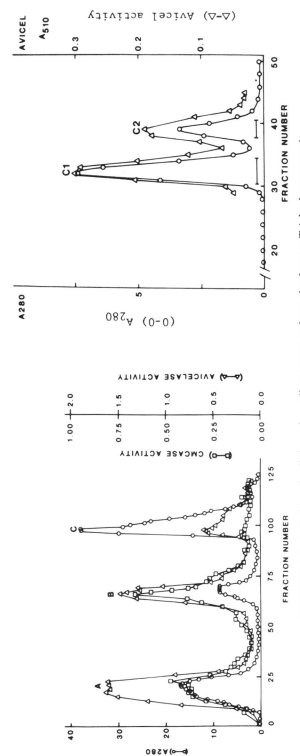

FIGURE 4-8. Isolation of cellulose-degrading enzymes from the fungus *Trichoderma reesei* by ion-exchange chromatography on DEAE Sepharose CL-6B (left) at pH 5.0 and (right) rechromatography of component C at pH 3.5. For details, see Section 4.9.1. (Reproduced with permission from Ref. 165.)

Component C, the most acidic one, was collected, lyophilized, and dissolved in 0.05 M ammonium acetate, pH 3.7, and applied on a column (2.6 × 30 cm) of DEAE Sepharose CL-6B in the same buffer. This time the column was eluted with a 700-ml linear gradient of ammonium acetate at pH 3.7 from 0.05 to 0.3 M, and component C was further separated into two cellobiohydrolases, both pure (Figure 4-8). Even though the two proteins coeluted at pH 5.0, they separated at the lower pH.

4.9.2   Isolation of Immunoglobulins from Egg Yolk[166]

Immunoglobulins can be obtained from immunized hens directly in the egg yolk in a high concentration. Drawbacks are the lower stability of chicken IgG, as compared to rabbit IgG, and the fact that they do not bind to protein A or G.

Egg yolk was collected and suspended in two volumes of 3.75% PEG in 10 mM sodium phosphate buffer, pH 7.5, containing 0.1 M NaCl. After centrifugation the IgG in the supernatant was precipitated by increasing the PEG concentration to 12%. The pellet was dissolved in 10 mM Tris-HCl buffer, pH 8.0, containing 0.05 M NaCl and applied to a column (2 × 15 cm) of DEAE Sephacel in the same buffer. The column was washed with 10 column volumes of starting buffer containing 0.5% Triton X-100 to elute lipid and unadsorbed material, followed by starting buffer without Triton X-100. It was then developed with a 150-ml linear gradient of 0.05 to 0.25 M NaCl in the Tris buffer. The single main peak contained the chicken IgG in 95% or better purity.

4.9.3   Isolation of α-Bungarotoxin from Snake Venom by Chromatography on CM Sephadex C-50[167]

The venom of the Taiwanese snake Krait, *Bungarus multicinctus*, contains a neurotoxin, α-bungarotoxin, widely used as a marker of the nicotinic acetylcholine receptor.

About 1 g of lyophilized venom was dissolved in 15 ml of 0.05 M ammonium acetate, pH 5.0, and applied to a column (2.5 × 85 cm) of CM Sephadex C-50 equilibrated in the same buffer. The column was developed with (1) 500 ml of starting buffer; (2) a 1,400 ml combined pH and ionic strength linear gradient of ammonium acetate, using 1 M buffer, pH 7.0, as the final buffer; (3) 800 ml of final buffer.

At least 14 components were resolved in the chromatography (Figure 4-9). α-Bungarotoxin eluted as component 3 in the first part of the gradient. It was obtained in a pure form after subsequent chromatography on CM-cellulose where it eluted in a gradient of ammonium acetate 0.05 M, pH 5.0, to 1 M, pH 7.0.

The volume of the CM Sephadex C-50 decreased by about 50% during the elution but apparently without any adverse effects. It is likely that a far shorter

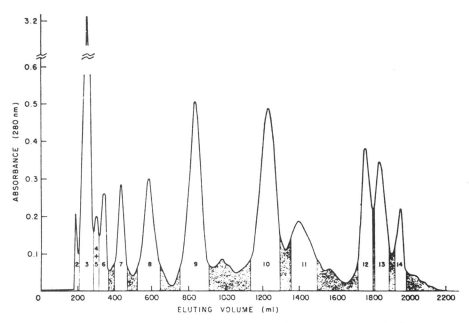

FIGURE 4-9. Fractionation of the venom of the Taiwanese snake, the krait *Bungarus multicinctus,* on CM Sephadex C-50. The chromatography is described in Section 4.9.3. (Reproduced with permission from Ref. 167.)

column would have been enough for the separation. It is also likely that rechromatography on the same column would have been as successful as the use of the cellulose ion exchanger for the second run. The chromatogram gives an indication of the complexity of snake venoms.

4.9.4   Separation of Six Monoacetyl Derivatives of a Cobra Neurotoxin by Chromatography on the Polymethacrylate Ion Exchanger Bio-Rex 70[25]

The functions of amino acid side chains in proteins can be studied by chemical modifications. In the chemical reactions used, several different side chains are normally derivatized, especially if they do not differ considerably. Meaningful interpretations of the results in such cases require that the protein derivatives are purified and assayed separately. A case in point is the acetylation of a neurotoxin from the cobra *Naja naja siamensis.* This toxin is a small basic protein with 71 amino acid residues in a single peptide chain cross-linked by five disulfide bridges. It contains five lysine residues and a free N terminus, all of which can react with acetic anhydride. An amount of reagent ([3H]acetic anhydride) corresponding to 1/12 of total amino groups produced a mixture of mostly mono derivatives that could be fractionated on the polymethacrylate ion exchanger Bio-Rex 70, despite the fact that they had an identical net charge.

The derivatized protein was dissolved in 0.02 M ammonium acetate, pH 6.7, and applied to a column (2 × 30.5 cm) of Bio-Rex 70 equilibrated in 0.20 M ammonium acetate (pH 6.7), pH 6.50. The column was first eluted with 0.05 M ammonium acetate (pH 6.7), then with a concave gradient of 0.05 to 1.4 M ammonium acetate (pH 6.7).

In the chromatography (Figure 4-10), underivatized toxin eluted last (peak 1), as expected since it has all its positive charges intact, preceded by six peaks of the six possible different monoacetyl derivatives. These were later identified by peptide mapping.[168] Minor amounts of di- and triacetyl derivatives are seen in the beginning of the chromatogram.

4.9.5 Hydrophobic Effects on the HPLC Cation Exchanger Bio-Gel TSK SP 5-PW[94]

Fasciculins (61 amino acids, 4 disulfides) from mamba snake venoms are noncompetitive inhibitors of acetylcholinesterase.[169,170] Fasciculin 2 (FAS2) was treated with nor-leucine methylester and a water-soluble carbodiimide. The hydrophobic group -NHCH(COOCH$_3$)CH$_2$CH$_2$CH$_2$CH$_3$ was thereby substituted into carboxyl groups, as well as into hydroxyl groups of tyrosines. Derivatives were isolated by HPLC on the cation exchanger Bio-Gel TSK SP-5-PW (7.5 × 75 mm) equilibrated with water (Figure 4-11). Buffer A: H$_2$O, B: 1.00 M ammonium acetate (pH 6.7). Program: 0- to 5-min elution with 15% B, 5- to 60-min linear gradient to 60% B, and 60- to 80-min linear gradient to 80% B. (A) Before and (B) after treatment with 0.5 M hydroxylamine (pH 7) to remove the substituent from tyrosine. Absorbance at full scale 0.4 in A and 0.08 in B.

Peaks A–E contained carboxyl derivatives, as they were also present after treatment with hydroxylamine. Modification neutralized the negative charge of carboxylates. A–E had a net charge of +5 (neutral pH) and an unmodified toxin (FAS2) of +4. Consequently, A–E eluted from a cation exchanger after

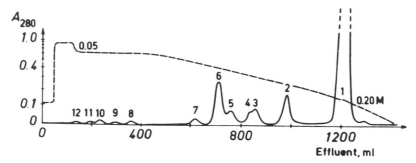

FIGURE 4-10. Isolation of acetyl derivatives of a neurotoxin from the cobra *Naja naja siamensis* on a column of Bio-Rex 70 using a gradient of ammonium acetate. The experiment is described in Section 4.9.4. (Reproduced with permission from Ref. 25.)

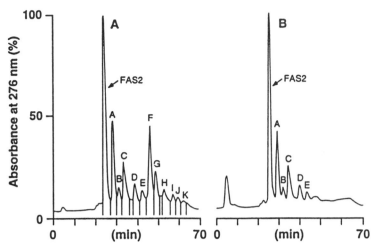

FIGURE 4-11. Isolation of derivatives of the acetylcholinesterase inhibitor fasciculin 2 by HPLC ion exchange on Bio-Gel TSK SP-5-PW (7.5 × 75 mm) equilibrated with water. The column was developed with a gradient of ammonium acetate (pH 6.7). The derivatives were obtained by modification with nor-leucine methylester and carbodiimide. Details are found in Section 4.9.5. (Reproduced with permissiom from Ref. 94.)

FAS2. A–E had the same net charge, and separation was due to differences in charge distribution (compare Section 4.9.4). F–K were tyrosine derivatives, as they were not detected after treatment with hydroxylamine. F was only a tyrosine derivative, as it eluted afterward as FAS2. Modification of tyrosine did not change the charge and F thus had the same net charge as FAS2 (+4). G–K were probably tyrosine derivatives of A–E and had a charge of +5. Modification made a tyrosine residue more hydrophobic, and the late elution was due to hydrophobic interactions with the ion exchanger, which were so strong that a molecule with a charge of +4 (derivative F) eluted after molecules with a net charge of +5 (A–E). Another example is given by fasciculin 1 and 2 that differs only by a single amino acid. FAS1 is more hydrophobic, with Tyr at position 47 instead of Asn in FAS2. The elution order is FAS2-FAS1 from BioGel TSK SP and from a reversed-phase C18 column, but FAS1-FAS2 from the more hydrophilic SP Sephadex C-25[170] and also from Bio-Rex 70.

4.9.6   Fractionation of Glycolytic Enzymes from Chicken Muscle by Consecutive Chromatographies on the HPLC Ion Exchangers Mono Q and Mono S[171]

Complex samples from natural sources often require the systematic use of several different selectivities to obtain pure proteins. The use of sequential anion and cation-exchange chromatography for the purification of creatine kinase from chicken muscle is a case in point. The sample was initially studied by electrophoretic titration curves (Figure 4-2).

A low ionic strength extract of chicken muscle was transferred to the starting buffer for the first step, cation exchange, by buffer exchange on Sephadex G-25. The sample was filtered (0.22 μm), and 800 μl was applied to a Mono S HR 10/10 column. The column was eluted with 0.05 M MES, pH 6.0, and a gradient from 0 to 0.5 M NaCl. Creatine kinase (CK) and phosphoglyceromutase eluted in the breakthrough peak, with several other enzymes being resolved in the gradient (Figure 4-12). The breakthrough peak was collected and transferred to the starting buffer for the anion-exchange step by gel filtration on Sephadex G-25. One milliliter of the sample was applied to a Mono Q HR 5/5 and eluted with 0.02 M Tris, pH 8.0, and a gradient from 0 to 0.5 M NaCl (Figure 4-12). Creatine kinase was obtained in good purity, well separated from other proteins.

This application demonstrates the ability to make an informed choice of conditions for ion-exchange chromatography by studies of the electrophoretic titration curves, as well as the power of modern high-performance ion exchange chromatography.

4.9.7 Large-Scale Purification of Human Serum Albumin and IgG by a Stepwise Elution Protocol[172]

Highly purified human blood plasma proteins are needed for a variety of scientific and therapeutic purposes. The usual procedure for large-scale fractionation of human blood plasma proteins is the Cohn cold ethanol procedure. An alternative method uses a combination of anion- and cation-exchange chromatography starting from cryoprecipitated and factor IX-depleted plasma.

After centrifugation and buffer exchange to 25 mM sodium acetate buffer, the pH is adjusted to pH 5.2 with acetic acid to allow the euglobins to precipitate. The plasma is then centrifuged and filtered before application to a column of DEAE Sepharose Fast Flow equilibrated in this buffer at a loading of approximately 35 g protein per liter ion exchanger. After eluting IgG under starting conditions, albumin is displaced by stepwise elution with the same buffer at pH 4.5 (Figure 4-13) and applied directly to a column of CM Sepharose equilibrated in this buffer. Highly purified albumin (>99% by cellulose acetate electrophoresis) is obtained by elution with 0.11 M sodium acetate buffer, pH 5.5.

In this example the chromatographic parameters were carefully optimized for high purity at large scale. Such optimization is not usually required at laboratory scale. If ion-exchange chromatography does not give a product of sufficient purity, additional chromatographic steps, such as gel filtration and reversed-phase chromatography, are used.

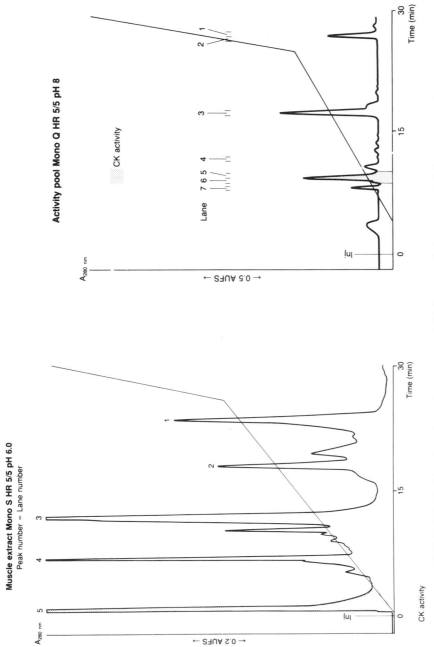

FIGURE 4-12. Fractionation of an extract of chicken muscle by sequential chromatography on Mono S HR10/10 and Mono Q HR 5/5. CK, creatine kinase. The electrophoretic titration curve for the sample is shown in Figure 4-2. Details are given in Section 4.9.6. (Reproduced with permission from Ref. 171.)

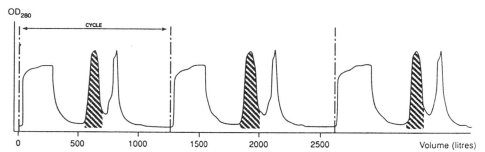

FIGURE 4-13. Large-scale fractionation of human plasma proteins by stepwise elution from DEAE Sepharose Fast Flow. Equilibration buffer: 0.020 M NaAc, pH 5.2 (IgG in breakthrough). Human serum albumin (HSA) eluted with 0.025 M NaAc, pH 4.5 (marked peak). Remaining proteins eluted with 0.15 M NaAc, pH 4.0. Column: BPSS 1000/150 (V_t 118 l). Flow rate: 120 cm/h. The HSA peak is directly applied to a CM Sepharose Fast Flow column for further fractionation. Final purification takes place by gel filtration on Sephacryl S-200 HR. (Data from Pharmacia Biotech AB, Uppsala, Sweden.)

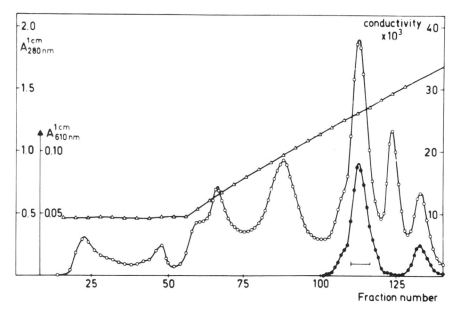

FIGURE 4-14. Hydroxyapatite chromatography of a ceruloplasmin-enriched fraction from ion exchange on DEAE-cellulose of human serum. The ceruloplasmin forms were monitored by absorption at 610 nm (filled circles). For details, see Section 4.9.8. (Reproduced with permission from Ref. 158.)

4.9.8   Separation of Human Ceruloplasmin Isoforms by Hydroxyapatite Chromatography[158]

Human ceruloplasmin, a blue copper glycoprotein, occurs in serum in two forms that differ in their carbohydrate content.[22] These forms could not be separated by ion-exchange chromatography, but hydroxyapatite was shown to be a good adsorbent for this purpose.

Defatted serum was diluted with 0.03 M potassium phosphate and 0.1 M NaCl (pH 6.8) and adsorbed to a column of DEAE-cellulose equilibrated with the same buffer, and ceruloplasmin was eluted with by a NaCl gradient to 0.45 M. The pooled ceruloplasmin peak (blue color), containing about 0.25 M NaCl, was dialyzed against 0.075 M potassium phosphate buffer (pH 6.8), applied on a column (2 $\times$ 75 cm) of hydroxyapatite prepared as described by Tiselius and co-workers,[156] and equilibrated with the 0.075 M buffer. The column was then developed with a 1,000-ml gradient of potassium phosphate up to 0.5 M at a flow rate of 12 ml/h. The proteins that coelute on DEAE-cellulose separate. The two ceruloplasmin forms contain about 11% (major form) and 7% (minor form) carbohydrate. All buffers contained 0.02 M ε-caproic acid to inhibit plasminogen activation.

ACKNOWLEDGMENTS

A special thanks to Dr. David Eaker, Professor Stellan Hjertén, and Dr. Göran Pettersson for sharing their great chromatographic experience and for many interesting discussions. The help of Dr. Kerstin Gunnarsson at the final stages of writing this chapter is very much appreciated.

4.11   REFERENCES

1. J. Bonnerjea, S. Oh, M. Hoare, P. Dunhill, *Biotechnology, 4,* 954–958 (1986).
2. S. Yamamoto, K. Nakanishi, R. Matsuno, *Ion-Exchange Chromatography of Proteins,* Chromatographic Science Series, Vol. 43, Marcel Dekker Inc., New York, 1988.
3. J. Ståhlberg, B. Jönsson, C. Horváth, *Anal. Chem., 64,* 3118 (1992).
4. J. Ståhlberg, B. Jönsson, *Anal. Chem., 68,* 1536–1544 (1996).
5. R.J. McColloch, Z.I. Kertesz, *J. Biol. Chem., 160,* 149–154 (1945).
6. S. Moore, W.H. Stein, *Adv. Protein Chem., 11,* 191–236 (1956).
7. S. Paléus, J.B. Neilands, *Acta Chem. Scand., 4,* 1024–1030 (1950).
8. W.H. Stein, S. Moore, *J. Am. Chem. Soc., 73,* 1893 (1951).
9. C.H.W. Hirs, W.H. Stein, S. Moore, *J. Biol. Chem., 200,* 493–506 (1954).
10. H.H. Tallan, W.H. Stein, *J. Am. Chem. Soc., 73,* 2976–2977 (1951).
11. H.H. Tallan, W.H. Stein, *J. Biol. Chem., 200,* 507–514 (1954).

12. H.A. Sober, E.A. Peterson, *J. Am. Chem. Soc., 76,* 1711–1712 (1954).

13. E.A. Peterson, H.A. Sober, *J. Am. Chem. Soc., 78,* 751–755 (1956).

14. H.A. Sober, F.J. Gutter, M.M. Wycoff, E.A. Peterson, *J. Am. Chem. Soc., 78,* 756–763 (1956).

15. W. Kopaciewicz, M.A. Rounds, J. Fasnaugh, F.E. Regnier, *J. Chromatogr., 266,* 3–21 (1983).

16. M.T.W. Hearn, A.N. Hodder, P.G. Stanton, M.I. Aquilar, *Chromatografia 24,* 769–776 (1987).

17. R.K. Scopes, E. Algar, *FEBS Lett., 106,* 239–242 (1979).

18. F.B. Rudolph, B.F. Cooper, J. Greenhut, in *Progress in HPLC,* H. Parvez Y. Kato, S. Parvez, eds., VNU Science Press, Utrecht, 1985, pp. 133–147.

19. R.K. Scopes, *Protein Purification: Principles and Practice,* 3rd ed., Springer-Verlag, New York, 1993.

20. R.E. Creighton, *Proteins: Structures and Molecular Properties,* 2nd ed., W.H. Freeman and Co., New York, 1993.

21. P.M. Hardy, in *Chemistry and Biochemistry of the Amino Acids,* G.C. Barrett, ed., Chapman and Hall, London, 1985, pp. 6–54.

22. T.H.W. Plummer, C.H.W. Hirs, *J. Biol. Chem., 238,* 1396–1401 (1963).

23. L. Rydén, in *Copper Proteins and Copper Enzymes,* R. Lontie, ed., Vol. III, CRC Publishers, Boca Raton, FL, 1984, pp. 37–100.

24. D.L. Brautigan, S. Ferguson-Miller, E. Margoliash, *J. Biol. Chem., 253,* 130–139 (1978).

25. E. Karlsson, D. Eaker, G. Ponterius, *Biochim. Biophys. Acta, 57,* 235–248 (1972).

26. P.G. Righetti, T. Caravaggio, *J. Chromatogr., 127,* 1 28 (1976).

27. P.G. Righetti, G. Tudoe, K. Ek, *J. Chromatogr., 220,* 115–194 (1981).

28. Y.J. Yao, K.S. Khoo, M.C.M. Chung, S.F.Y. Li, *J. Chromatogr., 680,* 431–435 (1994).

29. C.C. Chang, C.C. Yang, *Biochem. Biophys. Acta, 236,* 164–173 (1971).

30. K.D. Collins, M.W. Washabaugh, *Q. Rev. Biophys., 18,* 323–422 (1985).

31. M.W. Washabaugh, K.D. Collins, *J. Biol. Chem., 261,* 2477–12485 (1986).

32. F.E. Regnier, *Science 238,* 319–323 (1987).

33. M.T.W. Hearn, A.N. Hodder, M.I. Aguilar, *J. Chromatogr., 458,* 45–56 (1988).

34. J. Fausnaugh-Pollit, G. Thevenon, L. Janis, F. Regnier, *J. Chromatogr., 443,* 221–228 (1988).

35. C. Cerveñansky, Å. Engström, E. Karlsson, *Biochim. Biophys. Acta, 1196,* 1–5 (1994).

36. C. Cerveñansky, Å. Engström, E. Karlsson, *Eur. J. Biochem., 229,* 270–275 (1995).

37. C. Cerveñansky, R. Durán, E. Karlsson, *Toxicon, 34,* 718–721 (1996).

38. R. Drager, F.E. Regnier, *J. Chromatogr., 359,* 147–155 (1986).

39. N.K. Boardman, S.M. Partridge, *Biochem. J., 59,* 543–552 (1955).

40. M.T.W. Hearn, A.N. Hodder, P.G. Stanton, M.I. Aguilar, *Chromatographia, 24,* 769–776 (1987).

41. M.A. Rounds, F.E. Regnier, *J. Chromatogr., 283,* 74–85 (1984).

42. T. Koide, T. Ikenaka, *Eur. J. Biochem., 33,* 417–431 (1973).

43. R.D. Whitley, R. Wachter, F. Liu, N.H.L. Wang, *J. Chromatogr., 465,* 137–156 (1989).

44. I. Mazsaroff, S. Cook, F.E. Regnier, *J. Chromatogr., 443,* 119–131 (1988).

45. R.C. Hider, E. Karlsson, S. Namiranian, in *Snake Toxins: International Encyclopeadia of Pharmacology and Therapeutics,* A. Harvey, ed., Pergamon Press, New York, 1991, pp. 1–34.

46. S. Hjertén, K. Yao, K.O. Eriksson, B. Johansson, *J. Chromatogr., 359,* 99–109 (1986).

47. K. Yao, S. Hjertén, *J. Chromatogr., 385,* 87–98 (1987).

48. E.S. Parente, D.B. Wetlaufer, *J. Chromatogr., 288,* 391–411 (1984).

49. J. Withka, P. Moncuse, A. Baziotis, R. Maskiewicz, *J. Chromatogr., 398,* 175–202 (1987).

50. R.M. Chicz, F.E. Regnier, *J. Chromatogr., 443,* 193–201 (1988).

51. E. Boschetti, *J. Chromatogr., 658,* 203–236 (1994).

52. C.J.O.R. Morris, P. Morris, *Separation Methods in Biochemistry,* 2nd ed., Pitman, London, 1976, Chapter 5.

53. M.A. Smith, P.C. Gillispie, *J. Chromatogr., 469,* 111–120 (1989).

54. W. Kopaciewicz, F.E. Regnier, *Anal. Biochem., 133,* 251–259 (1983).

55. R.K. Scopes, *Anal. Biochem. 114,* 8–18 (1981).

56. H.A. Sober, G. Kegeles, F.J. Gutter, *J. Am. Chem. Soc. 74,* 2734–2740 (1952).

57. F.C. McIntyre, J.R. Schenck, *J. Am. Chem. Soc. 70,* 1193–1194 (1948).

58. C.L. Hoffpauir, J.D. Guthrie, *Textile Res. J. 20,* 617–620 (1950).

59. E.B. Astwood, M.S. Raben, R.W. Payne, A.B. Grady, *J. Am. Chem. Soc., 73,* 2969–2970 (1951).

60. H. Determan, N. Meyer, *Nature, 233,* 499–500 (1969).

61. W. Müller, *J.Chromatogr., 510,* 133–140 (1990).

62. R. Janzen, K.K. Ungers, W. Müller, *J. Chromatogr., 522,* 77–93 (1990).

63. A.E. Ivanov, V.P. Zubov, *J. Chromatogr., 673,* 159–175 (1994).

64. L.L. Lloyd, *J. Chromatogr., 544,* 201–217 (1991).

65. T. Hashimoto, *J. Chromatogr., 544,* 249–255 (1991).

66. F.E. Regnier, *Anal. Biochem., 126,* 1–7 (1982).

67. F.E. Regnier, *Chromatographia, 24,* 241–251 (1987).

68. M.A. Rounds, W. Kopaciewicz, F.E. Regnier, *J. Chromatogr., 362,* 187–196 (1986).

69. M.A. Rounds, W.D. Rounds, F.E. Regnier, *J. Chromatogr., 397,* 25–38 (1987).

70. H. Freiser, K.M. Gooding, *J. Chromatogr., 544,* 125–135 (1991).

71. T. Hashimoto, *J. Chromatogr., 544,* 257–265 (1991).

72. N.B. Afeyan, N.F. Gordon, I. Mazsaroff, L. Varady, S.P. Fulton, Y.B. Yang, F.E. Regnier, *J. Chromatogr., 519,* 1–9 (1991).

73. N.B. Afeyan, S.P. Fulton, F.E. Regnier, *J. Chromatogr., 544,* 267–279 (1991).

74. J.L. Liao, S. Hjertén, *J. Chromatogr., 457,* 175–182 (1988).

75. S. Hjertén, J.L. Liao, R. Zhong, *J. Chromatogr., 473,* 273–275 (1989).

76. J.L. Liao, R. Zhong, S. Hjertén, *J. Chromatogr., 586*, 21–26 (1991).

77. S. Hjertén, Y.M. Li, J.L. Liao, J. Mohammad, K. Nakazato, G. Pettersson, *Nature, 356*, 810–811 (1992).

78. J. Porath, L. Fryklund, *Nature, 226*, 1169–1170 (1970).

79. J. Porath, N. Fornstedt, *J. Chromatogr., 51*, 479–489 (1970).

80. P. Lundahl, Q. Ying, *J. Chromatogr., 544*, 283–304 (1991).

81. S.G. Mayhew, L.G. Howell, *Anal. Biochem., 41*, 466–470 (1971).

82. T. Fujita, Y. Suzuki, J. Yamauti, I. Takagahara, K. Fujii, Yamashita, T. Horio, *J. Biochem. (Tokyo), 87*, 89–100 (1980).

83. M.L. Heinitz, L. Kennedy, W. Kopaciewicz, F.E. Regnier, *J. Chromatogr., 443*, 173–182 (1988).

84. E. Peterson, H. Sober, *Methods Enzymol., 5*, 3–27 (1962).

85. K. Griebenow, A.M. Klibanov, *Proc. Natl. Acad. Sci. USA, 92*, 10969–10976 (1995).

86. D.J. Strydom, *J. Biol. Chem., 247*, 4029–4042 (1972).

87. C.C. Viljoen, D.P. Botes, *J. Biol. Chem., 248*, 4915–4919 (1973).

88. A. Adem, A. Åsblom, G. Johansson, P.M. Mbugua, E. Karlsson, *Biochim. Biophys. Acta, 968*, 340–345 (1988).

89. M. Jolkkonen, P.L.M. van Giersbergen, U. Hellman, C. Wernstedt, A. Oras, N. Satyapan, E. Karlsson, *Eur. J. Biochem. 234*, 579–585 (1995).

90. M. Jolkkonen, P.L.M. van Giersbergen, U. Hellman, C. Wernstedt, E. Karlsson, *FEBS Lett., 352*, 91–94 (1994).

91. A. Aneiros, I. García, J.R. Martínez, A.L. Harvey, A.J. Anderson, D.L. Marshall, Å. Engström, U. Hellman, E. Karlsson, *Biochim. Biophys. Acta, 1157*, 86–92 (1993).

92. C. Cerveñansky, Å. Engström, E. Karlsson, *Biochim. Biophys. Acta, 1199*, 1–5 (1994).

93. C. Cerveñansky, Å. Engström, E. Karlsson, *Eur. J. Biochem., 229*, 270–275 (1995).

94. C. Cerveñansky, R. Durán, E. Karlsson, *Toxicon, 34*, 718–721 (1996).

95. L. Söderberg et al., in *Protides of the Biological Fluids*, H. Peeters, ed., Vol. 30, Pergamon Press, Oxford, 1982, pp. 629–634.

96. W. Kopasciewicz, M.A. Rounds, J. Fassnaugh, F.E. Regnier, *J. Chromatogr., 266*, 3–21 (1983).

97. K.M. Gooding, M.N. Schmuck, *J. Chromatogr., 266*, 633–642 (1983).

98. M.A. Rounds, F.E. Regnier, *J. Chromatogr., 283*, 37–45 (1984).

99. K.M. Gooding, M.N. Schmuck, *J. Chromatogr., 296*, 321–328 (1984).

100. F.E. Regnier, *Methods Enzymol., 104*, 170–189 (1984).

101. FPLC *Ion Exchange and Chromatofocusing: Principles and Methods.* Pharmacia, Uppsala, Sweden, 1985.

102. M.T.W. Hearn, A.N. Hodder, M.I. Aquilar, *J. Chromatogr., 443*, 97–118 (1988).

103. M.T.W. Hearn, A.N. Hodder, M.I. Aquilar, *J. Chromatogr., 458*, 27–44 (1988).

104. W.R. Melander, Z. El Rassi, C. Horváth, *J. Chromatogr., 469*, 3–27 (1989).

105. A.N. Hodder, M.I. Aguilar, M.T.W. Hearn, *J. Chromatogr., 476*, 391–411 (1989).

106. A.N. Hodder, M.I. Aguilar, M.T.W. Hearn, *J. Chromatogr.*, *506*, 17–34 (1990).
107. A. Crestfield, W.H. Stein, S. Moore, *J. Biol. Chem.*, *238*, 2413–2419 (1963).
108. A. Crestfield, W.H. Stein, S. Moore, *J. Biol. Chem.*, *238*, 421–2428 (1963).
109. E. Karlsson, D. Eaker, L. Fryklund, S. Kadin, *Biochemistry*, *11*, 4628–4633 (1972).
110. E. Karlsson, H. Arnberg, D. Eaker, *Eur. J. Biochem.*, *21*, 1–16 (1971).
111. P. Marchot, A. Khélif, Y.H. Ji, P. Mansuelle, P. Bougis, *J. Biol. Chem.*, *268*, 12458–12467 (1993).
112. T. Arakawa, S.N. Timasheff, *Biochemistry, 21*, 6545–6552 (1982).
113. W.A. Jensen, J. Armstrong, J. De Giorgio, M.T.W. Hearn, *Biochemistry, 34*, 472–480 (1995).
114. W. Melander, C. Horváth, *Arch. Biochem. Biophys.*, *183*, 200–215 (1977).
115. J. Bello, D. Haas, H.R. Bello, *Biochemistry, 5*, 2539–2548 (1966).
116. H. Sigel, R.B. Martin, *Chem. Rev.*, *82*, 345–426 (1982).
117. H. Fang, M.I. Aguilar, M.T.W. Hearn, *J. Chromatogr.*, *729*, 67–79 (1996).
118. J. van Renswoude, C. Kempf, *Meth. Enzymol.*, *104*, 329–339 (1984).
119. K.H. Milby, S.V. Ho, J.M.S. Henis, *J. Chromatogr.*, *482*, 133–144 (1989).
120. *Ion Exchange Chromatography: Principles and Methods*, Pharmacia, Uppsala, Sweden.
121. Z. El Rassi, C. Horváth, *J. Chromatogr.*, *359*, 253–264 (1986).
122. Dj. Josic, W. Hofmann, W. Reutter, *J. Chromatogr.*, *371*, 43–54, (1986).
123. R.W. Stringham, F.E. Regnier, *J. Chromatogr.*, *409*, 305–314, (1987).
124. H. Hori, N. Tamiya, *Biochem. J.*, *153*, 217–222 (1976).
125. A.S. Arseniev, T.A. Balashkova, Yu.N. Utkin, V.I. Tsetlin, V.F. Bystrov, V.T. Ivanov, Yu.A. Ovchinnikov, *Eur. J. Biochem.*, *71*, 595–606 (1976).
126. A. Rousselet, G. Faure, J.C. Boulain, A. Ménez, *Eur. J. Biochem.*, *140*, 31–37 (1984).
127. F. von der Haar, *Eur. J. Biochem, 34*, 84–90 (1973).
128. S.S. Quadri, J.S. Esterby, *Anal. Biochem.*, *105*, 299–303 (1980).
129. R.K. Scopes, *Biochem. J.*, *161*, 253–263 (1977).
130. J.D. Karkas, J. Germershausen, R. Liou, *J. Chromatogr.*, *214*, 267–268 (1981).
131. G. Vanecek, F.E. Regnier, *Anal. Biochem.*, *109*, 345–353 (1980).
132. F.E. Regnier, *Methods Enzymol.*, *91*, 137–180 (1981).
133. J.M. Cooney, *Biotechnology, 2*, 41–55 (1984).
134. E.H. Cooper, R. Turner, J.R. Webbs, H. Lindblom, L. Fägerstam, *J. Chromatogr.*, *327*, 269–277 (1985).
135. J.-C. Janson, *J. Agr. Food Sci.*, *19*, 581–588 (1971).
136. R. Mhatre, W. Nashbabeh, D. Schmalzing, X. Yao, M. Fuchs, D. Whitney, F. Regnier, *J. Chromatogr.*, *707*, 225–231 (1995).
137. R.M. Bock, S.N. Ling, *Anal. Chem.*, *26*, 1543–1546 (1954).
138. B.M. Pogell, *Biochem. Biophys. Res. Commun.*, *7*, 225–230 (1962).
139. B.M. Pogell, *Methods Enzymol.*, *9*, 9–15 (1966).
140. R.K. Scopes, *Biochem. J.*, *161*, 265–277 (1977).

141. R.D. Davies, R.K. Scopes, *Anal. Biochem.*, *114*, 19–27 (1981).

142. E.A. Peterson, A.R. Torres, *Meth. Enzymol.*, *104*, 113–133 (1984).

143. A.W. Liao, Z. El Rassi, D.M. LeMaster, C. Horváth, *Chromatographia, 24*, 881–885 (1987).

144. D.H. Leaback, H.K. Robinson, *Biochem. Biophys. Res. Commun.*, *67*, 248–254 (1975).

145. G. Jayaraman, Y.F. Lei, J.A. Moore, S.M. Cramer, *J. Chromatogr., 702*, 143–155 (1995).

146. J.R. Whitaker, P.E. Granum, *Anal. Biochem.*, *109*, 156–159 (1980).

147. W.J. Waddell, *J. Lab. Clin. Med.*, *48*, 311–314 (1956).

148. P. Wolf, M. Maguire, *Anal. Biochem.*, *129*, 145–155 (1983).

149. R.K. Scopes, *Anal. Biochem.*, *59*, 277–282 (1974).

150. W.P. Neumann, E. Habermann, *Biochem. Z., 327*, 274–288 (1955).

151. H. Breithaupt, K. Rübsamen, P. Walsh, E. Habermann, *Naunyn-Schmiedebergs Arch. Pharmacol.*, *270*, 278–288 (1971).

152. R.A. Hendon, H. Fraenkel-Conrat, *Proc. Natl. Acad. Sci. USA, 68*, 1560–1563 (1971).

153. J. Fohlman, D. Eaker, E. Karlsson, S. Thesleff, *Eur. J. Biochem.*, *68*, 457-469 (1976).

154. C.T. Wehr, *Methods Enzymol.*, *104*, 133–154 (1984).

155. C.T. Wehr, *J. Chromatogr., 418*, 3–26 (1987).

156. A. Tiselius, S. Hjertén, Ö. Levin, *Arch. Biochem. Biophys. 65*, 132–155 (1956).

157. J.-C. Janson, J. Porath, *Methods Enzymol.*, *8*, 615–621 (1966).

158. L. Rydén, *FEBS Lett.*, *18*, 321–325 (1971).

159. P. Norberg, B. von Hofsten, *Biochim. Biophys. Acta, 220*, 132–133 (1970).

160. G. Bernardi, *Nature, 206*, 779–783 (1965).

161. G. Bernardi, T. Kawasaki, *Biochim. Biophys. Acta, 160*, 301–310 (1968).

162. T. Sato, T. Okuyama, S. Fuginuma, T. Ogawa, *Chromatography (Japan), 9*, 51–54 (1988).

163. T. Sato, K. Ohuchi, T. Okuyama, S. Fuginuma, *Chromatography (Japan), 9*, 171–172 (1988).

164. T. Kawasaki, *J. Chromatogr., 544*, 147–184 (1991).

165. R. Bhikhabhai, G. Johansson, G. Pettersson, *J. Appl. Biochem.*, *6*, 336–345 (1984).

166. S. Johansson, unpublished.

167. V.A. Eterovic, M.S. Herbert, M.R. Hanley, E.L. Bennet, *Toxicon, 13*, 37–48 (1975).

168. K. Balasubramaniam, D. Eaker, E. Karlsson, *Toxicon, 21*, 219–229 (1983).

169. C. Cerveñansky, F. Dajas, A.L. Harvey, E. Karlsson, in *Snake Toxins: International Encyclopedia of Pharmacology and Therapeutics*, A. Harvey, ed., Pergamon Press, New York, 1991, pp. 303–321.

170. E. Karlsson, A.L. Harvey, C. Cerveñansky, G.J. Kleywegt, M. Harel, I. Silman, J.L. Sussman, in *Enzymes from Snake Venoms*, G.S. Bailey, ed., Alaken, Inc., Fort Collins, CO, in press.

171. L. Fägerstam, L. Söderberg, L. Wahlström, U.B. Fredriksson, K. Plith, E. Walldén, in *Protides of the Biological Fluids*, H. Peeters, ed., Vol. 30, Pergamon Press, Oxford, 1982, pp. 621–628, and L. Fägerstam, unpublished.

172. Data from Pharmacia Biotech AB, Uppsala, Sweden.

5 Chromatofocusing

T. WILLIAM HUTCHENS
Department of Food Science and Technology
University of California
Davis, California 95616–8598, USA

Protein Purification: Principles, High-Resolution Methods, and Applications, Second Edition.
Edited by Jan-Christer Janson and Lars Rydén.
ISBN 0-471-18626-0. © 1998 Wiley-VCH, Inc.

5.1 INTRODUCTION AND BACKGROUND

Whether the goal is high-resolution separations at the analytical scale or the preparative scale isolation of biologically active macromolecules, often the most definitive separation techniques include exploitation of molecular charge properties. Thus, ion-exchange chromatography, because of its accommodating versatility and simplicity, is a frequently used separation technique in biochemistry. Similarly, electrophoretic focusing procedures, although generally more labor-intensive and cumbersome with larger sample volumes, allow macromolecular separations based on differences in net surface charge. Indeed, isoelectric focusing, as it is often referred to, is perhaps the most discriminating, high-resolution protein separation method available. During the period from 1977 to 1981, Sluyterman and colleagues offered a theoretical basis and practical means of realizing the most favorable attributes of both ion-exchange chromatography and isoelectric focusing in a single chromatographic focusing procedure.[1-5]

During chromatofocusing, proteins are separated as a result of the isocratic formation of "internal" pH gradients on ion-exchange columns (Figure 5-1). Because proteins of similar net surface charge may still vary in surface charge distribution, it is even possible to resolve proteins during chromatofocusing that are not well separated by isoelectric focusing.

One of the great benefits of chromatofocusing is the ease of operation. No gradient-forming devices or mixers are required. The pH gradient is formed with a single eluent, isocratically. In the case of descending pH gradients formed using an anion-exchange stationary phase, the upper limit or starting pH (e.g., pH 8) is determined by the stationary-phase equilibration buffer. The lower or limit pH is determined by titration of the mobile phase or chromatofocusing buffer to the desired end point (e.g., pH 4). The pH gradient slope is a function of the column bed volume, buffer capacity of the matrix, and chromatofocusing buffer concentration as well as the initial and final adjusted pH of the stationary and mobile phases, respectively. The protein-focusing zone moves more slowly down the length of the column than does the focusing buffer or eluent (Figure 5-2). Repetitive protein-binding events driven by buffer flow against the resistance of the stationary phase to change pH causes the focusing phenomena to occur.

As a powerful analytical and preparative separation technique, chromatofocusing is in its infancy. The purpose of this presentation is to provide enough

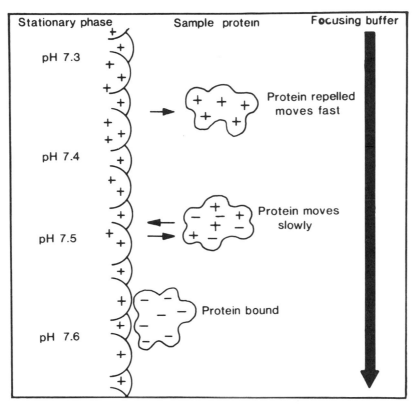

FIGURE 5-1. Illustration of the chromatofocusing process showing isocratic pH gradient formation, protein adsorption behavior, and migration of the focusing zone.

theoretical and practical guidelines to enable the successful use and further development of chromatofocusing as a high-resolution protein separation technique.

5.2 CHEMISTRY OF PROTEIN SURFACE CHARGE

The three-dimensional structure and overall surface molecular architecture of each different protein is unique. Most charged groups on proteins reside at the surface/aqueous interface. Locally, the immediate environment or milieu surrounding individual groups on the protein surface (e.g., His) may specifically affect the reactivity (e.g., pK_a) or chemical potential of that group. These surface reactive properties are influenced by both the mobile phase and the stationary phase during chromatofocusing.

The isoelectric point of a protein is defined as that pH in which the net charge is zero. Even though the majority of proteins (ca. 70%) examined to

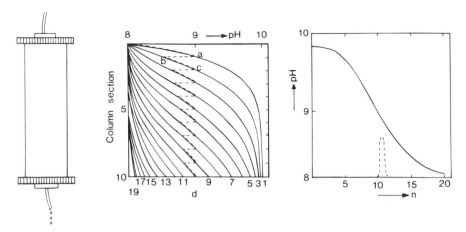

FIGURE 5-2. Formation of pH gradient in the chromatofocusing column. Each curve in the left diagram shows the gradient in the column (initial pH 10.0) after addition of a given number of unit volumes of focusing buffer (pH 8.0). The pH at the column outlet is given at the bottom of the left diagram and is used to construct the final chromatofocusing profile to the right. A sample protein with a pI of 9 in the column and the effluent is indicated by hatched lines. (Adapted from an illustration by Sluyterman and Wijdenes.[2]

date have isoelectric points in the range pH 5 to 8, considerable heterogeneity exists within this region. Several well-organized and useful compilations of protein molecular weights and subunit compositions[6] as well as apparent isoelectric points[7,8] have been published. It is important to emphasize that even though a protein with a net zero charge will not necessarily migrate in an electric field, it nevertheless possesses a given charge distribution that may be operative during chromatofocusing as the pH of the mobile phase alters the pH of the protein and the stationary phase.

5.2.1 Factors Influencing Protein Surface Charge

Proteins and peptides are amino acid polymers (often containing carbohydrate oligomers) with titratable functional groups varying in pK from approximately 3 to 13 (Table 4-2). It is the collective contribution of different buffering groups on the protein that allow it to be workably defined as relatively basic (pI > 7) or acidic (pI < 7). In buffers with pH values above its isoelectric point, the protein will have a net negative charge and interact with a positively charged matrix (e.g., anion-exchange resin) (cf. Figure 5-1). Below its isoelectric point an overall or net positive charge will result and the protein will of course not interact with positively charged structures. Most of the individual, titratable groups are located at the protein's surface, and a gradual change in pH will result in a continuously changing surface charge distribution. These changes can occur rapidly, in either direction, and are reversible. At pH values near the protein's isoelectric point, local charge distribution effects are important.

Several factors, both chemical and physical, can influence the protein surface charge. Intentional perturbations of this kind may include temperature, type, and concentration of ions (counterions, coions, and zwitterions) present, water structure (activity)-modifying reagents such as urea or ethylene glycol, and bound metal ions or other specific ligands, as well as detergents. In fact, the accurate determination of pH (see Section 5.6.2) is significantly influenced by solvent composition and temperature. Gelsema[9,10] and Ui[11] provide a more detailed consideration of these factors relative to the measurement of pH and pI. Unintentional influences on protein charge as a function of pH may include changes in protein quaternary organization (e.g., subunit dissociation or aggregation) and protein interaction with the stationary phase or polymeric ampholytes in the chromatofocusing buffer[12] or other macromolecules present. At a given pH, the mobile-phase ionic strength can influence protein surface charge and thereby its behavior, including interaction with other proteins and the stationary phase. (see Section 4.3).

5.2.2 Ionic Strength Effects and Protein Solubility

The solubility of a protein is partially a function of surface charge and thus can vary significantly with pH. Most proteins are least soluble at their isoelectric pH values due to minimized electrostatic repulsion. Thus, low ionic strengths also influence solubility at this pH (pI). The ionic strength of chromatofocusing buffers is best kept low to favor stationary phase–solute (i.e., protein) interactions. However, relative to isoelectric focusing in an electric field, chromatofocusing is, to a certain extent, independent of small changes in ionic strength.

Protein precipitation or aggregation at the isoelectric point is for many proteins a practical problem in chromatofocusing and has hampered the wide use of the technique. It is, however, not at all a compulsory reason to avoid the technique. Isoelectric precipitation is counteracted by the fact that proteins do not elute exactly at their pI. They need to have a slight net charge to adsorb to the ion exchanger. Increased ionic strength thus counteracts precipitation both because of the increased net charge needed for adsorption and because of the salt itself. A drawback is, however, that salt in the mobile phase adversely affects the pH gradient. This effect is considerably influenced by the choice and design of the stationary phase. Decreased protein load also diminishes the risk of precipitation. In fact, it should be noted that many proteins of considerably physiological interest occur in very low amounts and the concentrations needed for precipitation are never reached during chromatofocusing. Finally, if precipitation occurs the protein might still be recovered. For example, β-lactoglobulin is only retarded on the column. The protein precipitates but is dissolved continuously as the pH decreases below its isoelectric point.

5.3   STATIONARY-PHASE DESIGN

The choice of suitable stationary phases for chromatofocusing has increased with the availability of high-capacity, weak anion-exchange materials that are compatible with large biopolymers such as proteins. Those commercially available products listed in Table 5-1 are examples of both conventional and high-performance stationary phases that have been successfully employed for chromatofocusing.

The appropriate stationary phase for chromatofocusing depends very much on the intended application. In general, for linear pH gradient formation, it is a necessary and probably sufficient requirement for the stationary phase to have a significant and even buffering capacity over the pH separation range of interest. Depending on sample loading volumes and column capacity requirements, a percentage of permanent or fixed (nontitratable) charges can aid in separations over pH ranges outside the area in which the weaker groups titrate. Thus anion exchangers contain an amount of quaternary amines and cation exchangers sulfonates. A fundamental property of the stationary phase should be the lack of nonspecific attractions for the proteins or other solutes of interest. Such interactions vary among the several types of columns currently utilized for chromatofocusing due to basic differences in their chemical and physical nature. High porosity and rigidity are also important and have been achieved with several matrices (see Table 5-1).

The choice of polymer- or silica-based stationary phases for chromatofocusing depends on, among other things, the pH range of interest. The use of silica-based stationary phases is usually discouraged for applications requiring pH values above pH 8. However, there are exceptions depending on bonded phase chemistry. Other deciding factors include available equipment, cost, useful lifetime of the material, and preference.

5.3.1   Charge, Charge Density, and Capacity

pH gradients can be formed on ion exchangers over the range in which they titrate. To achieve a chromatofocusing effect, a descending pH gradient is formed internally on anion-exchange (cationic) stationary phases and, conversely, an ascending pH gradient is formed internally using suitable cation-exchange (anionic) matrices. Most practical experience, and this chapter, involves the use of weak anion-exchange resins (primarily mixtures of secondary and tertiary amines) for the development of descending pH gradients. The charge density (meq/m^2) of various anion-exchange matrices most often reflects the chemistry and capacity of the matrix to be derivatized.

The overall charge capacity (meq/mL gel) of a given stationary phase per unit bed volume is a property of particle size and porosity (i.e., surface area). Typical figures are 15 to 30 meq/mL, half of which are strong nontitratable groups. Not all of the stated charge (binding) capacity is necessarily available for protein due to geometrical constraints. In practical terms, the determina-

TABLE 5-1. Anion Exchange Stationary Phases Used for Chromatofocusing

	Immobilized Ligand or Bonded Phase	Stationary-Phase Composition	Particle Size (μm)	Average Pore Diameter	Reported Charge or Loading Capacity	pH Stability
Bakerbond WP-PEI (J. T. Baker, Inc.)	PEI	Silica	5, 15, and 40	250–300 Å	150–200 mg protein/g silica (40 μm)	pH 2–10
SynChroPake (SynChrom, Inc.)						
AX-300	PEI	Silica	6.5 and 7	300 Å	96 mg protein/g silica	<pH 8.5
AX-500	PEI	Silica	7	500 Å		<pH 8.5
AX-1000	PEI	Silica	10	1000 Å		<pH 8.5
Polybuffer Exchanger 94 (Pharmacia)	3° and 4° amines	Polymer Sepharose 6B	40–120	Porous	2–4 meq/100 ml gel	pH 3–12
Polybuffer Exchanger 118 (Pharmacia)	3 ° and 4° amines	Polymer Sepharose 6B	40–120	Porous	0.3–5 meq/100 ml gel	pH 3–12
Mono P (Pharmacia)	3° and 4° amines	Cross-linked hydrophilic	10	Porous	15–21 meq/100 ml gel	pH 2–12

tion of the stationary-phase working capacity is influenced by the choice of protein and interaction conditions (i.e., pH, ionic strength, temperature) as well as by nonspecific adsorption phenomena. Thus on Mono Q and Mono S (Pharmacia Biotech AB), 80 mg/mL bed volume of human serum albumin and 65 mg/mL of α-lactalbumin is adsorbed. For the much larger thyroglobulin the figure decreases to 25 to 30 mg/mL.

Finally, in addition to those stationary-phase charge groups with buffering capacity that participate in the generation of the internal pH gradient, additional permanent or fixed charges (e.g., quaternary amines) have been found useful in maintaining an even protein-binding capacity over a broad pH range.

5.3.2 Ionic Adsorption Properties

The ability of weak anion-exchange matrices to participate in electrostatic interactions with biopolymers is affected by pH as well as the type (i.e., pK_{a},), charge, and concentration of competing counterions. Thus, the operable pH range is best kept as narrow as is consistent with good separation. Generally, the initial adjusted pH of the anion-exchange column must be above (e.g., 0.5 to 1 pH unit) the pI of proteins to be adsorbed, so the buffering capacity of the stationary-phase ligand must be adequate at this pH. Similarly, the ionic strength of the chromatofocusing buffer must be less than that required to prevent the interaction of the designated protein(s) at the loading pH. These effects are variable and depend on the actual stationary-phase/mobile-phase system utilized. Table 5-2 gives recommended column equilibration buffers and elution buffers for developing pH gradients over several different ranges for use with Pharmacia Polybuffer Exchanger (PBE) and Mono P gels. In practice, both anionic and cationic as well as zwitterionic buffer constituents have been used successfully. Sometimes one good experimental effort provides acceptable results independent of accepted theory.

5.3.3 Buffering Capacity of the Stationary Phase

Anion-exchange columns and packing materials with suitable and even buffering capacity over a given pH range may be used effectively for chromatofocusing in that same pH range. The column or material in question can be titrated with acid (e.g., 0.5 to 2 N HCl) or base (e.g., 0.5 to 2 N NaOH) in the presence of elevated ionic strengths to calculate the buffering capacity (β) as originally outlined by Van Slyke.[14] Stationary phases with high and even buffering capacities over a fairly wide pH range have been developed specifically for chromatofocusing and are available commercially (see Table 5-1). Also, other cationic stationary phases developed for high-performance ion-exchange chromatography have been found to have excellent buffering capacities and pH gradient-generating qualities (Figure 5-3).

Certain columns have been carefully developed for a wide range of chromatofocusing applications, that is, pH ranges from pH 8 or 9 to pH 4. Yet, it is

TABLE 5-2. Buffer Substances for Focusing in Different pH Ranges

pH	Starting Buffer	Eluent	Dilution Factor	Approximate Volume (in Column Volumes)	
				Pregradient	Gradient
10.5–7	0.025 M triethylamine pH 11, HCl	Pharmalyte 8–10.5 pH 7.0 HCl	1:45	2.0	11.5
9–6	0.025 M ethanolamine pH 9.4, HCl	Polybuffer 96 pH 6.0, CH3C00H	1:10	1.5	10.5
8–5	0.025 M Tris pH 8.3, CH_3COOH	30% polybuffer 96 70% polybuffer 74 pH 5.0, CH_3COOH	1:10	2.0	10.5
7–4	0.025 M imidazole pH 7.4, HCl	Polybuffer 74 pH 4.0, HCl	1:8	2.5	11.5
5–4	0.025 M piperazine pH 5.5, HCl	Polybuffer 74 pH 4.0, HCl	1:10	3.0	9.0

These are based on instructions given in Pharmacia booklet (Ref. 13). Polybuffers are oligomeric ampholyte buffers titrating between pH 9 and 6 and pH 7 and 4 (PB 74).

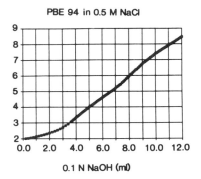

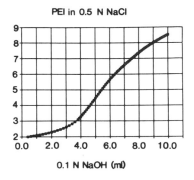

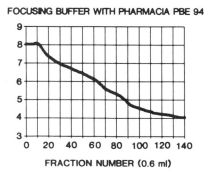

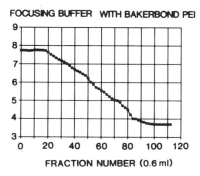

FIGURE 5-3. Titration curves of (1) a stationary phase specifically designed for chromatofocusing (Pharmacia Polybuffer Exchanger PBE 94) and (2) a silica-based high-performance anion-exchange stationary phase (J. T. Baker, Inc., Bakerbond WP-PEI). The characteristics of the pH gradients formed internally with a chemically defined focusing buffer are also illustrated for the two stationary phases.

difficult to maintain wide-range pH gradients that are linear over more than about 3 pH units. Narrow pH gradients (less than 2 to 3 pH units) may also be formed on columns with a buffering capacity limited to that same pH region. Thus, several other column types, as yet unexploited for chromatofocusing, may be found suitable for certain applications. It also needs to be emphasized that the stationary-phase buffering capacity and the pH gradient formed isocratically during separation need not necessarily be linear as long as a satisfactory or optimal separation of the proteins is obtained without it.

5.3.4 Column Volume and Geometry

The smallest possible chromatofocusing bed volume consistent with sample load requirements and resolution should be employed to minimize focusing buffer volumes and the time required for analysis. In general terms, the amount of buffering groups on the stationary phase and column volume defines the buffering capacity of the column. This determines the concentration and/or volume of focusing buffer required to generate a pH gradient of defined slope. Excessive column bed volumes also prolong column regeneration times. Good-quality analytical separations have been achieved using Pasteur pipettes or other narrow (e.g., 0.4 to 0.5 cm) diameter open columns packed with 2 to 4 ml bed volumes of Pharmacia Polybuffer Exchanger 94. An average chromatofocusing column is, however, 15 to 30 cm long. The pH gradient becomes less steep as it travels through the column, but after a certain distance further lengthening of the column does not change the result.

5.4 MOBILE-PHASE DESIGN

The single column eluent or mobile phase necessary to develop a particular pH gradient during chromatofocusing, like the mobile phase operative during isoelectric focusing, is actually a carefully designed set of buffering constituents. Collectively, these constituents make up one of the more complex buffer systems used in chromatography today. Relatively simple in concept, it is instructive to understand the mobile phase as an important variable contributing greatly to the efficiency and outcome of the desired separation.

5.4.1 Buffering Capacity ot the Mobile Phase

It was Van Slyke[14] who first described the concept of buffering capacity (β) and showed it to be an additive property of weak acids and bases in solution:

$$\beta = \frac{dB}{dpII}$$

where dB is the increment of base (B) added to the buffer solution and dpH is the incremental change in pH which results.

In the context of protein-focusing techniques, Svensson[15,16] first provided the theoretical calculations for using the buffering capacity of "carrier" ampholytes to create natural pH gradients in an electric field. The lack of simple ampholytes suitable for this purpose limited the use of isoelectric fractionation until Vesterberg[17] reported the synthesis of polyampholytes with the necessary isoelectric properties. The similarities of commercially available polyampholytes used for isoelectric focusing and chromatofocusing are not coincidental and the reader is urged to consult one of the many reviews on the subject (e.g., Ref. 18) for a more thorough appreciation of the synthesis and properties of polyampholytes.

Polyampholytes are designed to provide an even buffering capacity (β) over the pH range in which separation is to take place. Figure 5-4 illustrates the titration curve of Pharmacia Polybuffers 96 and 74. This curve represents

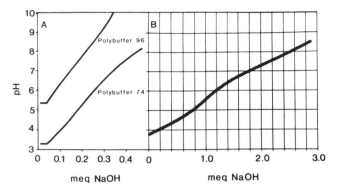

FIGURE 5-4. Titration curve of polymeric (polyampholyte) buffers (left) as well as of a simple nonpolymeric chromatofocusing buffer (right). Polybuffer 96 titrates between pH 9 and 6, whereas Polybuffer 74 titrates between pH 7 and 4.

the buffering action of titratable constituents with numerous successive pK_a values.

A generic alternative to oligomeric ampholyte focusing buffers is the use of chemically defined, nonpolymeric (so-called "simple") focusing buffers. Such universal or wide range buffer systems have existed for many years.[19–21] Several attempts have been made to extend this philosophy and formulate useful, "simple," focusing buffer systems for chromatofocusing.[22–26] Figure 5-4 also shows the titration curve of one such focusing buffer. Some examples of the utility of these focusing buffer systems are provided in Section 5.7.1 (also see Figure 5-5). Although earlier attempts to prepare simple, chemically defined, focusing buffers for electrophoretic focusing procedures[27–30] have been met with only partial acceptance, perhaps due to perceptions of electrofocusing theory (e.g., Ref. 18), the development of simple chromatofocusing buffer systems is expected to continue. There are presently no theoretical or practical limitations to the use and further improvement of nonpolymeric (i.e., very low molecular mass), chemically defined chromatofocusing buffer systems, especially for large-scale applications. For buffer use in general, Perrin and Dempsey[31] have published a useful book that provides a simplified overview of key considerations in buffer preparation and use.

In practice, pH gradient engineering for chromatofocusing is complicated by the contribution of stationary-phase effects.

5.4.2   Physical Properties of Chromatofocusing Buffer Constituents

A desirable chromatofocusing buffer will have sufficient and even buffering capacity over the pH range of interest. The ionic strengths have to be very low in order to give a separation according to isoelectric point, and thus to favor protein or solute interaction with the charged stationary phase during focusing. The light (UV)-absorbing properties should be minimal to permit

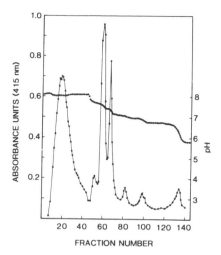

FIGURE 5-5. Fractionation of a commercial preparation of hemoglobin (15 mg) using the nonpolymeric focusing buffer described in Refs. 23 and 24 with Pharmacia Polybuffer Exchanger 94 as the stationary phase (1.0 × 5 cm). Two-minute fractions were collected at a flow rate of 0.5 to 0.6 mL/min. The average pH gradient slope (ΔpH/mL) from pH 7.75 to 6.75 was 0.01 to 0.02.

an adequate signal-to-noise ratio while monitoring the protein or solute elution (e.g., at 280 nm). The most successful chromatofocusing buffer should be composed of constituents that (1) are easily separated from most proteins and (2) do not interact with high affinity or irreversibly with the protein(s) being separated. In particular, the structure and/or function of the protein should not be practically or irreversibly compromised. The separation of proteins from polyampholyte chromatofocusing buffers is treated in more detail in Section 5.6.6.

The precise physical and chemical properties of oligoampholyte (oligomeric) focusing buffers are variable and poorly defined. The degree to which they interact with proteins remains the subject of considerable debate.[12,18] Even chemically defined, low molecular weight buffer constituents may interact with proteins.[28,29] Presumably, however, with defined chromatofocusing buffers these constituents can be identified and deleted or replaced if necessary. The degree to which these considerations pose a practical barrier must be determined on an individual base. There are strong arguments for the use and continued development of small, "noninteractive" focusing buffer systems.[22–30]

5.5   MECHANISM(S) OF SEPARATION

This section briefly evaluates existing theories of internal pH gradient formation on stationary phases, as well as the postulated protein separation mechanisms. The aim is to understand better the utility of given experiment variables for the solution of a given separation or analytical problem.

5.5.1 Theory of pH Gradient Formation

The initial assumptions (all-or-none model) and first approximation equations as originally outlined by Sluyterman and colleagues[2–5] provide a logical approach to the underlying principles generally utilized to explain the generation of internally developed pH gradients during chromatofocusing. They derived the following equation for final eluted pH (pH_f) when the two phases are mixed. The buffering capacity per defined unit (e.g., column length) of stationary phase is β_s and mobile phase is β_m

$$pH_f = \frac{(\beta_s)pH_s + (\beta_m)pH_m}{(\beta_s) + (\beta_m)} \tag{5-1}$$

In the simple case when β_s and β_m are assumed to be equal, Eq. (5-1) becomes

$$pH_f = \frac{pH_s + pH_m}{2} \quad (\beta_s = \beta_m) \tag{5-2}$$

That is, if the buffering capacities of the mobile phase and stationary phase are equal and additive, upon mixing, the resulting pH will reflect the simple average. To best illustrate the consequences of this assumption, Figure 5-6 shows the theoretical pH profiles generated using this equation. To begin, the pH of the stationary phase (pH_s) was adjusted to pH 8. The column (e.g., 100 individual units or sections) pH is indicated during the sequential addition of 100 aliquots of focusing buffer at pH 4, where each aliquot of focusing buffer is equal in buffering capacity to one unit or section of the column. The internal pH values of column sections 1, 10, 25, 50, and 75 are shown as a function of the number of focusing buffer aliquots added. Except for the extremes at either end, a series of linear pH gradients are developed.

These data reveal an important, experimentally verified, point. The slope of the pH gradient depends on the number of column sections for a given amount of focusing buffer. Also, if the buffering capacity of the mobile phase is increased relative to that of the stationary phase, the pH gradient will be greater in slope. Another important assumption for the model predicted by Eq. (5-2), aside from equal and additive buffering capacities for the stationary and mobile phases, is that of even (linear) and parallel buffering capacities over the pH ranges of interest. However, upon physical (electrostatic or hydrophobic) interactions between mobile-phase buffer constituents (e.g., polyampholytes) and the stationary phase, the functional buffering capacities (pK_a values) may be altered. Indeed, an unequal distribution of buffer constituents between mobile and stationary phases can result in uneven variations in the

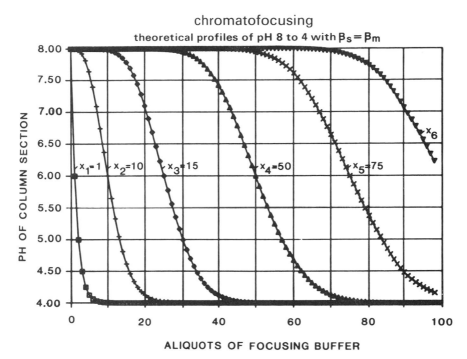

FIGURE 5-6. Theoretical pH profiles generated from pH 8 to pH 4 as a function of focusing buffer (volume) for each of 100 aliquots as described in the text. The pH gradient of the final column eluate is partially shown (far right). An equal buffering capacity of focusing buffer and stationary phase is assumed per unit volume. x_n is the number of column sections alternatively, column length.

column buffering capacity as a function of pH. So, while the buffering capacities of soluble buffering agents are additive, the buffering capacities of interacting mobile and stationary phase components may not always be additive. Nevertheless, it has been shown that the contribution of stationary-phase and mobile-phase buffering capacities alone can indeed be additive and account for the pH gradient formation predicted by Eq. (5-1). This was demonstrated using chemically identical (or similar), nonamphoteric buffering groups in both mobile and stationary phases.[25]

Despite the theoretical uncertainties involved, it has been shown that chemically defined mixtures of simple (nonpolymeric) buffer reagents,[22,23] nonamphoteric buffers,[25] and heterogeneous, polymeric ampholytes (e.g., Polybuffers 96 and 74) can be used to internally generate similar pH gradients.

5.5.2 Control of Separation Variables and Resolution

In practical terms, the most efficient means of controlling solute or protein separation efficiency (i.e., resolution) during chromatofocusing is by altering the shape and/or slope of the pH gradient generated. This may be accomplished for a given pH range by altering the stationary-phase bed volume (e.g., colum length) or changing the relative concentration of focusing buffer. Thus an increase in column length will only initially make the gradient more shallow, as explained earlier. The possibility of increasing the concentration of focusing buffer is also limited. A too high buffer strength that does not match the stationary phase will not give an even gradient. Its steepness will increase toward the end of the run.

Alternatively, the initial column equilibration pH (starting pH for gradient) and/or the final adjusted pH of the focusing buffer (limiting pH for gradient) may be adjusted to provide the desired pH gradient slope (ΔpH/mL). Protein resolution can be excellent with chromatofocusing. Protein peaks separated by 0.01 pH units have been reported.[26]

The specific elution pH of a given protein can also be affected by the inclusion of mobile-phase modifiers such as urea, metals, and detergents. These influence the way proteins in the sample behave in the separation.

5.5.3   Ampholyte Displacement Effects

Before the development of chromatofocusing, polyampholytes had been used to displace proteins from conventional anion exchangers. Ampholyte displacement chromatography, apparently first reported by Leaback and Robinson,[32] can be distinguished from chromatofocusing under specifically defined conditions. In particular, the formation of a pH gradient does not always appear to be necessary for the displacement of proteins when high concentrations of competing polyampholytes are used. Discussions have been presented to distinguish, theoretically and practically, chromatofocusing from ampholyte displacement chromatography.[33,38]

However, uncertainties remain as to the precise separation mechanism(s) operable in either case. Under the conditions thus far used for chromatofocusing, the distinction appears to be more apparent than real as these two protein separation processes do not necessarily have to be mutually exclusive. For example, as the pH approaches that point at which the protein interaction with the stationary phase becomes weaker, competitive binding of polyampholytes will act more effectively to displace the protein. Stated another way, as pH effects begin to decrease the affinity of protein for the stationary phase, competitive binding of ampholytes or other displacers may become a significant factor. The most basic polyampholytes (high pI values) elute first and others break through in order of their relative basicity. In practice, the separation mechanism may be irrelevant if purification is the desired end point. If the investigation is analytical or comparative in nature, unexpected elution behavior must be reconciled with the dominant elution mechanism.

5.5.4   Comparison with Isoelectric Focusing

It is important to emphasize that proteins may elute during chromatofocusing at pH values close to or well away from (e.g., 1 to 2 pH units) their isoelectric points. The several possible explanations for this phenomena are based on one of the major mechanistic differences between chromatographic focusing and electrophoretic focusing techniques, namely, the stationary phase.

Individual proteins may vary in their distribution and density of surface charge even if they are similar in their isoelectric points. Thus, differential interaction with the charged stationary phase is still possible. Because the slope of the titration curve for individual proteins (Δ surface charge/ΔpH) can vary significantly for a given difference between pI and pH, proteins may have markedly different affinities for the stationary phase. The interaction of a protein with a charged column matrix may itself alter the conformational response of that protein to changes in environmental pH. These processes must occur repeatedly during chromatofocusing because the focusing effect is the result of proteins repeatedly adsorbing to and desorbing from the charged matrix. It is during these interactions (at or near the pH of adsorption/desorption) that both concentration and composition of polyampholytes or salts may affect the elution of proteins differentially.

It has been postulated[2,3] that a Donnan potential between the interior and the exterior of the stationary-phase particles might decrease the elution pH of the proteins from chromatofocusing columns. During chromatofocusing on highly charged (positive), porous matrices, due to electrostatic repulsion, positively charged species in solution (including hydrogen ions, basic ampholytes, etc.) may be excluded and increase the pH of the buffer immediately surrounding the surface of the porous matrix.

To summarize, in contrast to electrophoretic focusing techniques, chromatofocusing relies on interfacial (solid/solution phase) molecular interactions that are influenced, in a collective manner, by such factors as pH, ionic strength, charge attraction and repulsion, and competitive binding reagents (displacers), as well as any "nonspecific," adsorption properties the stationary phase may have. It is therefore inappropriate to refer to elution pH as pI. However, it is not trivial to measure the pH on an isoelectric focusing gel, especially in the presence of various additives. Thus, it may also be questionable to equal the measured pH value and the actual pI.

5.6 ANALYTICAL AND PREPARATIVE CHROMATOFOCUSING TECHNIQUES

Chromatofocusing is a separation procedure useful for the analysis of protein surface charge heterogeneity as well as alterations in the surface charge properties of individual proteins at different stages of purification and in unfractionated biological fluids and extracts. However, in addition to the small-scale analysis and separation of proteins or protein isoforms, chromatofocusing can be useful on a larger scale for preparative protein purifications. Chromatofocusing, as a sample concentrating procedure, is particularly useful at early or intermediate stages in the protein isolation scheme. Since proteins are eluted from anion-exchange columns during the formation of descending internal pH gradients, the protein is not subjected to pH levels below that just required for elution.

It is also possible to collect fractions or individual protein peaks into pH neutralizing buffer solutions to preserve bioactivity (see Section 5.7.1).

The pH gradient formed during chromatofocusing can be monitored continuously along with sample elution during the separation experiment. Thus, gradient slope and shape may be modified as needed to alter separation efficiency (see Section 5.7.1). It is not always desirable to use linear pH gradients. In fact, nonlinear gradients can be predesigned to maximize the separation of given components.

5.6.1 Analytical to Preparative Scale Applications

Substantial increases in sample volume, sample protein concentrations and/ or column bed volume can often bring changes (e.g., pH gradient slope) not easily anticipated from analytical scale operations. Depending on sample volumes or quantities of protein needed, utilizing available high-performance stationary phases with rapid analysis times to make replicate "analytical" scale separations under familiar conditions may be more advantageous than altering column dimensions to achieve the desired end product. There are, however, no theoretical or practical reasons why chromatofocusing cannot be used on a large scale to address difficult separation problems.

Since chromatofocusing is indeed a focusing procedure, unlike zonal elution chromatography, a large part of the bed volume can be utilized during sample loading. Isocratic pH gradient formation greatly simplifies equipment requirements (i.e., controllers and mixers) for large volume applications. The continued development and commerical availability of suitable, inexpensive focusing buffers may further reduce the cost of large-scale chromatofocusing separations.

Large sample volumes can bring unexpected changes in protein elution behavior if sample contaminants or buffer constituents, even if present at the same concentrations as for analytical scale operations, build up in the sample loading zone. EDTA is an example of one common buffer constituent that causes no problems when present in minor quantities (small sample volumes) but can create serious problems with larger sample volumes. EDTA binds to most anion-exchange stationary phases used for chromatofocusing and elutes in a pH-dependent manner with absorbance at 280 nm,

5.6.2 pH Measurements

Particularly for analytical or comparative experiments by chromatofocusing, accurate and reproducible pH measurements are essential. Aside from pH electrode type and stability, the most immediate concerns are the temperature of separation and pH measurement (they should be the same), dissolved CO_2, and mobile-phase additives such as urea, glycerol, or anything else that affects the H^+ activity of aqueous eluents (see Refs. 9–11).

Flowthrough pH electrodes are convenient but may not be as accurate or reproducible as needed for sensitive analytical measurements. Of course, if the primary concern is resolution and separation, pH measurements are of less concern.

5.6.3   Sample Preparation and Column Loading

Proper sample preparation and column loading procedures will help increase resolution, sample recovery, and useful column lifetime. It is necessary to remove all sample components that are insoluble (e.g., precipitates) or immiscible (e.g., lipid layers or micelles) in the chosen sample and column equilibration buffer. Brief centrifugation at high speeds (10,000 to 15,000 g; 10 to 15 min) or filtration through inert, porous (0.22 or 0.45 μm) membranes (e.g., Gelman or Millipore) will normally suffice to ensure the application of fully soluble sample components if these procedures are performed at the temperature of chromatography. After equilibration of the sample into the column equilibration buffer, difficulties with protein solubility, when encountered, may be dealt with in several different ways. It is also possible to load the sample in focusing eluent buffer after that the gradient has developed on the column. Very diluted samples can of course also be used.

5.6.4   Solubility of Proteins

Sample protein solubility during chromatofocusing, depending on the protein, may become a necessary consideration. Protein solubility during chromatofocusing is most probably influenced by pH as well as by the concentration and type of salts (i.e., buffer and counterions) present (see Section 5.2.2). Several mobile-phase (focusing buffer) modifiers are helpful in minimizing solubility problems, and also increase resolution during protein separation. The low ionic strength column and sample equilibration buffers utilized during chromatofocusing sometimes induce protein aggregation even at the start pH, which may be well above (e.g., 0.5 to 2 or more pH units) the isoelectric point of the protein of interest. In this case, the addition of urea or nonionic detergent to the sample can sometimes prevent aggregation until protein adsorption to the stationary phase and/or gradient development begins and aggregation may no longer be a problem.

In fact, adding concentrations of urea from 1 to 6 M in the sample preparation buffer and both the column equilibration and focusing buffer is an effective, probably underutilized, means of optimizing sample solubility and resolution during chromatofocusing. Many proteins retain their tertiary and/or quaternary structure as well as their activity in relatively high concentrations of urea, especially at 4°C. Low concentrations of detergents may also be utilized to maintain protein solubility without compromising biological activity. Triton X-100, Nonidet P-40, and Tween 80 are all examples of nonionic detergents (available from Sigma) that have been used effectively during chromatofocusing. CHAPS (Cal Biochem) is a zwitterionic detergent prepared from cholic acid that has also been found useful in preventing aggregation and maintaining high recoveries during chromatofocusing. Low concentrations (4 to 10%, w/v) of zwitterions (e.g., betaine or taurine) and glycerol have been found to be effective solubilizing/stabilizing agents for some proteins during chromatofocusing.

5.6.5   Inclusion of Mobile-Phase Modifying Reagents

The mobile phase includes the chromatofocusing column equilibration buffer and the focusing buffer used to generate the internal pH gradient. Since the protein migration zone or focusing zone proceeds slowly down the length of the chromatofocusing column relative to the flow of buffer, the sample protein may be exposed to 10 to 20 column volumes of buffer before finally eluting. Thus, specific buffer constituents required to maintain protein structure (or solubility) and/or function must be included in the focusing buffer. Fortunately, most of them can be included. In fact, in contrast to focusing in an electric field where current (conductivity) is important, even relatively high concentrations of charged components can be included and maintained during chromatofocusing. A good example is the inclusion of 10 mM sodium molybdate in steroid receptor chromatofocusing buffers. The effect of molybdate on the stability of receptor structure and function is easily reversible but could be maintained during chromatofocusing. Other salts can be included (up to 20–30 mM) to influence protein solubility or prevent aggregation.

Urea, detergents, and glycerol have been especially useful in chromatofocusing buffers with "sticky" proteins.

There is a problem with mobile-phase modifiers when used to analyze surface charge heterogeneity or when comparisons of surface charge are being made. These additives can cause significant error in pH measurements. They can also cause major shifts in protein elution pH, which vary in a protein-dependent manner. Again, if purification is the only objective, elution pH may not be a major concern.

5.6.6   Sample Recovery

Some proteins can be encountered that are surprisingly resistant to pH-dependent elution during chromatofocusing. This depends very much on the properties of the individual protein and, just as much, on the properties of the stationary phase chosen for chromatofocusing. The number of different types of protein present (e.g., crude mixtures vs purified samples) often influences recovery simply due to the mass action of adsorption. Highly purified proteins can often be further purified or analyzed in the presence of carrier proteins to saturate so-called nonspecific adsorption sites. Alternatively, mobile phase additives (e.g., detergents) have also been effectively utilized to prevent adsorption.

If the protein being analyzed or purified is a receptor or other ligand-binding protein whose activity is determined by the association with ligand (e.g., steroid–receptor complex or metal–protein complex), it is important that the stationary phase have little or no affinity for the ligand in question. For example, [³H]estradiol was found to bind to SynChropak AX-300, AX-500, and AX-1000 silica-based anion-exchanged columns and elute in a pH-dependent manner.[39–42] However, no steroid interaction was evident using Bakerbond PEI (J. T. Baker) as the stationary phase.

The separation of proteins from focusing buffer constituents (most commonly polyampholyte buffers such as Pharmacia Polybuffers 74 and 96) can normally be achieved by size exclusion chromatography (e.g., Sephadex G-50 or G-75 Fine), precipitation with ammonium sulfate, hydrophobic interaction chromatography, or other types of affinity chromatography. However, certain polyampholyte properties[12,18] and reported difficulties in protein–polyampholyte separations[12,23] indicate that some caution should be used in defining adequate removal of focusing buffer constituents from the specific protein of interest.

5.6.7 Column Regeneration

In general, sample recovery and column maintenance require the routine use of high salt concentrations after completion of the pH gradient to help ensure quantitative elution of adsorbed material. A gradual rise in back pressure is an alert to a growing column contamination problem. Postanalysis column washing and/or column regeneration procedures are best performed with reversed flow, if possible. Typically, 1 to 2 M solutions of neutral salt (e.g., NaCl) should be adequate, although more chaotropic salts can be used, especially if hydrophobic adsorption is suspected. A useful practice is to utilize 1 to 2 M concentrations of equilibrating buffer adjusted to the upper or starting pH to facilitate column regeneration. The removal of more difficult proteins can often be accomplished with 8 M urea or 6 M guanidine–HCl. Lipoproteins and lipids can be removed with one or more of several different organic solvents. Ethylene glycol (e.g., 50%), diethyl ether, dimethylformamide, or dimethyl sulfoxide (e.g., 10 to 50%), acetonitrile (up to 100 ± 0.1% trifluoroacetate), ethanol, methanol, and other solvents can be used regularly. Care should be taken to ensure that all salts are washed away (Milli Q water) and that as many proteins as possible are removed before using organic solvents to avoid precipitation. Multiple injections of 0.1 M HCl and/or 0.1 M NaOH are also an effective cleaning procedure. Columns may be stored in 20% methanol as a preservative. Sodium azide should be avoided. The use of clean buffers (especially containers) will reduce the risk of bacterial contaminations of "clean" columns.

5.7   APPLICATIONS

5.7.1   Analysis of Surface Charge Heterogeneity and Purification of a Labile DNA Regulatory Protein: The Estrogen Receptor

The utilities of chromatofocusing are revealed by the examples provided below illustrating the characterization and purification of various steroid receptor protein isoforms. Because chromatofocusing is preparative even at the analytical scale and because the procedure is amenable to the inclusion of various structure-modifying reagents, it has been extremely useful in efforts to better define receptor, structures and their functional relationships.

The intracellular receptor protein(s) for the female sex hormone estradiol has been evaluated for surface charge heterogeneity and partially purified by chromatofocusing using several different types of stationary phases, including conventional open columns of polymeric anion-exchange gels[39,40] and high-performance polymeric[41] and silica-based2[3–26,39,42–44] anion-exchange matrices. During these investigations, both polyampholytes and nonpolymeric, chemically defined focusing buffers have been utilized to generate similar pH gradients and receptor elution profiles.[23,24,26] One goal of the original investigations[42] was to develop a rapid, high-resolution protein separation technique to better evaluate and purify the different forms of steroid receptor proteins known to exist in the presence and absence of (1) natural steroid ligand,[43] (2) receptor structure-stabilizing metal ions,[42,43] and (3) receptor structure modifying reagents known to induce exposure of the receptor's DNA-binding site.[44]

The steroid receptor proteins are identified by their ability to bind specific steroid ligands with high affinity. To extract and radiolabel the estrogen receptor proteins, endocrine target tissues (e.g., uterus or mammary glands) are homogenized in 2 to 4 volumes of buffer and centrifuged (>100,000 g, 60 min) to obtain particulate-free cytosol. The cytosol is then labeled with [^{3}H]estradiol-17β or [^{125}I]iodoestradiol-17β in the presence and absence of a radioinert, estrogen receptorspecific competitor (diethylstilbestrol) to identify receptor from nonspecific estrogen-binding proteins.[42] Cytosol preparations (10 to 20 mg protein/mL) containing what are considered to be biologically relevant concentrations of receptor protein (typically from as low as 5 to 10 fmol up to several hundred femtomole receptors per milligram cytosol protein) are equilibrated into chromatofocusing column equilibration buffer using minicolumns of Sephadex G-25. The chromatofocusing column equilibration buffer usually consists of 25 mM Tris-HCl, pH 8.0 (4°C), containing dithiothreitol (1 mM) and glycerol (10 to 20%, v/v). Where noted, 10 mM sodium molybdate, 6 M urea, detergent (CHAPS), and other receptor structure modifying reagents are included in the sample buffer, stationary-phase equilibration buffer, and focusing buffers. All procedures, including equilibration buffer and focusing pH determinations, are carried out at 0 to 4°C. Buffers and samples are filtered to remove particulate contaminants using Millipore

0.45-μm HAWP filters. Buffers are degassed before final pH adjustments and use. Sample aliquots of 0.5 to 2.0 mL (5 to 20 mg total protein) are usually loaded onto the column prior to initiation of the pH gradient. The formation of the internal pH gradient is initiated using either polyampholytes (Pharmacia Polybuffers 96 and 74) or other nonpolymeric, chemically defined focusing buffers[22-24] at flow rates of 0.5 to 1.0 mL/min. The pH of collected fractions (1.0 min) is determined immediately upon elution using a Corning Model 125 pH meter equipped with a microcombination calomel electrode. Flowthrough pH electrodes (Phoenix Electrodes, Houston, TX) have also been used. The protein elution profile is monitored at 254 or 280 nm using a Beckman Model 153 or Model 166 analytical flowthrough UV detector (8-μL flow cell) or other suitable spectrophotometer. The elution profile of labeled estrogen receptor proteins was monitored by liquid scintillation counting (30 to 38% efficiency) or gamma counting (60 to 80% efficiency).

The estrogen receptor elution profile shown in Figure 5-7 is typical of that obtained for the human uterine estrogen receptor chromatofocused in the

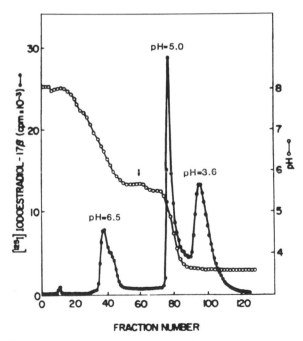

FIGURE 5-7. Open column chromatofocusing of human uterine estrogen-binding proteins on PBE 94. Cytosol was prepared from a fresh specimen of postmenopausal human uterus and labeled with 10 nM [³H]estradiol-17β. Termination of the labeling reaction and details of chromatofocusing on PBE 94 are described elsewhere.[39] Arrows mark initiation of the primary pH gradient [Polybuffers 96 and 74 (30:70) diluted 1:15, pH 4 at 0°C] and secondary eluent (Polybuffer 74 diluted 1:15, pH 3 at 0°C), respectively.

absence of structure-stabilizing or structure-destabilizing mobile-phase reagents. Open columns (0.7 cm i.d. × 9 cm) of PBE 94 were used with Polybuffers 96 and 74. Relatively small bed volumes of chromatofocusing gel were utilized to minimize separation times. The rather labile steroid receptor proteins are stabilized (reversibly) by the presence of relatively high concentrations of molybdate. Figure 5-8 illustrates an important property or chromatofocusing discussed earlier, namely, the ability to maintain continuous concentrations of mobile-phase modifies or stabilizers in both column equilibration and focusing buffers. It shows the elution behavior of human uterine estrogen receptor proteins chromatofocused using SynChropak AX-5OO high-performance anion-exchange columns in the absence and presence of 10 mM sodium molybdate. The practical use of simple, chemically defined focusing buffers is shown in Figure 5-9 for the separation of calf uterine estrogen receptor forms showing surface charge heterogeneity. Polyethyleneimine-derivatized high-performance silica-based columns (Bakerbond WP-PEI) from J. T. Baker Inc. were used for the experiments

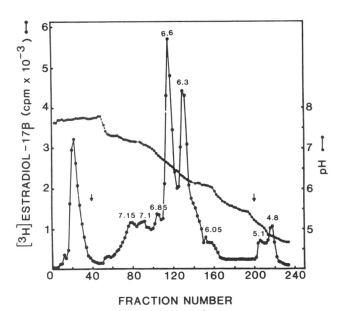

FRACTION NUMBER

FIGURE 5-8. Identification of molybdate-stabilized receptor species by HPLC chromatofocusing on AX-500 in the continued presence of molybdate, Human uterine cytosol was prepared and labeled with [125I]iodoestradiol-17β in the presence of 10 mM sodium molybdate. Receptor preparations (2 to 4 mg) were eluted from the AX-500 column with a biphasic pH gradient. The primary eluent was a 30:70 mixture of Polybuffers 96 and 74, diluted 1:10 with 20% glycerol containing 10 mM sodium molybdate and adjusted to pH 5.0. The secondary eluent (initiated at arrow) was Polybuffer 74, diluted 1:10 with 20% glycerol (no molybdate) and adjusted to pH 3.5. The recovery of activity in this representative experiment was 91%.

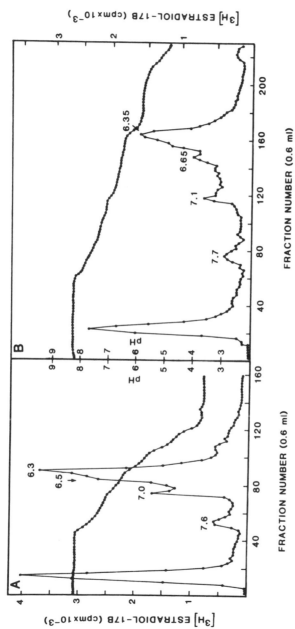

FIGURE 5-9. Surface charge heterogeneity of calf uterine estrogen receptor proteins evaluated by high-performance chromatofocusing. The internal pH gradients (O) were developed isocratically on Bakerbond WP-PEI columns (J. T. Baker, Inc.) using a nonpolymeric focusing buffer adjusted to pH 4 (see Refs. 23 and 24). The wide-range pH gradient (A) was generated using a single 25-cm PEI column. The narrow-range pH gradient (B) was generated using the same focusing buffer with two of the 25-cm PEI columns connected in series. In each case a 500-μl aliquot of labeled cytosol (22 mg protein/mL) was eluted at a flow rate of 0.6 mL/min and 1.0-min fractions were collected.

shown here with the defined focusing buffer. Since the estrogen receptor proteins have high-affinity binding sites from polyanions and are also relatively hydrophobic, one purpose of these experiments was to determine if polymeric ampholytes in the polybuffers previously utilized to evaluate receptor surface charge were themselves influencing receptor heterogeneity. These profiles also illustrate nicely how increased column buffering capacity can be used with a single focusing buffer to decrease the pH gradient slope and increase resolution. The average gradient slope (ΔpH/ΔmL) was decreased from approximately 0.1 to 0.03 over the separation pH range by doubling the column length. The use of this same nonpolymeric focusing buffer can be used to generate a narrow pH gradient on a polymer-based chromatofocusing column (Pharmacia PBE 94) to fractionate a commercial source of hemoglobin (Figure 5-5).

The unliganded form of any steroid receptor protein is even more labile than the steroid-bound form. The separation and recovery of unliganded estrogen receptor (aporeceptor) forms with apparently unaltered biological activity (i.e., steroid binding) were possible by collecting receptor fractions, directly into pH-neutralizing buffer solutions containing radiolabeled steroid.

Figure 5-10 shows how the inclusion of 6 M urea in both column equilibration buffers and focusing buffers was utilized to differentially affect the chromatofocusing elution properties of unfractionated cytosolic versus DNA affinity-purified calf uterine estrogen receptor forms. In this manner, by differential chromatofocusing, two previously uncharacterized receptor forms were identified, separated, and partially purified for further analysis.

pH gradient engineering can also be effectively utilized to maximally separate and purify protein isoforms. Affinity-purified estrogen receptors, indistinguishable by sedimentation or size exclusion analyses, were resolved into two major subfractions using a biphasic descending pH gradient.

5.7.2   Analysis and Preparative Scale Isolation of Heterogeneous Forms of Cyanobacterial Ferredoxin-NADP⁺ Oxidoreductase

Serrano[45] achieved nice results using chromatofocusing as both an analytical and a preparative tool to evaluate and separate molecular heterogeneous forms of ferredoxin-NADP oxidoreductase (FNR) (EC 1.18.1.2) from the nitrogen-fixing cyanobacterium *Anabaena* sp. strain 7119. The cells were grown in culture, harvested, washed, and stored at $-20°C$ before use. The cells were then thawed and disrupted by exposure to buffered solutions containing a 2% solution of the nonionic detergent Triton X-100. Duplicate batches of cells were processed in the presence and absence of the serine protease inhibitor phenylmethylsulfonyl fluoride (PMSF). All of the FNR enzyme activity was recovered in the supernatant liquid after centrifugation at 40,000 g for 2 min. The FNR (M_r 33,000 to 38,000) was purified by chromatography on DEAE-cellulose (Whatman DE-52) and 2′,5′ ADP-Sepharose 4B (Pharmacia).

Chromatofocusing was utilized to analyze molecular heterogeneity of the FNR purified in the presence and absence of PMSF. A column (1 $\times$ 25 cm)

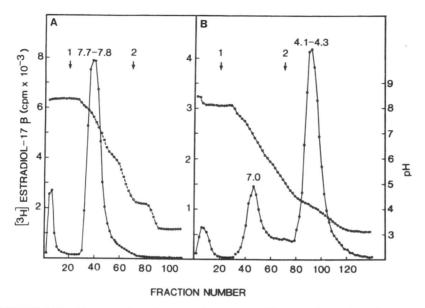

FIGURE 5-10. Chromatofocusing to evaluate urea effects on the surface charge properties of estrogen–receptor complexes purified by DNA affinity chromatography. Purified estrogen–receptor complexes were analyzed by chromatofocusing on PBE 94 in both the absence (A) and the presence (B) of 6 M urea. Nonspecific estrogen-binding components were absent after receptor purification by DNA affinity chromatography in the presence of 6 M urea. The specific elution pH of estrogen–receptor forms is indicated.

of Pharmacia PBE 94 was equilibrated with 25 mM piperazine–HCl buffer, pH 5.3. Samples of the purified FNR (3 to 13 mg in 5 mL) were dialyzed against the PBE column equilibration buffer before application. The focusing pH gradient was initiated by elution (12 mL/h) with 10-fold diluted Polybuffer 74 adjusted to pH 4 with HCl. As shown in Figure 5-11, a narrow (shallow slope) pH gradient was quite effective in the separation of (nicked) FNR enzyme generated by purification in the absence of the protease inhibitor PMSF. The one minor and three major peaks of FNR resolved (elution pH 4.3 to 4.6) were recovered in milligram quantities for subsequent characterization by SDS–PAGE, isoelectric focusing, and amino acid analysis. The high-resolution and preparative nature of chromatofocusing was clearly used in this example.

5.7.3   Isolation of Human Very-Low-Density Lipoproteins

In this example, Weisweiler and co-workers[46] utilized highly porous (>500 Å), monodisperse polymer particles (10 μm) developed by Ugelstad[47] and derivatized by Pharmacia (Mono P) for chromatofocusing to improve their separations of human very-low-density lipoproteins (VLDL). Serum was pre-

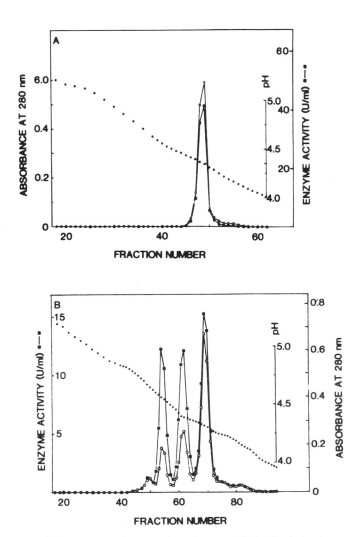

FIGURE 5-11. (A) Column chromatofocusing on PBE 94 of *Anabaena* sp. strain 7119 FNR purified in the presence of the protease inhibitor PMSF. A sample containing 3.5 mg of protein was applied to a PBE 94 column (1 × 25 cm), and the enzyme was eluted by using a pH gradient (●) generated with Polybuffer 74–HCl, pH 4.0 (7.5 μmol/pH unit/mL). Fractions of 2 ml were collected. Absorbance at 280 nm (○) and enzyme activity (■) were measured for each fraction. (B) Column chromatofocusing on PBE 94 of *Anabaena* sp. strain 7119 FNR purified in the absence of the protease inhibitor PMSF. A sample containing 13.5 mg of protein, purified from the same batch of cells used in A, was chromatographed under the conditions described for A.

pared from hypertriglyceridemic patients by a combination of low-speed centrifugation and ultracentrifugation (160,000 g for 24 h) at 4°C. VLDL was removed from the top of the supernatant, dialyzed against 0.01 M ammonium bicarbonate buffer (pH 8.6), and lyophilized. Lipids were removed by extraction with diethyl ether and ethanol/diethyl ether. The VLDL apolipoproteins were solubilized in the chromatofocusing equilibration buffer, which consisted of 0.025 M bis–tris (pH 6.3) containing 6 M urea. Sample protein (up to 10 mg) was applied to a 4-mL prepacked Mono P column (5 mm i.d. × 20 cm) in column equilibration buffer, and the descending pH gradient was developed with 30 mL of (10-fold diluted) Polybuffer 74 (pH 4.0) at a flow rate of 1 mL/min. Figure 5-12 shows the chromatofocusing elution profile of the VLDL apolipoproteins. Seven peaks were resolved and collected. Polybuffers and urea were removed by dialysis and gel filtration (Sephadex G-50 fine). Overall recovery was reported to be 78%. Individual VLDL protein fractions were identified by double immunodiffusion in 2% agar using monoclonal antibodies before further analysis by isoelectric focusing and SDS–PAGE. The speed of VLDL apolipoprotein separation by chromatofocusing (30 min) on MonoBeads (Mono P) was considered a significant improvement over previously utilized, more time-consuming procedures.

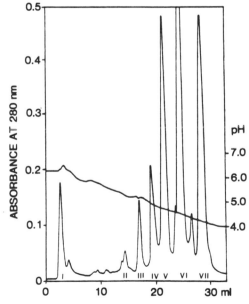

FIGURE 5-12. Fractionation of soluble VLDL apolipoproteins by chromatofocusing on 10-μm monodisperse beads. Apolipoproteins (approximately 8 mg protein) were applied to a prepacked column (pH 6.3) in 6 M urea (Mono P). The pH gradient was developed with 30 mL of 1/10 diluted polybuffer 74 (pH 4.0) containing 6 M urea. The absorbance at 280 nm and the pH of each fraction are as shown. Seven fractions were collected as indicated.

5.8 SELECTED OTHER APPLICATIONS

Steroid sulfotransferase from human fetal liver was purified approximately 1,800-fold by a combination of ion-exchange chromatography, chromatofocusing, and affinity chromatography.[48] A 90-kDa glucoamylase from *Aspergillus niger,* a glycoprotein, was purified by a combination of ion-exchange chromatography, chromatofocusing, and Con A affinity chromatography.[49] The major lignin peroxidase isoenzymes from the basidiomycete *Phanerochaete chrysosporium* were purified to high homogeneity by a one-step separation by chromatofocusing with a recovery of over 80%.[50] Chromatofocusing was used for the large-scale purification of thymosin *β*-4 and thymosin *β*-9 from bovine lung tissue (2 kg). The method developed is rapid and convenient and can be used for the isolation of thymosin *β*-4 and homologous peptides from various tissues.[51] A 32-kDa juvenile hormone from the hemolymph of the silkworm, *Bombyx mori* (pI 4.9), was purified by a combination of gel filtration, ion-exchange chromatography, chromatofocusing, and hydroxyapatite chromatography.[52] The *β*-hemolysin from *Staphylococcus aureus* was purified 38,000-fold to homogeneity in just two steps using glass bead chromatography and chromatofocusing.[53] Two novel chitinases from *Streptomyces griseus* were purified by a procedure based on ammonium sulfate precipitation, hydrophobic interaction chromatography, and chromatofocusing. Both enzymes had a molecular mass of approximately 27 kDa, but had different pI values (7.7 and 7.3).[54] Six glycoforms of plasminogen 2 were purified by a combination of lectin affinity chromatography and chromatofocusing on the basis of their sialic acid content.[55] Finally, an extracellular dextranase from *Penicillium notatum* was purified from a crude enzyme powder extract by sequential anion- and cation-exchange chromatography and chromatofocusing. Two highly purified fractions were obtained with molecular masses of 55.8 and 50.1 kDa and isoelectric points at pH 4.9 and pH 4.75, respectively.[56]

5.9 REFERENCES

1. L.A. AE. Sluyterman, J. Wijdeness, in *Proc. Int. Symp. Electrofocusing Isotachophoresis,* B.J. Radola, Graesslin, eds., de Gruyter, Berlin, 1977, pp. 463–466.

2. L.A. AE. Sluyterman, O. Elgersma, *J. Chromatogr., 150,* 17–30 (1978).

3. L.A. AE. Sluyterman, J. Wijdenes, *J. Chromatogr., 150,* 31–44 (1978).

4. L.A. AE. Sluyterman, J. Wijdenes, *J. Chromatogr., 206,* 429–440 (1981).

5. L.A. AE. Sluyterman, J. Wijdenes, *J. Chromatogr., 206,* 441–447 (1981).

6. D.W. Darnall, I.M. Klotz, *Arch. Biochem. Biophys., 166,* 651–682 (1975).

7. P.G. Righetti, T. Caravaggio, *J. Chromatogr., 127,* 1–28 (1976).

8. P.K. Righetti, G. Tudor, K. Ek, *J. Chromatogr., 220,* 115–194 (1981).

9. W.J. Gelsema, C.L. DeLigny, *J. Chromatogr., 130,* 41–50 (1977).

10. W.J. Gelsema, C.L. DeLigny, N.G. Van Der Veen, *J. Chromatogr., 140,* 149–155 (1977).

11. N. Ui, *Biochim. Biophys. Acta, 229,* 567–581 (1971).

12. L.S. Rodkey, *Protides Bio. Fluids, 34,* 745–748 (1986).

13. Chromatofocusing with Polybuffer and PBE, Pharmacia Biotech AB, Uppsala, Sweden, 1980.

14. D.D. Van Slyke, *J. Biol. Chem., 52,* 525–570 (1922).

15. H. Svensson, *Acta Chem. Scand., 15,* 325–341(1961).

16. H. Svensson, *Acta Chem. Scand., 16,* 456–466 (1962).

17. O. Vesterberg, *Acta Chem. Scand., 23,* 2653–2666 (1969).

18. P.G. Righetti, in *Laboratory Techniques in Biochemistry and Molecular Biology,* T.S. Work, R.H. Burdon, eds., Elsevier, Amsterdam, 1983.

19. H.T.S. Britton, R.A. Robinson, *J. Chem. Soc.,* 458–473 (1931).

20. H.T.S. Britton, R.A. Robinson, *J. Chem. Soc.,* 1456–1462 (1931).

21. D.A. Ellis, *Nature (London), 191,* 1099–1100 (1961).

22. R.L. Prestidge, M.T.W. Hearn, *Sep. Purif. Methods, 10,* 1–28 (1981).

23. T.W. Hutchens, C.M. Li, P.K. Besch, *J. Chromatogr., 359,* 157–168 (1986).

24. T.W. Hutchens, C.M. Li, P.K. Besch, *J. Chromatogr., 359,* 169–179 (1986).

25. T.W. Hutchens, *Prot. Bio. Fluids, 34,* 749–752 (1986).

26. T.W. Hutchens, C.M. Li, P.K. Besch, *Prot. Bio. Fluids, 34,* 765–768 (1986).

27. N.Y. Nguyen, A. Chrambach, *Anal. Biochem., 74,* 154–163 (1976).

28. C.B. Cuono, G.A. Chapo, *Electrophoresis, 23,* 65–75 (1982).

29. C.B. Cuono, G.A. Chapo, A. Chrambach, L.M. Hjelmeland, *Electrophoresis, 4,* 404–407 (1983).

30. M.T.W. Hearn, D.J. Lyttle, *J. Chromatogr., 218,* 483–495 (1981).

31. D.D. Perrin, B. Dempsey, *Buffers for pH and Metal Ion Control,* Chapman and Hall, London, 1974.

32. D.H. Leaback, H.K. Robinson, *Biochem. Biophys. Res. Commun., 67,* 248–254 (1975).

33. M. Page, M. Belles-Isles, *Can. J. Biochem., 56,* 853–856 (1978).

34. J.L. Young, B.A. Webb, D.G. Coutie, B. Reid, *Biochem. Soc. Trans., 6,* 1051–1054 (1978).

35. J.L. Young, B.A. Webb, *Anal. Biochem., 88,* 619–623 (1978).

36. J.L. Young, B.A. Webb, *Sci. Tools, 25,* 54–56 (1978).

37. J.L. Young, B.A. Webb, *Prot. Bio. Fluids, 27,* 739–742 (1979).

38. J.P. Edmond, M. Page, *J. Chromatogr., 200,* 57–63 (1980).

39. T.W. Hutchens, W.E. Gibbons, P.K. Besch, *J. Chromatog., 297,* 283–299 (1984).

40. A.H. Couffalik, M. Feldman, M. Platica, L. Toth, D. Dreiling, V.P. Hollander, *Cancer Res., 43,* 5235–5242 (1983).

41. FPLC ion Exchange and Chromatofocusing, Pharmacia Biotech AB (undated).

42. T.W. Huchens, R.D. Wiehle, N.A. Shahabi, J.L. Wittliff, *J. Chromatogr., 266,* 115–128 (1983).

43. T.W. Hutchens, H.E. Dunaway, P.K. Besch, *J. Chromatogr., 327,* 247–259 (1985).

44. T.W. Hutchens, C.M. Li, P.K. Besch, *Biochemistry, 26,* 5608–5616 (1987).

45. A. Serrano, *Anal. Biochem., 154,* 441–448 (1986).

46. P. Weisweiler, C. Friedl, P. Schwandt, *Biochim. Biophys. Acta, 875,* 46–51 (1986).

47. J. Ugelstad, L. Söderberg, A. Berge, J. Bergström, *Nature (London), 303,* 95–96 (1983).

48. T. Hondoh, K. Yokoyama, T. Suzuki, K. Hirato, T. Kadofuku, H. Saitoh, T. Yanaihara, *Endocr. J., 41* (Suppl.), S77–S84 (1994).

49. A.S. Vandersall, R.G. Cameron, C.J. Nairn, III, G. Yelenosky, G.R.J. Wodzinski, *Prep. Biochem. 25* (1–2), 29–55 (1995).

50. P. Ollikka, V.M. Leppanen, T. Anttila, I. Suominen, *Prot. Express. Purif., 6*(3), 337–342 (1995).

51. A. Roboti, E. Livaniou, G.P. Evangelatos, G. Tsoupras, O. Tsolas, D.S. Ithakissios, *J. Chromatogr. B, 662*(1), 27–34 (1994).

52. K. Kurata, M. Nakamura, T. Okuda, H. Hirano, H. Shinbo, *Comp. Biochem. Physiol. B, 109*(1), 105–114 (1994).

53. C. Herbelin, B. Poutrel, *J. Microbiol. Methods, 21*(2), 163–171 (1995).

54. M. Mitsutomi, T. Hata, T. Kuwahara, *J. Ferm. Bioeng., 80*(2), 153–158 (1995).

55. S.R. Pirie-Shepherd, E.A. Jett, N.L. Andon, S.V. Pizzo, *J. Biol. Chem., 270*(11), 5877–5881 (1995).

56. M. Pleszczynska, J. Rogalski, J. Szczodrak, J. Fiedurek, *Mykol. Res., 100*(6), 681–686 (1996).

6 High-Resolution Reversed-Phase Chromatography

MILTON T. W. HEARN

Department of Biochemistry and Molecular Biology
Center for Bioprocess Technology
Monash University
Clayton, Victoria 3168, Australia

Protein Purification: Principles, High-Resolution Methods, and Applications, Second Edition.
Edited by Jan-Christer Janson and Lars Rydén.
ISBN 0-471-18626-0. © 1998 Wiley-VCH, Inc.

6.1   INTRODUCTION

Spectacular advances in the development of chromatographic methods for the analysis and purification of polypeptides and proteins have been made since the early 1980s. To a large extent, these developments have been catalyzed by the recognition that the hydrophobic effect can provide important selectivity options not available to other separation methods that are based on differences in solute net charge or size. A further fundamental advantage is that chromatographic separations based on hydrophobic interactions have many features in common with the interplay of hydrophobic effects evident in the protein–lipid interactions found with biological membranes or the stabilization of the secondary and tertiary structure of polypeptides and proteins through intramolecular domain interactions. As a consequence, a natural nexus exists between polypeptide and protein conformation, insertion of proteins into nonpolar biomembranes and peptide, or protein interactions with hydrophobic ligands, irrespective of whether these nonpolar surfaces and ligands have a biological (i.e., lipid-like) or a chemical (i.e., *n*-alkyl-like) origin.

In the other modes on interactive chromatography, such as ion-exchange, immobilized metal ion chelate, or liquid–solid polar adsorption chromatography, interactions between the protein and adsorbent involve the participation of strong polar forces, either electrostatic, hydrogen bonding or coordination bonding in character. Separation selectivity in these modes of adsorption chromatography is thus mediated by differences in the net charge and charge anisotropy of the solute, by the availabilty of structural units that can act as Lewis acids or Lewis bases and accept or donate nonbonding electron pairs, or due to the accessibility of hydroxyl, thiol, amino, carboxlic, or amido groups which can act as hydrogen bond donors or acceptors. In addition, when the separation process only involves weak van der Waals-type forces or nonbonding steric interactions between proteins and nonpolar (or hydrophilic polar) surfaces, different and unique aspects of protein structure contribute selectively to the separation and are revealed as size- or volume-related phenomena typified in size exclusion separations.

Hydrophobic interactions are now commonly accepted to be of great biological significance in the determination of the tertiary and quaternary structural hierarchy of proteins and also in the dynamics of protein motion in solution. Similarly, the hydrophobic effect has important ramifications in the way polypeptides and proteins fold and are presented to biological membranes. Most proteins in their native, biological functional state possess hydrophobic regions as part of their surface topography. Extended regions of the surface of a typical membrane protein or an amphipathic polypeptide involve these hydrophobic patches. There these hydrophobic regions play major roles in determining the orientation of the protein in the membrane phospholipid bilayer, in the transduction of biological signals across the membrane boundaries, and in protein–protein aggregative phenomena leading to coated pits and clathrin-like strucures. Even with water-soluble globular proteins, a considerable proportion of the surface is also hydrophobic in the sense that the atoms in these regions are not able to form in aqueous solutions hydrogen bonds with the surrounding water molecules (see also Chapter 7). Specific protein–protein recognition phenomena involved in the assembly of multisubunit proteins, in multimeric enzyme complexes, or in intracellular organelles are all believed to be consequences of the hydrophobic effect. Further, solvent- and salt-induced precipitations of proteins are also mediated to a large extent through this effect. Is it little wonder then that much effort has been devoted to mimic these biological processes in protein fractionation methods so that hydrophobic interactions can be exploited in well-designed chromatographic systems?

This chaper is devoted mainly to the requirements for high-resolution separations of polypeptides and proteins by reversed-phase high-performance liquid chromatography (RP-HPLC). Recent advances in both the theory and the practice of RP-HPLC as it applies to the fractionation of these biomolecules are critically assessed. Selected examples of proteins purified by RP-HPLC methods are also used to illustrate different chromatograhic stratagems that can be employed. Polypeptide size has been deliberately limited by selecting examples with molecular weights greater than ca. 3,000. The reader is directed elsewhere for reviews on the separation by HPLC techniques of amino acids and small peptides, including the important area of RP-HPLC mapping of enzymatic digests of proteins.[1–5]

Historically, the chromatographic separation of polypeptides and proteins via hydrophobic mechanisms has evolved from studies with chemically modified soft polymeric gels. As early as 1950, the concept that polar components could be separated through the participation of weak, nonpolar interactions between the solute and a nonpolar stationary phase was established by Boscott,[6] Boldingh,[7] and Howard and Martin.[8] These latter investigators coined a descriptive acronym for this separation mode—*reversed-phase chromatography (RPC)*—in contradistinction to the normal or *polar-phase chromatography mode* in vogue at that time. Here, as in paper chromatography, the solutes elute in increasing order of polarity. In contrast, the distribution processes in RPC result in the elution of solutes in *decreasing* order of polarity. Thus, in RPC the solutes migrate in decreasing order of net charge, extent of ionization, and hydrogen-bonding capabilities.

At an early stage of its development, RPC adsorbents were made from cellulosic and other types of polymeric matrices, which were adsorptively coated with nonpolar stationary-phase components such as long chain alkanes. These systems lacked long-term stability and exhibited low resolution with most polypeptides and proteins. With the advent of chemically bonded *n*-alkylagaroses and similar materials,[9–11] difficulties associated with the rapid loss of the nonpolar ligands from the adsorbent surface were mostly circumvented. However, it was with the emergence of HPLC media, especially the meso- and macroporous chemically bonded, hydrocarbonaceous supports of small particle diameter, typically in the range of 5 to 10 μm, with narrow particle and pore size distributions, that enabled RPC to become the viable, rapid, high resolution method for the analysis and purification of polypeptides and proteins that biologists and other end users exploit with such vigor today (see, for example, Figure 6-1).

Silica-bound hydrocarbonaceous stationary phases are now in wide use and, for many applications, represent the preferred form of high-performance adsorbent for the isolation and purification of peptides, polypeptides, and proteins. However, it can be noted that the development of a range of additional support materials, based either on other types of inorganic oxides such as zirconia[12,13] or titania[14] or, alternatively, on a variety of organic polymeric materials involving polystyrene or polystyrene divinylbenzene copolymer combinations,[15,16] have opened new selectivity avenues for RP-HPLC applications with biopolymers. The current preference for the use of silica-based reversed-

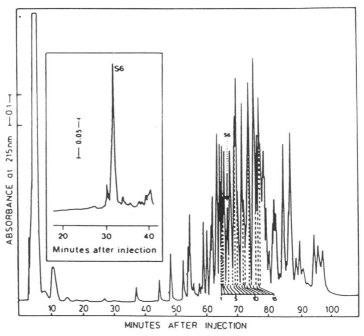

FIGURE 6-1. Resolution of proteins from whole rat liver ribosomes by reversed-phase HPLC. A 6 *M* guanidinium hydrochloride extract of whole ribosomes (200 μg) was separated on a Bakerbond C8 column using a 90-min linear gradient from 0.1% TFA in water (A) to 0.1% TFA–acetonitrile–water (50:50, v/v) (B) at a flow rate of 1 mL/min. A small amount of ^{32}P-labeled ribosomal protein S6 was added to the extract. The elution position of the protein S6 (and the ^{32}P-labeled S6 under these conditions were coincidental) is denoted by the arrow. (Insert) Fraction 4 was rechromatographed on a Bakerbond C8 column using a linear gradient of 45 to 100% B over 60 min. After these two chromatographic procedures the protein S6 was homogeneous, as revealed by N-terminal sequence analysis. (Adapted from Ref. 17.)

phase adsorbents is due mainly to their excellent mechanical stability, important in the HPLC mode, and their fairly robust chemical stability over a wide variety of elution conditions, provided the pH of the eluent is kept below about pH 7.5.

With RPC, considerable versatility exists for changes in eluent composition, thus allowing an extremely wide variety of substances to be resolved. Excellent selectivities can be achieved for polypeptides and proteins separated on chemically bonded n-alkylsilicas and other modern hydrocarbonaceous supports.[17] With many 5- or 10-μm n-alkylsilicas now in use, it is possible to achieve column efficiencies for polypeptides and proteins between 2,000 and 20,000 theoretical plates per meter, with short separation times on the order of minutes, and peak capacities approaching 250. This situation can be contrasted with conventional open column systems packed with soft gel matrices where efficiencies in the region of 100 to 200 theoretical plates per meter are typically obtained, with separation times on the order of hours or even days. This combination of higher efficiencies, more rapid separation times, and greater versatility in elution conditions has resulted in the more classical modes of RPC, such as low-pressure partition chromatography or soft gel hydrophobic interaction chromatography, being essentially displaced since the mid-1980s by higher performance adsorbents, often operating at higher linear flow velocities and in the gradient elation mode at room or slightly elevated temperatures. However, the underlying principles and concepts for the hydrophobic interaction chromatography (HIC) and reversed-phase chromatography (RPC) have a common basis, exploiting in each case the change in eluent surface tension to affect control over the chromatographic resolution in either the isocratic or the gradient elation mode.

6.2 CHARACTERISTICS OF STATIONARY AND MOBILE PHASES: SAMPLE PROPERTIES AND COLUMN DESIGN

6.2.1 Properties of the Stationary Phase

A typical stationary phase in RPC consists of 5- to 10-μm-diameter porous beads of silica, covered completely with covalently bound hydrophobic n-alkyl chains. The retention of proteins on a chemically modified hydrocarbonaceous silica-based adsorbent depends on both hydrophobic (solvophobic) and silanophilic effects. There has been considerable discussion in the scientific literature on the interplay between these effects.[18-22] Controversy still persists on the optimal stationary-phase characteristics of an "ideal" RPC adsorbent for the separation of polypeptides and proteins. An important reason limiting this debate is that insufficient systematic data exist on a number of crucial stationary-phase parameters. These parameters include the uniformity of surface coverage with n-alkyl chains, ligand density, the method used in the bonding treatment of the silica matrix, and preparation history and type of the silica itself. However, several schools of thought[23,24] now favor the use of the more bulky bis(isopropyl)n-alkylchlorosilanes and similar protected n-alkyldialkylchlorosilanes or the bis alkyldisilazines for the surface modification with subsequent end capping of the residual silanol groups to ensure that the surface assumes a well-covered if not necessarily homogeneous character. Generally, not all of the nonpolar surface area is accessible to a polypeptide or protein, once it enters the pore. This is due to solute–ligand conformational constraints, surface tortuosity, ligand chain-length compression, solute solvation, or other diffusion-controlled factors. However, accessibility of the polypeptide or protein to surface silanol groups leads to peak tailing or greatly enhanced retention behavior if the influence of these silanol groups is not adequately mashed.

Over the years, a large variety of nonpolar functional groups have been bonded to silica matrices, both as monolayers and as polymeric layers. The dimethylpropyldimethyloctyl, dimethyloctadecyl, and diphenyl phases are, however, normally used in polypeptide and protein separations. For compendia of available phases, see Refs. 5, 16, and 23–25. The application of some common RP-HPLC adsorbents are tabulated in Table 6-1.

Large-pore silicas with specific surface areas below 50 m²/g are very suitable support materials, provided a narrow pore distribution exists (see Section 6.3.2 for a further explanation of the rationale for the choice of these adsorbents). High coverage 5- to 10-μm particle diameter, 25- to 50-nm nominal pore diameter, n-alkylsilicas are currently the most popular. As is becoming increasingly apparent, these wide-pore support materials are not necessarily more suitable than smaller pore or even nonporous support materials in every case of protein separation. It is now known[26] that the retention dependencies (but not relative retention) of polypeptides and protein on solvent composition are essentially independent of pore size at low sample loadings of the solutes. With alkylsilicas of different porosity but identical alkylchain length and ligand coverage, the S parameters of polypeptides and proteins are essentially constant, although the extrapolated log k_o and log k_w values are different (the meaning of the S, log k_o, and log k_w terms are explained in Section 6.3.2). Important differences in terms of capacity and kinetics that are evident with stationary phases of different porosities relate to the accessibility of surface area and film and surface interaction rate constants differences, features adequately predicted by theory.

It has been widely assumed in the literature that the unsatisfactory retention behavior of polypeptides and proteins seen with some narrower-pore-diameter $(10 < p_d < 20$ nm$)$ n-alkylsilicas is due to the physical entrapment of the biosolutes arising from restricted diffusion in the pore chambers. Moreover, an assumption commonly held by biologist who employ RP-HPLC techniques as part of their investigations but with incomplete understanding of the underlying separation science concepts that conditions which favor high resolution will not be synonymous with conditions that engender optimal recovery of polypeptides and proteins. This paradox is not unique, however, to reversed-phase separations and has parallels also with ion exchange and several other adsorptive modes of chromatographic resolution of polypeptides and proteins. In fact, this controversy has been largely resolved with the development of improved models for the interpretation of the mechanisms of protein folding and unfolding at liquid–liquid or liquid–solid interfaces as well as by the development of reversed-phase adsorbents that show improved performance from the points of view of stability, isothermal regularity, and knowledge on the effects of surface coverage. For example, 10-fold differences in surface area per gram of packing material are commonly found between small and large pore n-alkylsilicas. As the bonding chemistries, as well as the extent of surface coverage, often differ between small- and large-pore materials, the outcomes are RPC adsorbents of significantly different surface interaction characteristics. Hence, eluents with significantly higher elutropic strength may be needed to achieve the same absolute retention, or k' value, with small-pore alkylsilicas ($p_d < 10$ nm) compared to large-pore alkylsilicas ($p_d > 30$ nm). Under such conditions these more forceful eluents may not be unacceptable as protein solvents and can cause impairment of the elution development and recovery. The solution is then to employ either chromatographic systems of smaller phase ratio, nonporous adsorbents or eluents that contain additives that change the surface tension of the mobile phase without causing irreversible conformational changes with the proteins or polypeptides.

The n-alkyl chain length above C_6 has been found to have only a small influence on relative retention if the ligand densities are of a similar magnitude. More significant differences have been observed with shorter chain length n-alkyl ligands (e.g., C_3 and C_4) compared with the longer C_8–C_{18} ligands. Detailed studies[27,28] on the molecular dynamics and flexibility of the C_4 to C_{18} ligands have indicated that the geometry of the ligand, when immobilized, significantly affects the interaction with polypeptides and proteins, including their orientation and interaction affinities. Such behavior can affect recoveries. For example, with some proteins with $M_r > 20,000$, recoveries can be improved when a n-octyl or a n-butyl ligand is substituted for an n-octadecyl ligand.[29,30]

The chemical characteristics of the parent silica and the pretreatment history have significant effects on selectivity. The presence of trace amounts of adsorbed metal ions, the water content of the parent silica prior to chemical modification, and the relative Bronsted acidity of the silica surface all ultimately affect the resolution in RP-HPLC separations of polypeptides and proteins. In order to minimize the free silanol group content, a variety of strategies have been attempted. These include vigorous end capping, sequential bonding, and combining two monofunctional alkylsilanes of different chain length or branching.

Theory predicts that improvement in chromatograhic performance will be obtained as the surface coverage becomes more uniform. This can be approached, for example, through the use of chlorotrimethylsilane in combination with chloro-n-alkyldimethylsilane. Such an approach has been employed with ProRPC stationary phases.[29] Several other options are available to de-

TABLE 6-1. Selected Examples of Polypeptides and Proteins Purified by Reversed-Phase HPLC

Column	Hydrophobic Chain	Mobile Phase	Protein Purified	Ref.
μBondapak C18	n-Octadecyl	0.1% TFA, 12–70% CH$_3$CN	Acid phosphatase	20
Nucleosil C18	n-Octadecyl	0.1% TFA, EtOH-butanol-methoxyethanol	Aldolase	134
μBondapak C18	n-Octadecyl	5% formic acid 40–80% EtOH	Bacteriorhodopsin	135
Nucleosil C18	n-Octadecyl	50 mM KH$_2$PO$_4$, 10–50% 2-methoxyethanol	Bacteriorhodopsin	137
LiChrospher C8	n-Octyl	400 mM Pyr-formate, 0–40% nPrOH	Bovine serum albumin	136
μBondapak alkylphenyl	n-Propylphenyl	1% TEAP, 10–50% CH$_3$CN	C-apolipoproteins	138
Aquapore RP300	n-Octadecyl	0.1% H$_3$PO$_4$, 10 mM NaClO$_4$, 0–60% CH$_3$CN	Calmodulin	139
Ultrasphere C8	n-Octyl	10 mM TFA, 0–45% nPrOH	Carbonic anhydrase	30
Ultrasphere C3	n-Propyl	155 mM NaCl, pH 2.1, 0–75% CH$_3$CN	Carbonic anhydrase	30
μBondapak C18	n-Octadecyl	0.1% TFA, 0–60% CH$_3$CN	Chorionic gonadotropin	140
LiChrospher C4	n-Butyl	10 mM H$_3$PO$_4$, 0–45% nPrOH	Chymotrypsinogen	18
Bakerbond diphenyl	Diphenyl	0.1% TFA, 0–50% CH$_3$CN	Collagen chains	141
U-ODS	n-Octadecyl	0.2% HFBA, 0–50% CH$_3$CN	Epidermal growth factor	142
Nucleosil C18	n-Octadecyl	50 mM KH$_2$PO$_4$, 10–50% 2-methoxyethanol	Ferritin	137
Pharmacia ProRPC	n-Octyl	0.3% TFA, 39–50% CH$_3$CN	Globin chains	71
LiChrospher C4	n-Butyl	100 mM NH$_4$HCO$_3$, 0–50% CH$_3$CN	Growth hormone	143
μBondapak C18	n-Octadecyl	0.2% TFA, 0–50% CH$_3$CN	Histone proteins	144
LiChrospher C8	n-Octyl	0.8 M Pyr-1 M formic acid, 0–60% CH$_3$CN	Human fibroblast interferon	145

Column	Alkyl	Conditions	Protein	Ref.
Ultrasphere C3	n-Propyl	0.1% TFA, 0–50% CH$_3$CN	Inhibin, follicular	89
LiChroprep RP8	n-Octyl	0.9 M Pyr-acetate, 20–60% nPrOH	Interleukin-2	146
W-DP	n-Octadecyl	50 mM KH$_2$PO$_4$ 2-methoxyethanol-PrOH	Leukocyte interferon	147
N-ODS	n-Octadecyl	33 mM NaOAc, 10–50% EtOH, pH 5.2	α_2-Macroglobulin	148
μBondapak C18	n-Octadecyl	12 mM HCl, 35–80% EtOH	β_2-Microglobulin	149
unspecified C8	n-Octyl	0.1% TFA, 0–60% iPrOH	Ovalbumin	150
LiChrospher C4	n-Butyl	10 mM H$_3$PO$_4$, 0–45% nPrOH	Papain	95
LiChrospher C18	n-Octadecyl	50 mM Tris–HCl, 0.1 mM CaCl$_2$, 0–70% iPrOH	Parvalbumin	151
Ultrasphere C3	n-Propyl	155 mM NaCl–HCl, 0–50% nPrOH	Phosphorylase	31
LiChrospher C8	n-Octyl	0.1% TFA, 0–50% iPrOH	Platelet-derived growth factor	152
μBondapak C18	n-Octadecyl	0.1% H$_3$PO$_4$, 10–75% CH$_3$CN	Proinsulin	87
Various	Various	0.1% TFA, CH$_3$CN-PrOH	Rhodopsin	151
Ultrapore RPSC	n-Octadecyl	0.1% TFA, 10–80% CH$_3$CN	Ribosomal 50S proteins	17
Synchropak RPP	n-Octadecyl	0.1% TFA, 15–75% CH$_3$CN	Ribosomal proteins	17
ToyaSoda TMS250	Trimethyl	0.2% TFA, 25–75% CH$_3$CN	Sendai viral proteins	134
LiChrospher C18	n-Octadecyl	15 mM H$_3$PO$_4$, 0–60% CH$_3$CN	Thyroglobulin	153
Bakerbond C18	n-Octadecyl	100 mM NH$_4$HCO$_3$, 0–50% CH$_3$CN	Thyrotropin and subunits	119

crease the magnitude of the S and log k_0 parameters, such as branched chain phases[23,24,31,32] and hydrophilic coverage phases, sequentially bonded with nonpolar groups.[33-35]

6.2.2   Different Mobile Phases

All of the common water-soluble organic solvents have now found use in the separation of polypeptides and proteins by reversed-phase HPLC (see Table 6-1), although their efficacy in terms of elutropic strength and protein recovery differs considerably. Organic solvent modifiers not only affect the gross properties of the mobile phase (through changes in surface tension, dielectric constant, viscosity, etc.) but also interact in specific ways with the proteins themselves. At concentrations below 0.1% (v/v), alcohols and other dipolar organic solvents can stabilize the structure of some proteins (e.g., chymotrypsinogen), whereas at higher concentrations these solvents usually induce either reversible or nonreversible structural deformation.[36-41] The denaturing effects of most organic solvents can be attributed to regional disruption of the hydrophobic interactions between nonpolar side chains in the protein and perturbation of the hydrogen-bonding characteristics of the protein through disruption of the peptide backbone dipoles. Addition of an organic solvent modifier to an aqueous medium containing a protein in its native biological conformation will, in general, alter the hydration structure of this protein. Whether this change leads to denaturation depends on the specific equilibria associated with each protein. Where mobile-phase-induced denaturation occurs, an increase in surface contact with the stationary phase is anticipated. The retention value k' (the protein elution volume expressed in the number of void volumes) becomes larger for proteins in the denatured (as well as many chemically modified states such as carboxymethylated) form with low pH eluents.[1,2,4,42]

The elutropic strength E_i of an aquo-organic solvent eluent increases with the content of the organic solvent (expressed in terms of a volume fraction) as confirmed by different experimental studies.[42-44] The relative retention or k' of a particular polypeptide or protein decreases in order of the following series of solvent modifiers methanol < ethanol < acetonitrile < 1-propanol, or 2-propanol at the same volume percentage. The solvent composition range over which the log k' value for a particular polypeptide or protein remains small (that is, when the polypeptide or protein elutes close to the volume expected for pore size exclusion selectivity and without evidence of selectivity reversals) is considerably greater with 1-propanol than for acetonitrile. This dependency, coupled with the more rapid decrease in bulk surface tension at a higher water content, is advantageous for 1- and 2-propanol, compared to acetonitrile or methanol. The propanols give shorter relative retention, generally higher mass recoveries, and less sensitivity to n-alkyl chain length effects with silica-based packing materials of nominally equivalent matrix characteristics. However, exceptions to these trends can occur with the polymeric type of RP adsorbents or, for that matter, with multimeric proteins where changes in conformation induced by the organic backbone of the polymeric adsorbent can lead to subunit dissociation and denaturation.[45,46] Although the relative retention for polypeptides and proteins decreases as the elutropic strength of the organic modifier increases, it does not necessarily follow that the overall peak capacity or even regional resolution will improve with more effective desorbing solvents. For example, with gradients of the same slope, different peak capacities will occur when different organic solvents are employed as is evident with the acetonitrile- and 1-propanol-based elution profiles shown in Figure 6-2.

It is often advantageous to use ternary mixtures containing water and two organic solvents. When their compositions are chosen to maximize the slope of the log k' versus φ dependency, this optimization can lead to improved selectivity. With such systems, hydrophobic proteins such as cytochrome c oxidase subunits[47] may be eluted with a mobile phase of higher water content than can be achieved with either organic solvent. A similar strategy can be employed by taking advantage of the different influences of ionic modifiers such as ion-pair reagents (see below).

The addition to binary mobile phases of small amounts of polar and nonpolar solvents such as 0.1 to 0.5% (v/v) ethylene glycol, formamide, tetrahydrofuran, dioxane, butanol, n-pentanol, or the nonionic detergents is anticipated to decrease the mobile-phase surface tension as well as to influence the characteristics of the RP-adsorbent surface. These composite phenomena will affect both the magnitude of the multisite binding at the sorbent–protein interface and the accessibility of the hydrophobic regions on the solute surface. Typical of this behavior are the plots[48] of the dependence of the capacity factor of several polypeptides on the concentration of the detergent Brij 35 in the mobile phase. This dependency and the corresponding gradient profiles are shown in Figure 6-3 and Figure 6-4. When high adsorption isotherms are observed with binary mobile phases, the use of ternary solvent modifiers may provide a remedy of these adverse sorption–desportion events. If no attention is given to these situation, erratic k' values and peak shapes, symptomatic of "irreversible" binding, can arise. In many cases though, with strongly retained solutes, protein solubility will limit the use of a secondary solvent. Then, further options must be sought other than simply increasing the volume fraction of the primary or secondary organic solvent modifier. Different pH conditions or the use of polar or surface-active ionic modifiers should be contemplated immediately.

6.2.3 The Sample: Protein Solubility and Avoidance of Protein Denaturation

One feature of conventional n-alkylsilica stationary phases that has caused concern to many protein chemists has been the assumed propensity of these hydrophobic supports to cause protein denaturation under some operational conditions. Although the more naive investigator has tended to rush into hasty

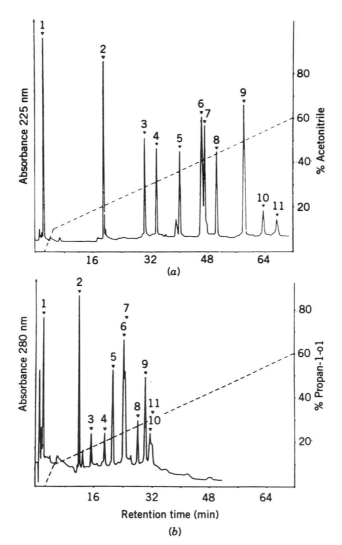

FIGURE 6-2. Comparative elution profiles using acetonitrile- or *n*-propanol-based gradients of the equivalent rate of change of organic solvent modifier concentration. Protein key: 1, tryptophan; 2, ACTH(1–24); 3, ribonuclease; 4, human calcitonin; 5, lysozyme, 6, bovine serum albumin; 7, bovine β-lactoglobulin; 8, lactoglobulin A; 9, bovine prolactin; 10, rat prolactin; and 11, rat growth hormone. Separations were carried out at 45°C, a flow rate of 1 mL/min on a Ultrapore C3 column [75 × 4.6 mm (i.d.)], and eluted with acetonitrile and a primary solvent of 0.155 *M* NaCl/0.01 *M* HCl (a) and 1-propanol and a primary solvent of 0.01 *M* TFA (b) using the gradient shapes indicated. (Adapted from Ref. 31.)

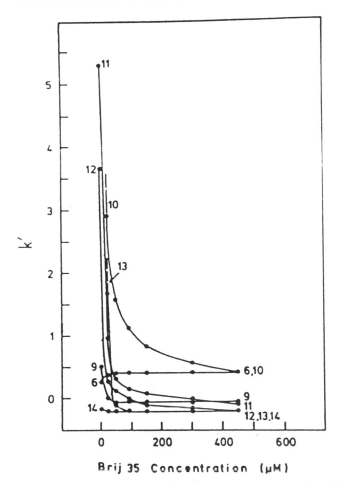

FIGURE 6-3. Dependence of the capacity factors of various polypeptides on the concentration of the nonionic detergent Brij 35 in the mobile phase. Chromatographic conditions: octadecylsilica, flow rate, 1.0 mL/min; temperature, 18°C; mobile phase, acetonitrile–water (30 : 70)–50 mM KH$_2$PO$_4$–15 mM H$_3$PO$_4$, pH 2.3, containing various concentrations of the detergent. Polypeptide key: 1, trityrosine; 2, angiotensin 1; 3, porcine glucagon; 4, bovine insulin B-chain; 5, bovine insulin; 6, cytochrome c; and 7, hen lysozyme. (Adapted from Ref. 92.)

conclusions about the generality of this phenomenon, practice teaches that denaturation and/or impaired mass recovery is not a general effect with high coverage n-alkylsilicas and aquo-organic solvent eluents of low to intermediate pH and buffer capacity. The opposite circumstance can be the case with denaturation as a consequence of inappropriately selected operation and handling conditions instead of the use of a RP-HPLC separation approach per se. Sufficient experimental data are now available with large polypeptides and

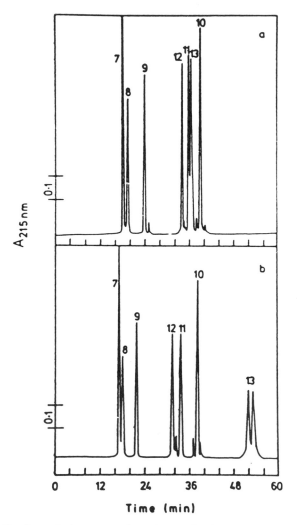

FIGURE 6-4. Gradient elution separation of several polypeptides on an octadecylsilica column in the absence (a) or presence (b) of 0.01% Brij 35 (w/v). Chromatographic conditions: column, 10-nm-pore diameter octadecylsilica; flow rate, 1.0 mL/min; initial conditions, (A) acetonitrile-water (15:85, v/v)–50 mM NaH$_2$PO$_4$–15 mM H$_3$PO$_4$, final conditions, (B) acetonitrile–water (50:50, v/v)–50 mM NaH$_2$PO$_4$–15 mM H$_3$PO$_4$. The detergent was added at 0.01% (w/v) to both initial and final conditions in b. Linear gradient from (A) to (B) over 60 min. Polypeptide key: 1, angiotensin 111; 2, angiotensin 11; 3, angiotensin 1; 4, bovine insulin; 5, bovine insulin B-chain; 6, cytochrome c; and 7, porcine glucagon. (Adapted from Ref. 92.)

proteins encompassing the molecular weight range of 5,000 to 65,000 to enable the following four scenarios with regard to the effect of the mole fraction, φ, of the organic solvent modifier on the retention behavior to be identified. Each scenario is time dependent and responsive to the manipulation of thermodynamic and kinetic variables such as temperature, ligand surface heterogeneity, or interaction rate constants.

The first scenario represents the optimal case history in which both the mobile-phase composition and the stationary phase do not cause observable changes in activity or recovery, for example, as found with serine proteases,[1,42] inhibins,[49] and several other protein hormones and growth factors.[50–52] Underlying this optimal behavior may well be subtle conformational effects, but these equilibrium effects do not impinge on the chromatographic behavior during the time required for the mass transport of the polypeptide or protein through the chromatographic bed.

The second scenario involves mobile-phase-mediated changes in conformation and is independent of the stationary phase surface per se, for example, such as observed for catalase,[53] when the mole fraction of organic modifier is $\varphi > 0.15$. These solvent-induced denaturation effects with some proteins in acidic low ionic strength eluents reflect surface tension-dependent changes in the solvated structure of the protein. These effects are thus not generally expected to respond to changes in incubation or dwell time at the stationary-phase surface. Experimentally, characterization of this effect is easy. From several small-scale pilot experiments, protein stability in the mobile phases of different composition is studied. The solvent combination that provides the maximal stability is then chosen as the initial mobile-phase composition for the chromatographic separation.

The third scenario involves ligand or support matrix-mediated changes in the conformation of the polypeptide or protein and is largely independent of mobile-phase composition, for example, as seen with pepsin[53] or β-lactoglobulin.[54] This effect responds to column dwell phenomena and is thus dependent on the stationary-phase surface area, porosity, and ligand density, as well as the elution time. Again, remedies for this effect have been developed, viz. short columns, steep gradients, and the use of nonporous stationary phases.[55,56]

Finally, the composite interplay of a particular mobile-phase composition with a particular stationary phase (scenario 4) may also give rise to protein denaturation, whereas a slight variation in the phase ratio or the eluent composition of the chromatographic system will result in acceptable behavior. The last situation represents the most troublesome circumstance for the novice who may become disenchanted with the RP-HPLC technique per se, although a perfectly suitable procedure might have been found if additional effort or imagination had been applied. The variation seen in the recovery of a particular protein with stationary phases of the same type of silica matrix, nominal chain length, and surface coverage from different commercial sources can often be ascribed to this last effect. Again, remedies and more rational approaches to overcome these difficult separations have been proposed.[57,58]

Methods[1,54–60] to rapidly characterize these effects in RP-HPLC have been established and new approaches have been developed to characterize the conformational changes with polypeptides and proteins on line. This latter technique, called chromatotopography, has provided[58,59] a rapid method to follow mobile-phase or stationary-phase-induced changes in the three-dimensional structures of polypeptides or proteins, provided the conformation intermediates have half-lives >10 times the time required for the polypeptide or protein to traverse a theoretical plate. The theoretical basis for evaluating these interconversions is now firmly established[56–61] for both analytical and semipreparative chromatographic separations (see also Section 6.6.6).

An alternative approach to overcome mobile-phase or stationary-phase-induced conformation changes in RP-HPLC while still utilizing the hydrophobic effect involves the use of salt-mediated hydrophobic interaction (HIC) stationary phases. These HIC adsorbents can be either of the ether-type[62] or the amide-type[63,64] ligand characteristics. Compared to n-alkylsilicas, the peak capacities of these hydrophobic interaction adsorbents are lower, typically by a factor of 5 to 10, and although bioactivity and mass recovery may be improved, the disadvantages associated with the handling losses during the subsequent desalting of the recovered sample must be overcome. This additional step frequently represents the greatest loss of material with micropreparative separations.

Despite more than 50 years of investigation involving isolation at the semipreparative, micro-, and ultramicroscale level of proteins and polypeptides by protein chemists and biochemists, it is surprising that postcolumn sample manipulation and handling still represent poorly investigated areas of development. The regeneration of bioactivity by proper attention to the collection technique, eluent neutralization, and concentration can permit an otherwise inactivated proteins to be reconstituted in active form (e.g., alkaline phosphates, horseradish peroxidase.[53]) Again what is required is a thoughtful strategy that accommodates the vagaries of the protein of interest rather than just a "hit- or -miss" attempt, totally dictated by a cursory application of one or two of the currently popular recipes. If this attention is not given to the problem, then peak shapes symptomatic of "irreversible" binding poor mass or biorecovery and low resolution performance can eventuate. The emergence of so-called intelligent instrumentation (e.g, the Pharmacia Smart system) or microhandling equipment based on minaturized membrane or microengineered devices offers the potential for practical solutions to some of these limitations.[65,66]

6.2.4   Column Design

Advantage can be taken of the high affinity and attendant multisite binding of polypeptides and proteins with n-alkylsilicas in the design of the column configuration. For example, a 10-fold increase in column length results in only a small improvement in resolution for many globular proteins, and such gains in chromatograhic performance are usually at the expense of protein recovery and column back pressure. As a consequence, columns with dimensions of 10 to 15 × 0.4 to 0.8 cm have gained popularity for both analytical and semipreparative separations, whereas columns with dimensions of 10 to 20 × 0.2 cm are increasingly favored for microbore applications. In contrast to high-performance ionexchange, process scale purification of proteins by reversed-phase procedures still remains undcrdcvclopcd. IIowever, the multigram/kilogram purification of polypeptides using such procedures has been realized[67,68] for several years by the pharmaceutical and biotechnology industries and similar procedures certainly find additional advocates in the food industry as well. The selection criteria for the use of columns of different dimensions in analytical and preparative RP-HPLC have been detailed in recent publications and the reader is referred to these literature sources.[69,70]

The large adsorption isotherms manifested by polypeptides and proteins with n-alkylsilicas have another practical benefit. Trace enrichment from very dilute solutions, with (ultimately) relatively high sample loadings, can be readily achieved, that is, loadings up to 50 mg of ribonuclease in volumes up to 500 ml did not significantly reduce the retention or recovery with a conventional analytical column (e.g., μBondapak Cl8, 25 × 0.4 cm). This feature permits the rapid processing of large volumes of a biological extract, particularly in situations where the chromatographic behavior of the desired component is already known.

6.3 BASIC THEORY

6.3.1 General Considerations

In the majority of the published literature on RPC separations of polypeptides and proteins the selection of chromatographic parameters has generally been made on empirical criteria. Typical conditions for the analytical or semipreparative separation on a macroporous n-octyl- or n-octadecylsilica support are as follows: a low pH organic solvent combination [usually encompassing the 0 to 50% (v/v) solvent], a relatively low ionic strength (usually in the range of 10 to 100 mM), a flow rate between 0.5 and 2.0 ml/min, and ambient temperature. Figure 6-5 illustrates a separation representative of these conditions, namely the resolution of globin chains from a heterozygous hemoglobin carrying a β-chain mutation (Pro^{36} → Thr^{36}).[29,71]

Both the nature of the solute–adsorbent interaction and the wide structural diversity encountered with even a relatively simple protein mixture require the use of gradient elution methods to ensure adequate elution times for all of the components. If the sample contains many substances of widely differing relative retentions, isocratic elution has generally been found to be ineffective.

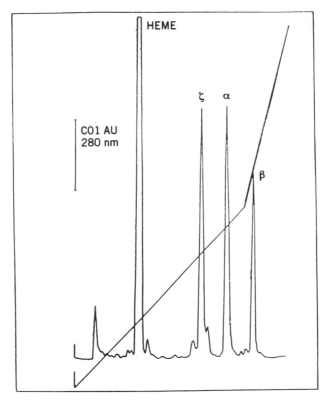

FIGURE 6-5. Separation of globin chains from a heterozygous hemoglobin carrying a β-36 (proline $\rightarrow$ threonine) chain mutation. Chromatographic conditions: column ProRPC HR 5/10, gradient, solvent A, 0.3% TFA in water–acetonitrile (61 : 39); solvent B, 0.15% TFA in water–acetonitrile (55 : 45); 60 min linear, flow rate, 0.2 mL/min. Detection, 280 nm. (Adapted from Ref. 71 with permission of authors and publishers.)

If isocratic conditions are chosen with complex mixtures, then components with short retention times tend to be poorly resolved or elute in the void volume, whereas strongly retained compounds, if they do elute, appear as asymmetric peaks with excessive band broadening. This chromatographic behavior is a direct consequence of the cooperative multisite interactions between the proteins and the heterogeneous stationary-phase surface. Here, gradient elution provides a simple and expedient solution to reduce separation times, decrease peak volumes, and improve peak symmetries.

The same variables that control retention, bandwidth, and resolution in isocratic elution are also relevant in gradient elution. Thus, isocratic retention data can be used to calculate corresponding gradient retention data and vice versa. Gradients can yield equal bandwidths for all compounds separated and roughly comparable values of retention expressed as the effective median capacity factor, $\bar{k}$. The gradient steepness parameters (i.e., the b terms[72]) can be easily varied to adjust bandspacing under gradient conditions as part of the general optimization of resolution.

Available evidence supports[73-75] the concept that the extent of retention for peptides and proteins on silica-based hydrocarbonaceous adsorbents follows their topographic surface polarity. That is, with water-rich eluents the polypeptide or protein solutes elute in order of increasing effective hydrophobicity. Various predictive methods (e.g., Refs. 76–80) have been developed for the estimation of polypeptide selectivity on n-alkylsilica and nonpolar polystyrene-based adsorbents. With small polypeptides of known sequence, up to 5,000 M_r reasonable correlations are found between predicted and experimentally observed retention behavior.[76,77,81,82] With larger polypeptides and proteins these approaches are much less reliable. Here, the database required for prediction does not yet adequately take into account amino acid positional effects or the unique changes in the secondary or tertiary structure of a particular protein that can arise under different chromatographic conditions. Inherent in most previous treatments has been the assumption that only a single, conformationally stable molecular species is involved in the distribution phenomena. In fact, it is highly likely that time-dependent molecular orientation and reorientation of all biopolymeric solutes occur at the stationary-phase surface. As a consequence, the overall adsorption–desorption behavior of these molecules on modern high-performance chromatographic supports often reveals retention and kinetic phenomena not obvious with conventional soft gel materials or, for that matter, by low molecular weight compounds. The recent development of sophisticated molecular modeling, molecular dynamics, and other computational approaches[27,28] offers a greatly enhanced capability to interpret these effects.

6.3.2 Relative Retention

Solute retention is frequently expressed in terms of the capacity factor, k' (also called k, see Chapter 2). The capacity factor is defined by the product of the phase ratio of the chromatographic system, Φ (equal to the ratio of the volumes of the stationary to the mobile phase V_s/V_m), and the adsorption (distribution) coefficient, K_D,

$$k' = \Phi \cdot K_D \tag{6-1}$$

The k' can also be expressed in terms of the retention time, t_R (or volume, V_R), and the column void time, t_0 (or volume V_0), such that

$$k' = \frac{t_R - t_0}{t_0} = \frac{V_R - V_0}{V_0} \tag{6-2}$$

or from the column parameters

$$k' = \frac{t_R \cdot v}{L} - 1 \tag{6-3}$$

where v is the linear velocity of the eluent and L is the column length. Typical values for k' are $1 < k' < 20$, where $k' = 1$ implies no interaction with the solid phase but complete permeation through the porous particulate system.

Experimental evidence has indicated that retention mechanisms based solely on the participation of loose nonbonding or van der Waals interactions do not adequately describe all the different types of separation effects seen when polypeptides or proteins interact with hydrophobic adsorbents. Alternative models, derived from thermodynamic, mechanistic, and nonmechanistic arguments, have been advanced.[44,83–85] In all of these theoretical treatments, it is recognized that the overall retention is the composite of several energy terms. These energy terms accommodate the participation of size exclusion as well as interactive processes involving hydrophobic interactions and coulombic and hydrogen-bonding interactions. Usually, chromatographic conditions can be selected so that "regular" reversed-phase retention behavior dominates (i.e., contributions to retention from polar effects are not significant and k' decreases monotonously as the volume fraction, φ, of an organic solvent modifier, such as acetonitrile or n-propanol, increases). Achieving the appropriate compromise between the primary distribution process and secondary equilibria mediated by coulombic effects or other types of polar interaction is the key to high resolution and frequently to high recovery.

For a specific column, the capacity factor k' for a specific solute can be expressed in terms of eluent composition, temperature, and pH. In the absence of slow competing conformational equilibria, the retention of polypeptides and proteins on n-alkylsilicas has been found to follow the following linearized form of the retention equation:

$$k'_i = k_{\text{sec, }i} + k_{\text{w, }i} \cdot e^{-S\varphi} + k_{\text{o, }i} \cdot e^{-D(1 - \varphi)} \tag{6-4}$$

where the three terms are, in order, the size exclusion term, the solvophobic term, and the silanophilic term. Here k_w and k_o are the capacity factors of the solute, i, in neat water and at the final organic solvent modifier concentration, respectively. The parameters S and D are solute- and condition-dependent variables.[44,83] As the mole fraction of the organic modifier increases, k will decrease, reach a minimum at the φ value where

$$\varphi = (\log k_{\text{w},i} + \log k_{\text{o},i} - D)/(S - D) \tag{6-5}$$

and then increase up to the point of solute precipitation. Figure 6-6 shows a schematic representation of the dependency of k' on φ. As the magnitude of the S and D variables increases, then the elution window over which realistic

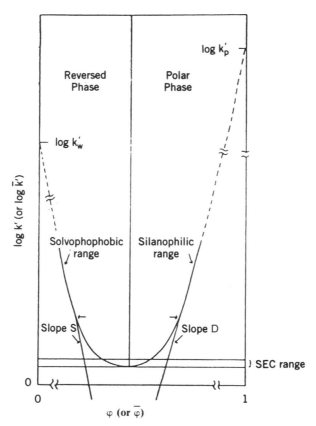

FIGURE 6-6. Schematic representation of the interplay between size exclusion and solvophobic and silanophilic phenomena on the retention dependency between the logarithmic capacity factor, log k', and the mole fraction, φ, of the organic solvent modifier. The extrapolated intercepts of log k' at $\varphi = 0$ and $\varphi = 1$ (dashed lines) are indicated as log k'_w and log k'_0, respectively. The tangents to the curves at a specified k' value, for example, log $k' = 2.5$, are given by S and D, respectively. The log k'_{sec} range corresponds to the retention time of t_R (hydrodynamic) to t_R (total inclusion) for noninteractive solutes of very large and small molecular weights, respectively. For convenience, the value of t_{sec} of most polypeptides can be set at the value of t_0 of the column when log k'_w is large.

k' values can be achieved will progressively narrow. Experimental observations are in good agreement with this anticipated behavior.[85,86]

6.3.3 Isocratic Elution

Over the retention range consistent with reasonable reversed-phase chromatographic practice (i.e., over the range $1 < k' < 20$), the relationships between organic modifier volume fraction and retention can be approximated to

$$\log k_i = \log k_{i,o} - S\varphi \tag{6-6}$$

where φ typically has the limits $0 < \varphi < 0.5$ (i.e., between 0 and 50% organic solvent). Since the slope parameter S is solute and condition dependent, evaluation of the S value dependencies provides crucial information for both isocratic and gradient optimization. Both nonmechanistic and thermodynamic treatments predict an inverse relationship of the log k' on φ over the range of mobile-phase compositions consistent with reversed-phase behavior. Experimental verification of these relationships has been provided by various studies (e.g., Refs. 26,44,86–89).

Isocratic retention can thus be classified in various ways as summarized in Figure 6-7. A typical plot of log k' versus φ for a protein is shown by curve (a) where only a narrow range in solvent percentages (or solvent concentrations) exists between the φ value at which the protein first begins to migrate in a column and the final φ value at which k' effectively becomes k_{sec} (i.e., the polypeptide or protein is not retarded at all). Generally, steep dependencies exist between log k' and φ, resulting in the S values of polypeptides and proteins typically being >20. In contrast, the S values of small peptides are usually in the range of 2 to 10.

6.3.4 Gradient Elution

Because of the interrelated molecular and chromatographic complexities, the separation of polypeptides and proteins by reversed-phase HPLC is typically carried out by gradient elution. The gradient shape can be described by the steepness parameter b, where

$$b = S \, \Delta\varphi \, t_o/t_G \tag{6-7}$$

where $\Delta\varphi$ is the change in φ during the gradient of duration, t_G. An alternative expression for steepness is

$$b = S \cdot V_m \cdot \frac{d\varphi}{dt} \cdot F^{-1} \tag{6-8}$$

which interrelates the flow rate F, the column volume, V_m, and the rate of change of φ with time.

For linear solvent strength gradients, that is, with gradients that result in constant b values for all compounds eluted, the effective k' value can be expressed, using the Snyder gradient elution model,[90] by $\bar{k}$. This term, $\bar{k}$, corresponds to the time taken for the solute to traverse halfway along the column. Under such elution conditions and when the value of t_{sec} approaches that of t_o, then $\bar{k}$ can be approximated by

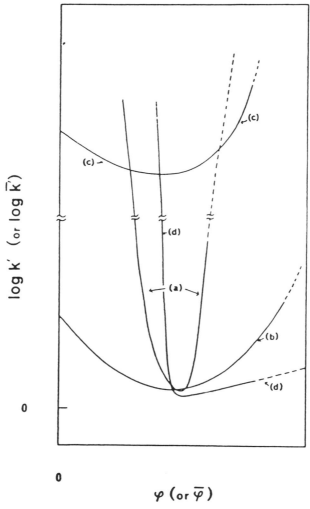

FIGURE 6-7. Schematic representation of the retention dependency of log k' on φ for solutes of different molecular weights and hydrophobicity. (a) Both the S and D slopes are large, and the corresponding log k'_0 and log k'_w values are also large. This behavior is typical of many polypeptides and proteins in water-organic solvent systems of low pH with some n-alkylsilicas. (b) The S and D slopes are small, with corresponding small log k'_w and log k'_0 values. This behavior is typical of small peptides. (c) The S and D slopes are small, but log k'_w and log k'_0 values are large due to the incorrect choice of chromatographic selectivity. (d) The S slope is large but the D slope remains small. This behavior has been observed with fibrous proteins and some proteins eluted with hydrogen-donating solvents. Other permutations have been documented (i.e., small S slope but large D slope) (see, for example, Refs. 21, 86, 87, 91, and 115). Generally, S increases with molecular weight and for smaller polypeptides takes the approximate form $S = 2.99 \cdot Mw.^{0.21}$

$$\bar{k} = 1/1.15b = F\, t_G/1.15\, \Delta\varphi\, S\, V_m \qquad\qquad (6\text{-}9)$$

It can thus be seen that changes in the magnitude of b or the $\bar{k}$ value are equivalent to changes in the average φ value (equal to $\bar{\varphi}$). Values of b for each component can be readily calculated for a particular chromatographic system and specified polypeptide or protein mixtures from multiple gradient runs with different t_G, F, or $\Delta\varphi$ values. From these computed b value data, values of S, $k_{i,o}$, $\bar{k}$, and φ can be calculated and plots of $\bar{k}$ versus $\bar{\varphi}$ (formally equivalent to the plots of k versus φ) generated. This derivation in turn allows a gradient retention time of a polypeptide or protein for any defined b value to be predicted.[91-94]

6.3.5 Peak Capacity of Reversed-Phase System

Changes in the b, $\bar{k}$, or $\bar{\varphi}$ values will have profound effects on band spacing between eluted polypeptides and proteins and will dramatically influence the average resolution of the system. The situation whereby the resolution of several polypeptides with small selectivity differences changes when the gradient slope is changed is probably familiar to many investigators (see Figure 6-8 for one such example). With solutes of different S values, variation in resolution and even selectivity reversals can be anticipated from the above treatment when gradients of different rates of change of organic modifier ($d\varphi/dt$) or flow rate are employed. For convenience, average resolution can be defined in terms of peak capacity PC (not to be confused with the capacity factor k) such that

$$PC = t_G/4\sigma_t = t_G\, F/4\sigma_v \qquad\qquad (6\text{-}10)$$

With well-packed columns, containing 4- to 7-μm particles maximally bonded with C8 or C18 chains, peak capacities approaching 250 can be achieved with appropriate gradient steepness parameters. These values equate with a peak width of ca. 500 μL for a component eluting at 60 min at a mobile-phase flow rate of 2 mL/min.

The resolution equation in gradient elution takes the familiar form, namely

$$R_S = \tfrac{1}{4}(\alpha - 1)N^{1/2}(\bar{k}/(1 + \bar{k}) \qquad\qquad (6\text{-}11)$$

and

$$RS = \left(\frac{k_1}{k_2} - 1\right)\left(\frac{k}{1 + k}\right)\sqrt{L}\bigg/\frac{1}{4\sqrt{H}} \qquad\qquad (6\text{-}12)$$

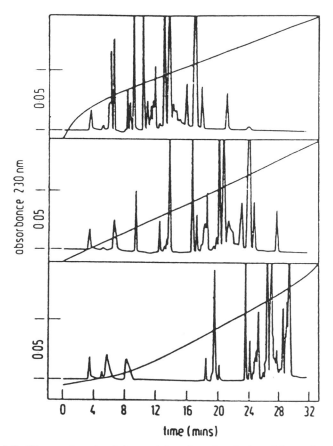

FIGURE 6-8. Change in selectivity and bandspacing induced in a chromatographic separation by variation of the gradient steepness parameter, b. Profiles show the separation of the tryptic peptides of hemoglobin A eluted a 0.1% H_3PO_4–water–acetonitrile gradient.

where α is the selectivity factor (ratio of $\bar{k}$ values) for band pairs when the bands traverse the column midpoint and N is the theoretical plate number at the same point. Hence, for a given chromatogram when constant selectivity is maintained, maximizing the peak capacity is synonymous with maximizing the average resolution. Both PC and R_S will thus show inverse dependencies on the b term with maximum values of PC and R_S favored by smaller values of b, that is for a given set of solutes (with defined S values) and predetermined gradient values ($\Delta\varphi$), by larger t_G values (Figure 6-9). With very long t_G values, a compromise may be necessary in the sense that peak height and detection sensitivity, both inversely proportional to bandwidth, will also decrease as the k' or $\bar{k}$ values of the eluting band increase. Higher peak detection limits are thus favored with steep gradients but obviously resolution will suffer.

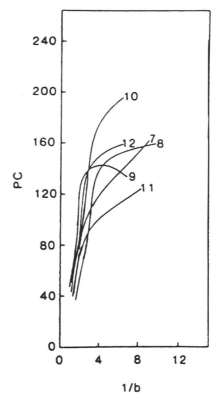

FIGURE 6-9. Plot of peak capacity, PC, versus the reciprocal of the gradient steepness parameter. As the b values increases, the PC decreases. However, the converse is only true at very small b values when anomalous kinetic effects occur, which dramatically influence the bandwidth of the migrating peak. (Data adapted from Ref. 94.)

6.3.6 Kinetic Effects

Up to this point, the discussion has assumed that polypeptide and protein retention in isocratic or gradient modes involves only a single, conformationally stable molecular species. The retention of proteins at n-alkylsilica surfaces is obviously far more complex than implied by this assumption. The reversible and nonreversible deformation of the conformation of polypeptides and proteins in water/organic solvent combinations at the hydrocarbonaceous interface must certainly be considered. The participation of other secondary equilibria processes mediated by mobile-phase or stationary-phase characteristics is also likely to occur (i.e., subunit dissociation, self-association, or aggregation, depending on the solubility coefficients). In many cases, these effects will be solute specific and time dependent. Inappropriate kinetics associated with these equilibria processes can lead to band broadening, at times bizarre in terms of peak skew, asymmetry, or degeneracy. The participation of some secondary equilibrium processes such as ionization or ion-pair formation can usually be exploited to enhance selectivity through appropriate choices of mobile-phase conditions (i.e., by variation of pH or co-/counterion concentration). Solute conformational changes or changes in subunit association/dissociation are generally undesirable secondary equilibrium processes in chromatographic separations, particularly when the kinetics of these phenomena are not sufficiently rapid with regard to the time scale of the chromatographic separation process.

As an illustrative case, the first-order reversible binding of a protein, P, with native globular structure to a hydrocarbonaceous stationary phase and the concomitant first-order interconversion of the native form to an unfolded form, P*, can be considered. Such a situation probably applies to the RP-HPLC of many proteins, including ribonuclease,[95] α-chymotrypsinogen,[92] and papain.[96] The simplest of these interconversion cycles can be represented by[92,97]

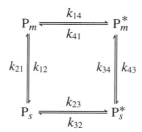

where the subscripts, m and s, refer to the mobile and stationary phases, respectively. According to this model, the concentration of the protein, P, can change biphasically depending on the chromatographic conditions and whether denaturation occurs in the mobile phase and/or at the stationary-phase surface. When the overall time constant for an interconversion cycle between P and P* is comparable to the time of mass transfer passage of P through the column bed, then an apparently homogeneous protein can elute as composite, skewed peaks with different retention times for the native, partially unfolded and fully unfolded forms. The choice of column configuration, elution velocity, and particularly the adsorbent characteristics (surface area, ligand density, pretreatment history, etc.) will influence these processes. The rate constant for interconversion of P and P* at the stationary phase of the order of 8 to 20×10^{-4}/sec at 20°C has been observed[91,96] for several proteins.

It can be recognized from the previous discussion that when such interconversion behavior exists, column efficiency, which reflects inter alia the adsorption/desorption kinetics, will depend not only on the usual chromatographic parameters such as flow rate, column length, and temperature, but also on the residence time the solute spends at the stationary-phase surface. In a typical polypeptide or protein separation on n-alkylsilicas, the plate height of a polypeptide or protein undergoing conformation interconversions is thus expected to be dominated by stationary-phase kinetics contributions.[3,98–101]

6.4   CHROMATOGRAPHIC TECHNIQUES

6.4.1   General Considerations

The goal of all high-resolution chromatographic separations is to achieve maximum selectivity with minimum zone broadening. Further, solute elution should occur rapidly, in high yield, and reproducibly. Few laboratories have access to the necessary silica-bonding technology or are directly involved in the preparation of specialty "in-house" hydrocarbonaceous silica-based adsorbents. Thus, success with complex polypeptide or protein separations on commerically available n-alkylsilicas depends to a very large extent on the ability to manipulate the mobile-phase composition. In some cases, selections of different commercial support materials, packed into columns of different configuration, can be used on a "trial and error" basis with a single mobile-phase composition. When details of the stationary-phase characteristics are poorly documented, such empirical approaches rarely shed much useful light on general trends to improve resolution and recovery. Obviously there are some physical and surface characteristics of n-alkylsilica stationary phases that may be considered more appropriate for polypeptide and protein separation than, for example, for the separation of biogenic amines. However, the same requirements for high-resolution separations (i.e., the same choice of particles with small diameters and narrow particle diameter distribution, uniform porosity, stability of the bonded ligand, suitable ligand coverage, appropriate column packing practices) that apply to the small molecule world are also pertinent to high-resolution separations of polypeptides and proteins.[102]

Similarly, appropriate selection of initial and final mobile-phase compositions is important. Here, variation of retention solely through changes in the water content of the eluent is but one option available. The chemical and physical properties of the organic modifier, the pH, the nature and concentration of ionic or nonionic additives and buffers, the temperature, and the flow rate can all be used to influence peak capacity, resolution, and overall recoveries in most separations. Each of these options can be exercised individually or in concert. The important goals of an RP-HPLC investigation can be summarized as method development that achieves (1) the optimal resolution R_S or peak capacity PC, (2) the smallest peak volume (equivalent to $4\sigma_v$), thus ensuring maximal detection sensitivity, (3) the shortest retention time (t_R or t_g) compatible with the determined resolution, and (4) the highest recovery.

The most effective strategy demands constant reappraisal of the limitations of all the separation and detection variables. This chromatographic diversity can generally be realized through the rational manipulation of remarkably few separation parameters. Clues for this selection of chromatographic conditions are provided by measurements on solubility and denaturation effects in eluents of different composition, from exploratory elution experiments carried out at the microanalytical scale, from evaluation of selectivity matrices, and from the effects on protein function or structure at the stationary-phase surface by changes in pH, ionization or pairing ion reagents. Except in the most straightforward cases, rigorous adherence to only a single chromatographic protocol cannot be expected to achieve the previously described goals.

6.4.2 Choice of Organic Solvent Composition and Gradient Design

When applied to strictly aqueous eluents, most polypeptides and proteins are strongly adsorbed to modern microparticulate reversed-phase packing materials. Because of the magnitude of the S and log k_o values for these solutes, gradient elution provides the most powerful routine method for working up a separation. Not only do gradient procedures permit rapid "scouting" of mobile-phase conditions for acceptable resolution, but they also permit ready assessment of solute "ghosting." The ghost is the residual amount of the polypeptide or protein that carries over on a subsequent blank gradient (i.e., running the same gradient without reinjection of the sample.[1–5,102]

Ideally, in exploratory runs, two different organic solvent modifiers should be used over the range 0 to 50% (v/v) with two different pH, ionic strength, or ionic modifier conditions. This approach gives background information on the nature of the separation. Based on the results of these four gradient experiments, resolution optimization is then attempted. As indicated in Section 6.2.1, resolution is dependent on the gradient steepness parameter, b, with peak capacity PC following an essentially inverse asymptotic dependence on b. With a defined n-alkylsilica support and mobile-phase condition, the average resolution of the system can be set, say at $R_S = 1.5$. An elution window corresponding to $k < 20$ for the desired component(s) is then generated through adjustment of the initial and final mobile-phase compositions. Band spacing can be subsequently varied by changing the steepness of the gradient. Hence, for very complex mixtures, where only a few components (often only one) are required to be purified, changes in b values and $\Delta\varphi$ can be exploited to achieve regional selectivity changes. Since the gradient steepness parameter b is related to the flow rate, F, as well as the rate of change of organic modifier $d\varphi/dt$, comparable resolution can be achieved at different flow rates or gradient rates provided $(d\varphi/dt)/F$ is held constant.

Commonly used gradient conditions such as a 1% change in organic modifier per minute at a flow rate of 1 to 2 mL/min for a standard stainless-steel analytical column (15 to 25 × 0.4 cm) should only be considered as initial points to establish the correct limits of these mobile-phase variables. Once appropriate F, $d\varphi/dt$, and k' or $\bar{k}$ values have been determined, significantly reduced separation times can be achieved using the procedures outlined earlier. When the initial solvent strength is close to the elution concentration, when the dwell time is small, and when the optimal value of the gradient steepness parameter b is achieved, then it has been the experience of this and other laboratories that optimal detection sensitivity, separation speed, and purification factors are realized. Often in these circumstances, a high recovery of mass and biological activity of globular proteins with molecular masses up to or in excess of ca. 60 kDa can also be achieved.

6.4.3   Choice of Ionic Strength and pH

Variation of the eluent pH regulates polypeptide and protein retention with n-alkylsilica and other types of reversed-phase adsorbents through several processes. For example, changes in the mobile-phase pH influence the extent of ionization of the solute and affect the ionization status of accessible silanol groups and other adsorbed ionizable groups present on the stationary-phase surface. Further, pH variations in the eluent will also affect protic equilibria of many ionizable components added to the mobile phase to enhance selectivity. Because of the chemical instability of n-alkylsilicas at high pHs, most reversed-phase separations use mobile phases with pH < 7.5, most commonly in the range pH 2 to 3. With the advent of polymeric organic matrices of large pore size, such as the TSK phenyl G5000 PW,[103] the pH limitations imposed by silica-based supports have been, to some extent, circumvented. However, a number of other reasons remain for the popularity of the low pH choice. These reasons can be summarized as follows:

1. At low pH condition, suppression of carboxyl group ionization occurs and the amino groups are essentially fully protonated. Thus, competing equilibria are suppressed and the solute can behave as a single, averaged ionized species.
2. The isoelectric points of most polypeptides and proteins are above these low pH values. For reasons not yet fully clarified, higher selectivity can be obtained when running below the pI value of a polypeptide or protein in reversed-phase systems.
3. Many alkanoic and perfluoroalkanoic acids have proved effective mobile-phase additives at low concentrations at these pH values. These compounds increase the solubility of polypeptides and proteins and lead to the participation of advantageous ion-pairing/dynamic ion-exchange type retention phenomena. All of these organic acids can be removed readily by lyophilization.
4. Low ionic strengths can be used at low pHs, facilitating easier recovery, better peak shape, and more reproducible retention.[2,11,67]

Several codicils, however, must also be applied when low pH conditions are contemplated for a specific RP-HPLC separation task. The possibility that a specific protein will be poorly soluble at or below its pI value must be taken into account when the pH value is chosen. Alternatively, if the protein is multimeric, the low pH conditions may cause dissociation of the subunits, resulting in the participation of multiple chromatographic peaks. Impaired solubility behavior will have dramatic effects on chromatographic performance, particularly evident in trace enrichment experiments or preparative separations where loss of resolution and blockage of the column may ensue. Similarly, pH-induced protein unfolding or subunit dissociation may occur at inappropriate pH values. In specific cases this effect can be utilized for the purification of the apoprotein moiety or individual subunits of a multimeric protein (i.e., apomyoglobin or the individual α- or β-subunits of the gonadotrophic glycoprotein hormones.[104,105] Reconstitution by the addition of the correct prosthetic group/cofactor or by subunit rehybridization experiments subsequent to completion of the RP-HPLC separation may lead to a useful purification of the native protein.

Even with relatively modest increases in the pH value of the mobile phase, significant band broadening of the eluted peaks with some polypeptides and proteins may nevertheless occur. Because of the existence of multiple optima in the dependency of retention versus pH, optimization of the resolution by random pH adjustments usually presents difficulties. Iterative window diagram procedures (based on simplex or other types of optimization algorithms[106,107]) have shown considerable promise for localizing optimal pH conditions with difficult separations. However, decreased resolution at higher pH values due to loss of column efficiency is often observed with polypeptides and proteins, particularly in the absence of suitable co- or counterions capable of stabilizing the secondary or tertiary structure of the solute in solution. High resolution separation of polypeptides and proteins with n-alkylsilicas solely by pH step or gradient adjustment under constant ionic strength conditions has not been employed extensively. Rather, the synergistic effects of suitable buffer additives and pH are usually combined (often inadvertently) through the use of so-called ion-pair or dynamic liquid–liquid ion exchange procedures.

6.4.4 Addition of Counterions: Hetaeric Chromatography

The introduction[43,108–110] of a variety of ionic modifiers during the late 1970s, many of them novel to peptide and protein purification, resulted in a new era in column liquid chromatography. The emergence of these high-resolution reversed-phase separations as viable techniques has undoubtably stimulated renewed interest in high-performance ion-exchange and affinity support media. During the ensuring years, solvophobic theory and the principles of secondary solution equilibria, as they apply to so-called "ion-pair" phenomena, have been successfully applied so that it is now possible to adequately rationalize the effect of many added co- and counterions. Following the proposal of Horvath and co-workers,[111] these complexing ions can be considered to be hetaerons and thus hetaeric chromatography denotes a technique in which the complexing ion is added to the mobile phase in order to affect selectivity of the chromatographic system by secondary equilibria.

Both nonpolar and polar hetaerons can be used to manipulate the RP-HPLC selectivity with polypeptides and proteins. Currently, more than 100 inorganic and organic anions and cations have been reported to be suitable as pairing ions in reversed-phase separations. Table 6-2 lists a selection of these species suitable for polypeptide or protein fractionation in reversed-phase systems. Retention and selectivity can be varied independently with these additives, with the extent of this variation being defined by the retention modulus or capacity modification factor. As a consequence, both increases and decreases in retention with constant (or varying) selectivity can be achieved with suitable hetaeric systems. Although the list of pairing ions continues to grow (for a compendium, see Ref. 43), two main classes have become favored,[112–126] that is, the bulky, surface-active ions of molecular weight ca. 200, such as the alkylsulfates and the quaternary alkylammonium salts, and the smaller, more polar ions with small adsorption isotherms for the hydrocarbonaceous stationary phase, such as phosphate, perfluoroacetate, bicarbonate, or tetrabutylammonium salts. Within each class, a further subdivision can be achieved based on whether the ion is volatile or removable by simple extraction techniques. Some controversy still surrounds the issue whether hetaeric chromatography with bonded reversed-phase adsorbents involves the retention of the complex of the solute-pairing ion, formed via ion–pair interactions in the bulk mobile phase followed by adsorption of the complex to the solvated nonpolar ligand, or by dynamic, stoichiometric ion-exchange processes involving transient interactions between the solutes and co- and counterion moieties adsorptively bound to the stationary phase. Certainly, many of the hetaerions now in popular use are surface active and exhibit stationary phase-binding properties as described by Langmuir or Freudlich isotherms. The conclusion has been reached that both mechanisms can operate simultaneously, permitting the polypeptide or protein to undergo a range of secondary equilibrium processes, each with its own kinetic constants and relative concentration distributions.[127,128]

6.4.5 Choice of Temperature, Flow Rate, and Column Pressure Drop

Temperature variation as a means of controlling the average resolution of polypeptides and proteins on reversed-phase columns has not been extensively investigated at this stage. Usually, ambient temperature is employed. Although hydrophobic interactions have been postulated to decrease with decreasing temperatures, often the capacity factors for polypeptides and proteins show the opposite effect with n-alkylsilicas. Higher temperatures favor more rapid mass transfer of these solutes, more rapid unfolding between native and disorganized species, and faster dynamics associated with ion and solvent interaction phenomena.[129] All of these effects result in improved efficiency, higher solute solubility (up to the critical temperature associated with gel-sol changes, helical-random coil transitions, or phase changes), and regional selectivity differences. These dependencies with temperature variation may well fall into

TABLE 6-2. Selection of Ionic Species which at pH < 7.0 Modify the Retention Characteristics of Polypeptides and Proteins on Chemically Bonded Hydrocarbonaceous Stationary Phases

Cationic Pairing Ion	Ref.	Anionic Pairing Ion	Ref.
Groups I and II inorganic cations	2, 67, 79, 103, 139	$H_2PO_4^-$	18, 42, 83, 103, 134, 138, 139
Pyridinium salts	79, 89, 84, 87, 92	Cl^-	30, 111, 112, 148, 151
NH_4^+	42, 89, 108, 113	ClO_4^-	2, 108
$CH_3N^+H_3$	42	SO_4^-	86
$C_3H_9N^+H_3$	42	HCO_3^-	42, 143
$C_{12}H_{25}N^+H_3$	42, 108, 109, 110	BO_3^-	115
$HOCH_2CH_2N^+H_3$	42	Tartrate	42
$C_6H_{13}N^+H_3$	42	HCO_2^-	43, 137
$(HOCH_2CH_2)_3N^+H$	28, 119	$CH_3CO_2^-$	89
$(C_2H_5)_3N^+H$	28, 120	$C_2H_5CO_2^-$	115
Morpholinium salts	93	$CF_3CO_2^-$	30, 42, 67, 113, 115, 116, 140, 141
N-Methylpiperidinium	70	$C_3F_7CO_2^-$	71, 120, 122
Piperazinium	70	$C_6F_{13}CO_2^-$	43, 67
TEMED	70	$C_4H_9SO_3^-$	43, 123
Triammonium propane	70	$C_5H_{11}SO_3^-$	43, 67
$(CH_3)_4N^+$	43, 108	$C_6H_{13}SO_3^-$	124, 125
$(C_2H_5)_4N^+$	67, 108	$C_7H_{15}SO_3^-$	124, 125
$(C_3H_7)_4N^+$	67, 108	$C_8H_{17}SO_3^-$	43, 115
$(C_4H_9)_4N^+$	75, 147	$C_{12}H_{25}SO_4^-$	124, 125
Cetylpyridinium	155	Camphor-10-sulfonate	123, 124, 125
$Benzl(CH_3)_3N^+$	110, 155	p-Toluene sulfonate	43, 67

This selection of ionic species is generally compatible with UV detection below 235 nm or, in the case of those reagents with an aromatic nucleus, with fluorometric or UV (above 235 nm) detection. The concentration of these compounds in the mobile phase is usually <5 mM or ca. 0.1%. (Adapted from Ref. 43.) See also Table 6–1 for selected applications.

the class of the so-called "entropy-compensated" effects,[130,131] which correlate with differences in the molecular shape of the solute molecules, as well as their orientation and dynamics at the liquid–solid interface. Moreover, the dependence of k' or $\bar{k}$ on temperature, T, can be expressed as

$$\ln k' = -\frac{\Delta G^{\circ}_{\text{assoc}}}{RT} + \log \Phi = -\frac{\Delta H^{\circ}_{\text{assoc}}}{RT} + \frac{\Delta S^{\circ}_{\text{assoc}}}{R} + \log \Phi \quad (6\text{-}11)$$

where $\Delta H^{\circ}_{\text{assoc}}$ and $\Delta S^{\circ}_{\text{assoc}}$ are the standard enthalpy and entropy change for the transfer of the solute to the stationary phase.

From plots of $\ln k$ versus reciprocal temperature (van't Hoff plots), it is possible to derive the $\Delta H^{\circ}_{\text{assoc}}$ values and hence the change $\Delta\Delta H^{\circ}_{\text{assoc}}$ and in $\Delta\Delta S^{\circ}_{\text{assoc}}$ values associated with polypeptide and protein unfolding at the stationary-phase surface.[41,130–131] One codicil that has arisen from studies with polypeptides has been the demonstration that although column efficiency improves with increasing temperature (ca. twofold for a 20°C increase), the average resolution or peak capacity in the gradient mode with solvent-compensated separations is lower at temperatures above 37 to 40°C than is the case with temperatures at 15 to 20°C.

Conventional chromatographic wisdom teaches that the column efficiency will depend on the linear flow velocity of the mobile phase and that this dependence can be visualized in the form of van Deempter or Knox plots. Improvement in column efficiency is thus anticipated as the flow rate is decreased. Changes in column and eluent parameters, which directly influence lateral diffusion of the solute at the mobile-phase–stationary-phase interface, have major effects on solute band broadening. This outcome can be explained by the fact that the reduced plate height (efficiency) is proportional to peak variance and is inversely proportional to particle diameter and column length. Because resolution changes only as the square root of column length, it is thus much more advantageous to manipulate eluent composition and flow rate in tandem.

A further factor that must also be considered in the design of column configuration is the pressure drop over the column bed given by

$$\Delta p = \frac{v \cdot \eta \cdot L}{\rho^* \cdot d_p^2} \quad (6\text{-}12)$$

where η is the eluent viscosity, v is the linear flow velocity of the eluent, and ρ^* is the column permeability. With silica-based hydrocarbonaceous phases, there is a linear relationship between Δp and v. The consequences of the dependencies inherent in Eq. (6-12) for scaleup should not be overlooked as they bear directly on the choice of column dimensions, particle diameter (d_p), and particle porosity. Although low flow rates are advisable in view of the slower diffusion rates of polypeptides and proteins, in gradient elution it is advisable to adjust the gradient slope so that relatively small b values are obtained. Proteins can show dramatic changes in their h versus v plots over very small k' ranges, with peak asymmetry factors typically larger for small-pore n-alkylsilicas than for large-pore n-alkylsilicas (see Figure 6-10). This behavior demands that adequate consideration be given to the total separation time as an integral part of the design of column configuration, as anticipated on the basis of the discussion in Section 6.2. Paradoxically, resolution may be increased with shallow gradients of low flow rate, but recovery can fall off disastrously under these conditions.

6.4.6 Detection and Sample Recovery

The availability of suitable detection methods also has significant bearing on the choice of the organic solvent modifier and, in general, the mobile-phase composition. Biological *in vitro* or *in vivo* response assays, radioligand assays, and radioimmuno assays can all detect, under suitable conditions, a specific polypeptide or protein at a level several orders of magnitude more sensitively than can be obtained by most spectrophotometric methods or by classical procedures of chemical analysis. All the popular primary organic solvent modifiers, methanol, ethanol, acetonitrile, and 1- or 2-propanols, that are used in RP-HPLC can be obtained with UV transparency at 210 to 215 nm of 0.005 au. At this wavelength the isobestic absorption of the polypeptide backbone occurs with molar adsorptivities essentially independent of conformation, ca. 10^3 M^{-1} per residue. As a consequence, with UV detectors of small cell volume (<5 μl) and short path length (<8 mm), the limits of detectability are ca. 20 ng/mL at a signal-to-noise ratio larger than 5. For most polypeptides and proteins the use of fluorescence detectors, being more selective, allows in specific cases a significant increase in the level of detectability (ca. 10- to 50-fold) since the detection is less susceptible to background perturbations due to solvent contaminants or other UV-absorbing artifacts. Where possible, endogeneous tryptophan fluorescence (215-nm excitation, 340-nm emission) can be monitored. Elsewhere, postcolumn derivatization will be required, but in these circumstances it must be remembered that even with a high derivatization efficiency, broader peaks (i.e., increased σ_v) and smaller peak heights (i.e., decreased $1/\sigma_v$) will always be obtained, thus decreasing the detectability of the solute.

Careful pre- and postchromatographic handling of samples is often required with labile proteins exposed to aqueous organic solvent mixtures at low pH values. A useful remedy in some cases, such as with prostatic acid phosphatase[118] and several monomeric and multimeric protein hormones,[94,119,120] involves the collection of the chromatographic fractions directly into tubes containing buffers of more appropriate pH. Deamidation, sequence-specific amino acid residue cleavage of proteins at acid labile sites, and rearrangement of aspartic acid residues to the corresponding α-aminosuccinimde group have all been observed following reversed-phase separations.[132]

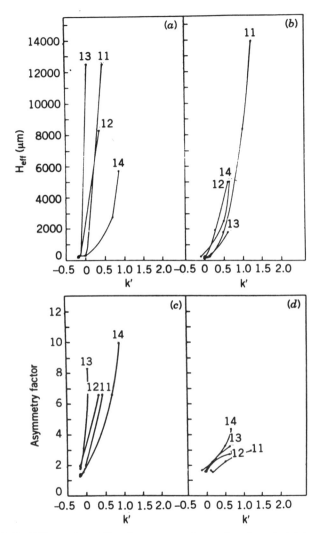

FIGURE 6-10. Effect of mobile-phase strength on the plate height equivalent, H_{eff} and peak asymmetry, a_s, for several polypeptides separated on a 7.3-nm pore size (a) or a 30-nm pore size (b) octadecylsilica at 2.0 mL/min. Polypeptides: 11, hen lysozyme; 12, sperm whale apomyoglobin; 13, porcine trypsin; and 14, bovine serum albumin. The divergence of H_{eff} and a_s, evident between the small- and the large-pore octadecylsilica, is particularly noteworthy and may indicate the fundamental reason why protein recoveries on large-pore silicas are often significantly higher than those obtained with small-pore n-alkylsilicas. (Adapted from Ref. 97.)

6.5 APPLICATIONS

One of the main application areas for reversed-phase chromatography of polypeptides and proteins is for structural studies, represented by two of the examples given in this section. Other areas selected from the extensive applications literature related to the conformational changes evident with proteins in RP-HPLC systems are described. Due to the extreme conditions often used in RPC, some proteins will not be eluted in their native, biologically active state. Proteins that retain their activities and native structures most often are the lower molecular weight proteins containing several disulfide bonds (e.g, with molecular masses <60 kDa). However, it is important to reflect that many of the protein hormones, growth factors, and immunomodulatory polypeptides and proteins that occupy such a prominent position in modern cell biology, molecular oncology, or molecular immunology have their origins in their purification and analysis by RP-HPLC procedures. Similar parallels can be found in numerous other areas of the biological sciences.

6.5.1 Ribosomal Proteins

Eukaryotic and prokaryotic ribosomes are composed of one large and one small subunit containing a large number of different proteins. Eukaryotic ribosomes contain about 80 proteins, including several phosphoproteins, of which about 33 belong to the small subunit and about 45 to the large subunit. The small and large subunits of the *Escherichia coli* ribosome contain 21 and 32 different proteins, respectively. The molecular weights range between 5,000 and 25,000 for the large subunit and between 8,000 and 27,000 (except S1, which is around 61,000) for the small subunit.

  Reversed-phase chromatography has proved extremely useful for the separation and purification of ribosomal proteins for structural studies from a variety of different cell types.[17,117,133] From the *E. coli* ribosome, 15 proteins and 23S proteins were isolated in sequencer purity using this technique.[117] Figure 6-11 shows a representative example of the separation of the 50S ribosomal proteins from *E. coli*. Similar procedures have been adapted for the separation of ribosomal proteins from a variety of other microorganisms and mammalian cells, including *Taq thermolytica*.[133]

6.5.2 Hemoglobin Varieties

The major hemoglobin, HbA, found in normal erythrocytes has the subunit structure $\alpha_2\beta_2$. The α-chain contains 141 amino acid residues and the β-chain contains 146 residues. There are more than 300 known point mutations in the human hemoglobin β-chain. About 100 of these single or multiple amino acid residue mutations give rise to symptoms with the individual, with the most common being fatigue due to stronger oxygen binding than in normal hemoglobin. Representative of the use of RP-HPLC procedures was the discovery in

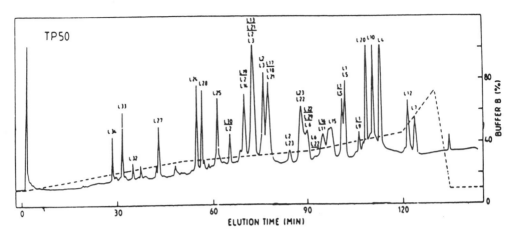

FIGURE 6-11. Purification of 50S ribosomal proteins from *E. coli* by reversed-phase HPLC on Ultrapore RPSC. Two milligrams of TP50 was injected into Ultrapore RPSC (5-μm particle size; 300-Å pore size; column size, 75 × 4.6 mm; 35°C; flow rate 0.5 mL/min). The eluent was buffer A, 0.1% TFA in water, buffer B, 0.1% TFA in acetonitrile. The gradient applied was 10% B to 30% B for 50 min; 30% B to 37% B for 40 min; 37% B to 50% B for 30 min; 50% B to 80% B for 10 min; 80% B to 10% B for 5 min; and reconditioning for 30 min at initial conditions. Detection measurements were made at 220 nm, range 0.64 au. The proteins were identified by two-dimensional polyacrylamide gel electrophoresis and microsequencing as indicated in the figure. (Adapted from Ref. 117 by permission of the authors and publishers.)

1984 of a new point mutation in Sweden. In this case, the β-36 Pro had been replaced by Thr. Figure 6-5 shows the separation of the hemoglobin chains of this individual by reversed-phase chromatography on an *n*-octyl-bonded silica adsorbent. Under the conditions of this experiment, the $\alpha_2\beta_2$ chains and the heme groups are completely dissociated. In this case, the individual was heterozygous with respect to the point mutation, and this coexpression explains why this individual had both normal and mutated chains present in the hemoglobin of the erythrocytes.

6.5.3 Conformational Behavior of Protein Hormones

A further representative example of the power of reversed-phase HPLC with the analysis of large polypeptides and proteins can be found in the application of this technique, not for the separation of different compositional species of different proteins, but in the resolution of different conformational intermediates and the ensuing studies on the mechanism of the interconversion of these conformers. A variety of elegant examples have been described in the literature, including studies on conformational effects of recombinant growth hormone and *N*-methyl recombinant growth hormone,[40] lysozyme,[46,131] trypsin,[1,3,100] ribonuclease A,[46,54] cytochrome c,[81] α-lactoglobin,[96,155] myoglobin,[3,46] the glycoprotein hormone α/β subunits,[51] insulins,[106,131] and a large array of amphipathic polypeptides[22,98] in the presence of the hydrophobic environment of the RP-HPLC milieu. The common characteristic of all these investigations is that the specific adsorption and desorption of the protein or polypeptide can be correlated with either solvent- or ligand-induced conformational changes. With some solvent systems (i.e., 1-propanol) the retention of the protein species decreased as the temperature was increased due to the greater flexibility of the hydrophobic ligands as well as structural rearrangement of the protein. In other cases, such as acetonitrile, a ligand surface-driven unfolding of the protein occurs, resulting in increased retention as the temperature is increased. This latter process can lead to enhanced denaturation of the protein and/or loss of the prothetic group which may, in appropriate cases, be associated with the specific protein (see, e.g., with myoglobin). These experimental methods provide a very important direct method to assess the apparent kinetics (from plots of log[peak area] versus time) of these transitions as well as the thermodynamics of the process (from plots of $\log k$ versus $1/T$, where k is the capacity factor or the median capacity factor and T is the absolute temperature). With increasing refinement of the adsorbent surface through the development of new classes of reversed-phase materials, including those that resemble more closely biological membranes, and new variations of on-line spectroscopic instrumentation to enable the direct monitoring of these conformational changes, the full potential of reversed-phase HPLC procedures in studies on protein conformational behavior will become increasingly recognized as one powerful manifestation of what is now increasingly being termed "molecular chromatography"—the study of the distinctive molecular events that characterize protein interactions with hydrophobic and other types of ligands in the HPLC mode.

ACKNOWLEDGMENT

The support of the National Health and Medical Research Council in Australia and the Australian Research Council of projects described in part in this chapter is gratefully acknowledged.

6.6 REFERENCES

1. M.T.W. Hearn, M.I. Aguilar, in *Modern Physical Methods in Biochemistry, Part B*. A. Neuberger, L.L.M. van Deenan, eds., Elsevier Science, 1988, pp. 107–142.
2. M.T.W. Hearn, in *High Performance Liquid Chromatography: Advances and Perspectives*, C. Horvath, ed., Academic Press, New York, 1983, Vol. 3, pp. 87–155.
3. M.T.W. Hearn, in *Chemical Separations: Separation Science and Technology*, C.J. King, J.D. Navratil, eds., Litarvan Press, CO, 1986, pp. 77–98.

4. R. Kellner, F. Lottspeich, H.E. Meyer, in *Microcharacterisation of Proteins*, VCH Press, Weinheim, 1994.

5. M.T.W. Hearn, in *HPLC of Proteins, Peptides and Polynucleotides*, VCH, New York, 1991.

6. R.J. Boscott, *Nature (London)*, *159*, 342 (1947).

7. J. Boldingh, *Experimentia*, *4*, 270 (1948).

8. G.A. Howard, A.J.P. Martin, *Biochem. J.*, *46*, 532 (1950).

9. S. Shaltiel, *Meth. Enzymol.*, *34*, 126 (1974).

10. Z. Er-el, Y. Zaidenzaig, S. Shaltiel, *Biochem. Biophys. Res. Commun.*, *49*, 383 (1972).

11. B.H.J. Hofstee, *Biochem. Biophys. Res. Commun.*, *50*, 751 (1973).

12. H.J. Wirth, M.T.W. Hearn, *J. Chromatogr.*, *646*, 143 (1993).

13. W.A. Schafer, P.W. Carr, *J. Chromatogr.*, *587*, 149 (1991).

14. U. Trudinger, G. Muller, K.K. Unger, *J. Chromatogr.*, *535*, 111 (1990).

15. G. Szepesi, in *Reverse-Phase HPLC*, VCH, New York, 1992, pp. 29–63.

16. K.D. Nugent, in *High Performance Liquid Chromatography of Peptides and Proteins*, C.T. Mant, R.S. Hodges, eds., CRC Press, Boca Raton, FL, 1991, pp. 279–287.

17. H.P. Nick, R.E.H. Wetherall, M.T.W. Hearn, F.J. Morgan, *Anal. Biochem.*, *148*, 93 (1985).

18. K.A. Cohen, K. Schellenberg, B.L. Karger, B. Grego, M.T.W. Hearn, *Anal. Biochem.*, *140*, 223 (1984).

19. B. Grego, M.T.W. Hearn, *J. Chromatogr.*, *266*, 75 (1983).

20. L.A. Witting, D.J. Gisch, R. Ludwig, R. Eksteen, *J. Chromatogr.*, *296*, 97 (1984).

21. J.D. Pearson, N.T. Lin, F.E. Regnier, *Anal. Biochem.*, *124*, 217 (1982).

22. A.W. Purcell, M.I. Aguilar, M.T.W. Hearn, *Anal. Chem.*, *65*, 3038 (1993).

23. J.J. Kirkland, J.L. Glajch, R.D. Fairlee, *Anal. Chem.*, *61*, 2 (1989).

24. R.E. Zhang, Z. Xie, X. Li, M.I. Aguilar, M.T.W. Hearn, *Anal. Chem.*, *63*, 1861 (1991).

25. C.T. Wehr, in *CRC Handbook of HPLC*, CRC Press, Boca Raton, FL, 1984, p. 31.

26. M.T.W. Hearn, B. Grego, *J. Chromatogr.*, *282*, 541 (1983).

27. I. Yarovsky, M.I. Aguilar, M.T.W. Hearn, *J. Chromatogr.*, *660*, 61 (1994).

28. I. Yarovsky, M.I. Aguilar, M.T.W. Hearn, *Anal. Chem.*, *67*, 1661 (1994).

29. G. Lindgren, B. Lundstrom, I. Kallman, K.A. Hansson, *J. Chromatogr.*, *296*, 83 (1984).

30. N.H.C. Cooke, B.G. Archer, M.J. O'Hare, E.C. Nice, M. Capp, *J. Chromatogr.*, *255*, 115 (1983).

31. M.J. O'Hare, M.W. Capp, E.C. Nice, N.H.C, Cooke, B.G. Archer, *Anal. Biochem.*, *126*, 17 (1982).

32. K.A. Cohen, J. Chazaud, G. Galley, *J. Chromatogr.*, *282*, 423 (1983).

33. Y. Kato, T. Kitamura, T. Hashimoto, *J. Chromatogr.*, *266*, 49 (1983).

34. J.L. Fausnaugh, E. Pfannkoch, S. Gupta, F.E. Regnier, *Anal. Biochem.*, *137*, 462 (1984).

35. M. Shin, N. Sakihama, R. Oshino, H. Sasaki, *Anal. Biochem.*, *138*, 259 (1984).

36. J.P. Harrington, T.T. Herskovits, *Biochemistry, 14,* 4972 (1975).

37. T. Askura, K. Adachi, E. Schwartz, *J. Biol. Chem., 253,* 6423 (1978).

38. T.T. Herskovits, H. Jaillet, *Science, 163,* 282 (1969).

39. W.A. Jensen, J. McD. Armstrong, J. DeGiorgio, M.T.W. Hearn, *Biochemistry, 34,* 472 (1995).

40. P. Oroszalan, S. Wicar, G. Tashima, S.L. Wu, W.S. Hancock, B.L. Karger, *Anal. Chem., 64* 1623 (1993).

41. M.I. Aguilar, M.T.W. Hearn, in *High Resolution Separation and Analysis of Biological Macromolecules,* B.L. Karger, W.S. Hancock, eds., *Meth. Enzymol.,* 1996, 270, 3–26.

42. M.T.W. Hearn, M.I. Aguilar, *J. Chromatogr., 359,* 31 (1986),

43. M.T.W. Hearn, in *Ion Pair Chromatography,* Marcel Dekker, New York, 1984, pp. 1–296.

44. X. Geng, F.E. Regnier, *J. Chromatogr., 296,* 1 (1984).

45. A.W. Purcell, M.I. Aguilar, M.T.W. Hearn, *J. Chromatogr., 593,* 103 (1992).

46. S. Lin, B.L. Karger, *J. Chromatogr., 499,* 89 (1990).

47. S.D. Powers, M.A. Lochrie, R.O. Poyton, *J. Chromatogr., 266,* 585 (1983).

48. M.T.W. Hearn, B. Grego, *J. Chromatogr., 296,* 309 (1984).

49. M.L. Tierney, N.H. Goss, S.M. Tompkins, D.B. Kerr, D.E. Pitt, R.G. Forage, D.M. Roberson, M.T.W. Hearn, D.M. DeKretser, *Endocrinology, 126,* 3266 (1992).

50. M.A. Chlenov, E.I. Kandyba, L.V. Nagornaya, I.L. Orlova Y.V. Volgin, *J. Chromatogr., 631,* 261 (1993).

51. J.B. Wu, P.G. Stanton, D.M. Robertson, M.T.W. Hearn, *Endocr. J., 136,* 410 (1993).

52. J.Frenz, W.S. Hancock, W.J. Henzel, C. Horvath, in *HPLC of Biological Macromolecules: Methods and Applications,* K.M. Gooding, F.E. Regnier, eds, Marcel Dekker, New York, 1990, pp. 145–160.

53. R. Janzen, K.K. Unger, H. Giesche, J.N. Kinkel, M.T.W. Hearn, *J. Chromatogr., 397,* 81 (1987).

54. X.M. Lu, K. Benedek, B.L. Karger, *J. Chromatogr., 359,* 19 (1986).

55. K.K. Unger, K.D. Lork, H.J. Wirth, in *HPLC of Proteins, Peptides and Polynucleotides,* VCH, New York, 1991, pp. 59–117.

56. T.W. Lorne, C.T. Mant, R.S. Hodges, in *High Performance Liquid Chromatography of Peptides and Proteins,* C.T. Mant, R.S. Hodges, eds., CRC Press, Boca Raton, FL, 1991, pp. 307–318.

57. J.W. Dolan, in *High Performance Liquid Chromatography of Peptides and Proteins,* C.T. Mant, R.S. Hodges, eds., CRC Press, Boca Raton, FL, 1991, pp. 23–30.

58. M.T.W. Hearn, in *Peptides: Chemistry, Structure and Biology,* J. Rivier, G. Marshall, eds., Escom, Leiden, The Netherlands, 1990, p. 415.

59. M.T.W. Hearn, M.I. Aguilar, *J. Chromatogr., 397,* 47 (1987).

60. M.T.W. Hearn, M.I. Aguilar, T. Nguyen, M. Fridman, *J. Chromatogr., 435,* 271 (1988).

61. M.T.W. Hearn, M.I. Aguilar, in *High Performance Liquid Chromatography of Peptides, Proteins and Polynucleotides,* M.T.W. Hearn, VCH, New York, 1989, pp. 345–370.

62. N.T. Miller, B. Feibush, B.L. Karger, *J. Chromatogr., 316,* 519 (1984).

63. H. Engelhardt, D. Mathes, *J. Chromatogr., 185,* 305 (1979).

64. A.J. Alpert, *J. Chromatogr., 359,* 85 (1986).

65. J. Liu, K.J. Volk, E.H. Kerns, S.E. Kohr, M.S. Lee, I.E. Rosenberg, *J. Chromatogr., 632,* 45 (1993).

66. D.E. Raymond, A. Manz, H.M. Widmer, *Anal. Chem., 66,* 2858 (1994).

67. M.T.W. Hearn, *Adv. Chromatogr., 20,* 1 (1982).

68. J. Rivier, R. McClintock, in *The Use of HPLC in Protein Purification and Characterisation,* A.R. Kerlavage, ed., John Wiley & Sons, New York, 1989, pp. 77–105.

69. G. Vigh, Z. Varga-Puchony, G. Szepesi, M. Gazdag, *J. Chromatogr., 386,* 353 (1987).

70. R.S. Hodges, T.W. Li, C.T. Mant, *J. Chromatogr., 444,* 349 (1988).

71. S.O. Jeppsson, L. Kallman, G. Lindgren, L.G. Fagerstam, *J. Chromatogr., 297,* 31 (1984).

72. A.W. Purcell, M.I. Aguilar, M.T.W. Hearn, *J. Chromatogr., 476,* 113 (1991)

73. M.C.J. Wilce, M.I. Aguilar, M.T.W. Hearn, *J. Chromatogr., 548,* 105 (1991).

74. M.I. Aguilar, K.L. Richards, A.J. Round, M.T.W. Hearn, *Peptide Res., 7,* 207 (1994).

75. R.A. Houghten, J.M. Ostresh, *J. Chromatogr., 386,* 223 (1987).

76. M.J. Wilce, M.I. Aguilar, M.T.W. Hearn, *J. Chromatogr., 632,* 11 (1993).

77. M.J. Wilce, M.I. Aguilar, M.T.W. Hearn, *Anal. Chem. 67,* 1210 (1995).

78. J.L. Meek, Z. Rossetti, *J. Chromatogr., 211,* 15 (1981).

79. S.J. Su, B. Grego, B. Niven, M.T.W. Hearn, *J. Liquid Chromatogr., 4,* 1745 (1981).

80. H. Stotzel, G.J. Hughes, *Biochem. J., 199,* 31 (1981).

81. K.L. Richards, M.I. Aguilar, M.T.W. Hearn, *J. Chromatogr., 676,* 17 (1994).

82. A.W. Purcell, M.I. Aguilar, M.T.W. Hearn, *J. Chromatogr., 711,* 61 (1995).

83. M.T.W. Hearn, B. Grego, *J. Chromatogr., 203,* 349 (1981).

84. T.J. Sereda, C.T. Mant, F.D. Sonnichsen, R.S. Hodges, *J. Chromatogr., 676,* 139 (1994).

85. J.G. Dorsey, K.A. Dill, *Chem. Rev., 89,* 331 (1989)

86. M.T.W. Hearn, B. Grego, *J. Chromatogr., 255,* 125 (1983).

87. M.T.W. Hearn, B. Grego, *J. Chromatogr., 266,* 75 (1983).

88. M.J. O'Hare, E.C. Nice, *J. Chromatogr., 171,* 209 (1979).

89. B. Grego, M.T.W. Hearn, *J. Chromatogr., 366,* 28 (1984).

90. L.R. Snyder, in *High Performance Liquid Chromatography,* C. Horvath, ed., Academic Press, New York, 1980, Vol. 1, p. 208.

91. M.T.W. Hearn, A.N. Hodder, M.I. Aguilar, *J. Chromatogr., 327,* 47 (1985).

92. M.T.W. Hearn, M.I. Aquilar, *J. Chromatogr., 392,* 33 (1987).

93. M.A. Stadalius, H.S. Gold, L.R. Snyder, *J. Chromatogr., 296,* 31 (1984).

94. M.I. Aquilar, A.M. Hodder, M.T.W. Hearn, *J. Chromatogr., 352,* 115 (1986).

95. S.A. Cohen, S. Dong, K. Benedek, B.L. Karger, *J. Chromatogr., 317,* 227 (1984).

96. R. Rosenfeld, K. Benedek, *J. Chromatogr., 632,* 29 (1993).

97. M.T.W. Hearn, B. Grego, *J. Chromatogr., 296,* 61 (1984).

98. A.W. Purcell, M.I. Aguilar, R.E.H. Wettenhall, M.T.W. Hearn, *Peptide Res., 8,* 545 (1995).

99. K.L. Richards, M.I. Aguilar, M.T.W. Hearn, *J. Chromatogr., 676,* 33 (1994).

100. M.T.W. Hearn, A.N. Hodder, M.I. Aguilar, *J. Chromatogr., 327,* 115 (1985).

101. F.E. Regnier, *Science, 238,* 319 (1987).

102. C. Jilge, R. Janzen, H. Giesche, K.K. Unger, J.N. Kinkel, M.T.W. Hearn, *J. Chromatogr., 397,* 71 (1987).

103. Y. Kato, T. Kitamura, T. Hashimoto, *J. Chromatogr., 292,* 418 (1984).

104. Y.F. Maa, C. Horvath, *J. Chromatogr., 445,* 71 (1988).

105. R.L. Patience, L.H. Rees, *J. Chromatogr., 352,* 241 (1986).

106. B.S. Welinder, in *High Performance Liquid Chromatography of Peptides and Proteins,* C.T. Mant, R.S. Hodges, eds., CRC Press, Boca Raton, FL, 1991, pp. 343–350.

107. J. Fekete, G. Morovjan, F. Csizmadia, F. Darvas, *J. Chromatogr., 660,* 33 (1990).

108. W.S. Hancock, C.A. Bishop, R.L. Prestidge, D.R.K. Harding, M.T.W. Hearn, *Science, 200,* 1168 (1978).

109. M.T.W. Hearn, W.S. Hancock, *Trends Biochem. Sci., 4,* 58 (1979).

110. M.T.W. Hearn, *Adv. Chromatogr., 18,* 59 (1980).

111. C. Horvath, W. Melander, I. Molnar, P. Molnar, *Anal. Chem., 49,* 2295 (1977).

112. A.F. Bristow, C. Wilson, N. Sutcliffe, *J. Chromatogr., 270,* 285 (1983).

113. E.C. Nice, M. Capp, M.J. O'Hare, *J. Chromatogr., 147,* 413 (1979).

114. J. Spiess, J. Rivier, C. Rivier, W. Vale, *Proc. Natl. Acad. Sci. USA, 78,* 6517 (1981).

115. B. Grego, M.T.W. Hearn, *J. Chromatogr., 336,* 25 (1984).

116. N.E. Tandy, R.A. Dilley, F.E. Regnier, *J. Chromatogr., 266,* 599 (1983).

117. R.M. Kamp, B. Wittman-Liebold, *FEBS Lett., 167,* 59 (1984).

118. M.P. Strickler, J. Kintzios, M.J. Genski, *J. Liquid Chromatogr., 5,* 1921 (1982).

119. P.G. Stanton, M.T.W. Hearn, *J. Biol. Chem., 262,* 1623 (1987).

120. H.P.J. Bennett, *J. Chromatogr., 266,* 249 (1983).

121. S.D. Power, M.A. Lochrie, R.O. Poyton, *J. Chromatogr., 266,* 585 (1983).

122. J.R. Walsh, H.D. Niall, *Endocrinology, 107,* 1258 (1980).

123. S. Terabe, K. Konaka, K. Inouye, *J. Chromatogr., 172,* 163 (1979).

124. M.T.W. Hearn, B. Grego, W.S. Hancock, *J. Chromatogr., 185,* 429 (1979).

125. M.T.W. Hearn, B. Grego, *J. Chromatogr., 218,* 497 (1981).

126. B. Grego, F. Lambrou, M.T.W. Hearn, *J. Chromatogr., 266,* 89 (1983).

127. C.T. Mant, R.S Hodges, in *High Performance Liquid Chromatography of Peptides and Proteins,* C.T. Mant, R.S. Hodges, eds., CRC Press, Boca Raton, FL, 1991, pp. 327–341.

128. C. Horvath, W. Melander, I. Molnar, P. Molnar, *Anal. Chem., 49,* 2295 (1977).

129. R.H. Ingraham, S.Y.M. Lau, A.K. Taneja, R.S. Hodges, *J. Chromatogr., 327,* 77 (1985).

130. A.W. Purcell, M.I. Aguilar, M.T.W. Hearn, *J. Chromatogr., 711,* 71 (1995).

131. R.A. Barford, B.J. Sliwinski, A.C. Breyer, H.L. Rothbard, *J. Chromatogr., 235,* 281 (1982).

132. P.E. Thompson, V. Cavallaro, M.T.W. Hearn, *Peptide Sci., 1,* 263 (1995).

133. I. Molnar, R. Boysen, P. Jekow, *J. Chromatogr., 485,* 469 (1989)

134. R. van der Zee, G.W. Welling, *J. Chromatogr., 244,* 134 (1982).

135. G.E. Gerber, R.J. Anderegg, W.C. Herlihy, C.P. Gray, K. Biemann, H.G. Khorana, *Proc. Natl. Acad. Sci. USA, 76,* 227 (1979).

136. R.V. Lewis, A. Fallon, S. Stein, K.D. Gibson, S. Udenfriend, *Anal. Biochem., 104,* 153 (1980).

137. W. Monch, W. Dehnen, *J. Chromatogr.,147,* 415 (1978).

138. W.S. Hancock, C.A. Bishop, A.M. Gotto, D.R.K. Harding, S.M. Lamplugh, J.T. Sparrow, *J. Lipid Res., 16,* 250 (1981).

139. K.J. Wilson, M.W. Berchtold, P. Zumskin, S. Klause, G.J. Hughes, in *Methods in Protein Sequence Analysis,* M. Elizinga, ed., Hamana Press, Clifton, NJ, 1982, p. 260.

140. G.J. Putterman, M.B. Spear, K.S. Meade-Cobun, M. Widra, C.V. Hixson, *J. Liquid Chromatogr., 5,* 715 (1982).

141. S.J.M. Skinner, B. Grego, M.T.W. Hearn, C.G. Liggins, *J. Chromatogr., 308,* 113 (1984).

142. A.W. Burgess, J.A. Knesel, L.G. Sparrow, W.A. Nicola, E.C. Nice, *Proc. Natl. Acad. Sci. USA, 79,* 5753 (1982).

143. B. Grego, G.S. Baldwin, J.A. Knessel, R.J. Simpson, F.J. Morgan, M. T.W. Hearn, *J. Chromatogr., 297,* 21 (1984).

144. L.R. Gurley, J.A.D. Anna, M. Blumenfeld, J.G. Valdez, R.J. Sebring, P.R. Dohahue, D.A. Prentice, W.D. Spall, *J. Chromatogr., 297,* 147 (1984).

145. S. Stein, C. Kenny, H.J. Freisen, J. Shively, U. Del Valle, S. Pestka, *Proc. Natl. Acad. Sci. USA, 77,* 5716 (1980).

146. R.A. Wolfe, J. Casey, P.C. Familletti, S. Stein, *J. Chromatogr., 296,* 277 (1984).

147. S.W. Herring, R.K. Enns, *J. Chromatogr., 266,* 249 (1983).

148. L. Sottrup-Jensen, T.M. Stepanik, C.M. Jones, P.B. Lonblad, T. Kristensen, D.M. Wierzbicki, *J. Biol. Chem., 259,* 8293 (1984).

149. V.L. Alvarez, C.A. Roitsch, O. Henriksen, *Anal. Biochem., 115,* 353 (1981).

150. W. Kopaciewicz, F.E. Regnier, *Anal. Biochem., 129,* 472 (1983).

151. M.W. Berchtold, K.J. Wilson, C.W. Heizmann, *Biochemistry, 21,* 6552 (1982).

152. C.N. Chesterman, T. Walker, B. Grego, K. Chamberlain, M.T.W. Hearn, F.J. Morgan, *Biochem. Biophys. Res. Commun., 116,* 809 (1983).

153. M.T.W. Hearn, B. Grego, *J. Chromatogr., 282,* 541 (1983).

154. G. Dasari, I.G. Prince, M.T.W. Hearn, *J. Biochem. Biophys. Meth., 30,* 217 (1995).

155. K. Benedek, *J. Chromatogr., 458,* 93 (1988).

7 Hydrophobic Interaction Chromatography

KJELL-OVE ERIKSSON
Department of Chemistry
Indiana University
Bloomington, Indiana 47405, USA

Protein Purification: Principles, High-Resolution Methods, and Applications, Second Edition.
Edited by Jan-Christer Janson and Lars Rydén.
ISBN 0-471-18626-0.   © 1998 Wiley-VCH, Inc.

7.1   INTRODUCTION

Hydrophobic molecules in an aqueous solvent will self-associate. This association is due to hydrophobic interaction. The hydrophobic interaction is of prime importance in biological systems. It is a major driving force behind the folding of globular proteins (e.g., formation of the molten globule state), in the association of protein subunits, and in the binding of many small molecules to proteins as in enzyme catalysis, regulation, and transport across surfaces. It is also responsible for the self-association of phospholipids and other lipids to form the biological membrane bilayer and the binding of integral membrane proteins.

In hydrophobic interaction chromatography (HIC) the hydrophobic interaction is utilized for the binding of proteins to adsorbents with hydrophobic ligands. Our present, rather detailed, knowledge of the protein three-dimensional structure has revealed that the surface of globular proteins can have extensive hydrophobic patches in addition to the expected hydrophilic groups. It is these hydrophobic regions that bind to hydrophobic ligands on the adsorbent, in media favoring hydrophobic interaction (e.g., an aqueous solution with a high salt concentration). Elution (and separation), according to differences in the strength of interaction between the proteins and the amphiphilic matrix, is generally brought about by decreasing the salt concentration of the eluent. In some cases a decrease of the solvent polarity is also needed.

The concept of separation proteins under HIC conditions was outlined by Tiselius already in 1948,[1] but the technique has been developed since the early 1970s. The first gels of practical use for HIC were of a mixed hydrophobic–ionic character.[2-4] Neutral adsorbents (alkyl and aryl ethers) were later prepared by Porath and co-workers[5] and Hjertén and co-workers,[6] the latter leading to the introduction of octyl- and phenyl-Sepharose. It was also Hjerte'n who suggested the now generally accepted name of the technique: hydrophobic interaction chromatography.[7] The term hydrophobic chromatography should be avoided, as it is the interaction between the solute and the adsorbent that is hydrophobic, not the chromatographic procedure.[8] The HIC technique has been adapted to the HPLC mode using the traditional gel material agarose,[9] as well as organic polymers[10] and silica-based matrices.[11] Alternative methods for immobilization of hydrophobic ligands, (e.g., attachment of alkyl sulfids to oxirane-activated agarose[12]) have been developed.

Salting-out chromatography and other types of chromatography related to HIC will be discussed briefly, as will other types of chromatographic adsorbents developed more recently.[13]

Adsorbents for reversed-phase chromatography (RPC) and HIC both contain hydrophobic ligands. In the RPC adsorbents the ligand density is much higher than in those used for HIC. Although the separation of both types of adsorbents is based on hydrophobic interaction, the mechanism on the molecular level is different. Whereas an RPC adsorbent can be regarded as a continuous hydrophobic phase, the ligands on a HIC adsorbent interact individually with the solutes. As a result, globular proteins very often denature when applied on RPC columns, which are therefore used mainly for peptides and small proteins. RPC also requires more drastic conditions for elution, such as a gradient of organic solvents, as compared to HIC. HIC thus has a more general field of application. RPC is described in Chapter 6.

In order to give the reader a general understanding of the principles of hydrophobic interaction chromatography, this chapter discusses the technique from both a theoretical and a practical point of view. Finally, some applications of the HIC technique will be described.

7.2 THE HYDROPHOBIC INTERACTION

7.2.1 Theory

Hydrophobic interactions in aqueous solvents are considered to be driven primarily by interactions within the solvent and to a lesser extent by interactions between the nonpolar solutes. However, in the case of protein folding, it has also been shown that van der Waals interactions between nonpolar amino acid side chains are significant.[14] It has also been shown that van der Waals interactions also play a role in chromatographic separations.[15] But to what degree van der Waals interactions influence the binding between proteins and ligands on a HIC gel is not known, and will thus not be discussed further in this chapter. For a detailed treatment of hydrophobic interaction, see Refs. 16–18. A short theoretical discussion follows, as a background to the HIC technique.

Water is a poor solvent for nonpolar solutes. Dissolving a nonpolar substance in water is thermodynamically unfavorable. Because of its hydrogen bonding capability, water has a unique structure, causing a high surface tension. A nonpolar solute forces the water to form a cavity in which the solute fits (Figure 7-1A). In the process, many hydrogen bonds between the water molecules are broken, but new hydrogen bonds are formed among the water molecules surrounding the cavity, leading to a negative change in enthalpy, ΔH. This change is due only to a small extent to the weak van der Waals interaction between the solute and the surrounding water. The increase in

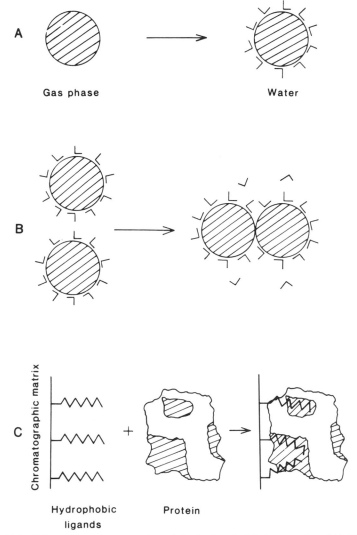

FIGURE 7-1. Schematic representation of the hydrophobic interaction. (A) Dissolving of a nonpolar solute in water. (B) The interaction of two nonpolar solutes in water. (C) A hypothetical representation of an interaction of ligands of an amphiphilic gel and hydrophobic surfaces on a protein. L represents water molecules.

the degree of order of solvent molecules near the solute explains the negative change in entropy, ΔS, that results.

The well-known expression

$$\Delta G = \Delta H - T\Delta S \tag{7-1}$$

describes the change in free energy, ΔG, of a process. A negative value of ΔG implies that the process is thermodynamically possible. The dissolution of a solute is thus favored either by a negative change in enthalpy or by a positive value of the entropy change. For the cavity-forming process earlier, the net change in free energy is positive, which makes it unfavorable.

In hydrophobic interaction (Figure 7-1B), the increase in entropy ($\Delta S > 0$) originating from water molecules leaving the more ordered structure around the nonassociated solutes for the more unstructured bulk water is the main driving force. The net decrease in cavity surface area that occurs when two nonpolar solutes associate explains why some water molecules go over to the bulk phase. The positive enthalpy term, ΔH, is smaller than the entropy term and does not influence the spontaneous association to a great extent.[19]

The interaction between an amphiphilic matrix and hydrophobic areas on a protein (Figure 7.1C) can be explained in the same way as the interaction between nonpolar solutes in water (Figure 7-1B).

Both entropy and enthalpy change with temperature for hydrophobic interactions. A theoretical treatment of the temperature dependence of the hydrophobic interaction is thus complicated, but one can conclude that the strength of hydrophobic interactions should increase with an increase in temperature, at least in the temperature range that is of interest for HIC.

Two factors of great importance for HIC are the type and concentration of salt used and additives that change the polarity of the solvent. The latter, exemplified by ethylene glycol, decrease the interaction between the HIC gel and proteins by changing the overall structure of water slightly toward a structure resembling an organic solvent.

The influence of different anions on hydrophobic interaction[20] follows the well-known Hofmeister (lyotropic) series.[21] Anions that promote hydrophobic interaction most are to the left in the series. The anions to the right—ClO_4^-, I^-, and SCN^-—are called chaotropic:

$$SO_4^{2-} > Cl^- > Br^- > NO_3^- > ClO_4^- > I^- > SCN^- \tag{7-2}$$

Also, cations influence the strength of interaction:

$$Mg^{2+} > Li^+ > Na^+ > K^+ > NH_4^+ \tag{7-3}$$

Melander and Horvath[22] have shown that the effectiveness of different salts in promoting hydrophobic interactions can be explained by their contribution to the surface tension of the solution. The formation of a cavity in water with a salt that gives a high surface tension needs a bigger input of energy than does the formation of a cavity with the same surface area in water containing a salt giving lower surface tension. The higher the salt concentration, the stronger the interaction. It is the salts that increase the surface tension most that give the strongest hydrophobic interaction.

Factors other than surface tension can also affect the hydrophobic interaction. Protein hydration and specific interactions between the protein and the salt ions seems to be factors that influence the strength of interaction.[23]

7.2.2   The Hydrophobicity of Amino Acids and Proteins

Although most hydrophobic amino acids are buried in the interior of globular proteins, and hydrophilic amino acids have a tendency to be exposed on the surface, some hydrophobic amino acids also appear on the surface. For globular proteins it has been shown that only about 20% of the amino acid side chains are totally buried.[24] The hydrophobicity of protein surfaces is the sum of the hydrophobicities of the exposed amino acids and part of the backbone. A discussion of protein surface hydrophobicity is thus based on the hydrophobicity of amino acids.

Two approaches have been used for the estimation of the hydrophobicity of amino acids. The first is based on direct measurements of the solubilities of individual amino acids in water and organic solvents.[25-28] A hydrophobicity scale for the different amino acids has thus been constructed on the basis of free energy of transfer for the amino acids from ethanol or dioxane to water.

The second approach is based on the empirical inspection of known protein structures.[29-33] Here, several hydrophobicity scales are based on, for example, the environment of the different amino acids, the fraction of amino acids that is buried in the protein, a side chain interaction parameter, or a fractional accessibility to the surrounding solvent of the different residues. For a comparative study of hydrophobicity scales published, see Ref. 34.

Table 7-1 is a comparison of two scales of hydrophobicity, the first based on amino acid solubility[25-28] and the second based on the fraction of amino acids buried within proteins[30] (average values obtained from 20 proteins with known structures). Some differences are striking. Proline is a rather hydrophobic amino acid, but its secondary structure-breaking properties make it appear in bends typically on the surface of proteins. Lysine is also classified as a rather hydrophobic residue in the scales based on solubility studies, although it is the most exposed of the amino acids. The four methylene groups of the lysine side chain disfavor the solubility of this amino acid in water, whereas the amino group with its hydrogen-binding capability has a strong tendency to be exposed on the surface of proteins.

To get more accurate values for the hydrophobicities of amphiphilic amino acids, one report suggests a scale based on the hydrophobicity of each individual atom.[35]

A linear relationship exists between the logarithm of the solubilities of hydrocarbons and the surface area they form in water.[36] The cavity area is the same as the accessible surface area of the hydrocarbons.

Protein surfaces are not smooth, but are rather rough and complex. Analyses of protein surfaces often use the method of Lee and Richards.[37,38] Figure 7-2 shows a part of a hypothetical protein surface.

TABLE 7-1. Hydrophobicity Scales for Amino Acids

Scale I (kcal/mol)		Scale II (Fraction Buried)	
Trp	3.77	Phe	0.87
Ile	3.15	Trp	0.86
Phe	2.87	Cys	0.83
Pro	2.77	Ile	0.79
Tyr	2.67	Leu	0.77
Leu	2.17	Met	0.76
Val	1.87	Val	0.72
Met	1.67	His	0.70
Lys	1.64	Tyr	0.64
Cys	1.52	Ala	0.52
Ala	0.87	Ser	0.49
His	0.87	Arg	0.49
Arg	0.85	Asn	0.42
Glu	0.67	Gly	0.41
Asp	0.66	Thr	0.38
Gly	0.10	Glu	0.38
Asn	0.09	Asp	0.37
Ser	0.07	Pro	0.35
Thr	0.07	Gln	0.35
Gln	0.00	Lys	0.31

Scale I is based on the solubility of the different amino acids, expressed as free energy of transfer for the amino acids from ethanol or dioxane to water.[25–28] Scale II is based on the fraction of the number of the different amino acids buried within proteins[30] (average values obtained from 20 proteins with known structures).

The accessible surface (see legend to Figure 7-2), with a water molecule as the probe, is used when protein surfaces are discussed. The nonpolar surface is defined as the area containing side chains with carbon and sulfur and main-chain carbon atoms (hydrogen atoms are not considered). The proportion of nonpolar surface area does not differ much among the globular proteins examined. A report on the subject gives figures of 41% for lysozyme, 48% for myoglobin, and 46% for ribonuclease.[37]

In one report,[39] 46 monomeric globular proteins with known structures were studied, and their surfaces and interior examined. This report also found that the proportion of hydrophobic surface area does not differ much among the proteins. With the definition of nonpolar compounds used in this report, the nonpolar fraction of the surfaces varied between 50 and 68%.

7.2.3 Interaction between Protein and the Hydrophobic Interaction Chromatography Adsorbent

No general and definite answer can be given to the question of how a protein surface should look to be able to interact with hydrophobic ligands on a HIC

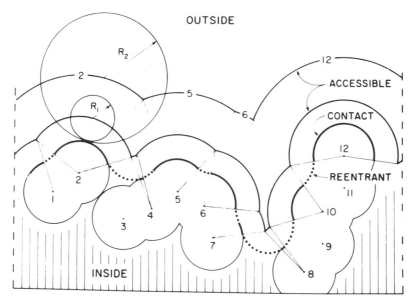

FIGURE 7-2. Possible molecular surface definitions. A section through part of a hypothetical protein is shown. Each atom (1 to 12) is represented by a sphere of its van der Waals envelope, and covalently bonded atoms are truncated. When a probe of radius R_1 is allowed to roll on the outside of the protein, having contact with the van der Waals surface of the different atoms, it will never contact atoms 3, 9, or 11. When choosing a probe with a larger radius R_2, even fewer atoms of the protein surface can make contact. These atoms are considered to be a part of the interior of the protein. There are several ways to measure the surface area. A straightforward procedure is to use the sheet defined by the locus of the center of the probe. This sheet or surface is called the accessible surface. Those parts of the molecular van der Waals surface that can actually contact the surface of the probe are called the contact surface. This provides a series of disconnected patches. The reentrant surface, also a series of patches, is the area on the protein with which the interior-facing part of the probe comes into contact when it is in contact simultanously with more than one atom. The sum of the reentrant and contact surfaces is called the molecular surface. (Reproduced from Ref. 38 with permission.)

adsorbent. It can, however, be concluded that more than one ligand on the adsorbent must be involved to get any adsorption,[8] so-called multipoint attachment.

Studies on the kinetics of the binding of a protein, phosphorylase b, to butyl-Sepharose have shown that the binding is a multistep reaction.[40] The rate-limiting step is not the collision between the protein and the amphiphilic adsorbent, but rather a slow conformational change or reorientation step of the protein on the HIC gel.

When α-chymotrypsinogen is activated by the cleavage of four peptide bonds, Ile-16 and Val-17 are buried and Met-192, Gly-193, and Arg-145 are exposed.[41] α-Chymotrypsin thereby becomes more hydrophilic than its zymogen. In HIC experiments, the enzyme elutes earlier than the zymogen,[42] as expected from this change.

The retention of seven different avian lysozymes has been studied on a HIC column,[23] and the effect of amino acid substitutions was investigated. Chromatographic retention is only affected by substitutions on the lysozyme surface opposite the catalytic cleft and not by substitutions close to it. Substitutions of hydrophobic as well as hydrophilic amino acid residues have an effect on the retention.[23] A similar effect was found when wild-type and genetically engineered subtilisin variants (from *Bacillus amyloliquefaciens*) were studied.[43] These observations indicate that the strength of interaction between a HIC adsorbent and a protein is not only determined by the proportion of the nonpolar surface area of a protein, but also by the distribution of the nonpolar areas on the surface of a protein.

Hydrophobic interaction chromatography is, in general, a mild method, due certainly to the stabilizing influence of salts, and recoveries are often high.[44] This can be illustrated by an experiment with model proteins chromatographed on an octyl-agarose column[9] (Figure 7-3). In the first chromatography the elution was started immediately after the application, whereas in the second the elution was delayed 16 h. No difference can be seen between the two chromatograms, indicating that these proteins did not alter their structures upon binding to the adsorbent (e.g., exposing their more hydrophobic interior), at least not irreversibly.

However, more labile proteins, upon contact with an HIC adsorbent, can change their structures. This has been shown for α-lactalbumin[45] (after removing the Ca^{2+} ions that normally stabilize the structure). A longer residence time on the column increased the size of a broad peak that eluted after the

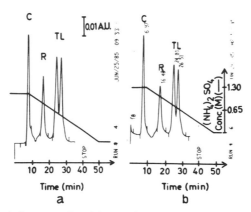

FIGURE 7-3. The influence of residence time of model proteins on a HPLC-HIC adsorbent (octyl agarose). (a) Elution started immediately after application and (b) elution delayed 16 h. (Reprinted from Ref. 9 with permission.)

native protein and consisted of several more or less unfolded species of α-lactalbumin. A rerun of the second peak again gave two peaks, showing that the unfolding is reversible. The higher the temperature, the larger the proportion of the α-lactalbumin that elutes with a partly unfolded structure.

The retention also increases with temperature. This is also the case for the more stable lysozyme,[46] but to a lesser extent. One concludes that the temperature effect in HIC is due to two factors: (1) the increase in hydrophobic interaction with temperature and (2) a temperature-dependent alteration of the structures of proteins, especially labile ones.

7.3 ADSORBENTS FOR HYDROPHOBIC INTERACTION CHROMATOGRAPHY AND ADSORBENTS RELATED TO HYDROPHOBIC INTERACTION CHROMATOGRAPHY

7.3.1 Adsorbents for Hydrophobic Interaction Chromatography

Many types of matrices are suitable for preparing adsorbents for HIC, but the most extensively used has been agarose. When the technique has been adopted to HPLC, silica and organic polymer resins have also been employed.

A number of commercially available HIC adsorbents are shown in Table 7-2 and some structures of HIC adsorbents are illustrated in Figure 7-4. All structures shown (except C) are octyl matrices, but any chain length can be chosen if one prepares the matrices oneself. The first (Figure 7-4A) was introduced by Shaltiel[3] and is based on cyanogen bromide activation[47] of an agarose gel, after which coupling with an alkylamine is performed. This coupling procedure creates ligands with a positive net charge. The structure in Figure 7-4A is the isourea derivative proposed by Kohn and Wilcheck.[48]

Figure 7-4 shows the B and C types based on the glycidyl ether (with an epoxide, oxirane, functional group) coupling procedure[6] (also used for the production of Octyl and Pheny Sepharose). This coupling method is widely used, so it will be described in some detail. Some glycidyl ethers are commercially available but they can also be prepared according to Ulbrich and co-workers.[49] The reaction scheme for the coupling of the ligand to the gel (usually agarose) is as follows:

$$
\text{M—OH} + \underset{\text{CH}_2\text{—CH—CH}_2\text{—OR}}{\overset{\text{O}}{\diagup\diagdown}} \xrightarrow[\text{catalyst}]{\text{BF}_3\text{Et}_2\text{O}} \text{M—O—}\underset{\text{CH}_2\text{—CH—CH}_2\text{—OR}}{\overset{\text{OH}}{|}} \quad (7\text{-}4)
$$

R is an alkyl or aryl group, and M is the matrix. The gel should be transferred to an organic solvent, as dioxane. This is done in steps (100-ml portions to 100-ml sedimented gel), on a Büchner funnel:

1. one washing with water–dioxane (4:1)
2. one washing with water–dioxane (3:2)

3. one washing with water–dioxane (2:3)
4. one washing with water–dioxane (1:4)
5. seven washings with dioxane

Transfer the gel to a reaction vessel equipped with a stirrer.

Add 100 ml dioxane to 100 ml of sedimented gel, and add 2 ml of a 48% solution of boron trifluoride etherate in diethyl ether. Stir for 5 min. Add 1 ml of glycidyl ether dissolved in 10 ml dioxane dropwise from a separatory funnel. The reaction takes about 40 min. After the reaction, transfer the gel back to water using the previous scheme, but in reverse order, and finally wash with water. The amount of ligand coupled to the gel can be controlled by varying the amount of the glycidyl ether added. The glycidyl ether coupling method produces gels that are charge free, and thus should have no other interaction with proteins than hydrophobic interaction. However, the phenyl group shown in Figure 7-4C, (and other aromatic groups) also has a potential for π–π interaction.

A recently introduced coupling method that leads to the structure shown in Figure 7-4D was used for coupling hydrophobic ligands to agarose gels used as an HPLC packing.[9,50] The agarose is first activated with γ-glycidoxypropyltrimethoxy silane in water. The immobilization of the ligands is then performed in the alcohol that is to be coupled to the gel. The resulting gel is noncharged and also contains a spacer, making the ligands more available for the proteins.

The ligand structure in Figure 7-4E was introduced by Maisano and coworkers.[12] The agarose is first activated with a bisepoxide, 1,4-butanediol diglycidyl ether, and is then coupled with an alkyl mercaptan. The gel, also charge free, contains a spacer arm. The ligand density can be regulated easily by varying the amount of alkyl mercaptan.

Polymeric coatings[51] are of great importance as stationary phases for HIC. Both silica and polymeric matrices can be derivatized this way. One drawback with polymeric coatings has been their slow mass transfer properties.[52,53] But modern polymeric phases have film thicknesses in the range of monolayer up to 5 nm,[54] so the difference in mass transfer velocity between polymeric and monomeric phases is small. Common polymeric phases used for HIC matrices are oligoethyleneglycol and polyether.

7.3.2 Ligand Density and Capacity of Hydrophobic Interaction Chromatography Adsorbents

The density of ligands is important for the strength of the interaction between the adsorbent and the proteins, as well as for the capacity. For the commercially available Octyl and Phenyl Sepharose gels, the ligand density is approximately 40 μmol/ml gel bed, corresponding to a degree of substitution of approximately 0.2 mol hydrophobic substituent per mole galactose.[55] The capacity of Phenyl

TABLE 7-2. Commercially Available Adsorbents and Columns for HIC (the list may not be complete)

Supplier	Product Name	Functional Group(s)	Remarks
A. Conventional packings			
J.T. Baker	HI-Propyl	Propyl	Silica, 15 and 40 μm, 300 and 275 Å
Bio-Rad	Methyl HIC	Methyl	Methacrylate, 50 μm
	t-Butyl HIC	Butyl	
Pharmacia	Phenyl Sepharose	Phenyl	Agarose, 34 μm
	Butyl Sepharose 4B	Butyl	Agarose, 45–165 μm
	Octyl Sepharose CL-4B	Octyl	Agarose, 45–165 μm
	Phenyl Sepharose CL-4B	Phenyl	Agarose, 45–165 μm
	Butyl Sepharose 4 FF	Butyl	Agarose, 45–165 μm
	Phenyl Sepharose 6 FF	Phenyl	Agarose, 45–165 μm
	HiLoad Phenyl Sepharose	Phenyl	Agarose, 34 μm, prepacked columns
Sigma		ω-amino C-2, C-3, C-4, C-5, C-6, C-8, C-10, and C-12. 4-Phenylbutylamine, 2,2-diphenylpropylamine. Alkyl: C-2, C-3, C-4, C-5, C-6, C-8, C-10, and C-12. Phenyl and trityl.	Agarose 4% CL or non-CL
SynChrom	SynChroprep	Propyl	Silica, 15 and 30 μm, 300 Å
TOSOHAAS	Ether-5PW	Oligoethyleneglycol	Polymer, 20–40 μm
	Phenyl-5PW	Phenyl	Polymer, 20–40 μm
	Ether 650	Oligoethyleneglycol	Polymer, 20–50 μm
	Butyl 650	Butyl	Polymer, 40–90 μm
	Phenyl 650	Phenyl	Polymer, 50–150 μm
B. HPLC packings			
J.T. Baker	HI-Propyl	Propyl	Silica, 5 and 15 μm, 300 Å
Biochrom Labs	Hydrocell C3 1000	Allyl	Polymer, porous
	Hydrocell C4 1000	Butyl	Polymer, porous
	Hydrocell C3 NP10	Allyl	Polymer, nonporous
	Hydrocell C4 NP10	Butyl	Polymer, nonporous

Company	Product	Ligand	Matrix
Bio-Rad	Bio-Gel MP-7 HIC	Methyl	Polymer, 7 μm, 900 Å
	Bio-Gel Phenyl-5PW	Phenyl	Polymer, 10 μm, 1,000 Å
Beckman	Spherogel-HIC		Silica, 5 μm, 300
Interaction	MCI GEL CQH3xs	Butyl, phenyl	Polymer, 10 μm
	Hydrophase HP-Butyl	Butyl	Polymer, 10 μm
Mitsubishi Kasei	MCI GEL HIC	Ether, butyl, phenyl	Polymer, 10 μm, wide pore
Pharmacia	Phenyl Superose	Phenyl	Agarose, 13 μm
	Alkyl Superose	Neophentyl	Agarose, 13 μm
Showa Denko	Shodex HIC	Phenyl	Polyhydroxymethyl acrylate, wide pore
Sigma	SigmaChrom HIC-phenyl	Phenyl	Polysaccharide CL, 12–15 μm
Supelco	Supelcosil LC-HINT	Diol	Silica, 5 μm, 100 Å
SynChrom	SynChropak	Propyl, hydroxypropyl, and pentyl	Silica, 6 μm, 300 Å
TOSOHAAS	Ether 5PW	Oligoethyleneglycol	Polymer, 10, 13, and 20 μm, 1,000 Å
	Phenyl 5PW	Phenyl	Polymer, 10, 13, and 20 μm, 1,000 Å
	Butyl-NPR	Butyl	2.5 μm, nonporous
YMC	YMC-Pack-HIC	CL polyamide-containing propyl ligands	Silica, 6.5 μm, 300 Å

C. Novel separation media

Company	Product	Ligand	Matrix
Millipore	MemSep chromatography cartridges		Continuous network of regenerated cellulose, derivatized, 1.2-μm flowthrough pores
Perseptive Biosystems	POROS PH	Phenyl	PS/DVB beads, covered by a CL polyhydroxylated polymer, large transecting throughpores
	POROS PE	Phenyl ether	
	POROS BU	Butyl	
	POROS ET	Ether	

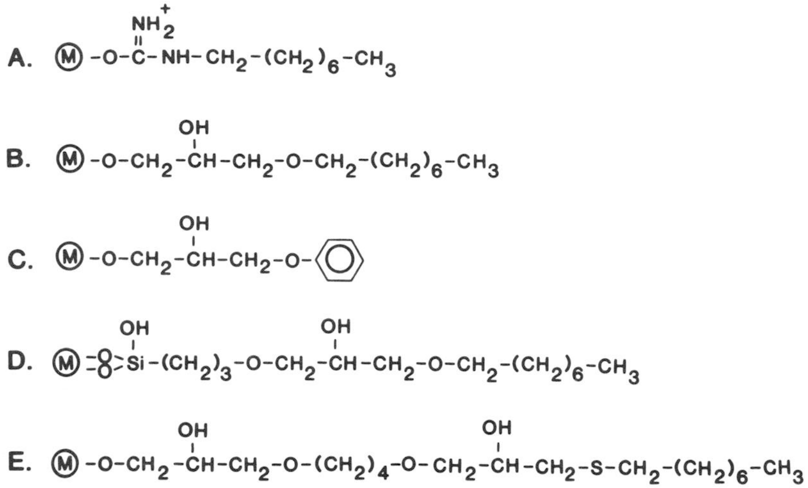

A. $\text{M} -\text{O}-\overset{\overset{+}{\underset{\|}{NH_2}}}{\text{C}}-\text{NH}-\text{CH}_2-(\text{CH}_2)_6-\text{CH}_3$

B. $\text{M} -\text{O}-\text{CH}_2-\overset{\overset{OH}{|}}{\text{CH}}-\text{CH}_2-\text{O}-\text{CH}_2-(\text{CH}_2)_6-\text{CH}_3$

C. $\text{M} -\text{O}-\text{CH}_2-\overset{\overset{OH}{|}}{\text{CH}}-\text{CH}_2-\text{O}-\bigcirc$

D. $\text{M} \overset{O}{\underset{O}{\ge}}\text{Si}-(\text{CH}_2)_3-\text{O}-\text{CH}_2-\overset{\overset{OH}{|}}{\text{CH}}-\text{CH}_2-\text{O}-\text{CH}_2-(\text{CH}_2)_6-\text{CH}_3$

E. $\text{M} -\text{O}-\text{CH}_2-\overset{\overset{OH}{|}}{\text{CH}}-\text{CH}_2-\text{O}-(\text{CH}_2)_4-\text{O}-\text{CH}_2-\overset{\overset{OH}{|}}{\text{CH}}-\text{CH}_2-\text{S}-\text{CH}_2-(\text{CH}_2)_6-\text{CH}_3$

FIGURE 7-4. Different hydrophobic ligands coupled to a gel matrix, M.

and Octyl Sepharose CL-4B in 0.01 M sodium phosphate buffer, pH 6.8, containing 1 M ammonium sulfate, is approximately 15 to 20 mg human serum albumin or 3 to 5 mg β-lactoglobulin per milliliter of gel.[55] Several procedures for determining ligand density have been employed,[56,57] including NMR, gas chromatography, and elementary carbon analyses. Ligand density determination for HIC adsorbents containing aromatic groups (e.g., phenyl) can also be made by derivative ultraviolet spectroscopy,[58] and adsorbents of the type shown in Figure 7-4E can be analyzed by sulfur determination.[12]

7.3.3   Other Types of Adsorbents Related to Hydrophobic Interaction Chromatography

At high salt concentrations, matrices used for gel filtration show adsorption of proteins.[59–62] The interaction, which is of a hydrophobic nature, can be used for the separation of proteins. Matrices such as cellulose, dextran, agarose, organic polymers, and modified silica can be utilized. The technique is called salting-out chromatography.

Spacers used in affinity chromatography are often of hydrophobic nature and can in some cases make a considerable contribution to the interaction between solutes and adsorbents.[63]

In addition to the hydrophobic interaction, HIC adsorbents with aromatic ligands also show the so-called $\pi-\pi$ interaction. This effect is explicitly utilized in the charge-transfer chromatographic technique,[64] using different kinds of ligands.

Another method related to HIC is the so-called thiophilic adsorption technique.[65,66] The ligand is divinylsulfone substituted with, for example, mercaptoethanol. Although the technique is related to HIC, the adsorption characteristics of the thiophilic adsorbents are different than those of HIC ones.

The binding of proteins to dye–ligand substituted adsorbents[67,68] is due partly (although far from predominantly) to the hydrophobic interaction. Also, binding to other ligands utilized in affinity chromatography can be partly of hydrophobic nature.

7.4 CHROMATOGRAPHIC TECHNIQUES

7.4.1 Introduction

Most proteins bind to HIC adsorbents, and the technique can thus be used in many purification procedures. HIC is based on a different separation principle from most other separation techniques and can thus, in combination with these, afford a high degree of purification. The mass recoveries of proteins are mostly excellent, as is the recovery of enzyme activity.

The high capacity of HIC adsorbents makes them suitable for use at an early stage in a purification scheme.[44] The similarities to salt precipitation, also based on protein surface hydrophobicity, make a combination of these two methods less useful. However, after an initial precipitation step to remove the most hydrophobic proteins, the usual second precipitation can be replaced by an HIC purification step, especially because HIC gives sharper separation than does salt precipitation.

Because of the rather high salt concentrations of the eluate from the HIC column, a subsequent ion-exchange chromatography step is not possible without an intermediate desalting step. The reverse is easier to perform. A HIC step can, of course, be followed directly by gel filtration, whereby the salt and other additives are automatically removed.

HIC adsorbents can also be used for batch procedures, making them suitable for purification of proteins on a large scale.[44] Rigid adsorbents allowing high flow rates can be used for large-scale chromatography, as can new types of chromatography media (e.g., membrane-based and perfusion materials) (Section 7.4.4)

7.4.2 Choice of Adsorbents

Adsorbents used for HIC should preferably be charge free. At low ionic strengths, the protein can interact with the positively charged amino groups on HIC gels with the structure shown in Figure 7-4A. In some cases, this can lead to an irreversible binding of the protein to the gel.[7,69]

The strength of the interaction between a protein and hydrophobic ligands on a HIC adsorbent increases with the increase in length of the ligand[56] (alkyl types of ligands). Ligands containing between 4 and 10 carbon atoms are suitable for most separation problems. For proteins with poor solubility in buffers of high salt concentration (e.g., membrane proteins) HIC adsorbents with rather long ligands are recommended.[8] A phenyl group has about the same hydrophobicity as a pentyl group, although it can have a quite different selectivity compared to pentyl ligands due to the possible $\pi-\pi$ interaction. Aromatic groups on protein surfaces can interact specifically with the aromatic ligands. The oligoethyleneglycol phases are intermediate in hydrophobicity, between butyl and phenyl.

7.4.3   HPLC-HIC

The adaption of HIC, as well as other classical protein purification techniques, to the high-performance mode has been made possible by the production of beads of small and uniform size. The smaller the bead size, the better and faster the separations can be done. This was already realized by Martin and Synge in 1941.[70] Smaller bead sizes limit the number of matrices that can be utilized successfully, as smaller beads have a higher flow resistance that leads to higher pressures. Only rigid matrices can be used as high-performance packings. High-performance HIC packings have been based on silica[11] and organic polymers[10] as well as on the traditional gel material agarose.[9,71] The capability of HPLC-HIC is illustrated in Figures 7-5 and 7-6, which show separations of model proteins and serum proteins on a pentyl-agarose column.

On columns packed with nonporous materials, separations can be achieved in 2 to 3 min.[72-74] The nonporous HIC matrices are of great value for quality

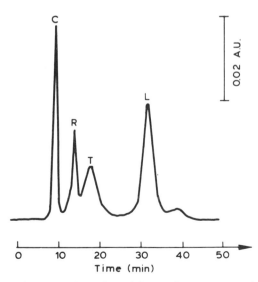

FIGURE 7-5. Isocratic separation of model proteins on a pentyl-agarose column in the HPLC mode. (Reprinted from Ref. 9 with permission.).

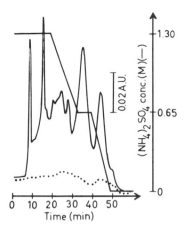

FIGURE 7-6. High-performance HIC of human serum on a pentyl-agarose gel. Elution with a negative salt gradient. (Reprinted from Ref. 9 with permission.)

control, process monitoring, and other applications that need rapid scanning of complex samples.

7.4.4 Novel Separation Media

In conventional chromatography, proteins (and other solutes) move to the outer surface of the beads by rapid convective flow through the column bed. Transport of the solutes to the inner surface (within the pores) occurs by diffusion. This process is slow, especially for conventional packing materials. By utilizing nonporous packing materials, this shortcoming is overcome, but at the cost of losing capacity. Nonporous packing materials are thus restricted to analytical applications. Some new approaches have been taken to overcome this problem. One such approach is to utilize a unified bed of through pores, made from regenerated cellulose[75] or from (polymerized) derivatives of acrylamide.[76,77] A second approach is to produce particles with large enough transecting pores to allow a convective flow through the particle itself.[78,79] These materials have been utilized for different types of protein separations and are now available as HIC supports (Table 7-2).

7.4.5 Conditions for Adsorption

Adsorption of proteins to a HIC adsorbent is favored by a high salt concentration, but due to differences in the strength of interaction between the adsorbent and different proteins, the concentration of salt needed for adsorption can vary considerably. The concentration of salt used for adsorption should be below the concentration that precipitates different proteins in the sample that should be chromatographed. For HIC gels with a ligand density similar to

Phenyl and Octyl Sepharose, the salt concentration is usually between 0.75 and 2 M (100% saturation is 4.05 M) with ammonium sulfate or 1 and 4 M with sodium chloride. For the most hydrophobic proteins the salt concentration can be lower. The salt is usually dissolved in a buffer solution with a concentration of 0.01 to 0.05 M.

Different salts give rise to differences in the strength of interaction between proteins and the HIC adsorbent. The strength of interaction as well as the capacity follows the series[20]:

$$Na_2SO_4 > NaCl > (NH_4)_2SO_4 > NH_4Cl > NaBr > NaSCN \qquad (7\text{-}5)$$

Na_2SO_4, NaCl, and especially $(NH_4)_2SO_4$ are the most utilized salts in HIC. Besides the differences in the strength of interaction, the different salts also show some additional selectivity.[9]

The pH of the buffers used in HIC experiments has a decisive influence on the adsorption of proteins to the adsorbent[9] (Figure 7-7). Some proteins with high pI values bind strongly to HIC gels at elevated pH values, although

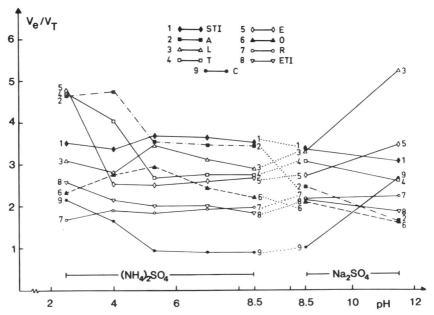

FIGURE 7-7. The pH dependence of the interaction between proteins and an octyl-agarose gel expressed as V_e/V_T (V_e is the elution volume of the different proteins and V_T is the elution volume of a nonretarded solute). Elution is by a negative linear gradient of salt. The model proteins used were STI, soy trypsin inhibitor; A, human serum albumin; L, lysozyme; T, transferrin; E, enolase; O, ovalbumin, R, ribonuclease; ETI, egg trypsin inhibitor; and C, cytochrome c. (Adapted from Ref. 9.)

no general trend in the interaction strength and protein pI values has been observed. Because the change in retardation with pH is large for most proteins, it can be worthwhile to test different pH values for adsorption. The only limitation is the stability of the protein to be purified and the stability of the chromatographic matrix (e.g., silica is not stable at high pH).

As discussed earlier, the temperature dependence of HIC is not simple, although, generally, a decrease in temperature decreases the interaction. Labile proteins should be chromatographed at low temperatures.

It is important that the column is equilibrated with the same buffer as the sample. If the sample is opalescent, it should be centrifuged or filtered before application to the column. The sample volume can be large because the proteins generally become bound to the column before elution starts.

7.4.6 Elution

Elution, whether done stepwise or with a gradient, can generally be achieved in three different ways:

1. Changing the salt concentration: The elution of solutes, in the order of increasing hydrophobicity, is accomplished by decreasing the salt concentration.
2. Changing the polarity of the solvent: A decrease in the interaction is achieved by adding solvents such as ethylene glycol or (iso)propanol. Up to 80% ethylene glycol may be used, whereas propanol should be used at lower concentrations. The addition of polarity-decreasing agents can be made after the salt has been removed from the column or concomitantly with the decrease of salt concentration.
3. Adding detergents: Detergents work as displacers of the proteins. They have been used mainly for the purification of membrane proteins by HIC.

In theory, the elution of proteins from HIC adsorbents can be accomplished by changing the salt species to a chaotropic ion (e.g., SCN^-), but this should be avoided due to the protein structure breaking properties of these ions. A change in pH may also be used for elution, but it is (so far) impossible to predict how the strength of interaction between a protein and HIC adsorbent will be affected upon changing the pH.

7.4.7 Regeneration of Hydrophobic Interaction Chromatography Adsorbents

HIC adsorbents have a long lifetime. For the regeneration, and storage, of the different HIC adsorbents available, the recommendations from the various manufacturers should be followed. For Sepharose based packings, it is recommended that the adsorbents be washed with 6 M guanidine hydrochloride to remove the strongest adsorbed proteins. After washing, the gel can be equilibrated with starting buffer and used immediately for the next run. If detergents have been used, a regeneration procedure involving washing with different alcohols has to be applied.[55] An adsorbent such as methacrylate can be sanitized in 1 M NaOH.

7.5 APPLICATIONS

7.5.1 Purification of an Acid Phosphatase from Bovine Cortical Bone Matrix: Elution by Decreasing Salt

In the purification of an acid phosphatase that displays phosphotyrosyl protein phosphatase activity, several purification methods have been employed, including HIC.[80]

Bovine long bones (tibia) were washed carefully to remove nonbone tissues and blood cells. Bone cubes were ground to powder, which was then homogenized and extracted. The extract was subjected to CM Sepharose ion-exchange chromatography, cellulose phosphate affinity column chromatography, Sephacryl S-200 gel filtration chromatography, and HIC.

For the last step, HIC on a Phenyl Sepharose column (Figure 7-8) was used. Active fractions from the gel filtration column were pooled and adjusted to 30% saturation with ammonium sulfate by slowly adding solid ammonium sulfate at 4°C and stirring gently for at least 2 h (buffer: 100 mM sodium acetate, pH 6.5). The enzyme fraction was then applied to the Phenyl Sepharose column equilibrated with sodium acetate buffer containing ammonium sulfate at 30% saturation. The chromatography was done at room temperature. The column was washed with this buffer and then eluted with a negative

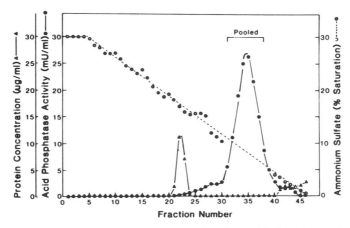

FIGURE 7-8. Phenyl Sepharose chromatography of an acid phosphatase from bovine cortical bone matrix. (Reprinted from Ref. 80 with permission.)

salt gradient, from 30% saturation to buffer without salt. The active fractions were concentrated and pooled. The phosphatase was stable for at least 4 months when stored at 4°C. The recovery of the enzyme in the HIC step was 92% and the purification 13-fold. The enzyme was pure according to SDS–gel electrophoresis.

7.5.2 Purification of Human Pituitary Prolactin: Elution by Decreasing Polarity

In the isolation of human pituitary prolactin, HIC on Phenyl Sepharose has been utilized.[81] The entire preparation was performed at 5°C, and the starting material was frozen human pituitary glands. The glands were homogenized, and extraction was done at an elevated pH. The first chromatographic step was gel filtration on Sepharose CL-6B at pH 9.8. The fractions from the gel filtration column that contained prolactin were pooled and chromatographed on a Phenyl Sepharose column (Figure 7-9). A sample containing 200 to

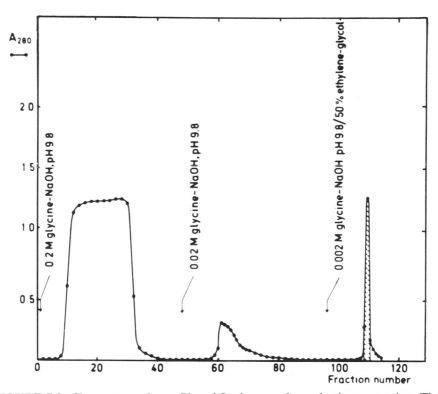

FIGURE 7-9. Chromatography on Phenyl Sepharose of a prolactin preparation. The hatched area represents the prolactin-containing fractions. (Reprinted from Ref. 81 with permission.)

400 mg of protein was applied to the column (3.2 × 25 cm). The column was equilibrated with 0.2 M glycine/NaOH buffer (pH 9.8), and the elution was carried out at pH 9.8 by a stepwise decrease in the buffer concentration (to 0.02 M glycine/NaOH) and, finally, by inclusion of ethylene glycol (50%, v/v). Fractions of about 15 mL were collected, and the flow rate was 45 mL/h. The prolactin activity was eluted by the buffer containing ethylene glycol, and recovery was 95%. The purification of the prolactin was completed by an additional gel filtration (Sephadex G-100 Superfine) and an ion-exchange step on DEAE Sepharose CL-6B.

7.5.3   Purification of Phospholipase C from *Trypanosoma brucei:* *Elution by Detergent*

In the purification of phospholipase C from *T. brucei,* HIC was used at an early stage.[82]

Trypanosomes were lysed, and the membranes were solubilized and extracted with *n*-octyl glucoside-containing buffers. A fraction containing the phospholipase activity was precipitated with an equal volume of saturated ammonium sulfate solution. After centrifugation, the supernatant (50 mL, containing 8.7 mg protein) was applied to a Phenyl Sepharose column (1.2 × 14 cm) equilibrated with 50% saturated ammonium sulfate in 25 mM sodium succinate, pH 6.0. A linear gradient (100 ml) with decreasing salt concentration from 50% saturation to buffer without ammonium sulfate was applied, followed by further washing with 50 mL buffer. The phospholipase was eluted with buffer containing detergent, 1% CHAPS (3-[(3-cholamidopropyl)-dimethylammonium]-1-propanesulfonate). The chromatogram, Figure 7-10, shows that most of the protein eluted in a sharp peak after introducing the detergent. The phospholipase, however, eluted in a broad peak. The trailing fractions of the activity peak, which contained about 70% of the activity but

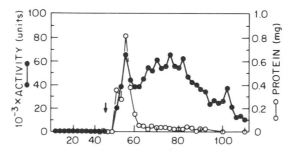

FIGURE 7-10. HIC of phospholipase C from *Trypanosoma brucei.* The vertical arrow shows the point at which elution with 1% CHAPS was initiated. Fraction numbers are shown on the abscissa; note the change in scale after fraction 44. (Reprinted from Ref. 82 with permission.).

relatively little protein, were pooled. The yield in the HIC step was 62% and the purification 22-fold. The HIC step was followed by an ion-exchange step (CM Sephadex C-25) and gel filtration chromatography (Sephacryl S-200).

7.5.4 Exchange of Detergents Bound to Membrane Proteins[83]

Unfortunately, with most intrinsic membrane proteins, no single detergent is usually well suited for all the different steps in the purification scheme. Methods for detergent exchange are therefore needed. One method is Phenyl Sepharose-mediated detergent-exchange chromatography. The alkyl detergents, lauryl maltoside, octyl glucoside, and dodecyl sulfate, were each successfully exchanged for Triton X-100, Triton N-101, or Nonidet P-40 present in a solution of cytochrome c oxidase, a mixture of inner mitochondrial membrane proteins, or a mixture of erythrocyte membrane proteins. The following description is one of an exchange of lauryl maltoside for cytochrome c oxidase-bound Nonidet P-40 (Figure 7-11).

A 0.5 × 10-cm bed of Phenyl Sepharose was presaturated with 13 ml of 10 mM lauryl maltoside in a pH 9.0 buffer at an ionic strength of 0.01 (Tris–HCl containing 0.1 mM EDTA), followed by 8 ml of 2 mM lauryl maltoside in the same buffer; a 0.6-ml protein sample was applied that contained 3 mg of protein/ml and 26 mM Nonidet P-40 in the above buffer to which enough lauryl maltoside was added to make the concentration 2 mM. Elution was affected by the above buffer containing 2 mM lauryl maltoside, and fractions were collected and analyzed. At pH 9 and an ionic strength of 0.01, only 5 to 15% of the Nonidet P-40-solubilized cytochrom c oxidase complex was bound to the column that had been saturated with lauryl maltoside. Less than 0.1% of the original amount of Nonidet P-40 remained in the complex after the detergent exchange.

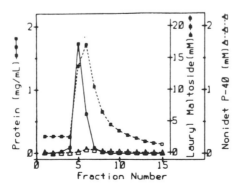

FIGURE 7-11. Exchange of lauryl maltoside for protein-bound Nonidet P-40. (Reprinted from Ref. 83 with permission.)

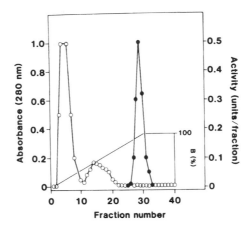

FIGURE 7-12. HPLC-HIC on a Synchropak propyl column of an acylpeptide hydrolase from bovine lens tissue. ○, absorbance; ●, activity. (Reprinted from Ref. 84 with permission.)

7.5.5 Purification of an Acylpeptide Hydrolase from Bovine Lens Tissue in the HPLC Mode: Elution by Decreasing Salt

In the purification of a bovine lens acylpeptide hydrolase, anion-exchange and hydrophobic interaction chromatography were employed.[84]

Bovine lenses were extracted with a 50 mM Tris–HCl buffer, pH7.5. The extraction, as well as all other steps in the purification, was performed at 4°C. After centrifugation the supernatant was used for further purification. The extract was subjected to ion-exchange chromatography on a DEAE-cellulose column. Active fractions were combined and concentrated in 20 mM sodium phosphate buffer, pH 7.0. A concentration of 4.0 M solid NaCl was added to the enzyme solution, and precipitated proteins were removed by centrifugation. A Synchropak propy column (250 × 4.1 mm) was used for the HPLC-HIC chromatography (Figure 7-12). Two-milliliters of the sample was applied to the column, and the elution was performed with a decreasing salt gradient, from 4.0 M (solvent A) to 0 M NaCl (solvent B), in a 20 mM sodium phosphate buffer, pH 7.0. Fractions (1 mL) were collected at a flow rate of 1 mL/min. Fractions were analyzed for enzyme activity and active fractions were pooled. A rechromatography was performed under similar conditions. The recovery of the enzyme in the HIC step was 47% and the purification was 20-fold.

7.6 REFERENCES

1. A. Tiselius, *Ark. Kemi. Mineral Geol., 26B,* 1 (1948).

2. R.J. Yon, *Biochem. J., 126,* 765 (1972).

3. Z. Er-el, Y. Zaidenzaig, S. Shaltiel, *Biochem. Biophys. Res. Commun., 49,* 383 (1972).

4. B.H.J. Hofstee, *Anal. Biochem., 52,* 430 (1973).

5. J. Porath, L. Sundberg, N. Fornstedt, I. Olsson, *Nature, 245,* 465 (1973).

6. S. Hjertén, J. Rosengren, S. Påhlman, *J. Chromatogr., 101,* 281 (1974).

7. S. Hjertén, *J. Chromatogr., 87,* 325 (1973).

8. S. Hjertén, in *Methods of Protein Separation,* N. Catsimpoolas, ed., Plenum, New York, 1976, Vol. 2, Chapter 6.

9. S. Hjertén, K. Yao, K.-O. Eriksson, B. Johansson, *J. Chromatogr., 359,* 99 (1986).

10. Y. Kato, T. Kitamura, T. Hashimoto, *J. Chromatogr., 360,* 260 (1986).

11. J.L. Fausnaugh, E. Pfannkoch, S. Gupta, F.E. Regnier, *Anal. Biochem., 137,* 464 (1984).

12. F. Maisano, M. Belew, J. Porath, *J. Chromatogr., 321,* 305 (1985).

13. J. Porath, *J. Chromatogr., 376,* 331 (1986).

14. P. Privalov, in *Protein Folding,* T.E. Creighton, ed., W.H. Freeman, New York, 1992, p. 83.

15. J. Ståhlberg, B. Jönsson, C. Horvath, *Anal. Chem.,* 64, 3118 (1992).

16. C. Tanford, The Hydrophobic Effect: Formation of Micelles and Biological Membranes, Wiley, New York, 1973.

17. T.E. Creighton, *Proteins: Structures and Molecular Properties,* W.H. Freeman, New York, 1984.

18. P.K. Ponnuswamy, *Prog. Biophys. Mol. Biol., 59,* 57 (1993).

19. S. Lewin, *Displacement of Water and Its Control of Biochemical Reactions,* Academic, New York, 1974, p.71.

20. S. Påhlman, J. Rosengren, S. Hjerte'n, *J. Chromatogr., 131,* 99 (1977).

21. F. Hofmeister, *Arch. Exp. Pathol. Pharmakol., 24,* 247 (1888).

22. W. Melander, C. Horvath, *Arch. Biochem. Biophys., 183,* 200 (1977).

23. J.L. Fausnaugh, F.E. Regnier, *J. Chromatogr., 359,* 131 (1986).

24. M.F. Perutz, J.C. Kendrew, H.C. Watson, *J. Mol. Biol., 13,* 669 (1965).

25. C. Tanford, *J. Am. Chem. Soc., 84,* 4240 (1962).

26. J.M. Zimmerman, *J. Theor. Biol., 21,* 170 (1968).

27. Y. Nozaki, C. Tanford, *J. Biol. Chem., 246,* 2211 (1971).

28. D.D. Jones, *J. Theor. Biol., 50,* 167 (1975).

29. P. Manavalan, P. Ponnuswamy, *Nature (London), 275,* 673 (1978).

30. D.H. Wertz, H.A. Scheraga, *Macromolecules, 11,* 9 (1978).

31. C. Chotia, *J. Mol. Biol., 105,* 1 (1976).

32. W.R. Krigbaum, A. Komoriya, *Biochim. Biophys. Acta, 576,* 204 (1979).

33. G.D. Rose, A.R. Geselowitz, G.J. Lesser, R.H. Lee, M.H. Zehfus, *Science, 229,* 834 (1985).

34. J.L. Cornette, K.B. Cease, H. Margalit, J.L. Spouge, J.A. Berzofsky, C. DeLisi, *J. Mol. Biol., 195,* 659 (1987).

35. D. Eisenberg, A.D. McLachlan, *Nature, 319,* 199 (1986).

36. R.B. Hermann, *J. Phys. Chem., 76,* 2754 (1972).

37. B. Lee, F.M. Richards, *J. Mol. Biol., 55,* 379 (1971).

38. F.M. Richards, *Annu. Rev. Biophys. Bioeng., 6,* 151 (1977).

39. S. Miller, J. Janin, A.M. Lesk, C. Chotia, *J. Mol. Biol., 196,* 641 (1987).

40. H.P. Jennissen, *J. Colloid. Interface Sci., 111,* 570 (1986).

41. S.T. Freer, J. Kraut, J.D. Robertus, H.T. Wright, N.H. Xuong, *Biochemistry, 9,* 1997 (1970).

42. J.L. Fausnaugh, L.A. Kennedy, F.E. Regnier, *J. Chromatogr., 317,* 141 (1984).

43. R.M. Chicz, F.E. Regnier, *J. Chromatogr., 500,* 503 (1990).

44. R.K. Scopes, *Protein Purification, Principles and Practice,* Springer, New York, 1982.

45. S.-L. Wu, A. Figueroa, B.L. Karger, *J. Chromatogr., 371,* 3 (1986).

46. S.-L. Wu, K. Benedek, B.L. Karger, *J. Chromatogr., 359,* 3 (1986).

47. R. Axén, J. Porath, S. Ernback, *Nature (London), 214,* 1302 (1967).

48. J. Kohn, M. Wilchek, *Appl. Biochem.Biotechnol., 9,* 285 (1984).

49. V. Ulbrich, J. Makes. M. Jurecek, *Collect. Czech. Chem. Commun., 29,* 1466 (1964).

50. S. Hjertén, K. Yao, Z.-H. Liu, D. Yang, B.-L. Wu, *J. Chromatogr., 354,* 203 (1986).

51. M. Petro, D. Berek, *Chromatographia, 37(9/10),* 549 (1993).

52. G. Schomburg, *LC-GC, 6(1),* 37 (1988).

53. S.A. Wise, L.C. Sander, H.-C.K. Chang, K.E. Markides, M.L. Lee, *Chromatographia, 25,* 473 (1988).

54. M. Hanson, K.K. Unger, G. Schomburg, *J. Chromatogr., 517,* 269 (1990).

55. Pharmacia Fine Chemicals, Octyl Sepharose CL-4B and Phenyl Sepharose CL-4B.

56. J. Rosengren, S. Påhlman, M. Glad, S. Hjerte'n, *Biochim. Biophys. Acta, 412,* 51 (1975).

57. B.-L. Johansson, I. Drevin, *J. Chromatogr., 346,* 255 (1985).

58. B.-L. Johansson, I. Drevin, *J. Chromatogr., 391,* 448 (1987).

59. J. Porath, *Nature (London), 196,* 47 (1962).

60. L.G. Hoffmann, P.W. McGivern, *J. Chromatogr., 40,* 53 (1969).

61. N. Sakihama, H. Ohmori, N. Sugimoto, Y. Yamasaki, R. Oshino, M. Shin, *J. Biochem., 93,* 129 (1983).

62. K. Adachi, *Biochim. Biophys. Acta, 912,* 139 (1987).

63. S. Shaltiel, in *Methods Enzymol.,* W.B. Jakoby, M. Wilchek, eds., Academic Press, New York, 1974, Vol. XXXIV, p. 126.

64. J. Porath, *J. Chromatogr., 159,* 13 (1978).

65. J. Porath, F. Maisano, M. Belew, *FEBS Lett., 185,* 306 (1985).

66. T.W. Hutchens, J. Porath, *Anal. Biochem., 159,* 217 (1986).

67. R.L. Easterday, I.M. Easterday, *Adv. Exp. Med. Biol., 42,* 123 (1974).

68. R.K. Scopes, *J. Chromatogr., 376,* 131 (1986).

69. L. Hammar, S. Påhlman, S. Hjertén, *Biochim. Biophys. Acta, 403,* 554 (1975).

70. A.J.P. Martin, R.L.M. Synge, *Biochem. J., 35,* 1358 (1941).

71. B.-L. Johansson, Ö. Jansson, *J. Chromatogr., 363,* 387 (1986).

72. R. Janzen, K.K. Unger, H. Giesche, J.N. Kinkel, M.T.W. Hearn, *J. Chromatogr., 397,* 91 (1987).

73. Y. Kato, T. Kitamura, S. Nakatani, T. Hashimoto, *J. Chromatogr., 483,* 401 (1989).

74. S. Hjertén, J. Mohammad, K.-O. Eriksson, J.L. Liao, *Chromatographia, 31(1/2),* 85 (1991).

75. J.R. Gerstner, R. Hamilton, S.M. Cramer, *J. Chromatogr., 596,* 173 (1992).

76. J.-L. Liao, R. Zhang, S. Hjerte'n, *J. Chromatogr., 586,* 21 (1991).

77. S. Hjertén, Y.-M. Li, J.-L. Liao, J. Mohammad, K. Nakazato, G. Pettersson, *Nature (London), 356,* 810 (1992).

78. F.E. Regnier, *Nature (London) 350,* 634 (1991).

79. A.I. Liapis, M.A. McCoy, *J. Chromatogr., 599,* 87 (1992).

80. K.-H.W. Lau, T.K. Freeman, D.J. Baylink, *J. Biol. Chem., 262,* 1389 (1987).

81. P. Roos, F. Nyberg, L. Wide, *Biochim. Biophys. Acta, 588,* 368 (1979).

82. D. Hereld, J.L. Krakow, J.D. Bangs, G.W. Hart, P.T. Englund, *J. Biol. Chem., 261,* 13813 (1986).

83. N.C. Robinson, D. Wiginton, L. Talbert, *Biochemistry, 23,* 6121 (1984).

84. K.K. Sharma, B.J. Ortwerth, *Eur. J. Biochem., 216,* 631 (1993).

8 Immobilized Metal Ion Affinity Chromatography

LENNART KÅGEDAL

Research and Development
Pharmacia Biotech AB
S-751 82 Uppsala, Sweden

Protein Purification: Principles, High-Resolution Methods, and Applications, Second Edition.
Edited by Jan-Christer Janson and Lars Rydén.
ISBN 0-471-18626-0. © 1998 Wiley-VCH, Inc.

8.1   INTRODUCTION

In 1975, Porath and co-workers published a paper entitled "Metal chelate affinity chromatography, a new approach to protein fractionation."[1] The principle of the suggested method was based on differences in the affinity of proteins for metal ions bound in a 1:1 complex of iminodiacetic acid (IDA) immobilized on a chromatographic support. The principle was not new—it had been suggested already in 1961 by Hellferich who named it *ligand exchange chromatography*[2] (LEC), and the technique has been widely used since then.[3] However, Porath's paper showed for the first time that immobilized metal ions could be used with advantage to fractionate and purify proteins. As is demonstrated in this chapter, the technique has become an important tool for the isolation of many proteins.

In the past, the term *metal chelate affinity chromatography* (MCAC) has been the accepted term for LEC when used in biochemistry. Other names have also been used [e.g., "metal chelate interaction chromatography" (MCIC)]. Following a proposal by Porath and co-workers,[4,5] this chapter uses the term *immobilized metal ion affinity chromatography* (IMAC).

A comprehensive review of LEC was published in 1977 by Davankov and Semechkin.[3] The theory they presented is, to a large extent, also relevant for IMAC and the reader is referred to them for a more complete coverage of the literature up to 1977. Some general aspects of the technique as applied to proteins have been reviewed by Porath and Belew,[4] Lönnerdal and Keen,[6] Sulkowski,[7] Porath and Olin,[8] Porath,[9] Rassi and Horvath,[10] and Chaga.[11] Mathematical modeling of IMAC has been done by Vunnum and co-workers.[12]

Most proteins can form complexes with metal ions. Many of these are multidentate complexes (chelates) and allow for the purification of the proteins by IMAC. The strength of the complexes formed varies from protein to protein, which, in many cases, gives rise to the high specificity of IMAC.

The chromatographic sorbent used in IMAC (see scheme in Figure 8-1) consists of a suitable chromatographic support to which a metal-chelating substance (B) has been attached by a leash or linkage group (A). The structure of the complex formed when metal ions are added must be such that some coordination sites are left free for the binding of solvent or solute molecules (ligands). Alternatively, the complex should be able to rearrange itself to allow incoming ligands to participate in the formation of chelates or complexes with the metal ion. Solvent or buffer molecules will occupy "free" coordination sites of the metal in the absence of ligands with higher affinity for the metal ion.

Muzzarelli and co-workers[13] have suggested the following definition of LEC: "Ligand exchange chromatography is based on the principle that a molecule or ion, which is part of a complex fixed on a support, can be released because a different molecule or ion enters to form a more stable complex, or because the complex collapses when the medium is altered." The definition may be somewhat limited compared to that proposed by Davankov and Semechkin (Ref. 3, p. 314), but it gives a sufficiently clear basis for the understanding of IMAC as reviewed in this chapter.

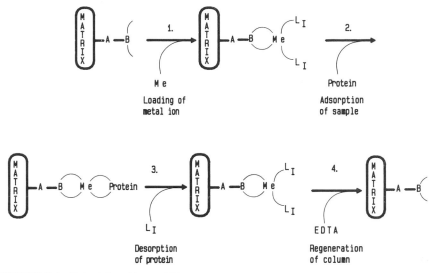

FIGURE 8-1. Principle of immobilized metal ion affinity chromatography. A, linkage (spacer) group; B, chelating group; Me, metal ion; L_I, solvent or buffer molecule.

Some of the special features of IMAC of proteins can be summarized as follows:

- Exposure of certain amino acid residues (histidine, cysteine, tryptophan) on the "surface" of the proteins is required for the adsorption of proteins.
- The steric arrangement of the protein chain plays an important role, which means that molecules with closely similar properties, with respect to charge, molecular size, and amino acid composition, but with differences in their secondary and tertiary structure, can be separated.
- Simple ionic adsorption and other complicating factors can be suppressed or modified by buffers of high ionic strength.
- Binding is influenced by pH. Low pH causes elution of adsorbed substances. Exceptions to this are known.[14]
- Several elution techniques are available (pH gradient, competitive ligands, organic solvents, chelating agents).
- IMAC is a general technique for purifying proteins. Metalloproteins do not bind specifically at their metal coordination sites, but rather through amino acid residues exposed at the protein surface.
- Clearance of viruses has been shown to be effective in a number of cases.[15]

IMAC is an established technique for the purification of proteins on a laboratory scale and is beginning to find industrial applications. Its acceptance may have been rather slow initially, but its usefulness and versatility have been unequivocally demonstrated sufficient to convince biochemists that it is a valuable addendum to the arsenal of preparative methods.

Because of the number of factors influencing the IMAC process, the apparent complexity of the method may have added to the reluctance of biochemists to adopt the method. Although there are still several question marks left concerning the exact mechanisms and chemical structures involved, there is sufficient information available for a systematic optimization of preparative experiments. For example, it is possible to modify adsorption and desorption by the correct choice of chelating group, metal ion, pH, and buffer constituents.

The specificity of IMAC obviously depends on the exploitation of the combined effects of primary structure (occurrence and position of a limited number of metal binding amino acid residues) and secondary and tertiary structure (exposure of certain amino acids on the protein surface). It may well be that the degree of proximity of the metal-binding amino acids is also important.

8.2 METAL CHELATE GELS

8.2.1 Choice of Chromatographic Support

Basically, the requirements of the support in affinity chromatography of biological molecules also apply to IMAC (Ref. 16, p. 20). Ideally the support should:

- Be easy to derivatize.
- Exhibit no unspecific adsorption.
- Have good physical, mechanical, and chemical stability.
- Be of high porosity to provide easy ligand accessibility.
- Permit high flow rates.
- Be stable to eluants including, for example, denaturing additives.
- Allow regeneration of column without deterioration of the gel bed.
- Provide a stable gel bed with no shrinking/swelling during the chromatographic process.

Beaded agarose is the support predominantly used for IMAC of proteins, and published protocols in most cases describe the use of Sepharose® 4B or Sepharose 6B to which IDA had been coupled by the original bisoxirane method described by Porath and co-workers.[1]

Silica-based IDA gels have been constructed for use in preparative work.[17-20] Gels for use in HPLC and FPLC™ have been prepared using hydrophilized resins,[21,22] cross-linked agarose[23,24] (10 μm), and silica[25] (5 μm) using IDA as the chelating ligand.

Membranes have also been used as supports for chelating groups.[26]

8.2.2   Chelating Ligand and Strength of Chelates; Capacity for Metal Ions

IMAC in general can be carried out using various types of ion exchangers.[3] In IMAC of proteins, however, only chelating groups have been used to fix the metal ion to the support, the reason being that metal ions bind more strongly to chelating groups. The binding energy of transition metal cations, as calculated by Schmuckler,[27] is 2 to 3 kcal/mol with ordinary cation exchangers and 15 to 25 kcal/mol with chelating groups. Table 8-1 lists the formation constants for 1:1 chelates of some chelating compounds. Compounds I and II have a structure similar to that of the chelating group in IDA Sepharose, and compound III is a close analog of the so-called TED group, introduced by Porath and Olin.[8] Studies of the formation constants of the IDA groups of Dowex A-1 indicated that formation constants of the same order of magnitude are obtained for analogous compounds free in solution or bound to a chromatographic support.[3] The relative stability of complexes formed with an IDA derivative of cellulose and divalent metal ions is in the order Cu(II) > Ni(II) > Zn(II) ≥ Co(II) >> Ca(II), Mg(II), as shown by Horvath and Nagydiosi,[28] which agrees well with data in Table 8-1. They also concluded that Ca(II) and Mg(II) ions do not form chelates with the IDA group, as their presence did not affect the shape of the potentiometric acid–base titration curve.

TABLE 8-1. Formation constants[a] (log K) for 1:1 complexes of chelating compounds[b] and metal ions

No.[c]	Formula	Ca^{2+}	Fe^{2+}	Fe^{3+}	Co^{2+}	Ni^{2+}	Cu^{2+}	Zn^{2+}
I	CH_3—N, CH_2COOH (×2)	3.8	6.3		7.6	8.7	11.1	7.6
II	CH_2—N with OH and two CH_2COOH	4.8	6.8		8.1	9.4	11.7	8.5
III	CH_2COOH / CH_2COOH, N—CH_2—CH_2—N, CH_2CH_2OH / CH_2COOH	8.2	12.2	19.5	14.5	17.1	17.5	14.6

[a] From critical stability constants Vols. 1–4, E. Martell, R. M. Smith, Plenum, New York London, 1974–1977.

[b] In the chelate the ligand is present in its ionized form.

[c] I, N-Methyliminodiacetic acid; II, N-(hydroxymethyl)iminodiacetic acid; and III, N-(hydroxyethyl) ethylenediaminetriacetic acid.

The fixing of the IDA–ligand to the support has, in many cases, been affected by Porath's original method of coupling to agarose into which reactive epoxy groups have been introduced by reaction with 1,4-bis(2,3-epoxypropoxy)-butane.[1] The method has the dual advantage of providing a chemically stable ether linkage to the ligand and a 12-atom spacer. A 3-atom spacer is obtained if activation of the agarose matrix is carried out with epichlorohydrin.[8] In chelating Sepharose Fast Flow (Table 8-2), a 7-atom spacer is used. The length of the spacer has been shown to have a certain impact on the chromatographic performance of the IMAC adsorbent.[29]

Results obtained by Hansson and co-workers[23] show that the metal ion capacity of the IDA gel influences the retention of proteins (Figure 8-2). IDA derivatives of Superose® 12 carrying varying amounts of the chelating moiety were used. The resolution of the sample proteins was also affected, indicating that the extent of the influence on retention is dependent on the properties of the proteins.

Commercially available column packings for IMAC of proteins are listed in Table 8-2.

8.3   FACTORS INFLUENCING ADSORPTION AND DESORPTION

8.3.1   Chelate Structure and Metal Ion

The exact structure of the stationary complex depends on the metal ion used and on the composition of the eluent buffer. According to Davankov and co-workers, Zn^{2+} and Cu^{2+} will leave one and Ni^{2+} three sites for solvent or buffer molecules when the three-dentate IDA is used.[3] Others seem to assume higher coordination numbers than four for the Zn^{2+} and Cu^{2+} ions.[30]

Figure 8-3 shows the postulated planar structure of the chelate formed between Cu^{2+} and Chelating Sepharose Fast Flow, including the linkage group.[3] One coordination site is occupied by a water molecule (or possibly by some buffer constituent) that will be substituted by a sample molecule during the course of the chromatographic experiment.

Porath and Olin[8] have used N,N,N-(tricarboxymethylethylene)diamine coupled to Sepharose (TED Sepharose). Because the TED group is a five-dentate chelating group, it can occupy five of the six coordination sites of the Ni^{2+} ion. Consequently, Ni^{2+}-TED Sepharose is a weaker adsorbent than the corresponding IDA column,[8,31,32]; this is demonstrated by, for example, the adsorption on Ni^{2+}-IDA Sepharose of proteins that pass through a Ni^{2+}-TED Sepharose column. Hochuli and co-workers[33] have used the Ni^{2+} chelate of a nitrilotriacetic acid derivative (NTA) for proteins containing adjacent histidine residues.

O-Phosphoserine (OPS), a tri-dentate chelating group, has been found to afford different selectivites as compared to IDA.[34] Carboxymethylated aspar-

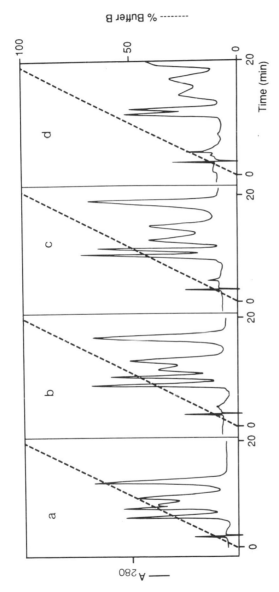

FIGURE 8-2. Cu^{2+} IMAC of proteins on IDA Superose 12 gels with different metal ion capacity.[23] Buffers: A, 20 mM sodium phosphate and 1 M NaCl, pH 7.2; B, 20 mM sodium phosphate and 1 M NH$_4$Cl, pH 7.2. Column: 5 × 50 mm (HR 5/5 Pharmacia Biotech AB). Sample: 50 μL of a solution consisting of cytochrome c (peak 1), lysozyme (peak 2), β-lactoglobulin (peak 3), lens lectin (peak 4), and α-lactalbumin (peak 5) dissolved in half-diluted buffer A. Flow rate: 0.6 mL/min. Conditioning of columns before sample application: Elution with water, 5 mL; 0.25 M Cu$_2$SO$_4$, 5 mL; water, 5 mL; and buffer B, 20 mL. Elution of proteins: From 100% A to 100% B in 20 min. Cu^{2+} capacity of packed gels (μmol/mL): a, 18; b, 26; c, 32; and d, 43.

TABLE 8-2. Commercially Available Chromatographic Supports for IMAC

Designation	Manufacturer	Description
Chelating Sepharose 6B	Pharmacia Biotech AB, Uppsala, Sweden	Agarose
Chelating Sepharose Fast Flow	Pharmacia Biotech AB, Uppsala, Sweden	Cross-linked agarose
Chelating Superose	Pharmacia Biotech AB, Uppsala, Sweden	Cross-linked agarose
HiTrap® Chelating	Pharmacia Biotech AB, Uppsala, Sweden	Cross-linked agarose, prepacked columns, 1 and 5 mL
Immobilized iminodiacetic acid gel	Pierce, Rockford, IL, USA	Cross-linked 4% beaded agarose
Immobilized iminodiacetic acid gel	Pierce, Rockford, IL, USA	TSK HW-65F
Iminodiacetic acid-agarose	Sigma Chemical Corp., St. Louis, MO, USA	Agarose
Iminodiacetic acid-epoxy activated Sepharose 6B	Sigma Chemical Corp., St. Louis, MO, USA	Cross-linked agarose
Zink chelate affinity adsorbent	Boehringer Mannheim, Germany	Agarose
TSKgel Chelate-5PWb	Tosoh, Japan	Polymer resin
"Nickel-NTA resin"	QIAGEN Inc., Chatsworth, USA	NTA bound to Sepharose CL-6B and precharged with Ni^{2+}
His · Bind Resin	Novagen, Madison, WI, USA	IDA bound to Sepharose
"Ni^{2+} immobilized resin column"	Invitrogen, CA	
Fractogel EMD Chelate 650(S)	E. Merck, Darmstadt, Germany	Polymer support with polymers extending from the support and carrying the chelating groups
TALON™ Metal Affinity Resin	CLONTECH Laboratories,	Tetradentate chelator precharged with Co^{2+} for his-tagged proteins
POROS MC	PerSeptive Biosystems	Hydrophilized polystyrene-divinyl-benzene particles

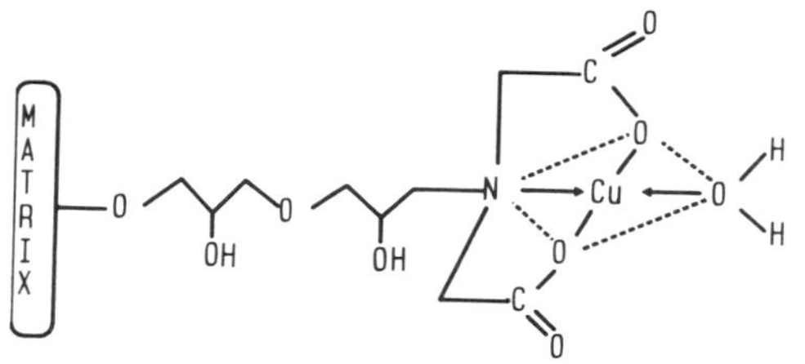

FIGURE 8-3. Postulated planar Cu^{2+} chelate with iminodiacetic acid as chelating group.

tic acid (CM-Asp)[9] and NTA mentioned earlier are tetra-dentate chelating groups.

It is reasonable to assume that the sorption of ligands from the sample will cause rearrangement of the stationary complex (Ref. 3, p. 326), which in turn may alter the stability of some of the stationary complexes. Such a supposition is supported by the observation that when a protein sample is applied to a column of chelating gel incompletely charged with copper ions (i.e., the lower part of the column is still white), the blue zone of the copper complex will move down the column to some extent.[35]

Cu^{2+}, Ni^{2+}, Zn^{2+}, and Co^{2+} form a series of metal ions with decreasing binding strength to IDA as shown by Porath and others.[4,7,36] Other ions such a Cd^{2+} may be used in a similar way.[16]

Ca^{2+} and Fe^{3+} seem to represent special cases. Ca^{2+} has been used for the purification of *Dolichos biflorus* seed lectin[37] and factor VIII:c coagulant activity.[9] The lectin has very high affinity for Ca^{2+} and was eluted from the column using EDTA. The special character of the interaction of Fe^{3+}–IDA gels with proteins is reflected in the requirement for special elution modes (see Section 8.3.3).

8.3.2 Structure of Protein or Peptide

The amino acid residues responsible for the adsorption of proteins in IMAC have been discussed in a number of papers,[5,36,38–45] yet there are several questions left open. According to Porath and co-workers, some studies with amino acids indicated that histidine and cysteine would be most likely to cause strong interactions with metals in IMAC.[1] Tryptophan, which contains the indole structure, was also thought to contribute to the binding. Porath later demonstrated that Fe^{3+} bound to IDA Sepharose has a rather specific affinity for tyrosine-containing proteins.[40]

The finding by Sulkowski[45] that there is an abrupt change in the adsorption of bovine serum albumin to a Ni^{2+}-IDA gel around pH 6.5 supports the notion that histidine residues are involved in the sorption process in this case. Above pH 8 the α-amino group of the amino terminal has an affinity for Cu^{2+}-IDA.[46,47]

Desorption of ribonuclease with an imidazole gradient was found to be influenced by the pH of the eluent, a higher pH requiring a lower imidazole concentration.

Using the α-chymotrypsinogen A and α-chymotryspin pair and glycosylated and nonglycosylated ribonuclease, Sulkowski[45] has shown that the microenvironment of the histidine residues influences the retention of proteins significantly.

Sulkowski and co-workers have used IMAC as a tool for the study of the surface topography of interferons.[36] The same group has also used some model proteins to establish the influence of various potential complex-forming amino acid residues on the behavior of proteins in IMAC. Their results showed that for angiotensin I, angiotensin II, and pancreatic ribonuclease A, the strength of binding to different metals decreases in the following order: $Cu^{2+} >$ $Ni^{7+} > Co^{2+} > Zn^{2+}$. The difference between the strength of binding to Co^{2+} and Zn^{2+} was rather small. In the absence of tryptophan and cysteine on the protein surface, an increasing number of histidine residues causes stronger retention. The pancreatic trypsin inhibitor, which lacks histidine, tryptophan, and cysteine on the protein surface, was adsorbed only onto Cu^{2+} at pH 7.4, "presumably due to coordination via amino groups."

In a study by Hammacher and co-workers,[24] IMAC was used in the structural characterization of platelet-derived growth factor (PDGF) purified from human platelets.

PDGF is the general name for any one of several dimers that differ in composition of the two homologous polypeptide chains denoted A and B. The chains are linked by disulfide bridges into homodimers of either the PDGF-AA or the -BB type or a heterodimer of the PDGF-AB type. Depending on the cell type used as a source of PDGF, the purified product is either a single isoform (PDGF-AA, PDGF-AB, or PDGF-BB) or a mixture of different isoforms. The A and B chains differ in the content of histidine, B having one and A three residues. When applied to IMAC (Cu^{2+}-IDA), PDGF-BB was not adsorbed, whereas PDGF-AA was retained on the column. The unknown sample, PDGF purified from human platelets, was partly retained. The retained material eluted before the A homodimer and was concluded to be a heterodimer (PDGF-AB).

Using carboxypeptidase A as a model protein, Muszynska and co-workers[31] have studied the possibilities of employing IMAC for the characterization of metalloproteins and their preparation in metal-free forms.

In general, it seems that whether a protein is a metalloprotein has no bearing on the binding or nonbinding of the protein to the chromatographic support in IMAC. For example, Lönnerdal and co-workers[6] showed that Fe-lactoferrin, Cu-lactoferrin, and apolactoferrin exhibited identical sorption and desorption properties; the Fe-lactoprotein was recovered from a Cu^{2+} chelate gel in its initial iron-saturated form.[6]

Nucleoside diphosphatase bound to Zn^{2+} and Cu^{2+} chelate gels exhibited almost full catalytic activity,[42] thus indicating that the binding site is separated from the catalytically active site.

When the amino acid compositions of the proteins to be fractionated are known, the following classification adapted after Sulkowski,[7,45] with respect to histidine and tryptophan residues, may be used to predict results:

Presence of Histidine or Tryptophan on "Surface" of Protein	Metal Ions Providing Adsorption
No His/Trp	—
One His	Cu^{2+}
More than one His	Cu^{2+} (stronger adsorption), Ni^{2+}
Clusters of His	Cu^{2+}, Ni^{2+}, Zn^{2+}, Co^{2+}
Several Trp, no His	Cu^{2+}

No foolproof rules can be set as yet. Even in the absence of the three amino acids considered to be most important for adsorption, namely histidine, tryptophan, and cysteine, adsorption or protein has been observed.[36] Free amino groups, which are not considered in Sulkowski's scheme, may, in many cases, play an important role, particularly in peptides.[46,47]

The fact that histidine residues are of such great importance in IMAC has been exploited very successfully in a mode of IMAC pioneered by Hochuli and co-workers[33] and Smith and co-workers.[48] Histidine-rich streches are introduced at either protein terminal by recombinant techniques. For example, the protein has, in many cases, been fused to a peptide consisting of six histidine residues. This modification of the protein ensures that the hybrid protein after expression in the appropriate microorganism can be efficiently adsorbed onto a chelating column and that the binding will be sufficiently strong to allow for a high degree of removal of other components before the protein is desorbed from the column. Adsorption can occur even if the sample consists of protein from inclusion bodies dissolved in a denaturing medium such as 8 M urea or 6 M guanidinium chloride. A number of papers have been published[49–70] that demonstrate that chelating groups such as IDA and NTA, in combination with a suitable metal ion (Cu^{2+}, Ni^{2+}, Co^{2+}, Zn^{2+}), can be used. A high degree of purity after only one chromatographic step has been claimed in several cases. Efficient proteolytic cleavage and removal of the affinity tail and the protease has also been demonstrated.[57]

The use of various metal ions with IDA as the chelating group has been studied in a series of experiments with recombinant protein A tagged at the C terminus with (HisGly)$_n$His[71] and expressed in *Escherichia coli*. The size of the peptide was varied with $n = 1, 2, 3, 4$, and 6. Isolation procedures were tested both under denaturing conditions with 6 M guanidinium chloride and under nondenaturing conditions. Under nondenaturing conditions the most suitable metal ion was found to be Zn^{2+} because of its relatively low affinity for native *E. coli* proteins. With Zn^{2+}, $n = 4$ or larger was required for binding of the fusion protein. In the presence of GuHCl, a metal ion with strong affinity for the protein of interest was required such as Cu^{2+} or Ni^{2+}.

Belew and co-workers[72] have made quantitative studies of the interaction between proteins and Cu^{2+}-IDA on chelating Sepharose Fast Flow using both frontal analysis and equilibrium-binding analysis. Closely similar results were obtained with the two methods. The dissociation constant (K_d) at pH 7.0 was 33 to 37 μM for lysozyme and 3.5 to 6.8 μM for ovalbumin. The measured binding capacities (L_t) were 6.8 and 1.7 μmol/ml gel for the same proteins. See also Hutchens and co-workers.[73,74]

8.3.3   pH, Types of Buffers, and Ionic Strength

There has been no comprehensive study of how buffers of different types affect sorption of proteins to Me^{n+}-IDA gels. Tables 8-3 and 8-4 give some indications how conditions can be varied to give a suitable strength of adsorption.

Fe^{3+}-IDA gels present a special case, as demonstrated by Andersson and Porath[14] and Muszynska and colleagues.[75] They have shown that phosphorylated proteins or peptides can be desorbed from such a gel by an ascending (sic!) pH gradient or by inclusion of phosphate in the elution buffer. Sulkowski[76] has extended the study to other types of proteins and has also shown that selective desorption can be affected with an increasing concentration of sodium chloride.

It is customary to include sodium chloride (0.1 to 1.0 M) in buffers used in IMAC. One effect of this is the suppression of ionic interactions between sample and the gel. Sulkowski[45] has, however, shown in some cases that the ionic strength of the eluent affects the retention of proteins on Cu^{2+}-IDA columns. Cytochrome c from horse heart (pI 10.6) eluted earlier in a pH gradient when ionic strength was high, whereas the opposite was observed with calmodulin (pI 4.1). The disparate behavior of these proteins was accounted for by assuming that the net charge of Cu^{2+}-IDA is negative.

High salt concentrations have also been used to suppress the protein–protein interactions during IMAC.[77,78,79] Such associations may make other chromatographic methods such as ion exchange difficult if not impossible, and the use of IMAC as an early step allows ion exchange or gel filtration to be used later in the purification scheme.

Wunderwalt and co-workers[80] used α_2-macroglobulin bound to Cu^{2+}-IDA Sepharose for the removal of endoproteinase from fluids in a "sandwich technique." They discovered that much less bound α_2-macroglobulin was removed from a column eluted with several column volumes of buffer of high ionic strength when sodium chloride was added to generate the ionic strength than when sodium phosphate was used.

8.3.4   Detergents and Other Additives

Detergents can, when necessary, be used with advantage in IMAC. Collagenase, a difficult enzyme to purify, was isolated with Zn^{2+} IMAC as one step of

TABLE 8-3. Behavior of Cohn IV Fraction Proteins from Human Serum in IMAC

Metal Ion	Buffers in Sample and Starting Buffer	Behavior of Proteins			
		Albumin	Transferrin	α_2-HS Glycoprotein	α_1-Lipoprotein
Zn^{2+}	50 mM Tris-HCl, pH 8.0, 150 mM NaCl	No retention	Retained	Retained	Retained
Cu^{2+}	100 mM Na acetate, pH 7.7, 500 mM NaCl	Complete retention	Retained	Retained	Retained
Cu^{2+}	20 mM phosphate, pH 7.7, 500 mM NaCl	Complete retention			
Cu^{2+}	50 mM Tris-HCl, pH 8.0, 150 mM NaCl	Partial retention			
Cu^{2+}	50 mM Tris-HCl, pH 8.0, 150 mM NH$_4$Cl	Major part unretained[a]	Retained	Retained (partial?)	Retained

[a] Albumin oligomers and a minor part of the albumin monomer were retained.

TABLE 8-4. Qualitative Influence of Various Factors on the Binding of Proteins in IMAC

System Parameter	Weaker Binding			Stronger Binding	
Stationary phase					
Metal ion	Ca^{2+}	Co^{2+}	Zn^{2+}	Ni^{2+}	Cu^{2+}
Chelating ligand					
Type	TED				IDA
Amount bound	Low				High
Mobile phase					
pH	5		7		8
Buffer ions	(EDTA)		Ammonia		Acetate
	Citrate		Tris		Phosphate
			Ethanolamine		

a purification scheme. A detergent, Brij 35, was used in all buffers.[81] Tween 80 was included in buffers used in a procedure for the isolation of plasminogen activator from human melanoma cells.[82] The use of detergent gave better yields than were obtained in an earlier isolation of the activator from human uterine tissue.[83] In both cases, Zn^{2+} IMAC was used as one of the steps, desorption being affected by an imidazole gradient.

In the purification of human factor XII, a trypsin inhibitor and surface-binding inhibitors were used in all buffers.[84]

Human placenta mitochondrial membrane proteins could be fractionated using 0.4% octaethylene glycol dodecyl ether in the eluents.[85]

A 1:1 mixture of ethanol–water was used to remove colored contaminants before eluting the adsorbed protein in the purification of albumin from a Cohn IV extract.[86]

8.4   CHROMATOGRAPHIC CONDITIONS

8.4.1   Planning of Experiments

For a protein with unknown properties regarding its binding to metal chelate columns, it is advisable to use Cu^{2+} and a neutral phosphate or acetate buffer containing 0.15 to 0.5 M NaCl for initial experiments. Ideally, total binding of the desired protein should occur with direct elution of contaminating proteins or the reverse. If inappropriate binding is obtained, a number of variations, as outlined in Table 8-4, can be tried.

A few experiments with variations of metal ion, pH, and buffer composition should be sufficient to establish suitable starting conditions. To vary the degree of substitution is more laborious. However, the amount of bound metal ions can be reduced by leaching the metal-primed column with a citrate buffer, at least when Cu^{2+} is used.[87]

The following section deals in some detail with practical aspects of IMAC for the optimization and systematization of experiments.

8.4.2 Loading of Metal Ion and Regeneration of Column

Metal ions such as Cu^{2+}, Zn^{2+}, Co^{2+}, Ni^{2+}, and Ca^{2+} can be loaded in neutral or weakly acid solutions to avoid hydroxy precipitates with, for example, Cu^{2+}. For Cu^{2+}, the saturation of the column can be followed simply by observing the blue color of the column. When Zn^{2+} is used, formation of a precipitate on the addition of Na_2CO_3 to the eluent indicates saturation of the column.

Some bleeding of metal ions from the column is likely to occur during most IMAC experiments. In many cases this does not matter, but the leakage can be suppressed if necessary by washing the primed column with the final eluent before equilibration with the starting buffer (the buffer in which the sample will be applied). Alternatively, the column can be charged with metal ion to only 70 to 90% of its maximum capacity; this will cause migrating metal ions to be captured by chelating groups at the bottom of the column, equivalent to having an unloaded small column in series with the separation column, and to extracting metal ions from the eluate with a small amount of chelating gel after the chromatographic run.

As a general rule it may be advisable to strip columns of metal ions between runs using chelating agents, such as EDTA, and to recharge with metal ion before the next experiment. Stripping and recharging are essential when ions of weak complexing strength, such as Zn^{2+} and Co^{2+}, are used with descending pH gradient protocols.

With Cu^{2+}- and Ni^{2+}-IDA columns, many chromatographic cycles can, at least under certain conditions, be performed without regeneration of the column. The experiment depicted in Figure 8-2c was repeated 20 times without regeneration of the column. The last chromatogram was indistinguishable from the first one (not shown).[23]

8.4.3 Sample Application

One obvious advantage of IMAC is that it can be used to concentrate a protein from very dilute solutions. It may, however, be necessary to equilibrate the sample carefully with the starting buffer by, for example, dialysis to remove components (e.g., amines) that may interfere in the adsorption step. Depending on the properties of the samples, salts, urea, detergents, glycol, and so on may be included in the sample and elution buffers. Such additives seem to have no or only a minor influence on the adsorption of the protein, but may serve to improve performance and yield. It is advisable to use 0.15 to 1.0 M NaCl or other salts in all buffers to get consistent results.

8.4.4 Elution Modes

After washing away unbound material, bound substances are recovered by changing conditions to favor desorption. A gradient or stepwise reduction in pH to 3 or 4 is often suitable. Alternatively, competitive elution with a gradient of increasing concentration of, for example, ammonium chloride, glycine, histamine, histidine, or imidazole may be used. A combination of ascending gradients of pH and concentration of ammonium chloride was found to give the best result in one case.[92] It has been pointed out[7] that when, for example, imidazole is used for elution, the columns should be saturated with imidazole prior to adsorption of the sample protein, and that imidazole should be included in the starting buffer. This is not always essential for good results, however, as shown in the purification of plasminogen activator.[83]

A third method of elution is to include a chelating agent such as EDTA in the eluent. In this case, all adsorbed proteins will be eluted indiscriminately along with the metal ion. In all cases, high ionic strength should be maintained.

8.5 AREAS OF USE OF IMMOBILIZED METAL ION AFFINITY CHROMATOGRAPHY

The predominant use of IMAC to date has been in the isolation and purification of proteins and peptides, as discussed in this chapter. The chromatography of nucleotides, dinucleotides, and related compounds on metal chelates has also been shown to be possible.[93,94] Pyrimidines show little interaction with the metal ions, but purines are resolved on Cu^{2+} chelates. The technique is potentially very useful for the large-scale purification of these compounds.

Of great interest is the demonstrated removal and inactivation of viruses in the IMAC process.[95]

The technique discussed earlier has also been used in the study of the structure of proteins, whereby the exposure of certain amino acid residues on the surface of the proteins can be explored, and for the immobilization of proteins to be used as active solid-phase enzymes[42,96] or affinity adsorbents.[80]

Table 8-5 lists some data from published purifications. Care has been taken to select examples exploiting various modifications of IMAC rather than showing only very typical procedures. The list, which is not exhaustive and does not include applications presented in Section 8.6, demonstrates that a wide range of adsorption and desorption buffers can be used.

8.6 APPLICATIONS

8.6.1 Purification of Copper, Zinc Superoxide Dismutase from Human Erythrocytes

A two-step purification protocol for the purification of copper, zinc superoxide dimutase (Cu,Zn SOD) has been designed by Weselake and colleagues.[97] The first step is ion-exchange chromatography of a filtered lysate from human erythrocytes on a column of DEAE-Sepharose CL-6B from which the Cu, Zn SOD-containing fraction is eluted in 10 mM potassium phosphate, pH 6.4, containing 100 mM NaCl.

An IDA Sepharose 6B gel was prepared according to Porath.[1] It contained 25 to 30 μmol of IDA per milliliter of sedimented gel as determined by the titration of immobilized carboxyl groups of the immobilized IDA. A 4 × 5 cm (diameter) column was packed and charged to saturation with 50 mM copper sulfate. A second column, 2 × 5 cm, packed with uncharged IDA gel was connected after the charged column to serve as a scavenger of any copper ions leached from the first column.

The columns were equilibrated with 10 mM potassium phosphate, pH 6.4, containing 100 mM NaCl and charged at a rate of 150 mL/h with the Cu,Zn SOD solution obtained from the ion-exchange step. The columns were then washed sequentially at a flow rate of 150 mL/h with

- 10 mM potassium phosphate, 1.0 M NaCl, pH 6.4, 3500 mL
- 10 mM potassium phosphate, pH 6.4, 400 mL
- 10 mM sodium acetate, pH 5.0, 300 mL
- 20 mM sodium citrate, pH 5.0 (desorption of Cu,Zn SOD)

Fractions containing Cu,Zn SOD obtained with the citrate buffer were collected as shown in Figure 8-4, concentrated to 30 mL by pressure ultrafiltration, dialyzed against distilled water, freeze-dried, and stored dessicated at −20°C. The volume of the collected fraction was 365 mL.

The specific activity of the purified Cu,Zn SOD was 3,800 U/mg. The overall purification factor was 2,000 and the yield was 58%. The purification in the IMAC step was 60-fold and the yield 63%.

Attempts to use higher concentrations of citrate buffer and/or lower pH resulted in sharper elution profiles but with concomitant release of impurities.

The purified Cu,Zn SOD was analyzed by SDS–gel electrophoresis, gel filtration, and isoelectric focusing and was found to be essentially pure. The authors concluded that

- The release of SOD from the copper ion column was related to the chelating properties of the citrate ion
- Cu^{2+} IMAC appears to be a simple and rapid procedure for purifying human erythrocyte Cu,Zn SOD, which avoids solvent and heat treatment steps
- The procedure did not appear to deactivate SOD by removing copper from the active site as judged from the retained enzymatic activity

8.6.2 Purification of Human Plasma α_2-Macroglobulin and α_1-Antitrypsin

A fine example of plasma protein purification by IMAC was given by Kurecki and co-workers[98] in their preparation of α_2-macroglobulin (α_2M) and α_1-

TABLE 8-5. Purification of Proteins by IMAC: Selected Examples

Protein	Metal Ion(s)	Starting Buffer	Elution Buffer	Remarks	Ref.
Plasma amyloid P component	Zn^{2+}	20 mM Tris–HCl, 150 mM NaCl, 1 mM benzamidine, pH 8.0	Linear gradient of histidine, 0–30 mM, in starting buffer	Dialyzed sample from Ba^{2+} citrate and $(NH_4)_2SO_4$ precipitation.	88
Thiol proteinase inhibitor	Cu^{2+}	25 mM Tris–HCl, 100 mM NaCl, pH 8.2		"Negative" chromatography. Flowthrough fraction used in next step.	89
Hu IFN-γ	Ni^{2+}	20 mM Tris–HCl, 150 mM NaCl, pH 8.5	100 mM Na acetate, 1.0 M NaCl, pH 4.5	22° gave better resolution and recovery. Typical recovery >90%. Very dilute sample could be used.	90
Granule proteins	Cu^{2+}	40 mM tris, 5 mM phosphate, 500 mM NaCl, pH 8.2	Phosphate-acetic acid gradient, pH 7.7–2.8, containing 500 mM NaCl	High ionic strength eliminates aggregation.	78
Nucleoside diphosphatase	Cu^{2+} Zn^{2+}	20 mM Tris–HCl, pH 7.5	10 mM L-histidine, 10 mM maleate, pH 6.6, or with EDTA	Enzyme is active when bound to chelating gel.	42
Dolichos biflourus seed lectin	Ca^{2+}	50 mM Tris-acetate, 500 mM NaCl, pH 8.2	10 mM EDTA in starting buffer	No desorption with 100 mM glycine–HCl, pH 2.2. Subunit with carbohydrate-binding activity binds to gel.	37
Collagenase (pig synovial)	Zn^{2+}	25 mM Na borate, 150 mM NaCl, pH 8.0	50 mM Na acetate, 800 mM NaCl, pH 4.7	Cu^{2+} and Ni^{2+} failed.	81
Ovalbumin	Fe^{3+}	100 mM acetate, pH 5.0	Gradient from A to B A: 50 mM Mes–NaOH, pH 5.7 B: 50 mM Pipes–HCl, pH 7.2		14

Protein	Metal ion	Binding buffer	Elution	Chelating group / tail	Ref.
Pepsin	Fe^{3+}	100 mM Na acetate, pH 5.0	100 mM Na acetate, 20 mM K phosphate, pH 5.0		14
Pituitary growth factor	Cu^{2+}	10 mM phosphate, 500 mM NaCl, 20% glycol, 1 mM imidazole, pH 7.4	Gradient with starting buffer containing 1–10 mM imidazole		91
Human 68-kDa (U1) ribonucleoprotein antigen	Ni^{2+}	5 mM Tris–acetate, 6 M urea, 0.5 M NaCl, pH 7.2	pH step gradient, 6 M urea and Tris–acetate, desorption in pH 5.0 buffer	Chelating group: IDA Histidine tail: $(His)_6$ at C terminus	50
HIV reverse transcriptase	Ni^{2+}	20 mM Tris, pH 8.0, 500 mM NaCl, 1 mM PMSF, 1 mM benzamidine, 10 mg/L leupetin, 10 mg/L aprotinin	Stepwise increase of imidazole concentration in starting buffer	Chelating group: IDA Histidine tail: Various histidine-rich peptides at N terminus	51
Upstream stimulatory factor	Ni^{2+}	20 mM Tris-HCl, pH 7.9, with several additives	Increasing concentrations of imidazole and KCl	Chelating group: IDA Histidine tail: $(His)_6$ at C terminus	52
TEM-β-lactamase	Zn^{2+}	50 mM phosphate, 1.0 M NaCl, pH 7.5	Stepwise reduction of pH in starting buffer	Chelating group: IDA Histidine-rich tail: Angiotensin I	54
Phenacrylic acid decarboxylase	Ni^{2+}	100 mM phosphate, 200 mM NaCl, 0.1 mM PMSF, pH 8	Stepwise reduction of pH in starting buffer	Chelating group: NTA Histidine tail: $(His)_6$ at N terminus	55
Penicillin-binding protein	Zn^{2+}	50 mM phosphate, 500 mM NaCl, 7 M urea, pH 8.0	Linear gradient of histidine, 0–300 mM, in starting buffer	Chelating group: NTA Affinity tail: Heptapeptide containing four histidines	56
Rat zinc finger protein Kid-1 construct	Ni^{2+}	50 mM Tris, pH 8.0, 150 mM NaCl, 6 M urea, 20 mM imidazole	1 M imidazole in starting buffer	Chelating group: NTA Affinity tail: $(His)_6$ at C terminus	58
GroES	Ni^{2+}	100 mM phosphate, 10 mM Tris-HCl, 6 M guanidin–HCl, pH 8.0	100 mM phosphate, 10 mM Tris-HCl, 8 M urea, pH 8.0	Chelating group: NTA Affinity tail: $(His)_6$ at N terminus	60

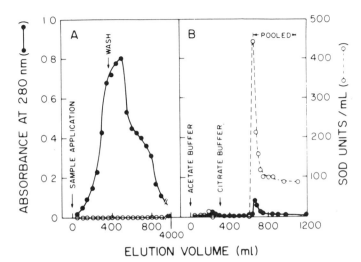

FIGURE 8-4. Cu^{2+} IMAC of Cu,Zn SOD. (A) Application of SOD solution (365 mL), washing with 3500 ml of 10 mM potassium phosphate buffer, pH 6.4, containing 1 M NaCl, 400 mL of same buffer without NaCl, and (B) with ca. 300 mL of 10 mM sodium acetate. Enzyme desorption with 20 mM citrate buffer, pH 5.0. Flow rate 150 mL/h. (Reproduced from Ref. 97 with permission.)

antitrypsin (α_1AT) (Table 8-6). They used dialyzed samples obtained by fractional ammonium sulfate precipitation. Large amounts of α_2M bound tightly to a zinc column at pH 6, allowing removal of contaminating proteins. At pH 5, α_2M eluted in a sharp concentrated peak (Figure 8-5). Their results show that a 2.5 × 14 cm zinc chelate column can accommodate at least 1.0 g of α_2M. An electrophoretically homogeneous product was obtained in good yield by this mild, two-step procedure.

α_1AT was isolated from the same batch of pooled human plasma (Figure 8-6). Because the major contaminating plasma protein, albumin, was not retained by the zinc chelate at the pH of the starting buffer, pH 8, this step was very efficient, giving a 20-fold purification by elution at pH 6.5. Two ion-exchange steps were required in addition to IMAC to obtain homogeneous α_1AT. According to the authors, this method increases the yield of α_1AT 2.5-fold as compared to other methods.

8.6.3   Purification of Albumin from a Cohn IV Fraction of Human Plasma

While purifying albumin from the Cohn IV fraction, Hansson and colleagues[86] carried out a study of the adsorption and desorption characteristics of albumin and contaminating proteins using zinc and copper chelates. The results, summarized in Table 8-3, showed that, for a given metal ion, the choice of buffer and salt in the elution medium governs the behavior of the proteins.

TABLE 8-6. Purification of α_1AT and α_2M [a]

Purification Step	Total Protein (mg)	Total Activity	Specific Activity	Recovery (%)	Purification (Fold)
α_1-Antitrypsin					
Plasma (1,000 mL)		665,000	16	100	1
Ammonium sulfate (50 to 80%)		612,000	28	92	2
Zn^{2+}-IDA gel		396,000	550	60	34
DE-52, pH 6.5		276,000	996	42	62
DE-52, pH 7.5	354	231,000	1,305	35	82
α_2-Macroglobulin					
Plasma (1,000 mL)		86,121	2.1	100	1
Ammonium sulfate (40 to 55%)		23,603	6.4	27	3
Zn^{2+}-IDA gel	342	18,605	61	22	29

[a] From Ref. 98 with permission.

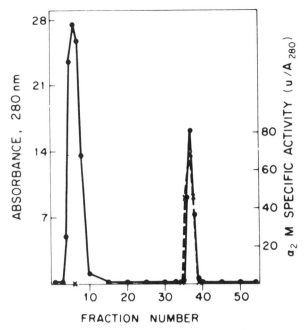

FIGURE 8-5. Purification of α_2-macroglobulin (α_2M) by Zn^{2+} IMAC. A 140-mL sample of α_2M in 50 mM sodium phosphate and 20 mM NaCl, pH 6.0, prepared by $(NH_4)_2SO_4$ precipitation from 1 L of human plasma was charged on a 2.5 × 14-cm Zn^{2+} chelate column equilibrated with 20 mM sodium phosphate and 150 mM NaCl, pH 6.0, at a flow rate of 100 mL/h. Elution was then begun at 50 ml/h with the same buffer, and 20-ml fractions were collected. At fraction 30 the buffer was changed to 20 mM sodium cacodylate and 150 mM NaCl, pH 5.0, and 11-mL fractions were collected. ●, A_{280}; ×, α_2M specific activity. The α_2M material in fractions 36 to 38 was homogeneous in polyacrylamide gel electrophoresis. (Reproduced from Ref. 98 with permission.)

Buffer constituents such as acetate and phosphate ions were found to favor the adsorption of the proteins, whereas the use of Tris or ammonium salts caused direct elution of the protein.

These findings were used to design the following purification procedure.

Sample:

An extract (65 ml) of a Cohn IV fraction containing 0.64 g of albumin and 0.76 g of other proteins, mainly transferrin.

Buffers:

A   0.1 M sodium acteate, 500 mM NaCl, pH 7.7

B   Ethanol–water, 1:9

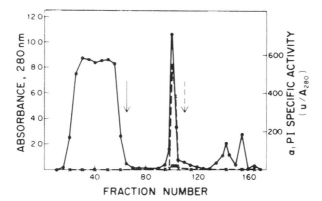

FIGURE 8-6. Purification of α_1-antitrypsin (α_1AT) by Zn^{2+} IMAC. A 700-mL sample of α_1AT in 50 mM sodium phosphate and 150 mM NaCl, pH 8.0, prepared from 1 L of human plasma by $(NH_4)_2SO_4$ precipitation was charged on a 2.5 × 90-cm Zn^{2+} chelate column equilibrated in the sample buffer at a flow rate of 100 mL/h. Elution was begun with the same buffer, and 22-mL fractions were collected. At fraction 65 the buffer was changed to 50 mM sodium phosphate and 150 mM NaCl, pH 6.5, and 12-mL fractions were collected. The column was stripped with 50 mM EDTA and 500 mM NaCl, pH 7.0, at fraction 110. ●, A_{280}; ×, α_1AT specific activity. Bar marks represent pooled fractions. (Reproduced from Ref. 98 with permission.)

C   Ethanol–water, 1:1
D   50 mM Tris-HCl, 150 mM NH$_4$Cl, pH 8.0
E   50 mM EDTA, 500 mM NaCl, pH 7.0

Colored impurities were eluted with aqueous ethanol before desorption of the albumin with a Tris buffer containing ammonium chloride.

Highly purified albumin was recovered with a 65% yield in one chromatographic step from the very crude and strongly discolored Cohn IV extract (Figure 8-7). Albumin in peak 3 of Figure 8-7a was 99.9% pure according to cellulose acetate electrophoresis.

8.6.4 Human Interferon-γ

Several types of interferons have been successfully purified by IMAC. The method is inexpensive and very efficient in isolating pure human fibroblast interferon (Hu IFN-β). Papers by Edy and co-workers[99] and Heine and co-workers[100,101] illustrate the methods. Using controlled porous glass in a batch procedure as a first step, they obtained interferon with an activity of 3.0 × 10^5 units/mg. From 44 mg of this they isolated ca. 7 μg of homogeneous Hu IFN-γ in one IMAC step using a Zn^{2+} chelate column. The specific activity was 1.7 × 10^9 units/mg and the yield in the IMAC step was 91.4%. Overall recovery was 52.6%.[101]

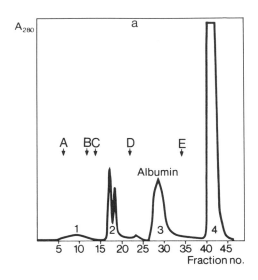

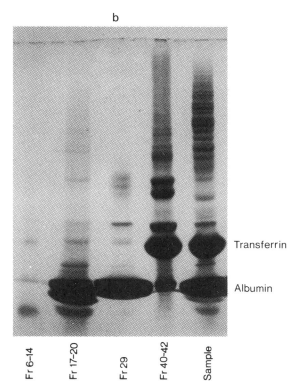

FIGURE 8-7. (a) Isolation of albumin from Cohn fraction IV extracts by Cu^{2+} IMAC. The copper-loaded column (K 16/20, Pharmacia Biotech AB, 65-mL gel bed) was charged with sample and eluted with eluents A–E as indicated in the chromatogram. See text regarding composition of sample and buffers. Fraction size was 11 mL and flow rate was 1.1 mL/min. (b) Polyacrylamide gel electrophoresis of fractions in a. Electrophoresis in a 4/30 gradient gel was carried out according to the manufacturer's

Of methodological interest is the use of the end buffer to wash the Zn^{2+} chelate column before equilibration with the starting buffer to remove excess Zn^{2+}.

The purification was performed at 4°C using a 1.5 × 16 cm column (K9/15, Pharmacia Biotech AB) packed with IDA Sepharose 6B prepared by epoxy coupling as described by Porath.[1] Before each run the column was regenerated. The flow rate was 15 to 20 mL/h. During elution steps, 1-mL fractions were collected and protein content, pH, and interferon activity determined.

Buffers:

A   20 mM phosphate, 1 M NaCl, 50 mM EDTA, pH 7.4
B   20 mM phosphate, 1 M NaCl, pH 7.4
C   100 mM sodium acetate, 1 M NaCl, pH 4.0
D   100 mM sodium acetate, 1 M NaCl, 1 mM ZnCl$_2$, pH 4.0
E   100 mM sodium acetate, 1 M NaCl, pH 5.9
F   100 mM sodium acetate, 1 M NaCl, pH 4.2
G   100 mM sodium acetate, 1 M NaCl, pH 4.0

Regeneration of column:

1   Removal of Zn^{2+} with A, 5 bed volumes.
2   Removal of EDTA with B, 5 bed volumes.
3   Equilibration with buffer C.
4   Introduction of Zn^{2+} with buffer D. Saturation tested by the formation of a precipitate when a drop of the eluate was mixed with a Na$_2$CO$_3$ solution.
5   Removal of excess Zn^{2+} with buffer C, 5 bed volumes.

Sample preparation:

6   Prepurification by adsorption to porous glass and dialysis against B. Final volume 100 mL.

Chromatography:

7   Application of dialyzed sample.
8   Wash with B, 1.5 bed volumes.
9   Wash with E, 5 bed volumes.

---

instructions (Pharmacia Biotech AB) after concentration of fractions in Minicon™ concentration cells (Amicon, Lexington, MA, USA). (Reproduced from Ref. 86 with permission.)

10   Elution with F, 1 bed volume.

11   Elution with G, 2 bed volumes.

The interferon peak fraction eluted at pH 5.2. Fractions with pH 5.6 to 4.2 contained interferon and were pooled. Table 8-7 summarizes the results from one typical run (simplified from Ref. 101).

8.6.5   Lactoferrin from Human Milk

The use of high ionic strength buffers to eliminate any nonspecific interactions with the gel is common in IMAC. High salt concentrations were used to prevent the association of lactoferrin from human milk[77] and granulocytes[78] with other proteins during IMAC (Figure 8-8).

8.6.6   Isolation of His-Tagged Glutathione *S*-Transferase Expressed in *Escherichia coli*

As discussed in Section 8.3.2, it has become a commonly used technique to introduce multiple histidine residues at either terminal of a protein to be expressed in a suitable microorganism. The high affinity of histidines to complexed metal ions is then exploited in an early IMAC purification step.

In the following example the target protein was glutathione *S*-transferase carrying six consecutive histidine residues at the C-terminal (GST-His$_6$). It was expressed in *E. coli* and secreted into the cytoplasmatic compartment during culturing. The vector used was pGEX-5X-1 and the strain of *E. coli* was XL1-blue MR.

The sample was the supernatant obtained after centrifugation of the cell homogenate and filtering through a 0.45-μm filter. The column was HiTrap Chelating 1 mL, a commercially available prepacked column (Pharmacia Biotech AB) containing Chelating Sepharose High Performance, a 34-μm support carrying IDA groups. All operations were carried out at ambient temperature. The chromatographic instrument was a FPLC system operated at a flow rate of 2 mL/min (312 cm/h) in all chromatographic steps. Desorption of the bound GST-His$_6$ was in this case done by a direct change to a buffer containing 300 mM imidazole.

TABLE 8-7. Purification of Human Fibroblast Interferon: Zn^{2+} IMAC Step

Fractions	Volume (mL)	Total Protein	Total Activity (units)	Specific Activity (units/mg)
Sample	100	44	13×10^6	3.0×10^5
Void + pH 7.4 wash	130	42	$<0.2 \times 10^3$	—
pH 5.9 wash	150	1.6	0.8×10^6	0.5×10^6
pH 5.6 to 4.2 eluate	6	0.0007	12×10^6	1.7×10^9

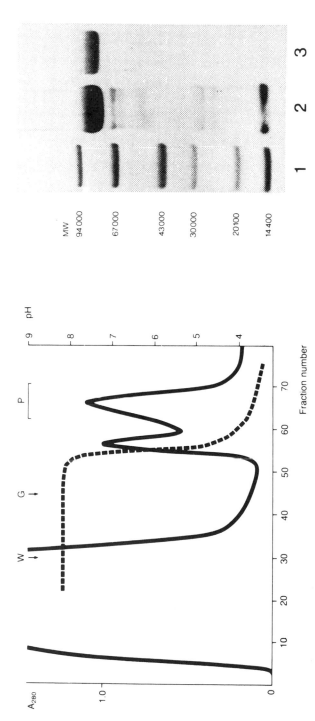

FIGURE 8-8. (Left) Isolation of lactoferrin by Cu^{2+} IMAC. Sample: 70 mL defatted, casein-free human milk equilibrated with the starting buffer. Column: C 10/20. Bed height: 16.3 cm, charged with Cu^{2+} to 10.3 cm from the top. Eluents: W, wash with starting buffer 50 mM Tris–acetate, pH 8.2, and 500 mM NaCl; G, development to final buffer 50 mM Tris–acetate and 500 mM NaCl, pH 2.8. Flow rate: 25 mL/h. Fraction size: 2.5 mL. (Right) Gradient gel electrophoresis in SDS. Gel: PAA 4/30. Lane 1, LMW calibration kit; lane 2, defatted casein-free milk; and lane 3, pooled material (work from Pharmacia Biotech AB).

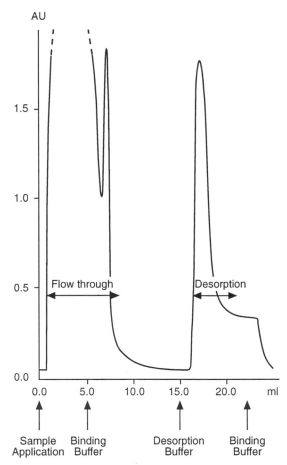

FIGURE 8-9. Purification of recombinant glutathione *S*-transferase fused to a hexahis-tidine peptide. See text for experimental details. The insert shows SDS gradient electro-phoreses of a low molecular weight calibration kit (lanes 1 and 7), starting material (i.e., cell extract used in the chromatography experiment) diluted 1 : 20 (lane 2), flowthrough fraction diluted 1 : 10 (lane 3), eluted glutathione S-transferase fraction diluted 1 : 5 (lane 4), diluted 1 : 10 (lane 5), nonrecombinant pure glutathione *S*-transferase, 0.2 mg/mL (lane 6) (work from Pharmacia Biotech AB).

Starting material and collected fractions were analyzed by SDS electrophoresis using PhastSystem® and PhastGel® Gradient 10–15 (Pharmacia Biotech AB). Silver staining was performed according to the manufacturer's standard protocol.

Buffers and solutions:

Metal salt   0.1 *M* NiSO$_4$ in distilled water.

Buffer A   20 m*M* sodium phosphate, 0.5 *M* NaCl, 20 m*M* imidazole, pH 7.4.

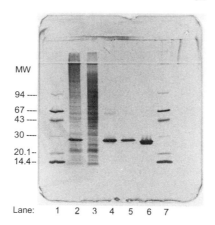

FIGURE 8-9. *(Continued).*

Buffer B   20 mM sodium phosphate, 0.5 M NaCl, 300 mM imidazole, pH 7.4.

Chromatography:

1   Washing of the column by elution with 5 mL distilled water.
2   Charging with Ni^{2+} ions using 0.5 mL of the salt solution.
3   Washing by elution with 5 mL distilled water.
4   Equilibration of the column with 10 mL of buffer A.
5   Application of 5-mL sample.
6   Removal of unbound or loosely bound material by elution with 10 mL of buffer A.
7   Desorption of GST-His$_6$ with 7 mL of buffer B.
8   Equilibration of the column with 5 mL of buffer A.

It is evident from Figure 8-9 that a high degree of purity of GST-His$_6$ was reached in only one step.

8.7   REFERENCES

1. J. Porath, J. Carlsson, I. Olsson, G. Belfrage, *Nature* (*London*), *258,* 598–599 (1975).
2. F. Hellferich, *Nature* (*London*), *189,* 1001 (1961).
3. V.A. Davankov, A.V. Semechkin, *J. Chromatogr., 141,* 313–353 (1977).
4. J. Porath and M. Belew, in *Affinity Chromatography and Biological Recognition,* I.M. Chaiken, M. Wilchek, and I. Parikh, Eds., Academic Press, San Diego, 173–190 (1983).
5. J. Porath, B. Olin, B. Granstrand, *Biochem. Biophys., 225,* 543–547 (1983).

6. B. Lönnerdal, C.L. Keen, *J. Appl. Biochem.*, *4*, 203–208 (1982).

7. E. Sulkowski, *Trends Biotechnol.* *3*, 1–7 (1985).

8. J. Porath, B. Olin, *Biochemistry*, *22*, 1621 (1983).

9. J. Porath, *Protein Expr. Purif.*, *3*, 263–281 (1992).

10. Z.E. Rassi, C. Horvath, in *HPLC of Biological Macromolecules: Methods and Applications*, K.M. Gooding, F.E. Regnier, eds., Chromatogr. Sci. Series, Marcel Dekker, Vol. 51, pp. 179–213 (1990).

11. G.S. Chaga, *Metal Ion Affinity for Purification and Immobilization of Enzymes*, Acta Univ. Ups., Comprehensive Summaries of Uppsala Dissertations from the Faculty of Science and Technology, 36, 1994.

12. S. Vunnum, S.R. Gallant, Y.J. Kim, S.M. Cramer, *Chem. Eng. Sci.*, *50*, 1785–1803 (1995).

13. R.A.A. Muzzarelli, A.F. Martelli, O. Tubertini, *Analyst*, *94*, 616–624 (1969).

14. L. Andersson, J. Porath, *Anal. Biochem.*, *154*, 250–254 (1986).

15. P.L. Roberts, C.P. Walker, P.A. Feldman, *Vox Sang.*, *67 (Suppl. 1)*, 69–71 (1994).

16. W.H. Scouten, Affinity chromatography, bioselective adsorption on inert matrices, in *Chemical Analysis, Vol. 59*, John Wiley & Sons, New York, 1981.

17. L. Fanou-Ayi, M. Vijayalakshmi, N.Y. Acad Sci., 413, 300–306 (1983).

18. F.B. Anspach, *J. Chromatogr. A, 672*, 35–49 (1994).

19. F.B. Anspach, *J. Chromatogr. A, 676*, 249–266 (1994).

20. F.B. Anspach, G. Altmann-Haase, *Biotechnol. Appl. Biochem.*, *20*, 313–322 (1994).

21. Y. Kato, K. Nakamura, T. Hashimoto, *J. Chromatogr.*, *354*, 511–517 (1986).

22. M. Gimpel, K. Unger, *Chromatography*, *17*, 200–204 (1983).

23. K.-A. Hansson, G. Moen, L. Kågedal, unpublished results.

24. A. Hammacher, U. Hellman, A. Johnsson, A. Östman, K. Gunnarsson, B. Westermark, A. Wasteson, C.-H. Heldin, *J. Biol. Chem.*, *263*, 16493–16498 (1988).

25. Z.E. Rassi, C. Horvath, *J. Chromatogr.*, *359*, 241–253 (1986).

26. G.C. Serafica, J. Pimbley, G. Belfort, *Biotechnol. Bioeng.*, *43*, 21–36 (1994).

27. G. Schmuckler, *Talanta*, *12*, 281–290 (1965).

28. Z. Horvath, G. Nagydiosi, *J. Inorg. Nucl. Chem.*, *37*, 767–769 (1975).

29. P. Smidl, J. Plicka, I. Kleinman, *J. Chromatogr.*, *598*, 15–21 (1992).

30. C.B.H. Loung, M.F. Browner, R.J. Fletterick, *J. Chromatogr.*, *584*, 77–84 (1992).

31. G. Muszynska, Y.J. Zhao, J. Porath, *J. Inorg. Biochem.*, *26*, 127–135 (1986).

32. L. Andersson, *J. Chromatogr.*, *315*, 167–174 (1984).

33. E. Hochuli, W. Bannwart, H. Döbeli, R. Gentz, D. Stuber, *Biotechnology*, Nov., 1321–1325 (1988).

34. M. Zachariou, I. Traverso, M.T.W. Hearn, *J. Chromatogr.*, *646*, 107–120 (1993).

35. L. Kågedal, unpublished data.

36. E. Sulkowski, K. Vastola, D. Oleszek, W. von Muenchhausen, in *Affinity Chromatography and Related Techniques*, Veldhoven, The Netherlands, June 22–26, 1981. T.C.J. Gribnau, J. Visser, R.F.J. Nivard, eds., Elsevier Scientific Publishing Company, 1982, pp. 313–322.

37. C.A.K. Borrebaeck, B. Lönnerdal, M.E. Etzler, *FEBS Lett., 130,* 194–196 (1981).

38. M. Conlon, R.F. Murphy, *Biochem. Soc. Trans., 4,* 860–861 (1976).

39. M. Yoshimoto, M. Laskowski, Sr., *Prep. Biochem., 12,* 235–254 (1981).

40. J. Porath, *Pure Appl. Chem., 51,* 1549–1559 (1979).

41. M.F. Scully, V.V. Kakkar, *Biochem. Soc. Trans., 94,* 335–336 (1981).

42. I. Ohkubo, T. Kondo, N. Taniguchi, *Biochim. Biophys. Acta, 616,* 89–93 (1980).

43. E. Bollin, E. Sulkowski, *Arch. Virol., 58,* 149–152 (1978).

44. A. Gozdzicka-Jozefiak, J. Augustyniak, *J. Chromatogr., 131,* 91–97 (1977).

45. E. Sulkowski, in *Protein Purification: Micro to Macro,* UCLA Symposia on Molecular and Cellular Biology New Series, Vol. 68, R. Burgess, ed., Alan R. Liss, 1987, pp. 149–162.

46. L. Andersson, E. Sulkowski, *J. Chromatogr., 604,* 13–17 (1992).

47. P. Hansen, G. Lindeberg, *J. Chromatogr., 690,* 155–159 (1995).

48. M.C. Smith, T.C. Furman, T.D. Ingolia, C. Pidgeon, *J. Biol. Chem., 263,* 7211–7215 (1988).

49. P. Loetscher, L. Mottlau, E. Hochuli, *J. Chromatogr., 595,* 113–119 (1992).

50. H. Berthold, M. Scanarini, C.C. Abney, B. Frorath, W. Northemann, *Protein Expr. Purif., 3,* 50–56 (1992).

51. A.F. Vosters, D.B. Evans, W.G. Tarpley, S.K. Sharma, *Protein Expr. Purif., 3,* 18–26 (1992).

52. M.W. Van Dyke, M. Sirito, M. Sawadogo, *Gene, 111,* 99–104 (1992).

53. C.A. Franke, D.E. Hruby, *Protein Expr. Purif., 4,* 101–109 (1993).

54. R.R. Beitle, M.M. Ataai, *Biotechnol. Prog., 9,* 64–69 (1993).

55. M. Clausen, C.J. Lamb, R. Megnet, P.W. Doerner, *Anal. Biochem., 212,* 537–539 (1993).

56. C.Y.E. Wu, L.C. Blaszczak, M.C. Smith, P.L. Skatrud, *J. Bacteriol., 176,* 1539–1541 (1994).

57. P.A. Walker, L.E.-C. Leong, P.W.P. Ng, S.H. Tan, S. Waller, D. Murphy, A.G. Porter, *Bio/Technology, 12,* 601–605 (1994).

58. R. Witzgall, E. O'Leary, J.V. Bonventre, *Anal. Biochem., 223,* 291–298 (1994).

59. G. Chaga, M. Widersten, L. Andersson, J. Porath, U.H. Danielson, B. Mannervik, Prot. Eng., 7, 1115–1119 (1994).

60. U. Schön, W. Schumann, *Gene, 147,* 91–94 (1994).

61. S.A. Doyle, D.R. Tolan, *Biochem. Biophys. Res. Commun., 206,* 902–908 (1995).

62. A. Seidler, *Prot. Eng., 7,* 1277–1280 (1994).

63. L.-O. Essen, A. Skerra, *J. Chromatogr. A, 657,* 55–61 (1993).

64. M. Kashlev, E. Martin, A. Polyakov, *Gene, 130,* 9–14 (1993).

65. T. Lu, M. Van Dyke, M. Sawadogo, *Anal. Biochem., 213,* 318–322 (1993).

66. P.A. Feldman, L. Harris, D.R. Evans, H.E. Evans, *Biotechnol. Blood Proteins, 227,* 63–68 (1993).

67. T.K. Chataway, G.J. Barritt, *Mol. Cell. Biochem., 144,* 167–173 (1995).

68. A. Skerra, I. Pfitzinger, A. Plückthun, *Bio/Technology, 9,* 273–278 (1991).

69. V. Boden, J.J. Winzerling, M. Vijayalakshmi, J. Porath, *J. Immunol. Methods, 181,* 225–232 (1995).

70. D. Peters, R. Frank, W. Hengstenberg, *Eur. J. Biochem.*, *228*, 798–804 (1995).

71. Unpublished work, Pharmacia Biotech AB.

72. M. Belew, T.-T. Yip, L. Andersson, J. Porath. *J. Chromatogr.*, *403*, 197–206 (1987).

73. T.W. Hutchens, T.-T. Yip, J. Porath, *Anal. Biochem.*, *170*, 168–182 (1988).

74. T.W. Hutchens, T.-T. Yip, *Anal. Biochem.*, *191*, 160–168 (1990).

75. G. Muszynska, G. Dobrowolska, A. Medin, P. Ekman, J. Porath, *J. Chromatogr.*, *604*, 19–28 (1992).

76. E. Sulkowski, *Macromol. Chem., Macromol. Symp.*, *17*, 335–348 (1988).

77. B. Lönnerdal, J. Carlsson, J. Porath, *FEBS Lett.*, *75*, 89–92 (1977).

78. A.R. Torres, E.A. Peterson, W.H. Evans, M.G. Mage, S.M. Wilson, *Biochim. Biophys. Acta*, *576*, 385–392 (1979).

79. H. Kikuxhi, M. Watanabe, *Anal. Biochem.*, *115*, 109–129 (1981).

80. P. Wunderwald, W.J. Schrenk, H. Port, G.-B. Kresze, *J. Appl. Biochem.*, *5*, 31–42 (1983).

81. T.E. Cawston, J.A. Tyler, *Biochem. J. 183*, 647–656 (1979).

82. D.C. Rijken, D. Collen, *J. Biol. Chem.*, *256*, 7035–7041 (1981).

83. D.C. Rijken, G. Wijngaards, M. Zaal-De Jong, J. Welbergen, *Biochim. Biophys. Acta*, *580*, 140–153 (1979).

84. R.A. Pixley, R.W. Coman, *Affin. Chromatogr. Thromb. Res.*, *41*, 89–98 (1986).

85. Y. Kato, T. Kitamura, K. Nakamura, A. Mitsui, Y. Yamasaki, T. Hashimoto, *J. Chromatogr.*, *391*, 395–407 (1987).

86. H. Hansson, L. Kågedal, *J. Chromatogr.*, *215*, 333–339 (1981).

87. K.C. Chadha, P.M. Grob, A.J. Mikulski, L.R. Davis, E. Sulkowski, *J. Gen Virol.*, 701–706 (1979).

88. I. Ohkubo, W. Sahashi, C. Nanikawa, K. Tsukada, T. Takeuchi, M. Sasaki, *Clin. Chim. Acta*, *157*, 95–102 (1986).

89. H.C. Ryley, *Biochem. Biophys. Res. Commun.*, *89*, 871–878 (1979).

90. D.D. Coppenhaver, *Methods Enzymol.*, *119*, 199–204 (1986).

91. J.M. Rowe, S.F. Henry, H.G. Friesen, *Biochemistry*, *25*, 6421–6425 (1986).

92. W.P. Michalski, *J. Chromatogr.*, *576*, 340–345 (1992).

93. P. Hubert, J. Porath, *J. Chromatogr.*, *198*, 247–255 (1980).

94. P. Hubert, J. Porath, *J. Chromatogr.*, *206*, 164–168 (1981).

95. P.L. Roberts, C.P. Walker, P.A. Feldman, *Vox Sang., 67 (Suppl. 1)*, 69–71 (1994).

96. M. Kashlev, E. Martin, A. Polyakov K. Severinov, V. Nikitorov, A. Goldfarb, *Gene*, 130, 9–14 (1993).

97. R.J. Weselake, S.L. Chesney, A. Petkau, A.D. Friesen, *Anal. Biochem.*, *155*, 193–197 (1986).

98. T. Kurecki, L.F. Kress, M. Laskowski, *Anal. Biochem.*, *99*, 415–420 (1979).

99. V.G. Edy, A. Billiau, P. De Somer, *J. Biol. Chem.*, *252*, 5934–5935 (1977).

100. J.W. Heine, M. De Ley, J. Van Damme, A. Billau, P. De Somer, *Ann. NY Acad. Sci.*, *350*, 364–373 (1980).

101. J.W. Heine, J. van Damme, M. De Ley, A. Billau, P. De Somer, *J. Gen. Virol.*, *54*, 47–56 (1981).

9 Covalent Chromatography

JAN CARLSSON

Biochemical Separation Centre
Uppsala University
Uppsala Biomedical Centre
Box 577, S-751 23 Uppsala, Sweden

FRANCISCO BATISTA-VIERA

C. de Bioquimica, F. de Quimica e
Instituto de Quimica, F. de Ciencias
Gral.Flores 2124. Casilla de Correo 1157
Montevideo, Uruguay

LARS RYDÉN

Department of Biochemistry
Uppsala University
Uppsala Biomedical Centre
Box 576, S-751 23 Uppsala, Sweden

Protein Purification: Principles, High-Resolution Methods, and Applications, Second Edition.
Edited by Jan-Christer Janson and Lars Rydén.
ISBN 0-471-18626-0.   © 1998 Wiley-VCH, Inc.

9.1   INTRODUCTION

Chromatographic techniques for the isolation of proteins and other biological macromolecules, such as ion-exchange and hydrophobic interaction chromatography, are based on noncovalent interactions between the molecules to be separated and the adsorbent. Covalent chromatography,[1–3] however, as indicated by the name, involves the formation (and breaking) of covalent bonds between the solute and the stationary phase (Figure 9-1). As is usually the case in the chromatography of biologically active proteins, specificity and mild conditions are required. These requirements can, as shown later, be fulfilled.

Potential sites for chemical reactions in proteins are primarily the amino acid side chains with their various functional groups. In particular the amino-

FIGURE 9-1. General scheme for covalent chromatography. A protein containing group Y reacts with a solid phase containing a reactive group X. After washing out nonbound proteins, the bound protein is released by a low molecular weight compound RY.

and carboxyl functions have been used for the immobilization of proteins to insoluble polymers.

In these cases the chemistry has been designed to give stable bonds and it is in practice difficult to release the immobilized protein without destroying it. The only functional group for which conditions are available for the formation of a stable covalent bond that can also be split under mild conditions is the thiol group. Covalent chromatography, which has already been discussed as a potential technique, has thus, with a few exceptions,[3] been applied to the isolation of thiol-containing substances.

Patchornik and co-workers[4,5] have, however, designed methods for the reversible covalent attachment of methionyl and tryptophanyl side chains to activated polyacrylamide matrices. The chemistry involved requires long incubation times and low pH (dilute acetic acid) and the methods have not received general attention. They would probably be useful for the isolation of peptides containing methionine or tryptophan in special cases.

This chapter deals exclusively with covalent chromatography on thiol-activated solid phases provided with reactive disulfides or disulfide oxides. Thiol groups frequently occur in proteins and often participate directly in their functions as enzymes, hormones, receptors, and so on.[6,7] Thiol groups in proteins vary in reactivity and can also be introduced chemically. Both circumstances make covalent chromatography a widely applicable technique.

To exploit fully the possibilities and to avoid some pitfalls of covalent chromatography, a thorough knowledge of the chemical properties of the thiol group is necessary. This chapter, therefore, gives a short description of the main types of reactions in which thiols and some of their derivatives take part, with an eye to their relevance for the practical application of the technique. In particular, their reactions with the so-called reactive disulfides and the disulfide oxides utilized in covalent chromatography and reversible protein immobilization are treated in some detail.

This chapter also briefly describes some of the properties of thiol groups in proteins from various biological sources. Special emphasis is placed on how the reactivities of thiol groups change with the conditions used.

It is hoped that these introductory sections make the main part of the chapter—the properties of the gel derivatives used in covalent chromatography and the principles and practice of the chromatography itself—easy to follow. The chapter is concluded with descriptions of a few practical applications.

9.2 CHEMICAL PROPERTIES OF THIOL GROUPS

9.2.1 Ionization, Oxidation, Metal Ligation, and Alkylation

Thiol groups are normally the most reactive groups found in proteins and they can participate in a large number of reactions.[6-10] This is due to the high nucleophilicity of the corresponding thiolate ions, which exist at reasonable concentrations at neutral to weakly alkaline pH values. Thiols easily oxidize to give disulfides (R—S—S—R), which are comparatively unreactive and important for the stabilization of the protein tertiary structure:

$$2 \text{ R—SH} \rightarrow \text{R—S—S—R (disulfide)} + 2 \text{ H}^+ + 2 \text{ e}^- \qquad (9\text{-}1)$$

The rate of oxidation increases with the increasing concentration of thiolate ion, that is, with increasing pH. At low pH (pH < 5) and in the absence of strong oxidizing agents, the thiol group is stable and can tolerate several days of incubation even in the presence of air at room temperature.

Thiols are avid ligands for heavy metal ions. In some cases, as in the cupric ion complex, the thiolate oxidizes rapidly in the presence of oxygen. The presence of chelating agents, such as EDTA, is thus advisable in situations where oxidation is a potential problem.

Other thiol complexes, such as zinc and even more so the mercury complex, are stable. The latter was used in some of the first applications of covalent chromatography,[11] where the gel contained an immobilized organic mercury compound.

Thiols can dissociate into a proton and a thiolate anion:

$$\text{R—SH} \rightarrow \text{R—S}^- + \text{H}^+ \qquad (9\text{-}2)$$

Their pK_a values generally lie in the range of 8 to 10.5. Thiolate-dependent reactions thus proceed readily at pH 8 and above, and at moderate speed in the pH range of 6 to 8. Some representative pK_a values are given in Table 9-1. The described pH dependence applies for low molecular weight alifatic thiols and many protein-bound thiols. The latter, however, occasionally show a completely different rate pH dependence due to microenvironmental effects.

Thiols react with oxidizing agents via either of two routes[10]: (1) the monomeric oxidation pathway, which proceeds via two electron steps and turns the thiols into sulfenic (R—S—OH), sulfinic (R—SO$_2$H), and sulfonic (R—SO$_3$H) acids or (2) the dimeric oxidation pathway, which converts the thiols to disul-

TABLE 9-1. pK_a Values for Some Low Molecular Weight Thiol Compounds

Compound	pK_{SH}
Glutathione	9.2
β-Mercaptoethylamine	8.3
β-Mercaptoethanol	9.5
β-Mercaptopropionate	10.3
β-Mercapto N,N'-diethylamine	7.8

Only the dissociation constant of the thiol proton is given.

fides (R—S—S—R), which can then be further oxidized to disulfide oxides [thiolsulfinates (R—SO—S—R) and thiolsulfonates (R—SO$_2$—S—R), respectively], and finally to sulfinic and sulfonic acids. The end products of the two routes are thus the same. Depending on the oxidizing agents used, the reactions take place by one of these two routes.

Alkylation that leads to a stable thioether can be used to block a thiol group permanently. In the most common procedure, iodoacetate or its amide is used and a carboxymethyl derivative is formed[12]:

$$I—CH_2—COO^- + R—SH \rightarrow R—S—CH_2—COO^- + HI \qquad (9\text{-}3)$$

This reaction, like other nucleophilic displacement reactions with thiols, involves the unprotonated form (i.e., the thiolate ion) and the rate therefore increases with increasing pH.

9.2.2   Thiol–Disulfide Exchange

Thiol–disulfide exchange is a special form of alkylation (S-alkylation). This reaction can easily be reversed and is used in covalent chromatography. It will therefore be discussed in some detail.

The reaction is a two-step nucleophilic displacement in which a mixed disulfide is formed as an intermediate:

$$R_1—S—S—R_1 + R—SH \rightarrow R_1—SH + R_1—S—S—R \qquad (9\text{-}4a)$$

$$R_1—S—S—R + R—SH \rightarrow R_1—SH + R—S—S—R \qquad (9\text{-}4b)$$

The sum of these two reactions is

$$R_1—S—S—R_1 + 2\,R—SH \rightarrow 2\,R_1—SH + R—S—S—R \qquad (9\text{-}4)$$

The reaction can also be described as a redox process, as the oxidation state of the sulfur atoms changes in the direction of greater electron deficiency in the disulfide.

As in the alkylation reactions, the rate increases with increasing pH. At physiological conditions (pH 7 and 25°C), rate constants up to 600 M^{-1} min^{-1} are typical for reactions involving alifatic thiols and disulfides. The corresponding equilibrium constants are usually near unity. In general, therefore, a large excess of thiol must be used to reduce a disulfide and, vice versa, a large excess of disulfide is required for the oxidation of a thiol.

Thiol–disulfide exchange is often utilized for the reduction of both intra- and intermolecular disulfides in proteins. When low molecular weight thiols are used, it is important to note that the necessary excess at a given pH increases with the pK_a of the thiol. A commonly used thiol is β-mercaptoethanol (pK_a 9.5):

$$R-S-S-R + 2\ HO-CH_2CH_2-SH\ (excess) \rightarrow 2\ R-SH$$
$$+\ HO(CH_2)_2-S-S-(CH_2)_2OH \quad (9\text{-}5)$$

When one of the two isomeric thiol–diols, dithioerythritol (DTE) or dithiothreitol (DTT), is used for the reduction, the situation is quite different. The mixed disulfides formed from these compounds and an alifatic disulfide are unstable and undergo subsequent internal thiol–disulfide exchange, leading to the formation of an internal disulfide in form of a stable six-membered ring, which drives the reaction toward completion. Thus DTE or DTT will reduce an alifatic disulfide in equimolar amounts.

$$(9\text{-}6a)$$

$$(9\text{-}6b)$$

It should also be mentioned that thiol–disulfide exchange reactions play an important role in biochemical processes, such as the biosynthesis of proteins, the aggregation and polymerization of some proteins, possibly the regulation of the activities of some intracellular enzymes, effector–receptor interactions, and membrane transport.

9.2.3 Reactions with Reactive Disulfides

A particularly interesting thiol–disulfide exchange is that between an alifatic thiol and a so-called reactive disulfide (Figure 9-2). These are compounds in which the corresponding thiol forms are stabilized either by resonance or by thiol–thione tautomerism. This results in decreased nucleophilicity and a correspondingly low reactivity toward disulfides. Reactive disulfides can be either homogeneous or mixed. In the first case, the two halves of the molecule are identical. In the mixed types, one of the halves $(R-S-S-)$ is derived from an alifatic thiol (RSH, where R denotes the alifatic group).

An example is the reaction between an alifatic thiol and the reactive disulfide 2,2′-dipyridyldisulfide (2-PDS):

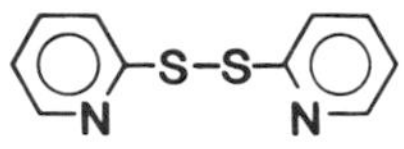
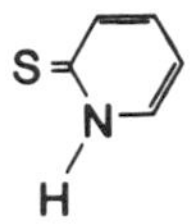

(Ia) (Ib)

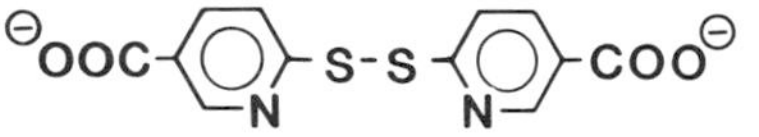
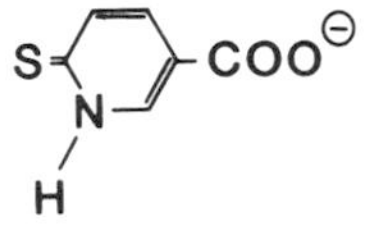

(IIa) (IIb)

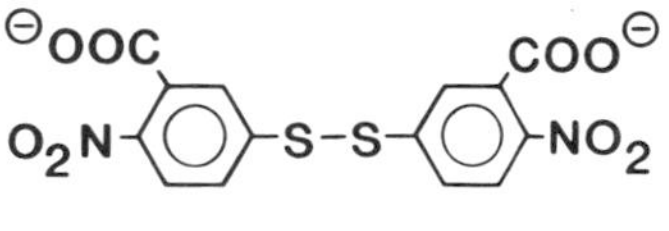
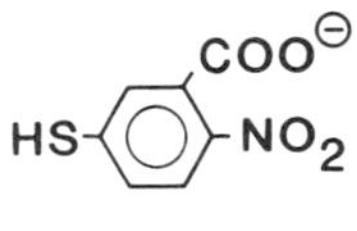

(IIIa) (IIIb)

FIGURE 9-2. Reactive homogeneous disulfides with their corresponding reduced forms: (Ia) 2,2′-dipyridyldisulfide, (Ib) 2-thiopyridone (λ_{max} 343 nm $\varepsilon = 8.08 \times 10^3$ M^{-1} cm $^{-1}$); (IIa) 5-carboxy-2-pyridyldisulfide; (IIb) 5-carboxy-2--thiopyridone (λ_{max} 344 nm, $\varepsilon = 10^4$ M^{-1} cm^{-1}); and (IIIa) 5,5′-dithiobis(2-nitrobenzoate), (IIIb) 2-nitro-5-mercaptobenzoate (λ_{max} 412 nm, $\varepsilon = 14.14 \times 10^3$ M^{-1} cm^{-1}).[3]

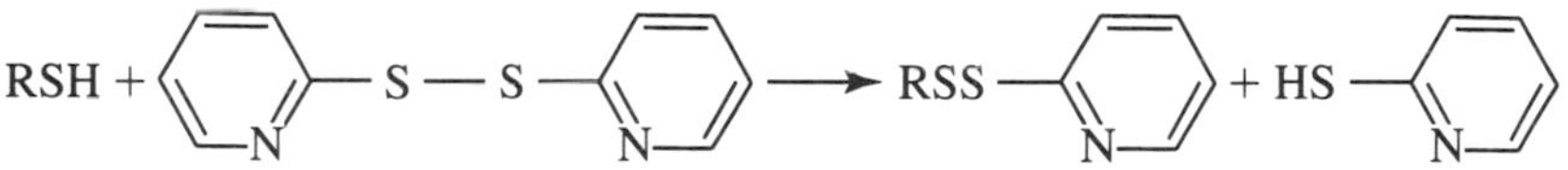

(9-7a)

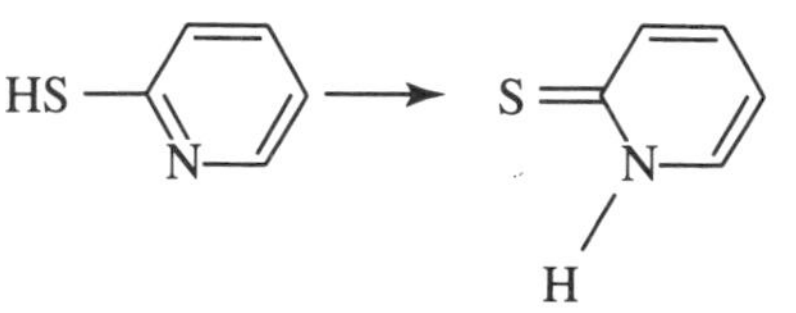

(9-7b)

thiol-form thione-form

In contrast to the thiol–disulfide exchange reaction involving an alifatic disulfide, where the equilibrium lies toward the "middle," these reactions are driven essentially to completion by formation of the thione form (K_{eq} around 3×10^4). An alifatic thiol can thus be quantitatively transformed into a mixed reactive disulfide concomitantly with the formation of equimolar amounts of thione.

If a mixed reactive disulfide is used, the disulfide formed will be an ordinary alifatic one:

$$\text{RSH} + \text{R}_1\text{S} - \text{S} - \bigcirc_N \longrightarrow \text{RSSR}_1 + \text{HS} - \bigcirc_N \qquad (9\text{-}8a)$$

$$\text{HS} - \bigcirc_N \longrightarrow \text{S} = \bigcirc_{\underset{H}{N}} \qquad (9\text{-}8b)$$

thiol-form   thione-form

Both reactions [Eq. (9-7) and Eq. (9-8)], which involve 2-thiopyridyl compounds, proceed at pH values where alifatic thiol–disulfide exchanges are slow or nonexistent. In fact, thiol–disulfide exchange with 2-pyridyldisulfides can be carried out at pH values in the range of 1 to 9. The low nucleophilicity of the alifatic thiol at acidic pH is compensated for by an increased electrophilicity of the 2-pyridyldisulfide as a result of the protonation of its ring nitrogen (pK around 3). It should be noted that in a mixed 2-pyridyldisulfide the sulfur atom on the alifatic side is the more electrophilic and will thus react with the incoming thiol sulfur to form the new disulfide, whereas the 2-thiopyridyl structure will leave as a thione. These facts explain the advantageous properties of the 2-thiopyridyl structure as a leaving group and why the binding step in covalent chromatography can proceed at low pH.

The thiol–disulfide exchange described with a mixed pyridyldisulfide produces an alifatic disulfide that is stable toward simple alifatic monothiols in excess at low pH. However, if DTT is used in the exchange, the intermediate form will rearrange to form the corresponding cyclic disulfide [Eq. (9-6b)] even at a low pH. The combined properties of the thiopyridyl compounds and DTT thus allow the formation of a new thiol over a wide pH range.

The properties of reactive disulfides have been exploited in a number of applications. The homogeneous types have been used for the titration of thiol groups in proteins.

The released thiones can be quantified easily by spectrophotometry, as their absorption maxima lie above the region where ordinary proteins absorb, and the molar absorptivity is above 8,000 (Figures 9-2 and 9-3). The most important application of the mixed type of reactive disulfide is in covalent chromatography. This is discussed in greater detail below.

9.2.4   Reactions of Thiols with Disulfide Oxides

Thiols react specifically and quantitatively with alifatic disulfide oxides (thiolsulfinates and thiolsulfonates)[10] at neutral or slightly alkaline pH. The reaction

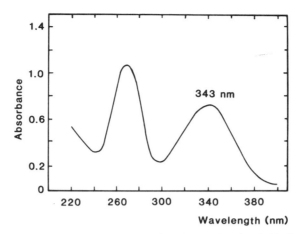

FIGURE 9-3. Spectrum of the reduced form (2-thiopyridone) of 2,2'-dipyridyldisulfide. The absorption of 343 nm fades at higher pH due to the loss of a proton (pK_a 9.5).

is faster when the incubation pH is around and above the pK_a value of the thiol group, as thiolate is the reacting species.

The comparatively stable thiolsulfonate reacts with 1 mol of thiol to give a mixed disulfide and a sulfinic acid in a thiol–disulfide-like exchange reaction [Eq. (9-9a)]. Thiolsulfinates, however, turn 2 moles of thiol per mole into two mixed disulfides when the sulfenic acid formed in a first reaction step oxidizes a second mole of thiol[13] [Eq. (9-9b)]

$$R-SH + R_1-SO_2-S-R_2 \rightarrow R-S-S-R_2 + R_1-SO_2H \quad (9\text{-}9a)$$

$$2\ R-SH + R_1-SO-S-R_2 \rightarrow R-S-S-R_2 +$$
$$R_1-S-S-R + H_2O \quad (9\text{-}9b)$$

These reactions also have practical applications. Alifatic thiolsulfonates [such as methylmethane-thiolsulfonate (MMTS)] have been widely used for the reversible blocking of thiol groups in thiol-dependent enzymes[14] and aromatic thiolsulfonates for the titration of thiol groups in proteins.[15]

9.3 THIOL-CONTAINING PROTEINS

9.3.1 Redox State in Biological Tissues

In tissues, the redox state within the cells is entirely different from that in the extracellular matrix and fluid. Inside cells a number of reactions are critically dependent on free thiol groups. Coenzyme A and lipoic acid are examples of essential thiol compounds. The intracellular environment must thus be kept in a reduced state. This is achieved by the cysteine-containing tripeptide glutathione (GSH) (Figure 9-4), which accounts for about 90% of all thiols in the cell and is present in total concentrations in the millimolar range. Glutathione serves as a scavanger of free radicals, oxidizing compounds, and other molecules that react with thiol groups.

The high concentration of reduced glutathione in the cell is maintained by new production through glutathione synthetase and by reduction of the disulfide form [oxidized glutathione (GSSG)]. The reduction is catalyzed by the NADPH-requiring enzyme, glutathione reductase. In an erythrocyte the ratio GSH/GSSG is about 500.

However, the extracellular environment is rich in oxygen, which is incompatible with the presence of exposed thiol groups.

9.3.2   Intra- and Extracellular Proteins

An overview of intracellular proteins,[6,7] which have been studied with regard to their content of thiol groups, shows that most of them have exposed thiol groups, whereas very few, if any, contain disulfide bridges. In many instances the thiol groups participate in the catalytic reactions and the alkylation of these groups leads to complete inactivation of the enzyme. This is in particular often the case of oxidoreductases and transferases. Sometimes the thiol groups serve as ligands to metal ions, as in alcohol dehydrogenase.

For extracellular proteins the reverse situation applies.[6,7] Thiol groups are exceptions and occur only when they are required for a special purpose, whereas disulfide bridges are common. Some examples are the extracellular hydrolases, such as the serine proteases trypsin, chymotrypsin, and several of the blood coagulation factors. All have several disulfide bridges and no thiol group. However, some intracellular proteases, such as the cathepsins, depend on thiols for their activity.

Small extracellular proteins, such as venom neurotoxins and the protease inhibitors, are especially rich in disulfide bridges, which are required for stabilization of the structure.[16] Similarly, many peptide hormones contain disulfides but none has a thiol group.

Exceptions to the rule are the plant proteases ficin, bromelain, and papain, which are extracellular thiol-dependent proteases. In these the thiol group occurs in an active site pocket where it is somewhat protected from oxidation. The mammalian plasma protein albumin sometimes carries a free thiol group (mercaptalbumin). This thiol group can trap thiol- or disulfide containing compounds in serum under the formation of mixed disulfides.

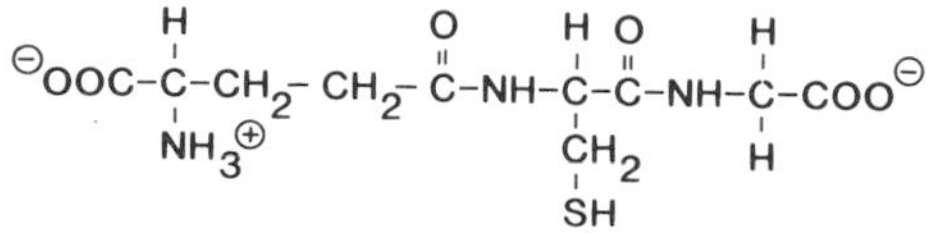

FIGURE 9-4. Structure of glutathione (γ-glutamyl-cysteinyl-glycine).

9.3.3 Demasking Protein Thiol Groups

The reactivity of a protein thiol depends on the conditions. Many proteins contain thiols that are buried in an hydrophobic environment and denaturants are needed for reaction with a hydrophilic reagent. In native human hemoglobin, two thiols per molecule react, whereas an additional four are unreactive.[17] The known three-dimensional structure of the protein indicates that these four are partly buried. In aldolase, one thiol per subunit reacts readily, but an additional three groups react if a small amount of detergent is added.[18] Still higher concentrations of detergent are required to abolish the enzymatic activity. In the native form of the copper enzyme ceruloplasmin, present in mammalian serum, no thiol group is accessible, but a slight modification— apparently the "nicking" of a single peptide bond—exposes one thiol.[19] This occurs without loss of activity.

In many cases, complete unfolding of the protein is necessary to allow all thiol groups to react. For example, in ceruloplasmin an additional three thiol groups are reactive in 8 M urea or 6 M guanidine containing EDTA to trap released copper.[19] This is also an example of a protein in which the thiols are liganded to a metal ion.

9.3.4 Reduction of Protein Disulfides

Thiols can be created in proteins by reducing disulfide bonds. This is usually done under conditions where all disulfides present will be reduced (use of denaturants and large excess of reducing reagent), but sometimes one can find conditions under which one or a few disulfide links can be reduced selectively. This is particularly true for interchain disulfides. In the lectin ricin, consisting of two peptide chains, and the immunoglobulins, containing two identical halves each with two peptide chains, the interchain disulfide links can be reduced selectively, without destroying the tertiary structure of the proteins, by a reducing agent such as DTT.[20,21]

In some cases an internal disulfide can be reduced without destroying the gross conformation or even the activity of the protein. One of the disulfides in bovine pancreatic trypsin inhibitor was reduced specifically by an equimolar amount of DTT in the absence of urea.[22] Most often one finds, however, that the reduction of intrachain bridges proceeds cooperatively in a zipper fashion.

9.3.5 Introduction of Thiol Groups

One can introduce thiol groups into proteins, thereby making them available to covalent chromatography. Several reagents are available for these thiolation reactions. The most commonly used is N-acetylhomocysteine thiolactone, which reacts with protein amino groups.[23] Reagents have also been designed for the introduction of protected thiol groups. One of them is N-succinimidyl-3-(2-pyridyldithio)propionate (SPDP).[24]

Protein — NH$_2$ + [pyridyl ring with N] — S — S — CH$_2$ — CH$_2$ — C($=$O) — O — N [succinimide ring with two O] →

(SPDP)

Protein — NH — C($=$O) — CH$_2$ — CH$_2$ — S — S — [pyridyl ring with N] + HO — N [succinimide ring with two O]   (9-10)

The 2-pyridyldisulfide introduced by this reagent is a mixed reactive disulfide and can be reduced under mild conditions, under which protein–disulfides are not split, by the use of an equimolar amount of DTT (an alifatic monothiol is not suitable, as this leads to the formation of a mixed disulfide). The 2-thiopyridone liberated can be determined spectrophotometrically. The pyridyldisulfide-substituted protein can also be immobilized on a thiol gel.

9.3.6   Derivatization of Thiol Groups

Thiol groups can be converted either to a blocked form (e.g., by alkylation) or to a reactive form by treatment with a reactive homogeneous disulfide. This chapter describes the latter of these possibilities, as it has been used to activate thiol proteins for covalent chromatography on a thiol matrix (see below).

The reaction that one wants to achieve is illustrated in Eq. (9-7). It is important to use an excess of 2,2'-dipyridyldisulfide as otherwise the newly activated thiol groups on the protein will be able to react with neighboring thiols to form disulfide links, either internally or intermolecularly, to give dimers or polymers of the protein.

The reaction can be performed either in ordinary buffers or in the presence of strong denaturants, in which case all thiols will become activated. The pH and other conditions are as described earlier. The reaction with a thiolated gel is discussed in a later section.

9.4   GELS FOR COVALENT CHROMATOGRAPHY

9.4.1   Principles

The most important application of the mixed reactive disulfides is as reactive groups in covalent chromatography. If a solid phase with a mixed reactive disulfide is incubated with a thiol-containing molecule, the latter becomes attached to the solid phase as a result of the thiol–disulfide exchange reaction. The disulfides shown in Figure 9-2 have been used with success for this purpose. Several others are likely to work as well. However, in most of the work reported so far, 2-pyridyldisulfide-substituted solid phases have been used. The principle of this type of chromatography is shown in Figure 9.5.

More recently, solid phases containing disulfideoxide groups (thiolsulfonates and thiolsulfinates) have been developed as alternative adsorbents for the reversible immobilization of thiols[25–27] (Figure 9-6).

Because of displacement of the electrons around the two sulfur atoms, these groups show high S reactivity. Thiol-containing molecules react with the more electrophilic of the two sulfur atoms (the unoxidized one) and become, as a result, immobilized to the solid phase by disulfide bonds. The pH range for thiol coupling is wide (5 to 8), but for most thiols the reaction seems to proceed faster at pH > 7.

Contrary to the case with the mixed reactive disulfides, the leaving groups (sulfinic or sulfenic acid) remain attached to the support (Figure 9-6).[26,27]

As with the reactive disulfide containing solid phases, thiol molecules immobilized on disulfide oxide solid phases can easily be released and recovered by reduction with alifatic low molecular weight thiols.

Any support material that can be properly derivatized is potentially usable. In fact, cross-linked dextran, agarose, cellulose, polyacrylamide, porous glass, and various silica derivatives have all been used as matrices in covalent chromatography. The same requirements as in other types of chromatography still apply: good packing and flow properties, low nonspecific adsorption, and so on. As in most affinity methods, beaded agarose gels are often preferred. They can be derivatized without changing their excellent properties for chro-

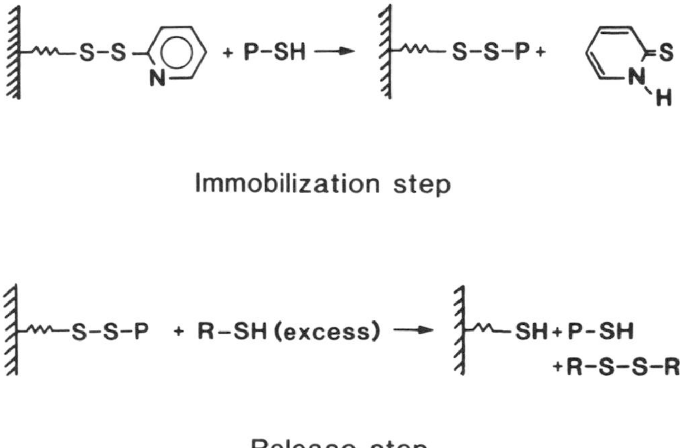

Immobilization step

Release step

FIGURE 9-5. Principle of 2-pyridyldisulfide-based covalent chromatography.

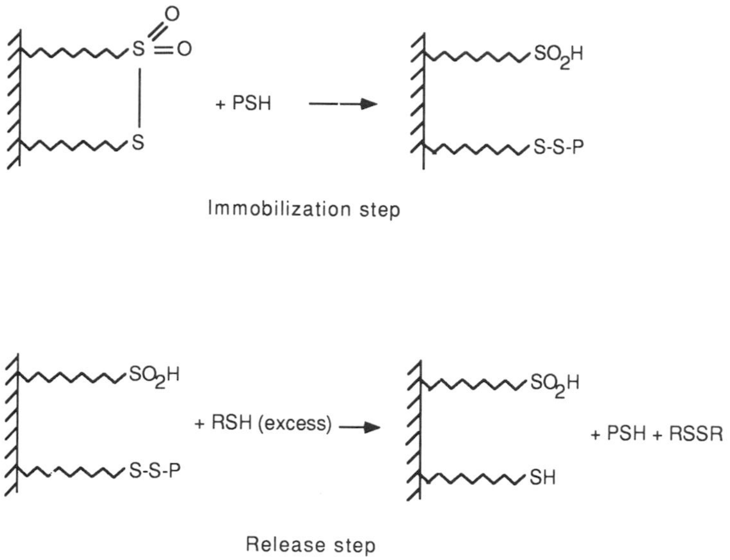

Immobilization step

Release step

FIGURE 9-6. Principle of thiolsulfonate-based covalent chromatography.

matography of proteins. They can also be transferred into nonpolar solvents without shrinking, which is a valuable property in some instances.

9.4.2 Introduction of Reactive Disulfides or Disulfide Oxides into Gels

This is usually done by first introducing thiol groups into agarose gels and subsequently converting the gel-bound thiols into 2-pyridyldisulfide or thiolsulfinate/thiolsulfonate groups. An early method for the introduction of thiol groups into agarose gels was the coupling[1,28] of glutathione to a cyanogen bromide-activated agarose gel:

$$\vdash O - C \equiv N + H_2N - \text{glutathione} - SH \longrightarrow$$

activated matrix

$$\vdash O - \overset{\displaystyle \overset{NH_2^+}{\|}}{C} - NH - \text{glutathione} - SH \qquad (9\text{-}11)$$

The presence of two carboxylate functions on the glutathione tripeptide (Figure 9-4) make the "arm" that links the thiol to the matrix negatively charged.

A second way to obtain a thiol-substituted agarose starts with the introduction of oxirane structures into the gel by the use of epichlorohydrin or bisepoxides.[29] The oxirane structures (the three-membered rings) are reacted with thiosulfate to form a Bunte salt derivative ($R-S-SO_3^-$), which is subsequently reduced to a thiol by an excess of a low molecular weight thiol.

$$\vdash OH + ClCH_2 - CH - CH_2 \longrightarrow \vdash O - CH_2 - CH - CH_2 \qquad (9\text{-}12)$$

matrix   epoxyactivated gel

$$\vdash O - CH_2 - CH - CH_2 + Na_2S_2O_3 \longrightarrow \vdash O - CH_2 - CH - CH_2 - SSO_3 \quad (9\text{-}13)$$

$$\vdash O - CH_2 - CH - CH_2 - SSO_3 + RSH \longrightarrow \vdash O - CH_2 - CH - CH_2 - SH$$

OH   OH

thiopropyl gel   (9-14)

In contrast to glutathione, the hydroxypropyl spacer contains no charged structures.

The methods described can also be used to introduce thiol groups into a number of other hydroxyl group-containing matrices. Thiol gels prepared as described earlier can be activated through different ways:

1. Converting gel-bound thiol groups into a mixed reactive disulfide, for example, by reacting them with an excess of 2,2'-dipyridyldisulfide [Eq. (9-15)].

$$\vdash O - CH_2 - CH - CH_2 - SH + \text{(pyridyl)} - S - S - \text{(pyridyl)} \longrightarrow$$

OH

$$\vdash O - CH_2 - CH - CH_2 - S - S - \text{(pyridyl)} + S = \text{(pyridyl)}$$

OH   H

activated thiopropyl gel   (9-15)

2-Pyridyldisulfide-based gels (either containing glutathione or 2-hydroxypropyl spacers) are stable and can be stored as suspensions at cold room temperature (+4°C) and pH 6 to 7 for several months. The use of sodium azide as a bacteriostatic should be avoided. The azide is a good nucleophile and reacts with the pyridyl disulfide group to form an immobilized labile sulfenylazide and thiopyridone, thereby consuming the active structures. In the presence of proper additives, such as dextran and galactose, the agarose derivatives can be lyophilized.

2. Creating disulfide oxide moieties. Oxidation of thiopropyl-agarose with hydrogen peroxide at acidic pH for 20 to 30 h converts thiol groups on the support (via disulfide and thiolsulfinate) into thiolsulfonate moieties[26] (Figure 9-7).

If gel-bound thiolsulfinates are preferred, the thiopropyl-agarose should be subjected to a two-step oxidation procedure.[27] The gel is first reacted with potassium ferricyanide at neutral pH. The formed disulfide moieties are then converted into gel-bound thiolsulfinate groups by controlled oxidation with a stoichiometric amount of monoperoxyphtalate at pH 5 to 7 (Figure 9-8).

The stability of thiolsulfinate/thiolsulfonate groups toward hydrolysis, especially at neutral and weakly acidic pH, is very high. In fact, solid phases containing these groups can be stored as suspensions at pH 5 and 4°C for at least 6 months, without a decrease in their thiol-binding capacity.

Contrary to the reactive disulfides, solid-phase disulfide oxides do not react with sodium azide in concentrations of 0.1% often used to prevent bacterial growth in gels.

9.4.3   Degree of Substitution and Practice Capacity of Gels

The degree of substitution of solid-phase, thiol-reactive groups obtained can differ considerably (Table 9-2) from one derivative to another and depends, to a large extent, on the original thiol content of the solid phase. Solid phases with both low and high thiol content can be prepared with the oxirane approach, depending on what excess of oxirane compound is used, whereas only a relatively low degree of thiol substitution can be obtained through the

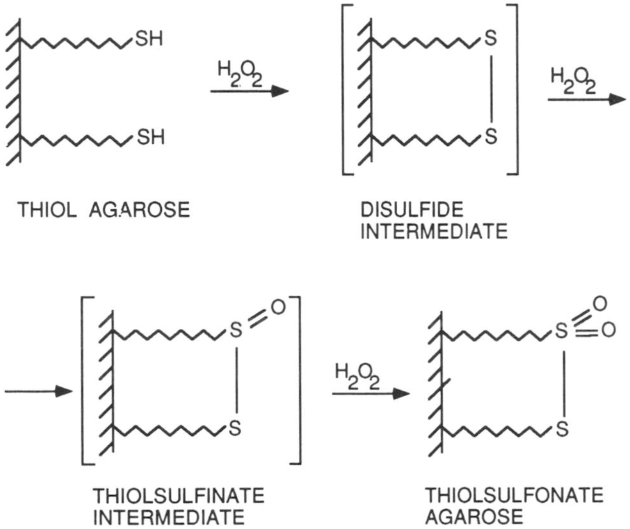

FIGURE 9-7. Preparation of thiolsulfonate agarose.

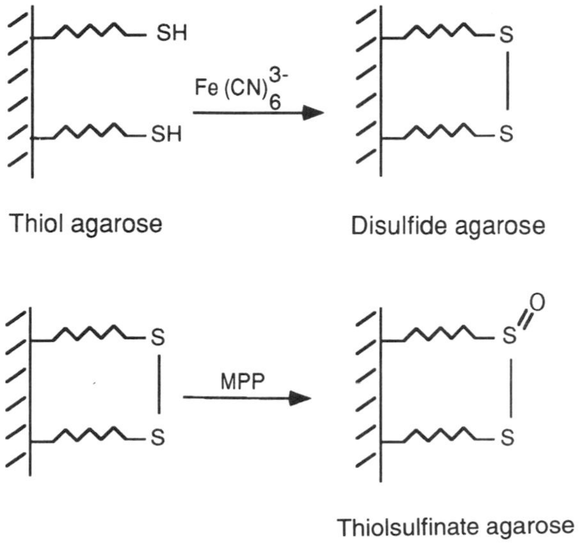

Thiol agarose   Disulfide agarose

Thiolsulfinate agarose

FIGURE 9-8. Preparation of thiolsulfinate agarose.

cyanogen bromide-based method. As the solid-phase, bound thiol groups can be more or less quantitatively converted into reactive disulfides[3] or disulfide oxide groups,[26,27] the final content of thiol-reactive groups is dependent on the original solid-phase thiol content, which can differ up to 100-fold between different derivatives. This, however, does not imply that the thiol-binding capacities of the solid phases will differ by the same factor. The figure for

TABLE 9-2. Capacities of Different Thiol-Reactive Gels

Gel Type	Degree of Substitution (μmol/mL gel)	Practical Capacity
2-Pyridyldisulfide-glutathione-agarose	1	2–3 mg HSA/mL gel (0.05 μmol/mL gel)
2-Pyridyldisulfide-hydroxypropyl-agarose	20	14 mg ceruloplasmin/mL gel (0.10 μmol/mL gel)
Thiolsulfonate-agarose	10	11 mg thiolated BSA/mL gel (0.17 μmol/mL gel)
Thiolsulfinate-agarose	17	12 mg thiolated BSA/mL gel (0.18 μmol/mL gel)
Thiolsulfinate-agarose	17	16 mg reduced Bence–Jones chains/mL gel (0.64 μmol/mL gel)

The degree of substitution refers to the content of gel-bound 2-pyridyldisulfide (2-Py$-$S$-$S$-$) or disulfide oxides ($-$SO$-$S$-$, $-$SO$_2$$-S-$) structures reactive toward low molecular weight thiol compounds. The practical capacity refers to the amount of protein that can be bound to the gels.[26,27,30] the theoretical thiol-binding capacity of a certain derivative is obtained by incubating an aliquot of it with a small excess of a low molecular weight thiol compound (e.g., glutathione) and, after thorough washing, quantifying the amount immobilized. This can be done easily by direct determination of the amount of released 2-thiopyridone by spectrophotometry for 2-pyridyldisulfide-based gels or by back titration (e.g., with 2,2'-dithiopyridine of unreacted free thiol in the case of disulfide oxide containing solid phases).

The practical capacity obtained in chromatography is in general much lower than the degree of substitution of thiol-reactive groups, especially for a highly substituted gel (Table 9-2). This is due to the fact that the binding of high molecular weight substances such as proteins is limited by the space available on the polymer rather than by the concentration of active groups. The amount of gel needed for a specific application should ideally be worked out in a pilot experiment.

9.4.4   Comparison of 2-Pyridyldisulfide and Disulfide Oxides as Solid-Phase Reactive Groups

Agarose-bound thiolsulfinate (disulfide monoxide) and thiolsulfonate (disulfide dioxide) groups show, as does the 2-pyridyldisulfide group, very high specificity for reaction with thiols.[25–27]

Their reactivity if the groups are available is high. Mercaptoalbumin, which contains a buried SH group, does not bind at all on thiolsulfonate-agarose but does to some degree to thiolsulfinate-agarose.[26,27]

This is different from 2-pyridyl disulfide-based agarose gels which bind virtually all thiol compounds capable of reaction with low molecular weight thiol-titrating reagents in solution. This property of thiolsulfonate-agarose can possibly be used for the separation of proteins with exposed thiol groups from those with buried thiols.

At high ionic strength the 2-pyridyldisulfide gels bind some proteins lacking thiol groups, especially immunoglobulins, through a noncovalent interaction. This has been ascribed to the so-called thiophilic adsorption interaction and has been used in chromatography.[31]

Disulfide oxides do not exhibit such thiophilic adsorption properties.

Perhaps the most important difference between disulfide oxides and 2-pyridyldisulfide is that when the former structures are used as solid-phase reactive groups, there is no release of any low molecular weight compound as a result of the immobilization of a thiol, as the formed sulfenic or sulfinic groups remain bound to the matrix. The reaction of thiol compounds with agarose-bound 2-pyridylsulfide groups, however, leads to the release of 2-thiopyridone. A disulfide oxide gel is thus the obvious choice in applications where unwanted thiols are to be removed from a solution.

However, the course of the immobilization on disulfide oxide gels can not be followed spectrophotometrically as no chromofore is released.

Solid-phase disulfide oxides have about the same stability as a function of pH as the corresponding 2-pyridyldisulfide derivatives. An additional advantage is that thiolsulfinate and thiolsulfonate gels are also stable in the presence of sodium azide.

The thiolsulfinate-agarose as the 2-pyridyldisulfide-agarose can, at least in theory, be regenerated an unlimited number of times. This is not possible with thiolsulfonate-agarose due to the formation of gel-bound, nonreducible sulfinate groups. This gel loses about 50% of its thiol-binding capacity after each cycle of use and regeneration and can thus in practice only be used a few times.

9.5 CHROMATOGRAPHIC TECHNIQUES

The recommendations and protocols presented in this section are focused on 2-pyridyldisulfide agarose derivatives as chromatographic material but should also be applicable for the disulfide oxide agarose derivatives after adaptation to the special characteristics of these derivatives, which are discussed in Section 9.4.4 (see also Refs. 25–27).

9.5.1 Preparatory Experiments

The authors recommend analyzing the thiol content of the sample for covalent chromatography in advance. This is to make sure that the capacity of the gel is not exceeded. Very often biological samples contain low molecular weight thiols such as glutathione. These should be removed by dialysis or gel filtration before the covalent chromatography is performed.

Thiol titration is conveniently done by spectrophotometric determination of the 2-thiopyridone released when a small amount of the sample (1 to 5 mg in 1 to 3 mL) reacts with 2,2'-dipyridyldisulfide (0.1 to 0.2 mM). The conditions can be chosen to suit the sample in question. Buffers between pH 3 and 8 (formate, acetate, phosphate, and Tris) in the concentration range from 0.05 to 0.4 M, with or without strong denaturants such as 8 M urea or 6 M guanidine–HCl, can be used. Under standard conditions (pH 7.5), a reaction time of a few minutes is usually enough for complete reaction. The addition of EDTA (5 mM) is recommended to trap transition metal ions, which might catalyze oxidation of the thiol groups.

A similar sample can also be reacted with the thiol-activated gel. In the case of activated thiopropyl agarose, 0.3 g of a swollen and equilibrated gel dried by aspiration on a filter is incubated batchwise with the same amount of protein as described earlier in 2 to 3 mL in a small tube that is closed and rotated end over end for the prescribed time (about 1 h). The tube is then centrifuged, and a spectrum is run on the supernatant. The amount of protein thiols bound by the gel can be calculated from the absorbance at 343 nm after the background level has been subtracted. It is important to run a spectrum and not just to read at a single wavelength because the background absorbance is sometimes considerable.

A small amount of the thiopyridone liberated upon incubation of the activated gel is sometimes released by mechanisms other than the binding of protein.[32] This "leakage" is of the order 0.02% of the active structures per hour in a buffer without denaturants, but is higher when high concentrations of urea or guanidine are present. The release corresponds to an absorbance at 343 nm of 0.004 per hour at pH 4.

The results of the preliminary experiments are used to determine proper conditions for the chromatography and the practical binding capacity of the gel. In many cases, especially with high molecular weight thiol proteins, this is not more than about 1% of the active structures present in a highly substituted 2-pyridyldisulfide-agarose gel (Table 9-2).

9.5.2   Binding of Sample Proteins

The coupling of the sample to the gel can be performed either batchwise (i.e., by suspending the gel in the sample solution) or columnwise (i.e., by letting the sample pass through a column packed with the gel equilibrated with the chosen buffer). The packing and dimensions of the column are not critical. It is, however, often better to use a long, thin column than a short, wide one and to adjust the flow rate such that the sample is in contact with the gel for the time chosen (at least 1 h in the standard procedure). The absorbance at 343 nm of the effluent can be used to estimate the amount of thiol groups that have reacted with the gel. If the reading at 280 nm is used to estimate the amount of nonbound protein, the contribution from released thiopyridone (Figure 9-3) has to be subtracted. The absorbance of 2-thiopyridone at 280 and 343 nm is roughly equal.

9.5.3   Washing

Unbound and nonspecifically adsorbed proteins should be washed off the column. The choice of washing buffers depends on the stability and intended use of the sample that is covalently bound to the column. Normally, a high ionic strength buffer is recommended to neutralize charge interactions (buffer containing 0.1 to 0.3 M NaCl). In the simplest case the same buffer is used for application, washing, and elution. If necessary, washing with detergents such as Triton and Tween can also be included. The washing should be monitored by measuring the absorbance of the effluent; 1 to 2 column volumes is usually sufficient. The washing operation is completed by equilibrating the column with the buffer to be used for the reductive elution.

· If the coupling is done batchwise, the washing can be performed either on a glass filter or in a column, the latter procedure being most convenient for a small amount of gel.

9.5.4 Reductive Elution

Reductive elution is normally done at pH 8. The low molecular weight thiol used to reduce the bound sample, as well as residual thiopyridyl structures, is mostly either 10 to 25 mM DTT or 25 to 50 mM 2-mercaptoethanol. If cysteine is used, it must be remembered that its oxidized form, cystine, which is less soluble than its reduced form, and may easily precipitate in the column after some time. As described previously, the elution can be followed by measuring the absorbance of the effluent at 343 nm.

Because the thiopyridyl groups are reduced much more easily, it is possible to avoid the contamination of eluted proteins by thiopyridone released from nonused groups on the gel by doing two separate reductive elutions. In the first the residual thiopyridyl structures are removed by an equimolar amount of reducing agent. After appropriate washing, the bound protein is then released by an excess of thiol. The first step can be performed at either alkaline or acidic pH.

In some cases, a series of solutions of thiols of different reducing power has been used to achieve the specific release of bound proteins (see Section 9.6.3).[33]

In the case of solid-phase disulfide oxides, recovery of the immobilized thiol molecule can be performed by reductive elution with alifatic low molecular weight thiols (e.g., 50 to 100 mM DTT in 0.1 M sodium phosphate buffer, pH 8.0).

9.5.5 Recovery of Thiol Proteins

When disulfide oxide agarose has been used as the chromatographic material, the bound thiol protein is the only compound released from the gel when it is treated with an excess of a low molecular weight thiol, as the simultaneously formed sulfenic or sulfinic acid groups are bound to the gel (Section 9.4.4). The released protein is eluted together with the excess of the reducing agent (and possibly a small amount of corresponding disulfide). In the case of 2-pyridyldisulfide agarose, a considerable amount of 2-thiopyridone, emanating from excess gel-bound reactive groups (if not removed in a separate step as described above), is also found in the eluate. The low molecular weight compounds should preferably be separated from the proteins before further handling, such as derivatization of thiol groups, activity measurement, and so on. This is done most easily by a gel filtration step. If it is possible to use a low pH buffer, the risk of oxidation of thiols is minimized. If the solutes are low molecular weight peptides rather than proteins, desalting is still possible by the use of a slightly hydrophobic gel, such as Sephadex LH-20, at an acidic pH. The peptides will then elute in the void volume, whereas thiopyridone, mercaptoethanol disulfide, and salts are retarded (Figure 9-9).[32]

The eluted material can also be recovered by lyophilization if appropriate volatile buffers are used. This is convenient if the eluted sample is recovered in a large volume (which is often the case). When lyophilization is to be performed, the authors recommend removing the unused thiopyridyl struc-

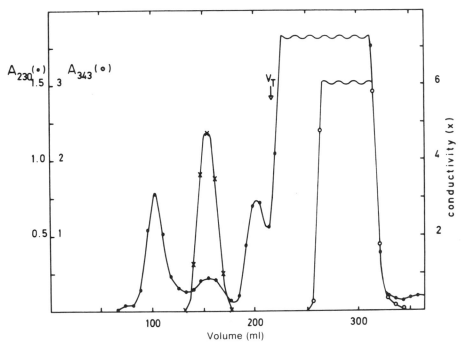

FIGURE 9-9. Chromatography of 20 mL of eluate from the reductive elution of an 11-mL bed of activated thiopropyl-agarose containing coupled thiol peptides from human ceruloplasmin on a column (3.2 × 27 cm) of Sephadex LH-20. Fractions of 7.3 mL were collected and analyzed for absorption at 230 nm (●), 343 nm (○), and conductivity (×). The fractions in the first peak (elution volume 100 mL), which contained the thiol peptides, were pooled and lyophilized. The Tris buffer eluted at about 150 mL, and the components with elution volumes of about 200, 240, and 290 mL are believed to be mercaptoethanol, mercaptoethanol disulfide, and thiopyridone, respectively. (Reproduced from Ref. 32 with permission.)

tures in a special reductive elution step as described earlier. If 2-mercaptoethanol and a volatile buffer salt such as ammonium acetate are used, all low molecular weight compounds will evaporate in the lyophilization step.

9.5.6 Reactivation of Thiol Gels

After chromatography the gel is in its thiol form and has to be reactivated before it can be reused. This is best done in a batchwise fashion, after washing the gel on a glass filter funnel. It is first incubated for 45 min with a 5 mM solution of DTT in 0.1 M sodium phosphate, pH 7.5, to reduce all alifatic disulfides, which might have been formed in the elution step, to thiols. Oxidized and excess reduced DTT and other low molecular weight thiols are then removed by washing with buffer.

The following steps differ depending on whether the thiol-reactive structure is a 2-pyridyldisulfide or a disulfide oxide. In the former case, the gel is incubated for 45 min with a saturated (1.5 mM) solution of 2,2′-dipyridyldisulfide in 0.1 M phosphate buffer, pH 8.0. With high-capacity gels (thiol content > 100 μmol/g dry derivative), it is necessary to use a 20 mM solution of 2,2′-dipyridyldisulfide to obtain complete reactivation. In this case, the reaction and the subsequent washing are done in buffer containing 20 to 30% ethanol in order to ensure that the reagent is dissolved. Excess reagent is finally removed by extensive washing. The reactivated gel can be stored in the same way as the fresh gel.

The presence of DTT in the eluates is conveniently checked by thiol titration with reagents such as 2,2′-dipyridyldisulfide or 5,5′-dithiobis(2-nitrobenzoate). Similarly, the 2,2′-dipyridyldisulfide can be assayed by adding a small amount of thiol compound followed by absorbance measurement at 343 nm.

In the case of a thiolsulfonate gel, it is practical to reactivate it only a few times (as mentioned earlier), at least twice, however, depending on the thiol group content of the starting thiopropyl-agarose.[26] Thiolsulfinate-agarose, however, can be regenerated, at least in theory, an unlimited number of times.[27] The recommended procedures for the original first synthesis and the reactivation of the reactive disulfide gels (in all cases starting from the thiol gels) are as follows:

For the thiolsulfonate gel, suction-dried thiol-agarose (15 g) is suspended in 0.2 M sodium acetate, pH 5.0 (45 mL). Hydrogen peroxide (30%) is added in aliquots under continuous shaking, 1.8 mL initially and 2.2 mL each after 30, 90, and 150 min. The incubation is then continued to give a total reaction time of 30 h. The oxidized gel is transferred to a sintered glass filter and washed with 0.1 M acetic acid until it is free from hydrogen peroxide. The activated gel is then equilibrated and stored in 0.2 M sodium acetate, pH 5.0.

For the thiolsulfinate gel, 15 g of suction-dried thiol-agarose is suspended in 30 mL 0.1 M sodium phosphate buffer, pH 7.0, and 0.1 M potassium ferricyanide is added in 1.0 mL-aliquots while shaking until the yellow color persists for at least 30 min. The gel is then thoroughly washed on a sintered glass filter with buffer: 1 M NaCl and 0.2 M sodium acetate, pH 5.0. The gel is then suspended in 30 mL of pH 5.0 buffer containing the required stoichiometric amount of dissolved magnesium monoperoxyphtalate (0.5 mol/mol −S−S− groups, 0.5 mol of the magnesium salt corresponds to 1 mol of monoperoxyphtalate).

The suspension is incubated under shaking for 2 h at room temperature. The gel derivative is then thoroughly washed with 50 mM sodium acetate buffer, pH 5.0, 0.1 M acetic acid, and 0.2 M sodium acetate buffer, pH 5.0, and is finally stored as a suspension in this buffer.

9.5.7 Chromatography of Activated Proteins or Peptides

Covalent chromatography can also be carried out in the reverse fashion to that just described. The sample is then treated with a homogeneous reactive disulfide and subsequently run on a gel containing immobilized thiol groups. In this approach it is important to remove excess reagent (e.g., by gel filtration) after the activation step.

The coupling should be done with a low-capacity gel to minimize unwanted thiol–disulfide exchange reactions on the gel, which could lead to immediate release of the newly linked protein. Such side reactions can be diminished by performing the coupling at a slightly acidic pH where alifatic thiol–disulfide exchange is minimized. The ensuing steps in the procedure are the same as in the conventional approach.

This reversed covalent chromatography is particularly useful for the isolation of peptides obtained by the proteolytic digestion of large proteins (see below).

9.6   APPLICATIONS

Covalent chromatography has found its most important use in the separation of thiol-containing molecules from nonthiols at an early stage in the fractionation of complex protein mixtures. Under certain conditions, a higher degree of specificity can be obtained, sometimes even allowing different thiol-containing molecules to be separated from each other. Moreover, the technique can be used to concentrate thiols from solutions in which they are present at very low concentrations. Another important application of thiol-reactive adsorbents is for the reversible immobilization of enzymes via disulfide bonds. Thiol enzymes containing exposed thiol groups can be immobilized directly, but if the thiol groups are buried or absent they can be provided through mild thiolation procedures (e.g., by use of SPDP, see earlier discussion).

These applications and others are illustrated by the following examples.

9.6.1   Isolation of Urease from Jack Bean[34,35]

Urease is an enzyme that catalyzes the hydrolysis of urea into ammonia and carbon dioxide. It is a thiol-rich protein consisting of six identical subunits noncovalently associated to an aggregate of molecular weight 500,000. Many of the thiol groups are nonessential and can be modified without loss of the urease activity.

The starting material was Jack bean meal, which is commercially available. The meal (60 g) was mixed with 300 mL of 0.05 M Tris–HCl buffer containing 36% ethanol, 0.1 M KCl, and 1 mM EDTA, pH 7.2. The mixture was stirred for 5 min at 28°C and filtered. The filtrate was centrifuged (500 g, 20 min). The supernatant (about 210 mL) was diluted to 300 mL with 0.05 M Tris-HCl buffer, pH 7.2, and the pH was adjusted to 7.2 by 0.5 M NaOH.

· A column with a total volume of 6.3 mL (1 × 8 cm) was prepared from activated thiopropyl-agarose, a high-capacity gel of the type described in Table 9-2. The column was equilibrated with 0.05 M Tris-HCl buffer, pH 7.2, containing 0.1 M KCl and 1 mM EDTA. The Jack bean meal extract (250 mL) was passed through the column at a flow rate of 20 mL/h. Most of the UV-absorbing material passed through the column unretained. The first 150 mL of eluate contained no urease activity. The activity gradually increased to that of the applied sample within the next 100 mL of eluate. The column was washed with the Tris-HCl buffer until the absorbance at 280 nm of the eluate was less than 0.04. The urease activity was released by eluting the column with 20 mM dithiothreitol (20 mL) dissolved in 0.05 M Tris-HCl buffer, pH 8, containing 0.1 M KCl and 1 mM EDTA.

The specific urease activity (units/mg dry material) of the eluted material (after the removal of low molecular weight substances on Sephadex G-25) was 167 times that of the starting material. This figure increased to 280 after an additional gel filtration on Sepharose 6B, which removed some high molecular weight material of low specific activity.

The capacity of the activated thiopropyl-agarose to bind urease active material was 5.1 mg/mL gel. The purified urease preparation was stable for several weeks when stored at 4°C. Before the reductive elution step the column could hydrolyze urea very efficiently when a solution of the substrate was passed through it. The column was thus an effective urease reactor.

9.6.2 Purification of Papain[28]

Papain is a protease that occurs in the latex of the tropical fruit *Carica papaya*. It is a single-chain protein with a molecular weight of about 23,500. Papain has a single thiol group that is essential for its activity. The following procedure worked both with a commercial crystallized papain preparation and an ammonium sulfate precipitate of dissolved dried papaya latex.

In the preparation based on the commercial enzyme, 200 mg of papain was dissolved in 0.1 M Tris-HCl, pH 8, containing 5 mM DTT. The reduction activated the enzyme by converting blocked active-site cysteines to the thiol form. Excess DTT was then removed on Sephadex G-25 equilibrated with 0.1 M Tris-HCl at pH 8 or 0.1 M sodium acetate at pH 4, both containing 0.3 M NaCl and 1 mM EDTA. The void material (usually containing 0.4 to 0.6 mol of thiol per mole of protein) in 100 mL of the Tris or acetate buffer was used as the sample for covalent chromatography. The preparation based on crude papain was obtained by dissolving 100 g of dried papaya latex containing 0.1 to 0.2 mol of thiol per mole protein in about 200 mL of either of the previously mentioned buffers.

Either sample was applied on a column (1.8 × 30 cm) of activated thiol-agarose (a low capacity gel, see Table 9-2). The column was eluted with the application buffer until the absorbance of the eluate at both 280 and 343 nm was less than 0.03. The gel was then equilibrated with 0.1 M Tris-HCl, pH 8, containing 0.3 M NaCl and 1 mM EDTA. The papain was eluted from the column with 50 mM L-cysteine in the same buffer. Fractions were read at 280 and 343 nm and were tested for esterolytic activity toward *N*-benzoyl-

L-arginine ethyl ester (BAEE). The fractions corresponding to the activity peak were pooled, and the protein was precipitated by the addition of 30 g of $(NH_4)_2SO_4$ per 100 mL of solution. The precipitate was redissolved in a minimum volume of pH 8 buffer, and the protein was separated from low molecular weight material by chromatography on Sephadex G-25 in 0.1 M KCl containing 1 mM EDTA. To prevent the formation of papain-L-cysteine mixed disulfide during the gel filtration, DTT was added immediately before application to the Sephadex G-25 column to a final concentration of 5 mM. Using these conditions, about 100 mg of pure papain with a thiol content of 1 mol per mole of protein was obtained.

The chromatography described earlier at pH 8 is an example of the separation of proteins on the basis of the presence of an exposed thiol. The specificity is even more pronounced at pH 4 when, due to microenvironmental effects, the thiol group of native papain reacts much faster with 2-pyridyldisulfide groups than does the thiol of denatured papain or low molecular weight compounds.[28] When a mixture of fully active papain (0.1 mM) and L-cysteine (up to approximately 5 mM) was subjected to covalent chromatography at pH 4, all of the papain reacted with the mixed disulfide gel and essentially all of the L-cysteine passed through the column.

9.6.3   Sequential Elution Covalent Chromatography[33]

Hillson succeeded in separating the two enzymes—protein disulfide isomerase (PDI) and protein disulfide oxidoreductase, also called glutathione-insulin transhydrogenase (GIT)—involved in the *in vivo* formation of protein disulfides by the reductive sequential elution of a thiopropyl-agarose column on which they had been immobilized. The starting material was a sample of partially purified protein disulfide isomerase from beef liver, which, apart from PDI, also contained several thiol-oxidoreductase activities. Samples were prepared by two different techniques, one involving ion-exchange chromatography (partially purified preparation) and the other ammonium sulfate precipitation (crude preparation).

The protein samples (25 to 125 mg at 10 mg/mL) were pretreated with 0.1 mM dithiothreitol in 50 mM Tris-HCl, pH 7.5, 25 mM KCl, 5 mM MgCl$_2$/ 1.25 mM EDTA/0.1 M NaCl (TKM/EDTA/NaCl buffer) at 30°C for 30 min. This gentle reduction unmasks any buried thiol groups that may be present as mixed disulfide. After reduction the sample was centrifuged and separated from DTT on a Sephadex G-25 column (2 × 25 cm) using the previously mentioned buffer. The reduced protein was then applied directly on a column of activated thiopropyl-agarose (30 to 45 g wet weight of high-capacity gel). The elution was then interrupted for 30 to 60 min at 30°C to allow binding to occur.

Alternatively, the sample was incubated batchwise with the gel for 16 h with gentle shaking, after which the gel was poured into a column. Batch incubation gave a higher level of coupling and is therefore the preferred sample loading method.

After application the column was cooled to 4°C and washed with TKM/EDTA/NaCl buffer to remove unbound and nonspecifically adsorbed protein. Bound proteins were then displaced from the gel by successive elution with different low molecular weight thiol compounds used in order of increasing reducing power: 20 mM L-cysteine, 50 mM glutathione, and 20 mM DTT, each in TKM/EDTA/NaCl buffer, pH 7.5. In each step one void volume of the reducing buffer was run into the column, which was then incubated at 30°C for 30 min to allow reaction to occur. Elution was then continued at 4°C at flow rates of 2 to 10 mL/h, followed by a wash with TKM/EDTA/NaCl buffer to rinse the column. Fractions of 5 mL were collected and monitored for protein at 280 nm and for displaced 2-thiopyridone at 343 nm.

In each step, fractions containing protein were pooled and solid DTT was added to a final concentration of 5 mM. The pooled fractions were incubated at 25°C for 30 min to reduce any mixed disulfide formed between protein and eluent, and the solution was then dialyzed extensively against TKM buffer, pH 7.5, at 4°C. The procedure resulted in preparations of four protein fractions (i.e., unbound protein washed through the column and material displaced by L-cysteine, glutathione, and DTT, respectively).

The breakthrough peak contained both PDI and GIT, probably due to overloading or incomplete reaction. PDI activity was found only in the cysteine fraction, with no associated GIT activity. The glutathione fraction showed no detectable activities except for a small amount of GIT in one run, and finally the DTT fraction contained GIT activity but no detectable PDI activity.

The degree of purification and the yields differed significantly, depending on whether the starting material was a partially purified (ion exchange) or a crude protein mixture (ammonium sulfate precipitate). In the latter case, the percentage yields of enzyme activities in the bound fractions were 70 to 98% of PDI activity in the cysteine fraction and 89 to 100% of GIT activity in the DTT fraction. Sequential elution covalent chromatography proved to be a powerful tool for the rapid isolation and separation of protein disulfide isomerase and protein disulfide oxidoreductase, which had not been achieved previously.

9.6.4 Purification of Thiol Peptides[36,37]

Covalent chromatography affords facile group-specific isolation of thiol peptides from protein digests. Svensson and co-workers isolated peptides from papain digests of reduced ribonuclease and of mercaptalbumin. Their approach was to react the proteins with 2,2′-dipyridyldisulfide before digestion and then to apply the activated peptide mixture to a column of reduced thiopropyl-agarose. The coupling to the column was carried out with 0.2 M ammonium acetate, pH 8.0. The reductive elution was done with the same buffer containing 50 mM 2-mercaptoethanol. The eluted peptide mixture was lyophilized, as only volatile buffer substances and mercaptoethanol had been used. The single thiol peptide from albumin was obtained directly in pure form.

In the procedure of Ryden and Norder,[23,29] the protein, human ceruloplasmin, was immobilized to the column via its thiol groups before the protease was added. This allowed the use of large amounts of protease and permitted the use of two proteases (pepsin and trypsin) sequentially to obtain subfragments of the original peptides. In these experiments, the reaction with the column was done at pH 4.0 in sodium acetate buffer and elution was done with Tris-HCl, pH 8.0, containing 50 mM mercaptoethanol. The thiol peptides eluted were purified further by gel filtration and HPLC. The recoveries in these experiments were around 70% for the coupling step.

Covalent chromatography is clearly the method of choice for the rapid isolation of thiol peptides from large proteins, which normally give very complex peptide mixtures upon digestion. Serum albumin contains 589 residues and human ceruloplasmin 1046.

9.6.5   Reversible Immobilization of β-Galactosidase (*Escherichia coli*)[38]

β-Galactosidase (lactase) catalyzes the hydrolysis of lactose into glucose and galactose, a reaction of great importance from a nutritional and technological point of view. The enzyme is used in the dairy industry to hydrolyze whey lactose, thereby solving a waste-handling problem. The glucose formed can then be fermented into more useful products such as ethanol.

Escherichia coli β-galactosidase is an oligomeric enzyme with a high content of cysteine (64 cysteine residues per tetramer). About one-fourth of them react with thiol reagents without affecting the enzymatic activity. It has been found that β-galactosidase can be immobilized on thiolsulfonate-agarose under mild conditions through its nonessential, exposed thiol groups.[38]

In the following experiment, β-galactosidase grade VIII from *E. coli* (Sigma, St. Louis, MO) and thiolsulfonate-agarose (with a degree of substitution of 300 μmol per gram dried gel) were used.

Aliquots of 2 g of suction-dried thiolsulfonate-agarose were incubated under gentle agitation for 16 h at 4°C with 7.5 mL of β-galactosidase solutions (containing 0.9 to 46 mg protein, respectively) in 0.1 M potassium phosphate, pH 7.0. The insoluble derivatives obtained were thoroughly washed with buffer and stored as suspensions at 4°C. Enzymatic activities of the conjugates were assayed with o-nitrophenyl-β-D-galactopyranoside in 0.1 M potassium phosphate, pH 7.5, and 3 mM MgCl$_2$. Active and stable conjugates with various contents of enzyme (8 to 114 mg protein per gram dried derivatives) were obtained. Thiolsulfonate-agarose gels bound the active enzyme selectively. The specific activity of immobilized β-galactosidase was up to 50% higher than that of the applied, commercial enzyme. The residual solid-phase thiolsulfonate groups could be reacted with various low molecular weight thiol compounds, with each compound giving rise to a specific microenvironment for the immobilized enzyme. Each derivative had different properties. Thus, blocking with glutathione led to improvements in the thermal stability for the immobilized β-galactosidase derivatives. These derivatives also showed excellent long-term stabilities; no decrease in enzymatic activity was noticed after storage for 18 months at 4°C in 0.1 M potassium phosphate, pH 7.0.

The immobilized enzyme was quantitatively released with 50 mM DTT in 0.1 M sodium phosphate, pH 8.5, which confirmed that binding was due to disulfide formation.

9.6.6 Characterization of Subunit Proteins[39,40]

The covalent chromatography of thymidylate synthase on activated thiopropyl-Sepharose 6B has been described.[39] This enzyme is in its native form a dimer with identical subunits, each containing a cysteine residue at the active site. The cysteine thiol groups did not react directly with activated thiol Sepharose 4B, but immobilization was performed via a splitable mixed disulfide bond introduced between the thiol-reactive solid phase and the catalytic thiol group(s) of the enzyme. The immobilization procedure allowed the isolation of enzymes with high specific activity from pools of pure thymidylate synthase, the activity of which had declined during storage. The decrease in activity took place even when pure thymidylate synthase was stored in the presence of 2-mercaptoethanol. By the covalent chromatography procedure, it was possible to separate the protein into two fractions, one with higher and the other with lower thymidylate synthase activity. The two fractions also showed distinct biochemical and biophysical properties.

The ability of the native enzyme to bind to the thiol-reactive adsorbent through more than one mixed disulfide bond was also studied. By restricting the length of the incubation time of the immobilization reaction, it was possible to limit the proportion of enzyme binding through more than one thiol group. Following this strategy, native thymidylate synthase could be immobilized via only one of its subunits, thus leaving the remaining active site thiol on the other subunit free and available for reaction with N-ethylmaleimide. The heterodimer was then eluted from the solid phase with 50 mM 2-mercaptoethanol. This novel approach, using covalent chromatography and selective chemical modification, is proposed as a new tool to study the subunit interactions involved in the catalytic and regulatory mechanisms of certain oligomeric proteins.[40]

9.7 REFERENCES

1. K. Brocklehurst, J. Carlsson, M.J.K. Kierstan, E.M. Crook, *Methods Enzymol. 34,* 531–544, Academic Press, New York, L.A.AE. Sluyterman, J. Wijdenes, *ibid.* pp. 544–547. A Ruiz-Carrillo, *ibid.* pp. 547–552. J.C. Nicolas, *ibid.* pp. 552–554. H.F. Voss, Y. Ashani, I.B. Wilson, *ibid.* pp. 581–591 (1974).
2. V.I. Lozinskii, S.V. Ragozhin, *Russian Chem. Rev.,* 49(5), 460 (1980).

3. K. Brocklehurst, J. Carlsson, M.J.K. Kierstan, in *Topics of Enzyme and Fermentation Biotechnology,* Vol. 10. A. Wiseman, ed. Ellis Harwood Limited, Chichester, 1985, pp. 146–188.

4. M. Rubinstein, Y. Schechter, A. Patchornik, *Biochem. Biophys. Res. Commun., 70*(4), 1257 (1976).

5. Y. Schechter, M. Rubinstein, A. Patchornik, *Biochemistry 16*(7), 1424 (1977).

6. M. Friedman, *The Chemistry and Biochemistry of the Sulfhydryl Group in Amino Acids, Peptides and Proteins.* Academic Press, London, 1972.

7. P.C. Jocelyn, *Biochemistry of the SH Group.* Academic Press, London, 1972.

8. S. Patai, *The Chemistry of the Thiol Group,* Parts 1 and 2. John Wiley & Sons, London, 1974.

9. Y.M. Torchinsky, *Sulfur in Proteins.* Pergamon Press, New York, 1981, pp. 6 and 7.

10. R.J. Huxtable, *Biochemistry of the Sulfur.* Plenum Press, New York, 1986, pp. 207–208.

11. L. Eldjarn, E. Jellum *Acta Chem. Scand., 17,* 2610 (1963).

12. C.H.W. Hirs, in *Methods Enzymol. 11,* Academic Press, New York, 1967, p. 199.

13. D. Barnard, E. Cole, *Anal. Chim. Acta, 20,* 540 (1959).

14. T.W. Bruice, G.L. Kenyon, in *Methods Enzymol, 47,* Academic Press, New York, 1977, p. 407.

15. J. Carnevale, K. Healey, *Anal. Chim. Acta, 140,* 143 (1982).

16. M.O. Dayhoff, *Atlas of Protein Sequences and Structure,* Vol. 5, National Biochemical Research Foundation, Washington, DC, 1978.

17. R. Cecil, M.A.W. Thomas, *Nature (London) 206,* 1317 (1965).

18. J.M. Nicolau, M. Bacila, *Arch. Biochem. Biophys., 129,* 357 (1969).

19. L. Rydén, D. Eaker, *FEBS Lett., 53,* 279 (1975).

20. S. Olsnes, A. Pihl, *Biochemistry 12,* 3121 (1973).

21. J.B. Fleishman, R.M. Pain, R.R. Porter, *Arch. Biochem. Biophys. Suppl. 1, 174* (1961).

22. W.K. Liv, J. Meienhofer, *Biochem. Biophys. Res. Commun., 31,* 467 (1968).

23. F.H. White, Jr., in *Methods Enzymol.,* Vol. 25 B, Academic Press, New York, 1972, p. 541.

24. J. Carlsson, H. Drevin, R. Axén, *Biochem. J., 173,* 723 (1978).

25. J. Carlsson, F. Batista-Viera *Biotechnol. Applied Biochem., 14,* 114 (1991).

26. F. Batista-Viera, M. Barbieri, K. Ovsejevi, C. Manta, J. Carlsson, *Appl. Biochem. Biotechnol., 31,* 175 (1991).

27. F. Batista-Viera, C. Manta, J. Carlsson, *Appl. Biochem. Biotechnol., 44,* 1 (1994).

28. K. Brocklehurst, J. Carlsson, M.P.J. Kierstan, E.M. Crook, *Biochem. J., 133,* 573 (1973).

29. R. Axén, H. Drevin, J. Carlsson, *Acta Chem. Scand., B29,* 471 (1975).

30. In Pharmacia Biotech AB Booklet. *Affinity Chromatography: Principles and Methods,* pp. 36–37.

31. S. Oscarsson, J. Porath, *Anal. Biochem., 176,* 330 (1989).

32. L. Rydén, H. Norder, *J. Chromatogr., 215,* 341 (1981).

33. D.A. Hillson, *J. Biochem. Biophys. Meth., 4,* 101 (1981).

34. J. Carlsson, R. Axen, K. Brocklehurst, E.M. Crook, *Eur. J. Biochem.,* **44,** 189 (1974).

35. J. Carlsson, I. Olsson, R. Axen, H. Drevin, *Acta Chem. Scand., B30,* 180 (1976).

36. A. Svensson, J. Carlsson, D. Eaker, *FEBS Lett., 73,* 171 (1977).

37. L. Rydén, D. Eaker, *Eur. J. Biochem., 132,* 241 (1983).

38. K. Ovsejevi, B. Brena, F. Batista-Viera, J. Carlsson. *Enzyme Microb. Technol., 17,* 151 (1995).

39. T.M. Bradshaw, R.B. Dunlap, *Biochim. Biophys. Acta, 1163,* 165 (1993).

40. T.M. Bradshaw, R.B. Dunlap, *Biochemistry, 32,* 12774 (1993).

10 Affinity Chromatography

JAN CARLSSON

Biochemical Separation Centre
Uppsala University
Uppsala Biomedical Centre
Box 577, S-751 23 Uppsala, Sweden

JAN-CHRISTER JANSON

Biochemical Separation Centre
Uppsala University
Uppsala Biomedical Centre
Box 577, S-751 23 Uppsala
Sweden

MARIANNE SPARRMAN

BioIndustry Division
Pharmacia Biotech AB
S-751 82 Uppsala, Sweden

Protein Purification: Principles, High-Resolution Methods, and Applications, Second Edition.
Edited by Jan-Christer Janson and Lars Rydén.
ISBN 0-471-18626-0.   © 1998 Wiley-VCH, Inc.

10.1 INTRODUCTION

All biological processes depend on specific interactions between molecules. These interactions might occur between a protein and low molecular weight substances (e.g., between substrates or regulatory compounds and enzymes; between bioinformative molecules—hormones, transmittors, etc.—and receptors, and so on), but biospecific interactions occur even more often between two or several biopolymers, particularly proteins. Examples can be found from all areas of structural and physiological biochemistry, such as in multimolecular assemblies, effector–receptor interactions, DNA–protein interactions, and antigen–antibody binding. Affinity chromatography (see Refs. 1–3 for general and earlier references) owes its name to the exploitation of these various biological affinities for adsorption to a solid phase. One of the members of the pair in the interaction, the ligand, is immobilized on the solid phase, whereas the other, the counterligand (most often a protein), is adsorbed from the extract that is passing through the column. Examples of such affinity systems are listed in Table 10-1.

In many cases, affinity chromatography is a very powerful method. This is particularly true when the protein of interest is a minor component of a complex mixture. The extraction of the vitamin B_{12} transport protein transcobalamin II from blood serum is given as an impressive example of the purification of 10 mg of active protein from 40 kg of plasma in a simple two-step procedure using a column with immobilized cobalamin as the ligand.[4]

The term *affinity chromatography* has been given quite different connotations by different authors. Sometimes it is very broad, including all kinds of adsorption chromatographies based on nontraditional ligands, in the extreme all chromatographies except ion exchange. Often it is meant to include immobilized metal ion affinity chromatography (IMAC), covalent chromatography, hydrophobic interaction chromatography, and so on. In other cases it refers only to ligands based on biologically *functional pairs,* such as enzyme–inhibitor complexes. This chapter uses the term not only to include functional pairs

TABLE 10-1. Examples of Biological Interactions Used in Affinity Chromatography

Ligand	Counterligand
Antibody	Antigen, virus, cell
Inhibitor	Enzyme (ligands are often substrate analogs or cofactor analogs)
Lectin	Polysaccharide, glycoprotein, cell surface receptor, membrane protein, cell
Nucleic acid	Nucleic acid-binding protein (enzyme or histone)
Hormone, vitamin	Receptor, carrier protein
Sugar	Lectin, enzyme, or other sugar-binding protein

but also the so-called *biomimetic ligands,* particularly dyes, whose binding apparently often occurs to active sites of functional enzymes, although the dye molecules themselves of course do not exist in the functional context of the cell.

Because affinity chromatography proper relies on the functional properties, active and inactive forms can often be separated. This is, however, not unique to affinity methods. Covalent chromatography (Chapter 9) can do the same thing when the activity depends on a functional thiol group in the protein. By affinity elution, ion-exchange chromatography (Chapter 4) is also able to separate according to functional properties. These are, however, exceptions to what is a rule for the affinity methods.

Affinity chromatography has proved to be of great value also in the fractionation of nucleic acids, where complementary base sequences can be used as ligands, and in the separation of cells, where cell surface receptors are the basis of the affinity. Its main use has, however, been in the context of protein purification.

A field that has been so successful that it is often treated separately is affinity based on antigen–antibody interactions, called immunosorption. Sometimes this is the only available route to the purification of a protein and is especially attractive when there is a suitable monoclonal antibody at hand. This technique will be dealt with in some detail in this chapter.

Very often the use of affinity chromatography requires that the investigator synthesizes the adsorbent. The methods for doing this, which are described later, are well worked out and are also easily adopted for those not skilled in synthetic organic chemistry. To further simplify the task, activated gel matrices ready for the reaction with a ligand are commerically available. The immobilization of a ligand can, in the best cases, be a very simple affair. In addition, immobilizations are just as easy for proteins as for small molecules.

A property that needs special consideration is the association strength between ligand and counterligand. If it is too weak there will be no adsorption, whereas if it is too strong it will be difficult to elute the protein adsorbed. It is always important to find conditions, such as pH, salt concentration, or inclusion of, for example, detergent or other substances, that promote the dissociation of the complex without destroying the active protein at the same time. It is often here that the major difficulties with affinity methods are encountered.

As in the example given earlier, ligands can be extremely selective, but they may also be only group specific. The latter type includes glycoprotein–lectin interactions, several dye–enzyme interactions, and interactions with immobilized cofactors. However, these interactions have also proved to be extremely helpful in solving many separation problems. Good examples are ligands that are group selective against immunoglobulins (e.g., staphylococcal protein A or streptococcal protein G).

This chapter first gives an overview of the various interactions that have been exploited in affinity chromatography. It will then follow with details of how to prepare an affinity adsorbent, along with the practicalities of the chromatography. Finally, applications are described for both the immobilization reactions and the chromatographies.

10.2 AFFINITY INTERACTIONS

A good affinity ligand should possess the following characteristics:

- The ligand must be able to form reversible complexes with the protein to be isolated or separated.
- The specificity must be appropriate for the planned application.
- The complex constant should be high enough for the formation of stable complexes or to give sufficient retardation in the chromatographic procedure.
- It should be easy to dissociate the complex by a simple change in the medium, without irreversibly affecting the protein to be isolated or the ligand.
- It should have chemical properties that allow easy immobilization to a matrix.

For any particular protein intended to be purified by affinity chromatography, there is often a choice of several different ligands. In addition to the obvious choice of using a monoclonal antibody of appropriate affinity, which is generally applicable to all immunogenic solutes, one may look for components of naturally occurring biospecific pairs such as enzyme–substrate (analogs), enzyme–cofactor (analogs), and enzyme–inhibitor complexes. For glycoproteins, there is the possibility of using immobilized lectins, and the latter are often isolated by adsorption to immobilized carbohydrates. Considerable interest has been focused on immobilized biomimetic dyes, which show a wide variety of specificities applicable to several groups of enzymes, plasma proteins, and other proteins such as interferons.

To adsorb a protein counterligand to an affinity gel, the binding constant K_a for the interaction needs, for most practical purposes, to exceed or equal to 10^5 to $10^6 M$ (corresponding to a dissociation constant K_D of 1 to 10 μM; the K_D is equal to the inverse of K_a). However, interactions in the order of millimolar to micromolar ($K_A = 10^3$ to $10^6 M$) will also, with a reasonable ligand density, cause retardation of the interacting protein. In these cases, isocratic elution chromatography with small sample volumes can sometimes be very useful. Note that heterogeneous immunoassays such as ELISA require higher association constants, and thus other detection methods are needed in these cases. This is why it is a risk to lose monoclonal antibodies that have lower affinities, but which are very useful for purification purposes. The high affinity interactions often require drastic and sometimes denaturing conditions for elution, (i.e., decrease of binding constant). Interactions with binding constants exceeding 10^{10} to 10^{11} M are sometimes impossible to use as the conditions required to dissociate the complex are often the same as those that unfold the protein.

Generally, ligands may be classified as either monospecific or group specific, each of which in turn may be divided into low molecular weight or macromolecular.

10.2.1   Monospecific Low Molecular Weight Ligands

This group includes ligands such as steroid hormones, vitamines, and certain enzyme inhibitors. The term *monospecific* refers to the fact that these ligands bind to a single or a very small number of proteins in any particular cell extract or body fluid. Thus, lysine binds only plasminogen from blood plasma samples[5] and vitamin B_{12} will bind only its transport proteins: intrinsic factor from pure gastric juices and transcobalamin II from plasma.[4]

Despite the high specificity, nonspecific adsorption may occur. This can be due to interaction with the ligand or with residues from the immobilization reaction or the spacer arm. One way to cope with this problem is to make a second adsorbent lacking only the ligand itself and to allow the desorbed material from the ligand-containing adsorbent to pass this under identical conditions. Another way is to use a specific displacer, (e.g., the ligand itself in soluble form) (see Section 10.5.4.3), followed by a more harsh, general displacement agent for the regeneration of the adsorbent.

Generally, monospecific ligands bind more strongly and require harsher eluents than group-specific ligands, which can usually be eluted under mild conditions. Examples of extremely strong binding are the steroids and steroid receptors, which have association constants in the range 10^8 to 10^{10} M. Here it is often impossible to find elution conditions that allow the protein to be recovered in native form. One possibility is to use steroid analogs that have lower binding constants as ligands. Another example of a monospecific low molecular weight ligand with a very high binding constant to its counterligand is biotin, which binds avidin[6] with a K_A of 10^{15}.

10.2.2   Group-Specific Low Molecular Weight Ligands

This is the largest group of ligands containing a wide variety of enzyme cofactors and their analogs. This group also includes biomimetic dyes, boronic acid derivatives, and a number of amino acids and vitamins. A representative list of group-specific ligands and their target proteins is given in Tables 10-2 and 10-3. The target proteins are most often enzymes and the most thoroughly studied are the NAD$^+$- and NADP$^+$-dependent dehydrogenases and kinases.

A large number of affinity chromatography adsorbents are based on group-specific ligands coupled to a variety of carrier matrices commerically available from several sources. Two of the most widely used adsorbents are N^6-(6-

TABLE 10-2. Examples of Group-Specific Low Molecular Weight Ligands and Their Target Proteins

Ligand	Target Proteins	Ref.
5′-AMP	NAD^+-dependent dehydrogenases	1
	ATP-dependent kinases	
2′,5′-ADP	$NADP^+$-dependent dehydrogenases	1
ATP	ATP-dependent kinases	1
NAD^+	NAD^+-dependent dehydrogenases	1
$NADP^+$	$NADP^+$-dependent dehydrogenases	1
Benzamidine	Serine proteases	7
Phenylboronic acid	Glycoproteins	8
Cibacron Blue F3G-A	See Table 10-3	9
Procion Red HE-3B	See Table 10-3	9

aminohexyl)-5′-AMP coupled to beaded 4% agarose and the biomimetic textile dye Cibacron Blue F3G-A coupled to cross-linked beaded 6% agarose (e.g., Blue Sepharose™ 6 Fast Flow). The 5′-AMP gel shows affinity for a variety of NAD^+-dependent dehydrogenases with a binding capacity of approximately 10 mg enzyme/mL gel. Despite its relatively broad specificity, very high purification factors may be obtained using specific elution protocols with either soluble cofactors or by ternary complex formation using a combination of cofactor and substrate.[10] Alternatively, when the ligand–enzyme association constants are sufficiently far apart, gradient elution with a soluble cofactor may result in adequate separation, as has been shown for LDH isoenzymes.

The blue dye ligand is an analog of adenylyl-containing cofactors. Consequently, the adsorbent can be used to purify a very wide range of enzymes requiring such cofactors,[11] including both NAD^+- and $NADP^+$-dependent enzymes, although it shows some selectivity for NAD^+-dependent enzymes. In this respect it resembles the 5′-AMP ligand. However, the blue ligand binds a wider range of proteins and has been used for the isolation of several quite disparate proteins, as shown in Table 10-3.

A red dye, Procion Red HE-3B, coupled to Sepharose CL-6B has been used for the purification of a variety of $NADP^+$-dependent enzymes and a number of other unrelated proteins such as interferon, inhibin, plasminogen, and dopamine-β-monooxygenase, suggesting that binding may depend not only on specific steric factors but also on ionic and hydrophobic interactions. A large variety of other dyes have been exploited as affinity ligands by Scopes.[12] It is also possible to use chemical modification of the textile dye structures to improve their specificities and affinities. Lowe and colleagues.[13] thus improved the interaction between horse liver alcohol dehydrogenase and Cibacron Blue F3G-A.

Another ligand type that appears to fit in the category of low molecular weight and group specific is the one described by Porath as the T gel (T =

TABLE 10-3. Examples of Proteins with Affinity for Two Broadly Specific Dye Ligands: Cibacron Blue F3G-A and Procion Red HE-3B

Cibacron Blue F3G-A	Procion Red HE-3B
Kinases and phosphatases	Dehydrogenases
Adenylate cyclase	Aldehyde reductase (NADP$^+$)
Adenylate kinase	Dihydrofolate reductase (NADP$^+$)
Amino acyl-tRNA synthetase	Glucose-6-phosphate dehydrogenase (NADP$^+$)
cAMP-dependent protein kinase	Glutamate dehydrogenase (NADP$^+$)
Creatine kinase	Glutathione reductase (NAD$^+$/NADP$^+$)
DNA polymerase	3-Hydroxybutyrate dehydrogenase (NAD$^+$)
Fructose diphosphatase	Isocitrate dehydrogenase (NAD$^+$)
cGMP-dependent protein kinase	Lactate dehydrogenase (NAD$^+$)
Nucleoside kinase	Malate dehydrogenase (NAD$^+$)
Phosphofructokinase	6-phosphogluconate dehydrogenase (NADP$^+$)
Phosphoglycerate kinase	
Phosphorylase A	Other proteins
Protein kinase	Carboxypeptidase G
Restriction endonucleases	Dopamine β-monooxygenase
Succinyl-CoA transferase	Inhibin
	Interferon
Dehydrogenases	3-Methylcrotonyl-CoA carboxylase
Alcohol dehydrogenase (NAD$^+$)	Plasminogen
Glutathione reductase	Propionyl-CoA carboxylase
Hydroxysteroid dehydrogenase	
Isocitrate dehydrogenase (NAD$^+$)	
Lactate dehydrogenase (NAD$^+$)	
Malate dehydrogenase (NAD$^+$)	
Phosphogluconate dehydrogenase	
Other proteins	
Albumin	
Blood coagulation factors II, IX	
Interferon	

thiophilic).[14] It contains sulfone and thio ether groups and shows high selectivity for immunoglobulins.

10.2.3 Monospecific Macromolecular Ligands

Specific protein–protein interactions are common and essential in biology. Examples include subunit interactions in quaternary structures, interactions in multienzyme complexes, and hormone–receptor protein interaction. Few of these have, however, been exploited in affinity chromatography. Exceptions include the binding of fibronectin to gelatin,[15] of antithrombin to thrombin and heparin,[16] a polysaccharide, and of the transferrin receptor to transferrin. The reader is referred to the relevant references for further information.

A group of specific protein–protein interactions that have large general significance is antibody–antigen binding, which is described in more detail below. A general description of immunological concepts is given in Chapter 14.

10.2.4 Immunoadsorbents

The high specificity of antibodies makes them extremely useful ligands for affinity chromatography, especially where the substance to be purified has no immediately apparent complementary-binding substance other than its antibody. Both antigens and antibodies can be used as affinity ligands, and the immobilized protein is known as an immunoadsorbent or immunosorbent. Immunoadsorbents can be used to purify soluble proteins and peptides, solubilized membrane proteins, viruses, and even whole cells.

The traditional immunoadsorbents based on polyclonal antibody preparations have largely been replaced by adsorbents based on monoclonal antibodies. Modern hybridoma technology allows highly specific antibodies to be obtained against a predefined antigen present at concentrations much less than 1% of total immunogen (see Chapter 14). By using suitable screening methods, it is possible to obtain rare hybridomas producing an antibody of virtually any desired specificity and affinity and which can be immobilized for use in purifying the antigen.

There are several advantages of using monoclonal antibodies for immunosorbents. For minor protein components, single-step purification factors of several thousandfold are possible, High to moderate affinities (K_D of 10^{-6} to 10^{-9} M) preserve the antigens (yields >90%) and increase the life span of the adsorbent, often to several hundred cycles. A pH of 2 to 3 will normally displace the antigen, and the capacity for binding is normally at least 10-fold higher than that of an adsorbent based on polyclonal antibodies. The binding capacity would probably be still higher if all monoclonals were immobilized via their F_c moieties. Uniform binding of the antigens allows sharp desorption peaks and consequently a high concentration of antigen in the eluate.

In principle, monoclonal antibodies also allow a constant supply of a highly uniform antibody, which gives rise to high reproducibility from batch to batch of immunoadsorbent. The degree of substitution (ligand density) is an important optimization factor.

One serious disadvantage with monoclonal antibodies is their high cost, which makes the use of a relatively small column with repetitive operation almost mandatory. The next disadvantage, which is shared with all adsorbents based on immobilized proteins, is the high risk of fouling and irreversible chemical denaturation, notably proteolytic degradation. General fouling by nonspecific adsorption of various biopolymers and lipids is best prevented by a preliminary purification step. Thus, to also prevent proteolytic attack, crude extracts should never be applied directly to columns packed with adsorbents based on monoclonal antibodies.

Monoclonal antibodies may be purified using a variety of tools. Generic methods are based on immobilized staphylococcal protein A or streptococcal protein G and, to some extent, on thiophilic adsorption.[14] Combinations of cation-exchange chromatography, hydrophobic interaction chromatography, and gel filtration have proved to be useful in many cases. However, because of the very wide distribution of isoelectric points and relative hydrophobicities among monoclonals, one is forced to develop tailor-made purification procedures in each individual case. A most useful handbook on the subject has been published,[17] and literature (including interactive software) is available from Pharmacia Biotech AB.[18]

Monoclonal antibodies have been covalently attached to several different matrices, using a variety of coupling methods. The use of 4% agarose (for process applications Sepharose 4 Fast Flow) and the NHS or CNBr method is normally recommended. The use of monoclonal antibodies in affinity chromatography has been thoroughly discussed by Goding.[19]

Chase and co-workers[20] have shown that there is a linear relationship between the binding capacity of an immunoadsorbent and the amount of immobilized antibody. The ability of a particular immunoadsorbent to bind low concentrations of antigen depends on the dissociation constant, K_D, of the immobilized antibody (Table 10-4). No effect on the antibody loading could be registered for the dissociation constant, but the kinetic properties of the adsorbent were substantially improved at low loadings. This means that the immunoadsorption experiments can be run at higher flow rates. However, the reduced capacity has to be compensated for by using larger bed volumes, which leads to larger volumes of wash and eluent buffers, dilution of desorbed antigen, greater risk for nonspecific adsorption to the matrix, and higher matrix costs. This is why immunoadsorbents with low antibody loading are primarily

TABLE 10-4. Influence of Dissociation Constant on the Adsorption of Low Concentration of Antigen to Immobilized Antibody

K_d of Immobilized Antibody (M)	M_r of Antigen						
	300	1,000	3,000	10,000	30,000	100,000	10^6
10^{-6}	0.3	1	3	10	30	100	1,000
10^{-7}	0.03	0.1	0.3	1	3	10	100
10^{-8}	0.003	0.01	0.03	0.1	0.3	1	10
10^{-9}	0.0003	0.001	0.003	0.01	0.03	0.1	1
10^{-10}	0.00003	0.0001	0.0003	0.001	0.003	0.01	0.1

The table shows the concentration of antigen (μg/ml) that results in a 50% utilization of the adsorption capacity of an immunoadsorbent when in equilibrium with the sample solution (from Ref. 20).

suggested for immunosubtraction procedures, (i.e., for the removal of low concentrations of known impurities from protein products).

Finally, it is appropriate to remember the most important general limitation of immunoadsorption as a tool for the isolation and purification of proteins. The immobilized antibody will only recognize and bind to the corresponding antigenic determinant of the actual protein. It will not discriminate between protein molecules that have been modified or partially degraded in other parts. This means that in the majority of cases other separation techniques have to be applied after the immunoadsorption step to remove molecules with possible immunogenic neodeterminant structures, irrespective of whether these molecules are biologically active or not.

10.2.5 Group-Specific Macromolecular Ligands: Lectins and IgG-binding Proteins

This group includes several ligands that have found widespread popularity, [e.g., lectins such as concanavalin (Con A) and lentil for the isolation of glycoproteins, staphylococcal protein A and streptococcal protein G for the purification of IgG, and calmodulin for the isolation of a wide variety of calcium-dependent enzymes]. An important member of this group is also the sulfated polysaccharide heparin, which is frequently used for the purification of several coagulation proteins and other plasma proteins in addition to a variety of enzymes and other unrelated proteins such as steroid receptors and virus surface antigens. Table 10-5 lists examples of proteins that have been shown to have affinity to immobilized heparin.

The ability of immobilized lectins to interact specifically with sugars makes them excellent tools for purifying both soluble and membrane-derived glyco-proteins and polysaccharides such as enzymes, hormones, blood plasma pro-teins, antigens, antibodies, and blood group substances. Table 10-6 lists the most commonly used lectins together with their specificities, Immobilized Con A has been the most widely used because of its specificity for the commonly occurring α-D-mannose and β-D-glucose and because the binding of soluble glycoproteins to the gel is easily reversed by the addition of low molecular weight sugars or sugar derivatives. Secretory glycoproteins that contain a large amount of N-acetylglucosamine are usually purified on immobilized wheat germ lectin. Immobilized lentil lectin has the same specificity as Con A but with a lower binding constant. This makes it more suitable for the purification of membrane glycoproteins that often possess very strong binding affinity to Con A.

10.2.6 Ligands Derived from Chemical and Biological Combinatorial Libraries

The rapid development in combinatorial chemistry and biology[22,23] offers a new approach for the design of affinity ligands.[24] Phage surface display peptide

TABLE 10-5. Proteins with Affinity to Immobilized Heparin[a]

Coagulation proteins	Enzymes that act on nucleic acids
Antithrombin III	Restriction endonucleases
Factor VII	RNA polymerase
Factor IX	RNA polymerase I
Factor XI	RNA polymerase III
Factor XII, XIIa	DNA polymerase
Thrombin	DNA ligase
	Polynucleotide kinase
Other plasma proteins	
Properdin	Protein synthesis factors
Complement Cl	Initiation factors
Complement factor B	Elongation factor (EF-1)
βIH	Ribosomes
Complement C2	
Complement C3	Receptors
Complement C4	Steroid receptors
C3b inactivator	Estrogen receptor
Inter-α-trypsin inhibitor	Androgen receptors
Gc globulin	Other proteins
Protein HC	Platelet factor 4
Fibronectin	β-Thromboglobulin
β_2-Glycoprotein 1	SV 40 tumor antigen
C-reactive protein	Hepatitis B surface antigen
Lipoprotein lipase	Trehalose phosphate synthetase
Hepatic triglyceride lipase	Hyaluronidase
Lipases	Collagenase inhibitor
Lipoprotein lipase	Collagenase inhibitor
Hepatic triglyceride lipase	
Lipoproteins	
VLDL, LDL	
VLDL, apoprotein	
HDLP	

[a] Adapted from Ref. 21.

libraries[25,26] and chemically synthesized peptidomimetic randomized libraries[27,28] have thus been used for the screening of ligands directed toward, for example, streptavidin[26] and chymotrypsin.[29] High-speed, parallel chemical synthesis techniques have been developed for the production of hundreds of novel organic compounds per day.[30] In this way, nonrandomized combinatorial chemical libraries, each containing more than 10,000 lead compounds, have been produced, and methods have been developed for the high-speed screening of potential affinity separation ligands directed toward a number of important target proteins.[31]

TABLE 10-6. Lectins and Their Sugar Specificities

Source of Lectin	English Name	Specificity
Dolichos biflorus	Anti-A lectin; horse gram	α-N-Acetyl-D-galactosamine
Bandeirea simplicifolia	Sunn hemp	α-D-Galactose
Ricinus communis	Castor bean	D-Galactose, N-acetyl-D-galactosamine
Canavalia ensiformis	Con A	α-D-Glucose, α-D-mannose
Helix pomatia	Snail	N-Acetyl-D-galactosamine
Lens culinaris	Lentil	α-D-Mannose, α-D-glucose
Pisum sativum	Pea	α-D-Mannose, α-D-glucose
Arachis hypogea	Peanut	Galactose, β-1,3-N-acetyl-D-galactosamine
Phaseolus vulgaris	Phytohemagglutinin	N-Acetyl-D-galactosamine
Glycine max	Soybean	N-Acetyl-D-galactosamine
Triticum vulgaris	Wheat germ	Tri-N-acetyl-D-glucosamine

10.3 PREPARATION AND EVALUATION OF AFFINITY ADSORBENTS

A general description of gels used in chromatography is given in Chapter 2. A list of commercially available matrices intended for ligand immobilization is provided in Table 10-7.

10.3.1 Choice of Matrix

As in all adsorption chromatography, an adsorbent with a large surface area per unit column volume is desirable to maximize the capacity of the affinity adsorbent. Hydrophilic gels with a high surface-to-volume ratio (Chapter 2) are very suitable as matrices. For affinity chromatography applications, the ideal gel material should meet the following characteristics:

- Macroporous to accommodate the free interaction of large molecular weight proteins with ligands that could themselves be proteins or other macromolecules.
- Hydrophilic and neutral to prevent the proteins from interacting nonspecifically with the gel matrix itself.
- Contain functional groups to allow derivatization by a wide variety of chemical reactions.
- Chemically stable to withstand harsh conditions during derivatization, regeneration, and maintenance.
- Physically stable to withstand hydrodynamic stress in packed beds and, when applicable, sterilization by autoclaving.
- Readily available at low cost to facilitate industrial applications.

TABLE 10-7. Commercially Available Activated Matrices

Ligand to be coupled	Functional group	Type of gel	Name of product	Name of manufacturer
Proteins	$-NH_2$	Beaded agarose with cyanate ester groups	CNBr-activated Sepharose 4B	Pharmacia Biotech
Proteins (peptides)	$-NH_2$	Beaded agarose with reactive ester on spacer	Activated CH Sepharose 4B	Pharmacia Biotech
Carbohydrates	$-OH$	Beaded agarose with epoxy (oxirane) groups on short spacer (low capacity)	Epoxi-activated Sepharose 6B	Pharmacia Biotech
Thiol compounds (e.g., proteins)	$-SH$			
Amines (peptides, proteins)	$-NH_2$			
Thiol compounds (thiol proteins and low mol. weight thiols)	$-SH$	Beaded agarose with reactive disulphide groups (short spacer) (high capacity)	Thiopropyl Sepharose 6B	Pharmacia Biotech
Amines and thiols (including proteins)	$-NH_2$ $-SH$	Beaded agarose with reactive sulfonic ester groups	Tresylactivated Sepharose 4B	Pharmacia Biotech
Aldehydes (low and high mol. weight)	$-CHO$	Beaded agarose with adipic acid hydrazide groups	Agarose-Adipic acid hydrazide	Pharmacia Biotech
Amines (esp low M_W type)	$-NH_2$	1,1'-carbonyldiimidazole activated 6% crosslinked beaded agarose, part. diam. 45–165 μm	Reacti-Gel (6X)	Pierce
Amines (esp low M_W type)	$-NH_2$	1,1'-carbonyldiimidazole activated Fractogel TSK, part. diam. 32–65 μm, frac. range: 50,000–500,000 MW	Reacti-Gel (HW-65F)	Pierce
Amines (esp low M_W type)	$-NH_2$	1,1'-carbonyldiimidazole activated Trisacryl GF-2000	Reacti-Gel (GF-2000)	Pierce

Target compounds	Reactive group	Description	Product name	Manufacturer
Amines (esp proteins)	—NH₂	Glutaraldehyde-activated Ultrogel (2% polyacrylamide, 2% agarose). Part. diam. 60–140 μm, Exd. limit: 3 × 10⁶ Daltons	Act-Ultrogel ACA 22	Biosepra
Amines (esp proteins)	—NH₂	Derivatized crosslinked agarose gel containing N-hydroxysuccinimide ester groups	AffiGel 10 Gel AffiGel 15 Gel	BioRad BioRad
Amines Thiol compounds (in principle also carbohydrates but matrix will hydrolyze under the harsh conditions necessary)	—NH₂ —SH	Hydrophilic polymer-based oxirane derivative groups part. diam. 32–63 μm	AF-Epoxy 650 Fractogel TSK	Merck
Amines (esp. low M_w)	—NH₂	Hydrophilic polymer-based derivative with imidazoyl-carbonate groups, part size: 32–63 μm	AF-CDI 650 Fractogel TSK	Merck
Amino acids, ketoacids, carboxylic acids	—COOH	Beaded agarose (4%) with (6 carbon) aminospacer	AH-Sepharose 4B	Pharmacia Biotech
Amino acids, ketoacids, carboxylic acids	—COOH	Beaded agarose (4%) with amino terminal (6-carbon spacer attached to matrix via highly stable ether bond thus more stable for leakage than AH-variety)	EAH-Sepharose 4B	Pharmacia Biotech

(continued)

TABLE 10-7. (*Continued*)

Ligand to be coupled	Functional group	Type of gel	Name of product	Name of manufacturer
Amino acids, peptides	—NH$_2$	Beaded agarose (6%) with (6 carbons) carboxylspacer	CH-Sepharose 4B	Pharmacia Biotech
Amino acids, peptides	—NH$_2$	Beaded agarose (6%) with carboxyl terminal or 6 carbon spacer attached to matrix via highly stable ether bond (thus more stable to leakage than CH-type)	ECH-Sepharose 4B	Pharmacia Biotech
Amino acids, ketoacids, carboxylic acids	—COOH	Amino terminal agarose gel with 6-atom, hydrophilic spacer	Affi-Gel 102	BioRad
Amino acids, ketoacids, carboxylic acids	—COOH	Aminoethylpolyacrylamide matrix	Aminoethyl Bio-Gel P-2 and P-100 gels	BioRad
Amines, amino acids, peptides	—NH$_2$	Carboxy terminal agarose beads without spacer arm	CM-Bio Gel A Carboxymethyl agarose	BioRad
Carboxyl cont. ligands (esp. low M$_W$)	—COOH	Hydrophilic polymer substituted with amino groups, part. size: 32–63 μm	AF-Amino 650 Fractogel TSK	Merck
Amino group cont. ligands (esp. low M$_W$)	—NH$_2$	Hydrophilic polymer substituted with carboxylic groups, part. size: 45–90 μm	CM-650 (M) Fractogel TSK	Merck

These characteristics point to gels based on polymers highly substituted with alcohol hydroxyls and thus to polysaccharides. Among the latter, the spontaneously gel-forming galactan *agarose* indeed possesses most of the characteristics of an ideal matrix for affinity chromatography. The major weakness of native agarose is its chemical and physical instability, which, however, has been largely compensated for by chemical cross-linking of the physically cross-linked so-called junction zones in the agarose gel structure[39] (see Chapter 2). Ever since its introduction in 1968 by Cuatrecasas, Wilchek, and Anfinsen,[33] 4% agarose has been the most popular matrix for affinity chromatography. A contributing reason for this popularity, in addition to the advantageous matrix properties as such, is that there were early simple and convenient coupling methods developed for agarose (see Section 10.4.1) and even commercially available preactivated matrices (Table 10-7).

Less frequently used gel matrices for affinity chromatography are cellulose, cross-linked dextran, polyacrylamide, and silica. To this group also belong the potentially interesting and commercially available matrices made of mixtures of polyacrylamide and agarose (Ultrogel) and polymerized tris(hydroxymethyl)acrylamide (Trisacryl, Biosepra, France). *Cellulose* has found its niche as a carrier for ligands in the affinity chromatography of oligonucleotides and nucleic acids. This area has been throughly treated by Schott.[34] Because cellulose is much more inexpensive than agarose, but used with the same immobilization methods, it is an alternative in large-scale industrial applications of affinity methods. Gels based on *cross-linked dextran and polyacrylamide* both suffer from the serious disadvantage of having too small pore diameters, which become still smaller after derivatization with affinity ligands. Also, fewer methods are available for immobilization on polyacrylamide.

In traditional low-pressure affinity chromatography systems, beads with a diameter of approximately 100 μm are usually standard. However, beads with diameters in the range 5 to 30 μm are used in so-called high-performance liquid affinity chromatography (HPLAC). In this variety of affinity chromatography, higher pressure drops are often required, which also means a demand for higher gel rigidity. This is why the first HPLAC applications were based on modified and derivatized porous *silica*.[35] The major reason for using smaller particles is to increase the chromatographic efficiency by decreasing the diffusion path lengths and increasing the interphase area between the stationary and the mobile liquids. The most serious drawback with silica-based stationary phases is their solubility at pH above 7.5, which prevents their regeneration and maintenance under alkaline conditions.[36] An alternative to silica for HPLC applications, which has also proven useful in HPLAC, is small-diameter agarose beads.[37] *Synthetic organic polymers,* highly substituted with alcohol hydroxyls and with adequate porosity and rigidity, should also present interesting matrices. Some of these are now commerically available (Table 10-7).

Some notable differences between the different matrices exist with respect to their chemical properties. The majority of immobilization methods depend on the presence of hydroxyl groups on the matrix and have been adapted for use in aqueous solvents. Agarose, however, also retains its macroporosity in organic solvents. Thus activation procedures that require an organic milieu can be performed as well as coupling of ligands not soluble in water. As most organic chemistry is based on work in apolar solvents, this means that a wealth of ligand immobilization methods is potentially available for beaded agarose.

10.3.2 Properties of Ligand

For the preparation of the affinity adsorbent the ligand should

- Be compatible with the solvents used during the coupling procedure
- Possess at least one functional group by which it can be immobilized to the matrix. [Commonly used groups include $-NH_2$ (amino), $-COOH$ (carboxyl), $-CHO$ (aldehyde), $-SH$ (thiol), and $-OH$ (hydroxyl)]
- Possess a functional group for coupling that is nonessential for its binding properties, (i.e., the binding properties of the ligand should not be adversely affected as a result of its immobilization)

Ligands of a high molecular weight type (e.g., proteins) with a large number of suitable functional groups can normally be immobilized without adversely influencing structure or function. In low molecular weight ligands the coupling of course results in a relatively large change in the molecule. If the affinity interaction decreases, a chemical modification of the ligand may be necessary to provide it with an appropriate functional group for immobilization.

The functional group used should permit the formation of a stable covalent bond so that the ligand is not released from the matrix. This is particularly important for small ligands where "single-point attachment" is often the case. For proteins, "multipoint attachment" between ligand and matrix is rather common. In such affinity adsorbents, the stability of each individual bond is less critical.

It is, of course, also essential that the ligand remains intact during the immobilization procedure and that it is sufficiently stable to allow the planned affinity chromatography to be carried out. This might be a problem when proteins are coupled at high pH. It is essential that the ligand reagent is as pure as possible and, in particular, does not contain substances with functional groups that can react competitively in the immobilization. Proteins should be subjected to gel filtration to remove low molecular weight substances such as ammonium sulfate.

10.3.3 Choice of Spacer Arm

Occasionally, an affinity adsorbent might show poor function due to low steric availability of the ligand. This rarely happens with high molecular weight ligands but may occur with low molecular weight ligands. The use of a "spacer arm" in many cases solves this problem. Commonly used spacer arms are aliphatic, linear hydrocarbon chains with two functional groups located at each end of the chain. One of the groups (often a primary amine, $-NH_2$) is attached to the matrix, whereas the group at the other end is selected on the basis of the ligand to be bound. The latter group, which is also called the terminal group, is usually a carboxyl ($-COOH$) or amino group, ($-NH_2$).

The most common spacers are 6-aminohexanoic acid [$H_2N-(CH_2)_6-COOH$], hexamethylene diamine [$H_2N-(CH_2)_6-NH_2$], and 1,7-diamino-4-azaheptane (3,3-diaminodipropylamine).[38-40] The spacer arm is introduced into the matrix by the same immobilization methods, which are described later for ligands.

Longer spacer arms can be introduced by first immobilizing a spacer arm with a terminal primary amine and then increasing the length of the arm by reaction with succinic anhydride.[39] Another possibility is to immobilize a spacer arm with a terminal carboxyl group and then increase its length by reaction with 1,7-diamino-4-azaheptane by aid of a condensation reagent (see later).

A drawback with the hydrophobic arms, especially the longer ones, is that they can give rise to unwanted nonspecific interactions. Polypeptides, particularly glycine oligomers, are examples of hydrophilic spacers. These, however, might bind proteins by nonspecific ionic interactions.[40]

Sometimes it is stated that a spacer should be used for ligands with a molecular weight of less than 5,000, but because of the risk of introducing nonspecific binding sites by side reactions in the gel during coupling, by the arm itself, or by both, the authors recommended first trying to prepare an affinity adsorbent by direct coupling of the ligand to the matrix. It should also be remembered that in several of the ligand immobilization methods described later, such as the bisepoxirane and glutaraldehyde method, the ligand will automatically be provided with a spacer as a result of its immobilization.[41,42]

10.3.4 Evaluation of the Prepared Affinity Adsorbent

Before an attempt is made to use the prepared adsorbent in affinity chromatography, one should always make sure that the ligand immobilization has succeeded and, if possible, determine the ligand density (degree of substitution as micromoles or milligrams of ligands per milliliter of affinity gel).

The analysis can be carried out in several ways. A simple method is "indirect evaluation" (i.e., the amount of immobilized ligand is calculated as the difference between the amount of ligand originally added to the matrix and the amount of ligand recovered in the liquid phase and pooled washings after finished coupling). If the ligand absorbs light of a suitable wavelength ($\lambda_{max} > 250$ nm) with an acceptable molar absorptivity ($\varepsilon > 5,000$ cm^{-1} M^{-1}), this analysis can simply be carried out with a photometer. This method, however, often gives erroneous values and should only be used for a rough estimation.

A very useful method is *elementary analysis*. This technique can be used if the ligand contains elements such as nitrogen, sulfur, halogen, or phosphorus, provided that these elements are not present in the matrix or become introduced into the matrix as a result of the activation and coupling procedures. Peptide and protein ligands can, after hydrolysis, be quantitatively determined by amino acid analysis. Another possibility is to label the ligand with a suitable γ-emitting radioisotope before coupling. In the case of small ligands having carboxyl or amino groups, acid–base titration is sometimes an easy method used for determining the ligand density.

Finally, activity determination for enzyme ligands can be used. It must be taken into account that immobilized enzymes often have changed kinetic properties due to steric and diffusional restrictions in the interaction with substrate (i.e., comparison with free enzyme might not be valid). An analysis of the *adsorption characteristics* of the affinity gel should also be performed.

The capacity of an affinity adsorbent is defined as milligrams or micromoles of counterligand that can be adsorbed per milliliter of sedimented gel. In a 90-μm average diameter beaded 4% agarose gel, the total surface area is approximately 5 m^2/ml bed volume, of which only about 8 cm^2 refers to the outer particle surface. (The external matrix surface area of a 4% agarose bead is only 2% of the surface area of a corresponding solid sphere.) The maximum theoretical binding capacity of a 60,000 molecular weight protein should thus be approximately 80 mg/ml. This value cannot be achieved for several reasons.

It is appropriate to distinguish between static and dynamic binding capacities, respectively. The *static capacity* is measured in batch experiments, which allow ample time for equilibrium to establish. The static capacity depends on the density of immobilized ligand and its availability for interaction with a particular protein. Some of the immobilized ligands might be inaccessible to a particular protein as a result of steric exclusion due to their location within the gel matrix. This is particularly true when small ligands are used for the binding of high molcular weight proteins. Thus, the functional binding capacity is often much lower than the nominal binding capacity as calculated from the measured ligand density.

The *dynamic capacity* of the affinity adsorbent is the binding capacity under operating conditions (i.e., in the packed affinity chromatography column during the sample application and washing procedures). Factors that influence the dynamic binding capacity are discussed in Chapter 2.

For affinity systems based on low molecular weight ligands that bind high molecular weight proteins, the matrix-bound affinity complexes sometimes sterically shield neighboring ligands from interacting with unbound protein molecules. In such cases, adequate binding capacity can be achieved at a substantially lower ligand substitution. In fact, a high degree of substitution should be avoided as it may cause undesired nonspecific adsorption. When the ligand and corresponding binding protein are of approximately the same size or, more unusually, when the ligand is much larger than the protein to be isolated, the problem of gel porosity primarily concerns the immobilized ligand. An alternative way to achieve a large surface area besides using beads of high porosity is to use smaller beads, as in HPLAC.

10.3.5 Storage of Affinity Adsorbent

The conditions for storage of the prepared affinity adsorbents of course depend on the stability of the matrix, the ligand, and the covalent bond by which the ligand is attached to the matrix.

Polysaccharide matrices such as beaded agarose hydrolyze with a significant rate at acidic pH < 4 and oxidize with formation of matrix-bound carboxyls at high pH > 9. Protein ligands may change their conformation and lose their activity as a result of exposure to extreme pH, high temperatures, and denaturing agents (organic solvents, urea, etc.). The commonly employed CNBr method leads to the formation of an isourea linkage,[43] which is split at a rather high rate through hydrolysis and aminolysis at alkaline pH (>8).

The immobilization procedure might decrease or increase the stability of the system. Several procedures, notably CNBr and epichlorohydrin coupling,[44] lead to the introduction of covalent cross-linkages into the matrix and thus render it more stable. Certain ligands may also be stabilized as a result of their immobilization. Thus aggregation and autodigestion that occur in solution with, for example, proteases might be prevented.

Normally the affinity adsorbent can be stored as a suspension in an appropriate buffer at physiological pH at 4°C for long periods of time. It is, however, advisable to add an antimicrobial substance such as sodium azide to prevent bacterial growth. For very labile adsorbents, lyophilization can be used to increase the storage time. To make sure that the beads will reswell properly on reconstitution with water or buffer, it is necessary to prevent them from irreversibly collapsing as a result of the lyophilization. Dextran or polyethylene glycol (PEG) is then added before lyophilization and, after reconstitution of the beads, washed away on a glass filter.

10.4 IMMOBILIZATION TECHNIQUES

In general, the immobilization procedure consists of three steps:

1. Activation of the matrix to make it reactive toward the functional group of the ligand
2. Coupling of the ligand
3. Deactivation or blocking of residual active groups by a large excess of a suitable low molecular weight substance such as ethanolamine

The activation normally consists of the introduction of an electrophilic group into the matrix. This group later reacts with nucleophilic groups, such as —NH₂ (amino), —SH (thiol), and —OH (hydroxyl) in the ligand. Alternatively, a matrix with nucleophilic groups can be used to immobilize a ligand containing an electrophilic group, although such an approach is less common (see Section 10.4.1.6). The activated structure is sometimes stable enough for the activated matrix to be isolated and stored until the coupling of ligand is performed. In other cases the coupling procedure has to be performed immediately after activation.

The ligand is either coupled directly to the activated matrix or the matrix is first provided with a spacer arm to which the ligand is subsequently attached. Coupling to the spacer arm is often performed in a one-step procedure by use of a condensation reagent that forms amide bonds between carboxyl and amino functions present in the ligand and spacer arm, but spacer arms containing terminal carboxyl group can also be activated in a separate step.

When affinity adsorbents are prepared, a bond as stable as possible should be established between the matrix and the ligand to prevent leakage of the ligand. In certain cases, however, it may be useful to have the ligand attached through a bond that is stable but can be cleaved when so desired. An example of such a bond is the aliphatic disulfide, which can be both formed and split under mild conditions by thiol–disulfide exchange reactions. Procedures for carrying out such reversible covalent immobilizations of ligands are described in Chapter 9.

An overview of the various immobilization methods to be presented is found in Table 10-8. The more useful of these are described according to matrix—agarose, polyacrylamide, and silica—and properties of the ligands. Methods used for agarose matrices can be used equally for other polysaccharide matrices, with the exception of methods requiring organic solvents.

In Section 10.6 gives several applications of the techniques in some detail.

10.4.1 Methods for Agarose and Other Polysaccharide Matrices

10.4.1.1 CNBr and CDAP Cyanylating Procedures   These methods are suitable for —NH₂-containing ligands, especially polypeptides and proteins.

The original CNBr technique as developed by Axen and co-workers[45,46] is a classical two-step method with activation and coupling. A water suspension of a polysaccharide (e.g., beaded agarose) is reacted with CNBr at a high pH (11 to 12), which leads to the introduction of cyanate ester and imidocarbonate groups into the matrix.

$$\text{Matrix} \quad \begin{array}{c} -\text{OH} \\ -\text{OH} \end{array} + \text{CNBr} \longrightarrow \begin{array}{c} -\text{O}-\text{C}\equiv\text{N} \\ -\text{OH} \end{array} \quad \text{or} \quad \begin{array}{c} -\text{O} \\ -\text{O} \end{array}\!\!>\!\text{C}=\text{NH} \qquad (10\text{-}1)$$

Matrix   Cyanate ester   Cyclic
                        (very reactive)   imidocabonate

The relative amounts of the two groups depend on the type of polysaccharide used. (e.g., the relative amount of cyanate esters is higher for agarose than for cross-linked dextran in which imidocarbonate is predominant). If the right conditions are used the activated matrix can be stored for a long time without significant loss of reactive groups either as a suspension or lyophilized. The cyanate esters are hydrolyzed at alkaline pH, whereas the imidocarbonates are converted to carbonates at acidic pH. Activated agarose is commercially available (Table 10-7).

Amino-containing ligands are covalently linked to the activated matrix in aqueous medium at close to physiological pH, 7 to 8.

$$
\begin{array}{c}
\text{O}-\text{C}\equiv\text{N} \\
\text{OH}
\end{array}
+ \text{H}_2\text{N}-\text{Ligand} \longrightarrow
\begin{array}{c}
\overset{\displaystyle \overset{+}{N}H_2}{\underset{\parallel}{}} \\
\text{O}-\text{C}-\text{NH}-\text{Ligand} \\
\text{OH}
\end{array}
$$

Isourea derivative   (10-2a)

The bonds formed between the ligand and the matrix are mainly of the isourea type. When the ligand reacts with the imidocarbonates, the products are N-substituted imidocarbonates as well as isourea derivatives. N-substituted carbamates also occur when the ligand reacts with cyclic carbonate [Eq. (10-2c)] formed by hydrolysis of the N-substituted imidocarbonates.

$$
\begin{array}{c}
\text{O} \\
\text{O}
\end{array}\!\!>\!\!\text{C}=\text{NH} + \text{N}_2\text{H}-\text{Ligand} \longrightarrow
\begin{array}{c}
\text{O} \\
\text{O}
\end{array}\!\!>\!\!\text{C}=\text{N}-\text{Ligand}
$$

Substituted
Imidocarbonate   (10-2b)

$$
\begin{array}{c}
\text{O} \\
\text{O}
\end{array}\!\!>\!\!\text{C}=\text{O} + \text{H}_2\text{N}-\text{Ligand} \longrightarrow
\begin{array}{c}
\text{OH} \\
\text{O}-\overset{\displaystyle}{\underset{\underset{\text{O}}{\parallel}}{\text{C}}}-\text{NH}-\text{Ligand}
\end{array}
$$

N-Substituted
carbamate   (10-2c)

The simplicity of the method, the fact that it works so well in combination with beaded agarose, a matrix with excellent chromatographic properties, and that it is mild enough for binding sensitive ligands such as proteins like antibodies and enzymes have made the CNBr technique by far the most used technique for the preparation of affinity adsorbents.

In the activation step the pH is kept constant either by addition of strong sodium hydroxide or by use of a buffer.[46,47] The high pH is needed to deprotonate the polysaccharide hydroxyl groups ($pK = 12$) to corresponding alkoxide

TABLE 10-8. Overview of Immobilization Procedures

Reagent for activation and coupling	Matrix	Functional group in matrix	Activation conditions	Activated structure	Ligand	Functional group in ligand	Coupling conditions	Ligand matrix bond	Refs.
CNBr (titration)	Polyols (esp. polysacch.)	—OH	aq. pH 11–12	Eq. 10-1	amines (esp. proteins)	—NH$_2$	aq. pH 7–8.5	isourea	46, 48
CNBr (buffer)	Polyols (esp. polysacch.)	—OH	aq./buffer	Eq. 10-1	amines (esp. proteins)	—NH$_2$	aq. pH 7–8.5	isourea	47
CDAP	Polyols (esp. polysacch.)	—OH	org./aq.	Eq. 10-4	amines (esp. proteins)	—NH$_2$	aq. pH 7–8.5	isourea	48
DSC (N,N'-disuccinimidylcarbonate)	(esp. agarose)	—OH	organic	Eq. 10-12	amines (esp. proteins)	—NH$_2$	pH 6–8	carbamate	51
CDI (Carbonyldiimidazole)	(esp. agarose)	—OH	organic		amines (esp. low MW)	—NH$_2$	pH 8–10	carbamate	52
Tosylchloride	(esp. agarose)	—OH	organic	Eq. 10-11	amines, thiols	—NH$_2$, —SH	pH 9–10	sec. amine thioether	50
Tresylchloride	(esp. agarose)	—OH	organic		amines, eg. proteins thiols	—NH$_2$, —SH	pH 8–9	sec. amine thioether	50
Bisoxiranes	polyols	—OH	aq. pH 13–14 aq. pH 8–10	Eq. 10-7	carbohydrates, amines thiols	—OH —NH$_2$, —SH	pH 11.5–13 pH 8–11	ether, thioether sec. amine	41
Epichlorohydrine	polyols	—OH	aq. pH 13–14	Eq. 10-9	carbohydrates, amines, thiols	—OH —NH$_2$, —SH	pH 11.5–13 pH 8–11	ether, thioether sec. amine	49

Method	Matrix	Reactive group	Conditions	Eq./Fig.	Compound	Reactive group	pH	Product	Ref.
DVS (Divinylsulphone)	polyols	—OH	aq. pH 13–14	Eq. 10-10	carbohydrates, amines, thiols	—OH, —NH$_2$, —SH	pH 10.5–12, pH 8–11	ether, thioether, sec. amine	44
Carbodiimides	polyols	—COOH, —NH$_2$	aq.	Eq. 10-13	amines carboxylates	—NH$_2$, —COOH	pH 5	amides	54–57
Esteractivated carboxyl	polyols	—COOH	organic	Eq. 10-14	amines (esp. proteins)	—NH$_2$	pH 5–9	amides	58
Reaction with matrix thiol	polyols	—SH	aq.		unsaturated compounds	—CH=CH—, =CO, —CNH	pH 8–10	thioether	57
Thiol-disulphide exchange	polyols	—SH	aq.	Eq. 10-15	thiols e.g. proteins	—SH	pH 2–9	disulphide	Chap. 9
Glutaraldehyde	polyamide	—CO NH$_2$	aq.	Eq. 10-16	amines (esp. proteins)	—NH$_2$	pH 7	prob. sec. amine	65
Hydrazine (acylazide)	polyamide	—CO NH$_2$	NaNO$_2$ in HCl	Fig. 10-2	amines (esp. low MW)	—NH$_2$	ph 7–9	amide	66
Hydrazine	polyamide	—CO NH$_2$	NaNO$_2$ in HCl	Fig. 10-13	aldehydes, ketones	—CHO, —CO	ph 7–9	amide	67, 68
Oxirane via silanization	silica, glass various	—Si—OH	aq. pH 6.5	Fig. 10-13	amines, thiols	—NH$_2$, —SH	pH 8	sec. amide	70
Isocyanide		—COOH, —NH$_2$, —COOH, =CO, —NC		Eq. 10-15	variety of comp.	—COOH, —NH$_2$, —COH, =CO	pH 6.5	amide	62, 64

ions that are sufficiently nucleophilic to react with the CNBr. The basic reaction medium causes the hydrolysis of CNBr to inert cyanate ions OCN, thus consuming more than 90% of the initially added amount of CNBr (Figure 10-1). The formed cyanate esters to a great extent also hydrolyze to inert carbamate groups and react with matrix hydroxyls to less active imidocarbonates.[48] This of course decreases the capacity of the activated matrix to bind ligand and introduces possible sites for nonspecific interactions in the affinity adsorbent.

In some cases the imidocarbonates (as well as the carbonates formed as a result of hydrolysis at acidic pH of the imidocarbonates) also act as covalent cross-linkages and stabilize the matrix mechanically and chemically without significantly changing the porosity. This is particularly useful when noncrosslinked agarose is used. Despite the side reactions, the original CNBr method and varieties of it have been widely used in many successful applications of affinity chromatography, which show that by performing the activation and coupling in an accurate way, reproducible results can be obtained. The yield in the reaction is poor. Less than 2% of the CNBr forms useful reactive groups. This does not present an economic problem, at least not in the small scale,

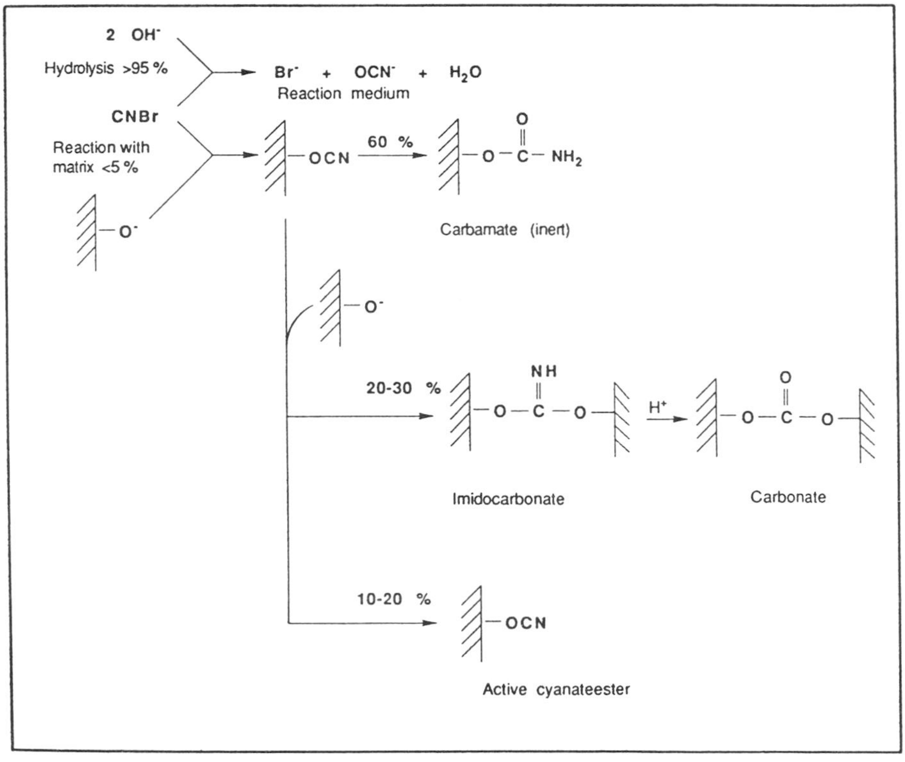

FIGURE 10-1. Reactions occurring when a polysaccharide matrix is activated by the classic CNBr-method at pH 11 to 12.[35]

as CNBr is an inexpensive chemical. What is more serious are the health hazards arising from dealing with large quantities of CNBr. Because of its toxicity and high vapor pressure, all work with CNBr should be carried out in properly ventilated fume hoods.

Kohn and Wilchek[48] have devised a method to increase the electrophilicity of CNBr by forming a so-called cyanotransfer complex by CNBr and certain bases such as triethylamine (TEA) and dimethylaminopyridine (DAP). The cyano transfer complex formed with TEA is not stable, but the one formed with DAP can be isolated as 1-cyano-4(dimethylamino)pyridinium bromide.

Dimethylaminopyridine, DAP   1-cyano - 4 -(dimethylamino) - pyridiniumbromide

These cyano transfer complexes are far more electrophilic than CNBr and are thus able to cyanylate the matrix hydroxyl groups to cyanate esters at a much lower pH than used in the conventional procedure. The hydrolysis of CNBr is thus avoided, as is the transformation of cyanate esters to other products. As a result the yield of cyanate esters improves dramatically to 20–80% depending on the conditions.

However, the activation reaction requires that a mixed organic solvent–water systems is used (e.g., acetone–water = 6:4); in a pure water system, irreproducible and low yields result. A low temperature, (0°C) typically also gives better results than room temperature.

The cyano transfer reaction is best suited for agarose beads that can be transferred to mixed solvents without shrinking. Note that the matrices will not be reinforced by covalent cross-linkages to the same extent as with the original CNBr method discussed earlier. Several cyano transfer complexes have been described. One of the more useful reagents is 1-cyano-(4-

dimethylamino)pyridinium tetrafluoro borate (CDAP),[48] which is also available commercially.

$$\left[N\equiv C - N^+ \diagdown \diagup N \diagup^{CH_3}_{\diagdown CH_3} \right] BF_4^- \qquad (10\text{-}5)$$

1 - cyano - 4 - (dimethylamino) -
pyridiniumtetrafluoroborate

CDAP is a quite stable and nonhygroscopic salt that can be safely handled on the laboratory bench in open vessels without health hazards. It can be stored as a solid at room temperature for long periods of time or dissolved in acetonitrile at $-20°C$ for weeks. It hydrolyzes rather slowly in 0.1 M HCl, but complete hydrolysis occurs in a few hours at pH 7.

In the coupling step, regardless of the activation method used, ligand amino groups react with the matrix-bound cyanate esters to form isourea bonds, [Eq. (10-2a)]. In the case of cross-linked dextran and cellulose, which are most conveniently activated by the conventional CNBr procedure, the majority of the activated groups are imidocarbonates which, in the coupling step, can be converted to both N-substituted isourea structures and N-substituted imidocarbonates, as mentioned earlier.

Isourea derivatives have pK values of about 9.5 and therefore are positively charged at neutral pH. The adsorbent thus becomes a weak anion exchanger. This does not usually present a problem. More serious is the fact that the reaction is reversible and the isourea bond can be cleaved (e.g., by hydrolysis at weakly alkaline pH and by aminolysis with low molecular weight amines[43]). It can in fact be demonstrated that single point attached ligands are released at a significant rate.

$$\begin{array}{c} NH_2^+ \\ \| \\ \diagup\!\!\!\!\diagup - O - C - NH - \text{Ligand} + H_2N - R \longrightarrow \end{array}$$

$$\begin{array}{c} NH_2^+ \\ \| \\ \diagup\!\!\!\!\diagup - OH + RNH_2 - C - NH - \text{Ligand} \quad (10\text{-}6) \end{array}$$

Thus, the CNBr technique is not the ideal immobilization method for such ligands. For ligands bound through multiple points such as polypeptides, the rate of release is not greater than for other commonly used immobilization methods and is more dependent on the stability of the matrix used.

10.4.1.2 *Bisepoxirane, Epichlorohydrin, and Divinylsulfone Methods*
Activations based on bisepoxiranes permit the immobilization of ligands containing hydroxyl, amine and thiol groups.[40] Especially useful is the possibility of coupling sugar ligands (e.g., mono- and oligosaccharides). An interesting characteristic is that the ligands will be provided automatically with a hydrophilic spacer arm. The reactions are described in Eq. (10-7).

This activation also introduces covalent cross-linkages in the matrix. Although this might lead to a certain decrease in porosity it is, at least in the case of agarose, advantageous as it increases the stability (e.g., thermostability) and rigidity of the gel.

The oxirane group is rather stable at pH values below 8. The activated matrix can be stored, therefore, as a suspension for prolonged periods of time until used. This, however, also means that rather high pH have to be used to couple hydroxyl (pH 11 to 12) and amino ligands (preferably pH > 9). The method is therefore not suitable for unstable ligands (e.g., many proteins). It should, however, be possible to couple thiol-containing proteins at lower pH values as the thiol (thiolate ion) is a better nucleophile than the other two functional groups. The poor reactivity of the oxirane groups also makes it difficult to eliminate residual activity (remaining groups) after coupling in the blocking step under conditions tolerable for the coupled ligand. Reactions with the commonly used deactivation reagent ethanolamine have to be performed at a rather high pH to be efficient. Thiol reagents such as mercaptoethanol will work at a lower pH, but can only be used if the ligand does not contain easily reduced disulfide bonds. Residual oxirane groups can be determined easily by reaction with sodium thiosulfate.[40] The sodium hydroxide formed is simply titrated with acid.

$$
\text{Gel}-O-\!\!\!\wedge\!\!\wedge\!\!-\overset{\displaystyle O}{\overset{/\quad\backslash}{CH-CH_2}} + Na_2S_2O_3 \longrightarrow
$$

$$
\text{Gel}-O-\!\!\!\wedge\!\!\wedge\!\!-\overset{\displaystyle OH}{\overset{|}{CH}}-CH_2-S-SO_3^- + OH^- + 2\,Na^+ \quad (10\text{-}8)
$$

Typically, a degree of substitution of 50 μmol of active groups per milliliter of gel can be obtained on beaded 6% agarose by varying excess bisepoxirane used in the activation.[40]

The most commonly used bisepoxiranes are 1,4-butanediol-bis(epoxypropylether), ethyleneglycol-bis(epoxypropylether), and 1,2:3,4-diepoxybutane.

The activation and coupling procedures with epichlorohydrin on polysaccharide gels are very similar to those used for bisepoxiranes.[49]

$$
\text{Gel}-O^- + Cl-CH_2-\overset{\displaystyle O}{\overset{/\quad\backslash}{CH}}-CH_2 \longrightarrow \text{Gel}-O-CH_2-\overset{\displaystyle O}{\overset{/\quad\backslash}{CH}}-CH_2 \quad (10\text{-}9a)
$$

Epichlorohydrine   Activated matrix

$$
\text{Gel}-O-CH_2-\overset{\displaystyle O}{\overset{/\quad\backslash}{CH}}-CH_2 + H_2N-Ligand \longrightarrow
$$

$$
\text{Gel}-O-CH_2-\overset{\displaystyle OH}{\overset{|}{CH}}-CH_2-NH-Ligand \quad (10\text{-}9b)
$$

Activation with epichlorohydrin leads to the introduction of gel-bound oxirane groups and of cross-linkages in the matrix. The properties of the activated gel are thus very similar to those obtained with the bisepoxiranes, except for the fact that the spacer introduced is shorter.

As with bisepoxiranes and epichlorohydrin, divinylsulfone (DVS) can also be used for the immobilization of amino-, hydroxyl-, and thiol-containing ligands to hydroxyl-containing matrices.[44]

$$\left\{ \!\!\!\! \text{—} O^- + H_2C =\!\!= CH \text{—} \overset{\displaystyle O}{\underset{\displaystyle O}{\overset{\|}{\underset{\|}{S}}}} \text{—} CH =\!\!= CH_2 \longrightarrow \right.$$

Divinylsulfone , DVS

$$\left\{ \!\!\!\! \text{—} O \text{—} CH_2 \text{—} CH_2 \text{—} \overset{\displaystyle O}{\underset{\displaystyle O}{\overset{\|}{\underset{\|}{S}}}} \text{—} CH =\!\!= CH_2 \quad (10\text{-}10a) \right.$$

$$\left\{ \!\!\!\! \text{—} O \text{—} CH_2 \text{—} CH_2 \text{—} \overset{\displaystyle O}{\underset{\displaystyle O}{\overset{\|}{\underset{\|}{S}}}} \text{—} CH =\!\!= CH_2 + H_2N \text{—} \text{Ligand} \longrightarrow \right.$$

$$\left\{ \!\!\!\! \text{—} O \text{—} CH_2 \text{—} CH_2 \text{—} \overset{\displaystyle O}{\underset{\displaystyle O}{\overset{\|}{\underset{\|}{S}}}} \text{—} CH_2 \text{—} CH_2 \text{—} NH \text{—} \text{Ligand} \quad (10\text{-}10b) \right.$$

Gels activated with DVS are more reactive than those activated with oxiranes. Ligand coupling can therefore be performed at 1 to 2 units lower pH. This is especially useful when coupling hydroxyl ligands such as sugars. However, the adsorbents formed are less stable at alkaline pH than those prepared by the oxirane method. Moreover, DVS is a highly toxic and expensive chemical.

10.4.1.3 *Organic Sulfonyl Chlorides: Tosyl and Tresylchloride Methods*
These methods are most suitable for the immobilization of amino- and thiol-containing ligands to beaded agarose.[50] The reactions are described in Eq. (10-11):

$$\left\{ \!\!\!\! \text{—} OH + Cl \text{—} \overset{\displaystyle O}{\underset{\displaystyle O}{\overset{\|}{\underset{\|}{S}}}} \!\!-\!\!\!\bigcirc\!\!\!-\!\! CH_3 \longrightarrow \right\{ \!\!\!\! \text{—} O \text{—} \overset{\displaystyle O}{\underset{\displaystyle O}{\overset{\|}{\underset{\|}{S}}}} \!\!-\!\!\!\bigcirc\!\!\!-\!\! CH_3 + HCl$$

Tosylchloride   Activated matrix   (10-11a)

$$\text{—NH—Ligand} + {}^-O_3S-\hspace{-0.5em}\bigcirc\hspace{-0.5em}-CH_3 \quad (10\text{-}11b)$$

Organic sulfonyl chlorides react with hydroxyl groups forming good leaving groups, sulfonates, that allow binding of nucleophiles directly to the hydroxyl carbon. This principle can be used for the coupling of low molecular weight ligands, as well as high molecular weight ligands such as proteins, to hydroxyl group-carrying matrices. The method works with both amino- and thiol-containing ligands that become immobilized to the matrix through stable $-CH_2-NH-$ and $-CH_2-S-$ linkages.

The activation has to be performed in nonaqueous solvents, preferably acetone. Thus only matrices that swell in the solvents can be activated. Beaded agarose is most often used, but cellulose and silica derivatives containing hydroxyl groups have also been activated successfully and used for the immobilization of a variety of ligands. Activated matrices can be stored for several weeks as suspensions in 1 mM HCl without losing their coupling capacity.

The coupling can be performed both in an aqueous solvent and in an organic solvent such as DMF. The latter is used when the ligand is not soluble in water.

The reactivity of the sulfonate ester formed is strongly influenced by the R group. Tosylates (R = $CH_3C_6H_5$) and especially tresylates (R = CF_3CH_2) seem to be the most suitable for the immobilization of ligands.[50] In fact, tresylates allow efficient immobilization even at neutral pH and at 4°C. Tosylated matrices, however, are less reactive and require coupling at pH 9 to 10.5 and are thus used with ligands that can tolerate such conditions. Apart from being less expensive than tresylchloride, tosylchloride also has the advantage of releasing a chromphore upon reaction. This means that the coupling reaction can be followed photometrically.

10.4.1.4 Methods Using N,N'-Disuccinimidylcarbonate (DSC) or Carbonyldiimidazol

A two-step method, using DSC, as the activating agent for the immobilization of amino-containing ligands to beaded agarose, has been described by Wilchek and Miron.[51]

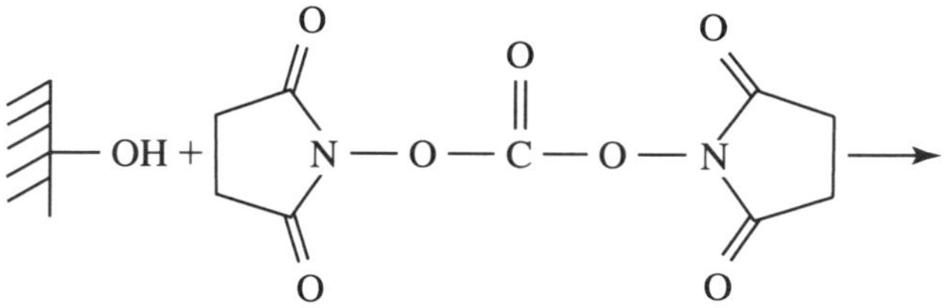

N,N' - disuccinimidyl carbonate (DSC)

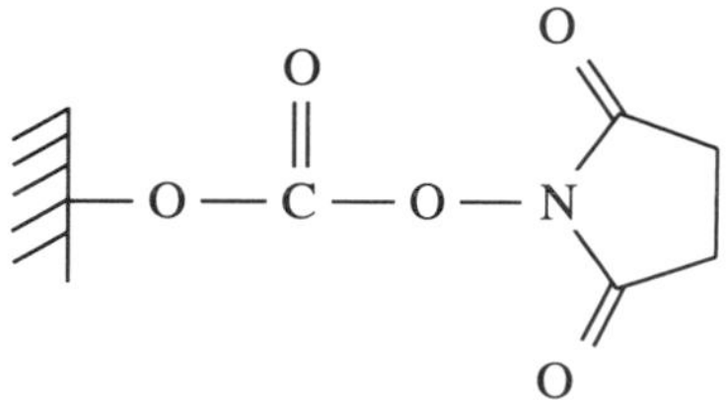

(10-12a)

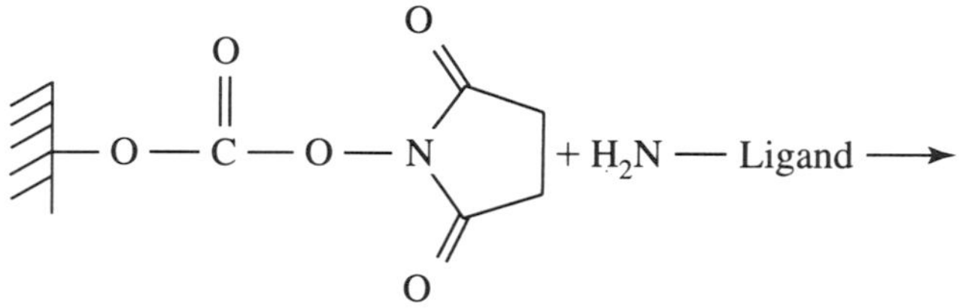

Activated matrix

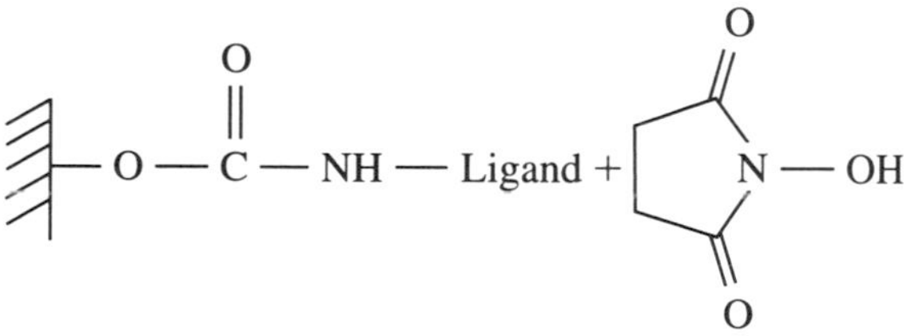

+ H₂N — Ligand ⟶

(10-12b)

N - Substituted carbamate   N - hydroxy succinimide

Hydroxysuccinimide carbonate groups are first introduced into the gel by reaction of its hydroxyl groups (especially primary ones) with DSC in organic milieu with a base catalyst such as TEA. The carbonate subsequently reacts with amines under the formation of carbamates. The coupling reaction runs both in aqueous systems under mild conditions (pH 6 to 8) and in organic solvent (if a base catalyst such as TEA is used). It can be used to immobilize amino-containing ligands of both high molecular weight (such as sensitive proteins) and low molecular weight types. Unlike the isourea bond formed in the CNBr methods, the carbamate bond is very stable and non-charged under conditions usually employed for affinity chromatography.

In aqueous systems the hydroxysuccinimide carbonate groups rapidly decompose by hydrolysis with regeneration of the gel hydroxyl groups and release of N-hydroxysuccinimide. The hydrolysis is faster at higher pH but proceeds at an appreciable rate even at neutral and weakly acidic pH. The activated gel should therefore be protected from water before it is mixed with the ligand solution. N-Hydroxysuccinimide has a λ_{max} at 261 nm with a molar extinction coefficient of 10,000 $M^{-1}cm^{-1}$.

The degree of activation of the gel can thus be determined photometrically after complete hydrolysis of the hydroxysuccinimide carbonate groups at high pH or after aminolysis with hydroxylamine. The hydroxysuccinimide groups also react to some extent with neighboring hydroxyl groups on the gel under formation of gel-bound carbonate groups. This seems to occur in conditions favoring a high degree of substitution (large excess of DSC). Despite their low reactivity, the carbonate groups formed might cause problems later on when the gel derivative is used as affinity adsorbent by inadvertently immobilizing reactive substances in the sample. The carbonate groups in the gel can be eliminated by prolonged exposure to alkaline pH (about 11), conditions that might be detrimental to the ligand.

The carbonyldiimidazol (CDI) reagent can also be used to activate hydroxyl-containing matrices.[52] The imidazoyl carbonate groups thus introduced into the matrix react with amino-containing ligands with the formation of carbamates. The activation has to be performed in an organic solvent, but the coupling can be run in aqueous systems. The imidazoyl carbonate groups are not as reactive as the groups introduced in the similar DSC method. Thus the ligand coupling has to be performed at a higher pH. The degree of activation (moles of imidazoylcarbonate groups per milliliter of gel) can be determined by keeping the activated gel at pH 3 for 4 hr. This treatment leads to hydrolysis of the reactive groups and to the release of imidazol, which can be determined by titration between pH 9 and 4.[53]

10.4.1.5 *Condensation Methods Based on Carbodiimides*

For a long time the carbodiimides have been used for synthesis of peptides. They were also among the first reagents to be employed in the synthesis of affinity chromatography adsorbents and are still among the most widely used. Using these reagents, stable amide bonds can be formed between a ligand that contains an amino group and a carboxyl-containing matrix (or vice versa) in a one-step procedure.

The reaction is performed by mixing the ligand with the matrix together with the reagent at slightly acidic pH (about 5) (pH adjustment with HCl and buffer).

Carboxylgroup   Carbodimide   Isourea ester
containing gel

(10-13a)

$$
\begin{array}{c}
\overset{O}{\underset{\|}{C}} \quad \overset{N-R}{\underset{\|}{}} \\
-C-O-C-NH-R' + H_2N-Ligand \longrightarrow
\end{array}
$$

$$
\overset{O}{\underset{\|}{C}}-NH-Ligand + R-NH-\overset{O}{\underset{\|}{C}}-NH-R' \quad (10\text{-}13b)
$$

N,N' - dialkylurea

The first step in the condensation is the addition of the carboxylate to either C $=$ N bond of the diimide, yielding the highly reactive and unstable O-acylurea (isourea ester). This, in turn, reacts chiefly with the amine to produce an amide and N,N'-dialkylurea. The major side reaction, the intramolecular rearrangement of the O-acylurea and formation of a stable N-acylurea, can be minimized if a large excess of amine-containing ligand is used. If that cannot be achieved, some of the resulting N-acylurea will be bound to the matrix.[39]

The reaction time is usually several hours and it is necessary to adjust the pH in the suspension during the first hour. Carbodiimides are relatively unstable compounds and must be handled with care because of their toxicity. The method is usually used to couple ligands to spacer arms with carboxyl or amino groups as terminal groups, but can of course also be used to immobilize ligands to any matrix containing amino or carboxyl groups.

Carbodiimides that have been used in the preparation of derivatives for affinity chromatography include 1-cyclohexyl-3-(2-morpholinoethyl)carbodiimide-p-toluenesulfonate (CMCL) and 1-ethyl-3-[3-(dimethylamino) propyl]carbodiimide hydrochloride (EDCL).[54–57]

An alternative way to activate carboxyl-containing matrices is to convert the carboxyl groups to reactive esters by reaction with N-hydroxysuccinimide in anhydrous medium in the presence of a carbodimide [e.g., dicyclohexylcarbodiimide (DCC)].[58] The matrix-bound N-hydroxysuccinimide ester reacts easily at pH 5 to 8 with amines with the formation of an amide and release of N-hydroxy-succinimide.

$$
-C-OH + HO-N \xrightarrow{\text{DCC}} -C-O-N \quad (10\text{-}14a)
$$

N - hydroxysuccinimide   Reactive ester   (10-14a)

$$(10\text{-}14b)$$

The ester is labile and undergoes hydrolysis in aqueous solutions, especially at a pH above 6. The activated gel should therefore be stored as a suspension in an anhydrous medium, (e.g., dioxane). For shorter periods (minutes to a few hours) it can be stored in water at pH 5 to 6, provided that there are no amines present. As the technique is more laborious than those based on a direct use of a condensation reagent it is only recommended for ligands containing both primary amines and carboxyl functions (e.g., proteins), where the use of a condensation reagent in the coupling would lead to unwanted inter- and intramolecular cross-linkages.

10.4.1.6   Methods for Thiol-Containing Matrices   So far two main principles for the immobilization of ligands to matrices have been discussed: (1) Introduction of electrophilic groups into the matrix that subsequently react with nucleophilic groups in the ligand, and (2) use of condensation agents to establish amide bonds between amino groups in the ligand and carboxyl groups in the matrix (or vice versa).

Another possibility is to immobilize a ligand by means of an electrophilic group (either already present or introduced prior to the immobilization) that reacts with a nucleophilic group in the matrix. Although not widely used, this approach has several merits. It is most easily applied to thiol-containing matrices. Thus, reaction of a thiol-containing matrix with alkyl or aryl halides and ligands containing $C = O$, and under certain conditions $C = C$ bonds, will lead to stable thio ether derivatives through nucleophilic displacement (in the first case) and addition (in the later case). Unsaturated compounds such as testosterone and estradiol have been attached to beaded thiol agarose using γ radiation.[59] Heavy metal ion-containing ligands can also be bound to a matrix via thiol groups by mercaptide formation.[60] Activated halides with the halogen α to a carbonyl group (as in iodoacetic acid) react smoothly at weakly alkaline pH with thiols. These reactions usually take place in aqueous or polar organic solvents under rather mild conditions, but for unactivated halides higher pH values have to be used. Estradiol and testosterone, mentioned earlier, require higher pH values.

Most matrices suitable for affinity chromatography can be provided with aliphatic thiol groups by simple organic chemistry. Several methods for the thiolation of polysaccharides, especially beaded agarose, are described in detail in Chapter 9 on covalent chromatography. Porous glass and silica are also easily substituted with thiol groups by their silanization with γ-mercaptopropyltrimethoxysilane.[48]

10.4.1.7 The Isocyanide Method Most of the ligand immobilization methods described earlier require that a reactive amino group is present in the ligand, although in some cases they also work with hydroxyl or thiol compounds. Much more flexible is the so-called isocyanide (isonitrile) method, which, despite the fact that it was described a long time ago, has not attracted much interest.[62] The method is based on the four-component condensation of amine, carboxyl, isonitrile and carboxyl compounds originally discovered and examined by Ugi and colleagues.[63]

The principle is outlined in Eq. (10-15):

$$R^1-\overset{\overset{\displaystyle O}{\|}}{C}-OH \quad H_2N-R^2$$

Carboxyl   Amine

$$+ \longrightarrow R^1-\overset{\overset{\displaystyle O}{\|}}{C}-N-\overset{\overset{\displaystyle R^2}{|}}{C}H-CONH-R^4$$

$$R^4-N\equiv C \quad H-\overset{\overset{\displaystyle O}{\|}}{C}-R^3$$

Isonitrile   Aldehyde

(10-15)

An ammonium ion structure is formed from an aldehyde or ketone and an amine. With the isocyanide, this structure forms a highly reactive intermediate that is very susceptible to addition reactions with nucleophiles, such as carboxylic and hydroxylic ions. A stable amide is finally produced by intramolecular rearrangement. Although the reaction appears complicated, the technique is, in practice, very simple to use for immobilization purposes.

The ligand to be attached may contain amino, carboxyl, aldehyde, ketone, or isocyanide functions. The matrix may contain any of the others. The two remaining functional groups are added to the reaction mixture as low molecular weight substances (e.g., aliphatic amines, carboxylic acids, aldehydes, ketones, isonitriles).

The reaction occurs in aqueous medium at pH 5 to 6. Matrices that have been used include cross-linked dextran and beaded agarose containing carboxyl groups, polyacrylamide, agarose substituted with amino groups or carbonyl groups, and a large number of insoluble polymers containing isonitrile groups.[62,64] Immobilized substances are proteins, peptides, amino acids, biotin, and steroids (which have been used as carbonyl compounds).

The coupling of a protein can be directed toward its amino or carboxyl groups using an excess of low molecular weight amine or carboxyl compound, respectively.

10.4.2 Methods for Polyacrylamide Matrices

The glutaraldehyde and hydrazine methods are suitable for matrices having amide groups, such as polyacrylamide, and for the immobilization of amino-containing ligands.[41,65] These methods do not work on polysaccharide matrices. The mechanism of the activation and coupling is not completely understood but is supposed to follow the scheme outlined in Eq. (10-16).

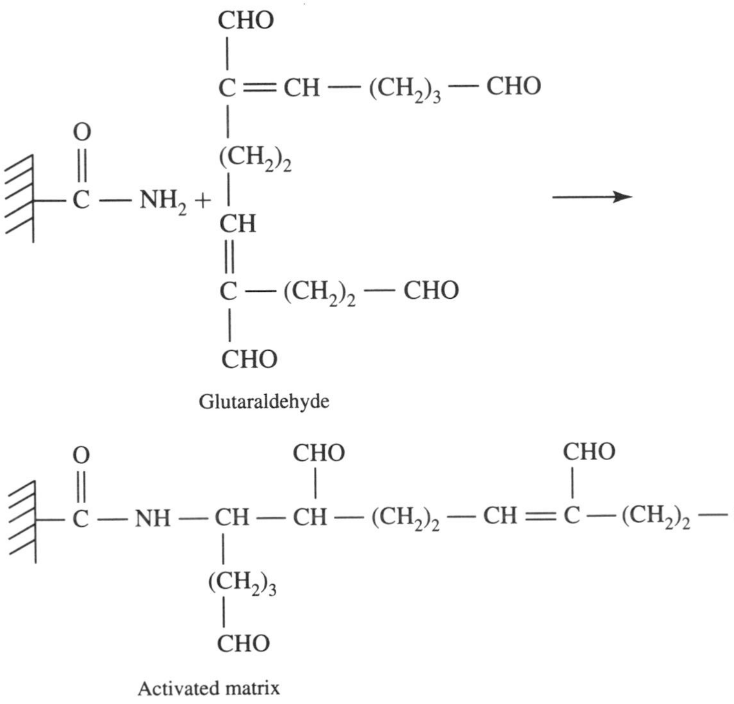

Glutaraldehyde

$$\text{(10-16a)}$$

Activated matrix

Activated matrix $+ H_2N$ — Ligand $\longrightarrow$

$$
\text{—}C \underset{\overset{\|}{O}}{} \text{— NH — CH} \underset{\underset{\text{CHO}}{\overset{|}{(CH_2)_3}}}{\overset{|}{}} \text{— CH} \underset{}{\overset{\overset{\text{CHO}}{|}}{}} \text{— (CH}_2)_2 \text{— CH} \underset{\underset{\text{Ligand}}{\overset{|}{NH}}}{\overset{|}{}} \text{— CH} \underset{}{\overset{\overset{\text{CHO}}{|}}{}} \text{— (CH}_2)_2 \text{— CHO (10-16b)}
$$

Immobilized ligand

The yields obtained in this immobilization depend on the ligand. Advantages with the technique are that a spacer arm is introduced as a result of activation, that the coupling can be carried out with good yields at pH around neutral, and that the bonds formed are stable enough to prevent leakage of ligands.

Hydrazine is a particularly useful reagent in combination with amide-containing matrices. When polyacrylamide is heated with hydrazine, hydrazide acrylamide (or polyacrylhydrazide) is formed. The hydrazide groups can be converted to reactive acylazides by treatment of the matrix with sodium nitrite in hydrochloric acid.[66] Amino-containing ligands can then be coupled to the activated matrix by the formation of stable amide bonds (Figure 10-2). It has been claimed that yields when coupling high molecular weight ligands are usually low, possibly because of a combination of instability of the acylazide group and steric hindrance. For low molecular weight ligands the technique seems to work well.

The hydrazide matrix can also be used for the immobilization of ligands containing aldehyde and ketone groups.[68] This reaction occurs at low pH (about 5) with formation of a hydrazone, which is then stabilized by reduction with alkaline sodium borohydride (pH about 9) (Figure 10-2). Although ligand immobilization by means of hydrazines is most simply performed on amide-containing matrices, it can also be used in combination with other matrices such as agarose and porous glass, provided they are modified in a suitable way.[68,69]

10.4.3 Methods for Silica Gels and Porous Glass Matrices

Although silica and porous glass beads show good dimensional stability and thus can cope with the often rather high pressures used in HPLAC, they do not fulfill several of the other requirements of an ideal affinity support (Section 10.3.1). Their hydrophobic character and content of negatively charged silanol groups can be changed by chemical modification. In addition to making the

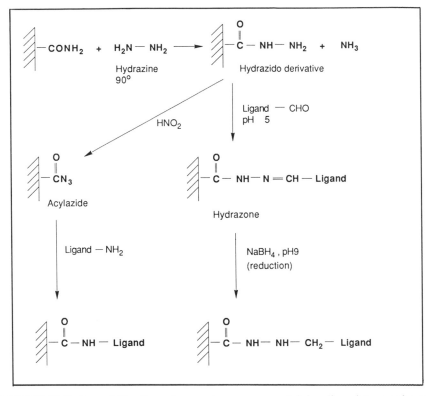

FIGURE 10-2. Immobilization of an amino group containing ligand to a polyacrylamide matrix by the hydrazine method.

matrix surface more hydrophilic and masking the acidic silanol groups, the modification should provide the matrix with functional groups suitable for the coupling of ligands either directly or after activation.

A commonly used procedure (Figure 10-3) is silanization with reagents such as γ-aminopropylsilane and γ-glycidoxypropylsilane[34]:

$$
\underset{\gamma \text{ - aminopropylsilane}}{H_3CO - \overset{\displaystyle OCH_3}{\underset{\displaystyle OCH_3}{\mid}} - (CH_2)_3 - NH_2}
\qquad
\underset{\gamma \text{ - glycidoxypropylsilane}}{H_3CO - \overset{\displaystyle OCH_3}{\underset{\displaystyle OCH_3}{\mid}} - (CH_2)_3 - O - CH_2 - CH - CH_2}
\qquad (10\text{-}17)
$$

Reaction with the first of these reagents gives an amino group functionalized matrix to which many affinity ligands are easily attached either directly by

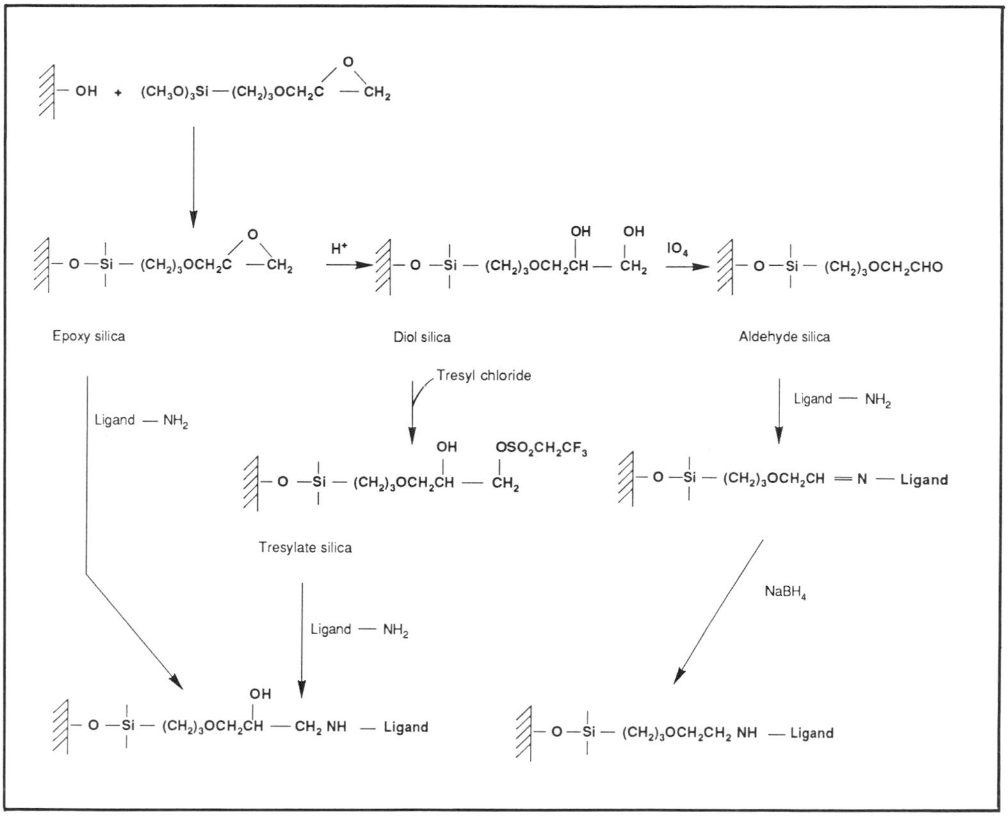

FIGURE 10-3. Immobilization of an amino group containing ligand to a silica matrix using activation with γ-glycidoxypropylsilane.

use of the condensation reagents discussed earlier or after further derivatization. The matrix may be converted to a carboxyl group containing matrix by reaction of the amines with a suitable anhydride. Unfortunately, these reactions turn the silica or glass into an ion exchanger.

Silanization with γ-glycidoxylpropylsilane leads to the introduction of oxirane groups into the silica or porous glass.[34] As discussed previously, this group reacts with nucleophiles with the formation of stable bonds and allows the coupling of ligands containing amino, hydroxyl, or thiol groups to the activated matrix. Because of steric shielding, only a small fraction of the oxirane groups are used up in the ligand immobilization. The excess groups may be hydrolyzed to diols by treatment with acid. These diols give the silica surface a hydrophilic character and decrease its tendency for unwanted protein binding.

10.5   CHROMATOGRAPHIC TECHNIQUES

Having immobilized the ligand or selected the ready-made affinity gel, the next step is chromatography. A typical separation by affinity chromatography consists of four stages: adsorption, washing, elution, and column regeneration. The general practical aspects for affinity chromatography are described in this section. For detailed practical recommendations on individual adsorbents, reference should also be made to the manufacturer's instructions.

10.5.1   Sample Preparation; Prefractionation

Many affinity separations constitute one-step purification procedures. However, it may be advantageous to include a preliminary step using precipitation or ion-exchange chromatography (see Chapter 1), as this removes some of the major contaminants, reduces the amount of material that must subsequently be processed, and improves the resolution and concentrating effects of the affinity step. This preliminary separation step is recommended when large amounts of sample are to be processed or when the substance of interest is in the presence of very large amounts of precipitating contaminants, such as lipoproteins and clotting factors. When a protein ligand is used in the affinity step, there is also a risk of causing proteolysis by proteolytic enzymes in the extract. This is diminished by prefractionation.

For efficient adsorption, the conditions (pH, ionic strength, etc.) of the sample and the column must allow efficient binding and preferably be optimal for binding. This can be achieved by dialysis of the sample, by desalting and exchange of the buffer by gel filtration, or by adding chemicals or adjustment of pH.

When the starting material consists of tissue, cell culture, fermentation product, or plant material, other steps, such as solubilization, homogenization, extraction, filtering, and/or centrifugation, are included in the fractionation scheme, as described in Chapter 1.

10.5.2   Column Operation

The column size is usually not a critical parameter. Rather the column size is governed by the capacity of the adsorbent and the amount of the substance to be purified. Capacities of affinity gels are generally high, and short, wide columns are often used to obtain rapid separations on beds of usually 1 to 10 ml of gel.

Typical capacities are 25 mg immunoglobulin/mL gel on Protein A Sepharose CL-4B, 10 mg lactate dehydrogenase/mL gel on AMP Sepharose, and more than 1 mg fibronectin/mL gel on Gelatin Sepharose 4B.

If the binding affinity of the ligand is markedly low ($K_D > 10^{-4}\ M$) the protein to be purified will not bind but will be retarded on the column (isocratic elution), and the separation of the desired protein from the unbound contaminants will become dependent on the column length. Longer columns are recommended. It is also advisable to use small sample volumes, (e.g., 5% of the total bed volume). Ligand density should be maximized and flow rates reduced in isocratic elutions.

Column packing should follow the usual precautions for chromatographic techniques. Use the recommended flow rates and pack evenly to ensure straight bands.

To achieve binding equilibrium, several approaches apply. Sometimes batchwise procedures are preferred with gentle stirring during both adsorption and desorption or only during the adsorption step. When columns are used they can be left with reduced or stopped flow to make prolonged contact possible to achieve adequate adsorption and desorption, respectively.

This chapter only gives general recommendations with respect to flow rates. The investigator has to compromise between the acceptable level of resolution, the time required for the separation, and the utilization of available adsorption capacity.

However, for sharp elution peaks and maximum recovery with minimal dilution of the purified protein, the lowest flow rate acceptable from a practical point of view should be used. This is especially important for adsorbents that bind several proteins and when competitive elution is used. Here the flow rate used is governed by the rate of dissociation from the ligand.

For standard low-pressure affinity chromatography flow rates 50 cm/h and higher have been used successfully. In HPLAC (see later) the improved mass transfer properties, giving higher dynamic capacities, allow the separation to be performed at much higher linear flow rates, with sharper elution peaks giving smaller elution volumes. Typical flow rates used in HPLAC are between 50 and 125 cm/h. In some cases, flow rates in excess of 200 cm/h have been used successfully in industrial scale applications.

10.5.3 Adsorption of the Sample; Washing

For efficient adsorption the column and sample must be equilibrated with a buffer reflecting the conditions optimal for binding. The volume in which the sample is applied is not critical, provided that the ligand binds specifically and effectively and that the total capacity of the column is not exceeded.

After sample application the column must be washed with several volumes of the starting buffer to remove all unbound material. The chromatography is usually monitored by UV-absorbance, and the washing step is finished when the original base line is reached. It is recommended to wash with high ionic strength buffers when nonspecific ionic binding is suspected. Nonspecific hydrophobic binding is less easily handled as the specific binding often depends on hydrophobic interactions.

10.5.4 Desorption

The principle of desorption is to change the binding equilibrium for the adsorbed substance from the stationary to the mobile phase. This can be achieved specifically or nonspecifically.

Ligand–protein interaction is often based on a combination of electrostatic, hydrophobic, and hydrogen bonds. Agents that weaken such interactions might be expected to function as effective nonspecific eluants. Careful consideration of the relative importance of these three types of interaction, as well as the degree of stability of the bound protein, will help in the choice of a suitable eluant. A compromise may have to be made between the harshness of the eluant (or the combination of eluants) required for effective elution and the risk of denaturing the proteins to be purified. Elution can be the most difficult stage of affinity chromatography, especially if the dissociation constant of the ligand is very low.

Broadening of desorption peaks is often a problem. This can be due to slow diffusion mass transport or to slow equilibration kinetics or a wide range of binding affinities in the system. The diffusion is usually not limiting in the HPLAC gels but might be so in the larger standard gel beads. Slow dissociation from the ligand can be a problem regardless of the gel. Peak shapes are often improved by reversing the direction of flow during desorption, as the distance in the column covered by the desorbed protein is minimized. This is called *reversed elution.*

Gradient elution often gives excellent results in affinity chromatography. A heterogeneity in the sample with regard to binding to the column is then used. For example, isoenzymes can often be separated by gradient elution[70] and glycoproteins can be separated according to affinity on immobilized lectins by a sugar gradient.[71]

10.5.4.1 Change in pH or Ionic Strength   The most frequently used method for eluting strongly bound substances nonspecifically is by decreasing the pH of the buffer[72,73]; sometimes an increase in pH can also be effective.[74,75] The chemical stability of the matrix, the ligand and the adsorbed substance determine how low the pH can go. This is usually around pH 2 to 4. It is important to neutralize the fractions as soon as possible after elution. This is most easily done by pipetting a small volume of buffer into each tube of the fraction collector in advance.

An increase in the ionic strength of the buffer elutes proteins bound by predominantly electrostatic interactions.[76,77] Such interactions typically dominate binding to dye columns. Usually 1 M NaCl is sufficient, but occasionally 2 or 3 M salt is required.[78] Continuous or stepwise salt gradients can resolve different proteins adsorbed to a particular dye column.[79]

10.5.4.2 Change In Polarity   When the binding is very strong and dominated by hydrophobic interactions, rather drastic methods of elution have to be used, such as reducing the polarity or including a chaotropic salt or denaturing agent in the buffer. This type of elution is typical for immunosorbents based on immobilized polyclonal antibodies.[80–82]

Commonly used chaotropic salts are KSCN, KCNO, and KI in the concentration range of 1 to 3 M; urea and guanidine HCl in moderate concentrations

(4 to 6 M) are sometimes preferred. The polarity can often be decreased enough to promote elution by including 20 to 40% ethylene glycol in the buffer, although sometimes much higher values are required.[83] It is effective as well as being mild and less likely to denature proteins.

When eluting most hydrophobic proteins, such as membrane proteins, decreasing the hydrophobic interactions with detergents is often preferred. Useful detergents are Lubrol, Nonidet P-40, or octylglycosides; Triton X-100 is disadvantageous because of its UV adsorption. The concentrations used are just below the CMC (Table 10-5).

A low concentration of a detergent is sometimes included during the entire purification process (see Chapter 1 and Ref. 84), Thus, nonspecific hydrophobic adsorption, as well as aggregation, can be suppressed, particularly in antigen–antibody purification or when handling membrane proteins. The detergents can also be left in the buffer when eluting with, for example, a moderate change in ionic strength or pH. An alternative is to increase the detergent concentration to achieve elution.

One example of an affinity system with very strong interactions is rat biotin-binding protein—biotin-AH-Sepharose.[85] This protein binds very tightly to the adsorbent and does not elute with a saturating concentration of biotin (0.004 M) or with other desorption methods such as elevated temperature (40°C), 8 M urea, 8 M KI, or low pH 3.6. However, the protein could be eluted by 3 M guanidine–HCl or by 2 M urea plus 0.004 M biotin in the equilibration buffer.

10.5.4.3 Specific Elution In specific elution, bound proteins are desorbed from the ligand by the competitive binding of the eluting agent either to the ligand or to the protein. Specific eluents are most frequently used with group-specific adsorbents as the selectivity is greatly increased in the elution step. Glycoproteins can be desorbed from lectin columns by elution with competing carbohydrates.[71,86] Desorption of a lectin from an immobilized carbohydrate with a competing carbohydrate is described in Section 10.6.2.1 and in Ref. 87. Specific elution has also proved to be effective when isolating receptors.[88]

Elution generally occurs at rather low concentrations of eluant (5 to 100 mM). A concentration of a single eluant or pulses of several different eluants can be used as well as gradients. Specific elution is often performed at neutral pH and is thus a mild desorption method causing little or no denaturation. If the eluting agent is bound to the protein, it can be dissociated by desalting on a gel filtration column or, when the binding is stronger, by dialysis.

10.5.5 Regeneration of Adsorbent

All affinity adsorbents should be reusable. The exact extent of reuse depends on the nature of the sample and the stability of the ligand and matrix with respect to the elution and cleaning conditions used.

The most important aspect in regeneration is to remove any material still bound to the adsorbent. In most cases it is sufficient to reequilibrate with several column volumes of starting buffer. If necessary, one can wash with buffers of alternately high and low pH and even include detergents and denaturing agents, depending on the stability of the ligand. For ready-made adsorbents the manufacturer's instruction should be followed.

10.5.6 High-Performance Liquid Affinity Chromatography

High-performance liquid affinity chromatography (for a review, see Ref. 34) combines the high specificity and selectivity of affinity chromatography with the speed of HPLC. As supports are used mechanically stable particles, provided with a noncharged, hydrophilic surface, which can withstand prolonged use in aqueous buffers at moderate or high pressures. The particles are usually considerably smaller than the ones used in standard low-pressure affinity chromatography. This is because smaller particles lead to improved efficiency due to shorter diffusion distances (see Chapter 2). Solid beads offer optimum conditions in this respect. However, to get adequate surface area, <5-μm average particle diameters are required.

HPLAC was introduced by Mosbach and co-workers[70] in 1978 by demonstrating separations on derivatized 10-μm macroporous silica beads. Since then other noncompressible porous particles such as highly cross-linked agarose, methacrylate, copolymers of ethylene glycol and methacrylate, and methacrylate vinyl copolymers have been employed as support materials. Supports substituted with a number of standard affinity chromatography ligands have been described for HPLAC. To provide maximum flexibility of application, supports with reactive or activable groups, to which desired ligands can be covalently bound, have been prepared as well.[89]

HPLAC is often performed using prepacked columns. Such columns containing supports with various immobilized ligands can be purchased from several manufacturers. Examples are Selecti-Spher-10™ concanavalin A, SelectiSpher-10™ protein A and SelectiSpher-10™ protein G (Pierce Chemical Company), and Protein A Superose®, Protein G Superose®, and Chelating Superose® (Pharmacia Biotech AB). HPLAC columns with several ligands such as biotin, Con A, and heparin are available from Showa Denko K.K., Durasphere (silica 7 μm). Silica substituted with Cibachrome blue F3GA, Con A, protein G, protein A, mellitin, and heparin are available from Alltech. TSK 5PW and ToyoPearl (ethylene glycol/methacrylate copolymers, 10 and 40 to 90 μm, respectively) substituted with Cibachrome blue F3GA, aminophenyl boronate, heparin, and protein A can be purchased from TosoHaas. Sigma Chrom affinity media (methacrylate 2 μm) are available substituted with lysine, N-acetylglucose amine, avidin, Cibachrome blue, Procion red, glutathione, Con A, lentil lectin, ricin lectin, wheat germ lectin, 5'-adenosine monophosphate, heparin, pepstatin, protein A, and protein G.

Also available are commercially prepacked preactivated columns such as NHS-activated Superose® (Pharmacia Biotech AB), SelectiSpher-10™ activated tresyl (Pierce Chemical Company) and Ultraffinity™ EP columns (Beckman Instruments), epoxy and divinylsulfon-activated Durasphere, TSK 5PW and Toyopearl with tresyl and epoxy groups, and Affyprep (Bio-rad) with N-hydroxysuccinimide. With these products the users can easily perform *in situ* immobilization of the desired ligand.

It is also possible to prepare HPLC adsorbents from, for example, silica and cross-linked agarose with the batch activation and immobilization methods described in Sections 10.3 and 10.4. The user should, however, be aware of the technical difficulties of obtaining well-packed HPLC columns from small diameter particle suspensions (see Chapter 2).

Small (<4 μm) nonporous particles have been used as an alternative to the larger (10 to 90 μm diameter) macroporous particles.[90] Thus amino functions were intoduced into nonporous, monodisperse, polystyrene beads (average particle diameter 3.7 μm) by nitration and subsequent reduction. Two affinity adsorbents were prepared by covalent coupling of the ligands p-aminophenyl-β-D-glucopyranoside and p-aminobenzamidine using the cross-linker hexamethylenediisocyanate and aminated polystyrene particles. The adsorbents were used for HPLAC of concanavalin A and trypsin, respectively.[90]

HPLAC has so far been used mainly in micropreparative and analytical applications. Thus cardiac myosin and actin have been purified in one step using salicylate immobilized to a preactivated Ultraffinity EP column.[91] Desorption was achieved by specific competitive elution and by increased ionic strength.

Human IgG oligosaccharides have been resolved by serial affinity chromatography on lectin columns.[92] This application demonstrated nicely the resolution of peaks obtained with small differences in k' during isocratic elution. Extremely fast (20 sec or less) analytical separations have been obtained on affinity packings using minicolumns (<2 cm length).[93] One example of an analytical application of HPLAC is the determination of plasmin and plasminogen in human blood using Toyopearl (a hydrophilic vinyl polymer resin) substituted with the trypsin inhibitor p-aminobenzamidine as the affinity medium.[94] The plasmin/plasminogen system plays an important role in fibrinolysis and its monitoring therefore is important in diagnosis, prevention, and therapy of various disorders such as thrombosis and myocardial infarction.[94]

The following considerations should be kept in mind when selecting a support to prepare a HPLAC adsorbent: When a suitable commercial HPLAC prepacked column or adsorbent cannot be found for a particular separation problem, it is always possible to prepare an adsorbent from an unsubstituted support. Thus, a large variety of well-characterized silica, agarose, and resin base matrices are available commercially as are a number of reagents for their derivatization (see earlier discussion). As previously pointed out, the major disadvantage of silica particles is their instability in alkali, which leads to degradation and ligand leakage.

Cross-linked polystyrene particles are available commercially as both solid and porous beads and are much more chemically stable than silica. However, these particles are very hydrophobic and their surface has to be modified to be more biocompatible to prevent denaturation or unspecific binding of proteins. The modification and introduction of the desired ligands can also be a rather lengthy and difficult task for individuals with ordinary skill in organic preparative chemistry.

The agarose-based beads (e.g., Superose and Sepharose High Performance) are reasonably biocompatible, and a number of easy and straightforward synthetic routes exist for their substitution with a large variety of ligands. However, even the most extensively cross-linked agarose beads are softer than both polystyrene and silica, which means lower flow rates through agarose-based columns and thus slower separation procedures. In addition, it is not possible to prepare agarose beads of the same degree of monodispersity as the best polystyrene particles.

10.6   APPLICATIONS

10.6.1   Immobilization of Ligands

The following sections present the experimental details of a number of commonly used procedures for the immobilization of different ligands to various matrices. Although most of the methods have been developed for beaded agarose as a matrix, they can often be used in combination with other polyol matrices, provided that they retain their structure in the solvents used during the activation and coupling steps.

10.6.1.1 Some General Advice   Before and after activation and after coupling, the matrix is usually washed and equilibrated with different buffer solutions. These washing steps are most easily performed on a glass filter funnel fitted on a Buchner flask (connected to a vacuum suction device). The matrix is suspended in a suitable volume of the desired medium, which is subsequently filtered off, or the reaction mixture is poured into the filter and the liquid with soluble components is sucked off. The matrix is resuspended in a new aliquot of medium and the liquid is again removed by filtration. This procedure is repeated until the unwanted soluble components are removed or when equilibrating the matrix with a new medium. Allow ample time in each step for the diffusion of soluble reagents and reaction products in and out of the beads. Often the filtration can be performed just by gravity flow or at a very small negative pressure. After each filtration a glass rod can be used to resuspend the matrix in a new portion of buffer.

Never let the matrix dry out completely on the filter during the filtration process (leave a few millimeters of liquid on top of the matrix surface) as this might lead to unwanted aggregation of the beads. This is especially critical when agarose is transferred from an aqueous to an organic medium such as acetone.

To obtain fast kinetics and efficient reactions, the activation, coupling, and deactivation should be carried out by suspending the nonreacted, activated, or coupled matrix in a minimum volume of liquid containing the reagent, ligand, and deactivation compound, respectively, to get a slurry that is dense but still can be agitated. As most beads, particularly agarose beads, are mechanically fragile, agitation with stirrers, fleas, and other harsh conditions should be avoided to minimize disintegration. For small volumes (up to 20 mL of suspension), end-over-end rotation can usually be employed by using sealed off tubes as reaction vessels (provided that there is no gas evolved as a result of the reaction). A shaking board can also be used.

When the coupling reactions are performed in a buffered medium, it is essential to select buffers that do not react with the activated matrix in competition with the ligand to be immobilized. Thus, amino-containing buffers should be avoided when the ligand is coupled through an aminofunction, carboxylic acids should not be used as buffers when the ligand is to be coupled via a carboxyl-group, and so on. When condensation agents such as carbodiimides are used, neither amines nor carboxyl-containing buffers can be used. In most cases the reagents used for activation are highly toxic and often volatile.

It is therefore recommended that all work be performed in a ventilated fume hood until the washed and drained activated gel is incubated with the ligand. After coupling of the ligand, residual activated groups should be deactivated by reaction with an excess of a suitable low molecular weight compound (ethanolamin, mercaptoethanol etc.) under the same conditions used for coupling the ligand.

The coupled matrix should then be washed very thoroughly to remove all noncovalently bound material. If possible, both alkaline and acidic buffers with varying ionic strength should be used. A tentative washing protocol could be

1. 0.1 M sodium phosphate buffer, pH 8
2. The same buffer with 0.5 M NaCl
3. 0.1 M sodium acetate buffer, pH 5
4. The same buffer with 0.5 M NaCl
5. The buffer in which the adsorbent is to be stored or used in the affinity chromatography

The procedure should, of course, be modified according to the special properties of the ligands. Thus, if an organic solvent has been used, the washing should start with this solvent. If necessary, an appropriate detergent can also be included in the washing buffers.

Until used, the prepared affinity adsorbents are best stored as suspensions at 4°C, the medium depending on the stability of ligand, matrix, and covalent bond employed. The suspension should also contain some antibacterial substance such as 0.02% sodium azide or Merthiolate.

10.6.1.2 The CNBr Method

The method described is essentially according to Kohn and Wilchek,[48] which is a modification of the original CNBr buffer procedure originally presented by Porath and co-workers.[47]

Activation step. The agarose gel (e.g., Sepharose 4B) is washed with distilled water. The drained gel (10 g) is then mixed with 10 mL distilled water and 20 mL of a 2 M sodium (or potassium) carbonate solution. The obtained suspension is cooled to 0°C using an ice bath. An approximately 10 M solution of CNBr in acetonitrile, DMF, or N-methylpyrrolidone is prepared by dissolving 1 g of CNBr in 1 mL of organic solvent. This solution is added, all at once, to the gel suspension under vigorous agitation. After exactly 2 min the reaction mixture is transferred onto a glass filter funnel and washed with ice-cold water until all CNBr is removed. This washing procedure should be done as quickly as possible so as not to lose reactive groups by hydrolysis. The activated agarose beads should be used immediately for coupling. This procedure gives a coupling capacity for low molecular weight amines of about 10 μmol/g drained agarose gel, which is about half the amount that can be obtained by the CNBr titration procedure.[46] For proteins the differences are less pronounced. The method has the advantage of being reliable and easy to perform compared with the titration method, which also requires automatic titration equipment.

Coupling step. The ligand to be coupled is dissolved in 0.1 M NaHCO$_3$, pH 8.3, containing 0.5 M NaCl. This is the most commonly used coupling buffer. Other buffers with lower ionic strength in the range of 8 to 10 (such as borate or phosphate buffers) can also be used. In this pH interval the aminogroups on the ligand are predominantly in the unprotonated and reactive form. The high salt concentration in the buffer (0.5 M NaCl) serves to minimize ionic protein–protein adsorption.

The amount of ligand added to the reaction mixture depends on what ligand density is desired. As a rule, a higher ligand concentration in the reaction medium leads to higher ligand content in the adsorbent. However, as discussed earlier, it might not always be advantageous to obtain a very high ligand content as this might have adverse effects on the affinity chromatography. For an efficient adsorbent, 1 to 10 μmol of low molecular weight ligand/ mL gel and 5 to 10 mg protein/mL gel for protein are recommended. These figures can be obtained by adding two to three times excess protein to the reaction mixture. A lower ligand concentration may in fact be more effective in, for example, immunoadsorbents, as it facilitates desorption.

The washed and drained activated gel is suspended in the ligand solution. If necessary, more buffer may be added to make a slurry that can be efficiently mixed. The reaction mixture is then agitated for 2 h at room temperature (22 to 25°C) or overnight at 4°C. A number of residual active groups may remain on the gel after coupling (this is particularly the case after immobilizing high molecular weight ligands such as proteins). These groups can usually be removed by hydrolysis by leaving the gel for 2 h with Tris-HCl buffer, pH 8, or by adding an excess of a small primary amine (e.g., ethanolamine, glycine, or glutamic acid). The obtained gel product is finally washed and stored as described earlier.

10.6.1.3 The CDAP Method[35]

The agarose gel (e.g., Sepharose 4B) is washed with water, then with acetone:water (3:7), and finally with acetone:water (6:4). The gel is drained for a moment by mild suction (see earlier discussion). Ten grams of drained gel is transferred into a 50-mL glass beaker and 10 mL of acetone:water (6:4) is added. This should give a dense but easily stirrable slurry. The reaction mixture is cooled at 0°C. The desired volume of CDAP (CDAP is commercially available from Sigma Chemical Company, St. Louis, MO) stock solution (see Table 10-9) is first added under vigorous stirring of the gel suspension, and after 30 sec, the corresponding volume of TEA solution is added dropwise over a period of 1 to 2 min.

The entire reaction mixture is then rapidly transferred into 200 mL of ice-cold 0.05 M HCl. (This is to hydrolyze and remove the pyridinium isourea derivative from the gel that is formed as a by-product in the activation. The active group, the cyanate ester, is stable toward dilute mineral acid and will thus not be affected.) The gel is allowed to sediment for 15 min, is then washed on a glass filter funnel with ice-cold water, and is used for coupling immediately. Coupling of ligand as well as deactivation is performed as described previously for the CNBr buffer method.

With the CDAP technique, much higher degrees of activation can be obtained and thus more ligand, especially if low molecular weight, can be coupled to the agarose beads.

TABLE 10-9. Amounts of CDAP and TEA Employed for the Activation of 10 g of Drained Agarose (Sepharose 4B)[48]

Degree of Activation	Approximate Coupling Capacity (μmol ligand/g gel)	0.1 g CDAP/mL Dry Acetonitrile (mL)	0.2 M Aqueous Solution of TEA (mL)
Weak	5	0.251	0.2
Moderate	15	0.752	0.6
Strong	30	1.50	1.2

10.6.1.4 The DSC Method   This method is essentially as that described by Wilchek and Miron.[51] Beaded agarose (10 g of wet gel) is dehydrated (washed) and mixed slowly under agitation with 1.5 mmol of succinimidyl carbonate DSC (can be obtained from Sigma Chemical Co.). A 1.5- to 2-fold molar excess (with respect to DSC) of base catalyst [either 0.38 mL TEA in 10 mL pyridine or 325 mg DAP (obtained commercially from Sigma) in 10 mL acetone slowly under agitation] is added to this suspension.

After agitating the suspension for 30 to 60 min at room temperature, the gel is washed successively with solutions of acetone, 5% acetic acid in acetone, methanol, and 1 mM HCl (4°C). If the gel is to be used within a few hours, it can be stored in 1 mM HCl. As a suspension in acetone at 4°C, the activated gel is stable for several weeks. The proper washing of the gel can be checked by diluting an aliquot of the methanol washings with 0.25 M NH$_4$OH. Using this treatment, the remaining reagent will turn into N-hydroxysuccinimide, which can be detected photometrically at 260 nm as described earlier.

With the previously described conditions, an activated gel containing 20 to 40 μmol of hydroxysuccinimide carbonate groups/g gel is obtained. The degree of substitution depends, among other things, on the excess DSC used in the reaction.

Coupling of amino-containing ligands and proteins is performed at pH 6 to 9 by mixing the activated gel, after filtering off the 1 mM HCl or acetone, with the ligand dissolved in either 0.1 to 0.2 M phosphate buffer, pH 7.5, or fresh solutions of 0.1 to 0.2 M NaHCO$_3$, pH 8.3, for 4 to 16 h at 4°C. A ligand density of 37 μmol/mL gel was obtained when aminocaproic acid was added to the activated gel suspension to give a final concentration of 0.5 M, and 2.5 mg of protein A/mL gel was bound as a result of adding 3.0 mg of protein/ mL gel. A higher degree of substitution is obtained by adding more protein to the activated gel. Deactivation is not necessary, as the N-hydroxysuccinimide carbonate groups are rapidly hydrolyzed at the conditions of coupling.

10.6.1.5 The Tresylchloride Method   This method is essentially as described by Nilsson and Mosbach.[50] A 10-mL gel is washed with 3 to 4 volumes of water, then sequentially with acetone : water mixtures (30 : 70, 60 : 40, 80 : 20), and finally with pure acetone: 10 to 20 mL per washing and three to five washings for each mixture.

The drained gel is suspended in 5 mL of dry acetone. Pyridine is added to twice the volume of the tresylchloride. Under agitation, 0.05 to 0.2 mL of tresylchloride is added dropwise (total addition time: 1 min). The reaction mixture is then agitated for another 10 to 15 min. Wash twice with 10 volumes of acetone and then wash sequentially with mixtures of acetone and 1 mM HCl (in water) (70 : 30, 50 : 50, 20 : 80) and finally with 3 volumes of 1 mM HCl.

The coupling is performed by suspending the gel in an equal volume of suitable buffer of pH 7.5 to 9.5 (e.g., 0.1 to 0.2 M carbonate or phosphate) containing the ligand. The reaction mixture is then agitated overnight at 4 or 25°C, depending on the stability of ligand. The gel is washed with the coupling buffer and is then resuspended in an equal volume of 0.1 M buffered ethanol-amine, pH 7.5 to 8.5, and agitated for an additional 3 to 5 h at 4°C or at room temperature. The obtained gel derivative is washed and stored as described earlier.

10.6.1.6 Immobilization with Water-Soluble Carbodiimides

Perform this method essentially as described in Ref. 9 (see also Refs. 54–57). An agarose derivative with carboxyl or amino groups substituted directly on the polysaccharide backbone or via a spacer is purchased or prepared. The gel should contain carboxyl groups if the ligand is an amine and amino groups if the ligand is to be coupled through a carboxyl group. Ten milliliters of the selected drained agarose derivative is washed with 3 to 4 volumes of 0.5 M NaCl and then with distilled water. The gel is then transferred to a reaction vessel containing 5 mL of 0.04 to 0.1 M ligand dissolved in water or organic solvent (dioxane, ethyleneglycol, ethanol, methanol, or acetone). The recommended concentration range will result in a molar excess, relative to groups on the gel, of low molecular weight ligands for which the method is recommended. Protein ligands are preferably immobilized by the technique based on activated matrix carboxyl groups, which is described later. The coupling is usually performed in unbuffered medium. In all cases, avoid using buffers containing amine, carboxyl, or phosphate groups.

Adjust the pH to 4.7 with 0.1 M HCl or 0.1 M NaOH. Add 5.2 mL of 0.1 M EDCL [1-ethyl-3-[3-(dimethylamino)propyl]carbodimide hydrochloride] or CMCL [1-cyclohexyl-3-(2-morpholinoethyl)carbodiimide-metho-p-toluene sulfonate] in water or mixed solvent. Up to 50% of organic solvents can be used when the ligand is poorly water soluble. Maintain the pH at 4.5 to 5 for 1 h (pH usually decreases under this time) by adding dilute NaOH. The reaction mixture is agitated overnight at room temperature.

A blocking reaction is not usually necessary when an excess of ligand is used, but can be carried out with ethanolamine or glucosamine in the case of carboxyl-containing agarose and with acetic acid when the matrix is an amino agarose derivative.

The gel is washed and stored as described earlier.

10.6.1.7 Immobilization to Activated Carboxyl Agarose

This is a modification of the procedure described by Cuatrecasas and Parikh.[58] Ten milliliters of carboxyl-containing gel is washed with deionized water and successively with 10 × 50 mL dioxane to remove all traces of water from the gel. The gel is then suspended in 15 mL of dioxane. N-Hydrodxysuccinimide is added (240 mg) and agitation is performed until this compound is completely dissolved. Cyclohexyl carbodiimide is then added (400 mg) and agitation is continued for another 2 h. The gel is then washed with 4 to 10 volumes of dioxane and four times with 1 volume of pure methanol (to eliminate the poorly soluble N,N'-dialkylurea derivative produced during activation). After washing the gel another three times with dioxane (or isopropanol) it can be stored as a suspension in a well-sealed vessel in the dark at 4°C until used. This procedure usually gives a degree of activation of about 12 μmol of ester groups/mL gel. The degree of substitution can be determined photometrically after the release of N-hydroxy succinimide as described in Section 10.4.1.4.

Before coupling of the ligand, excess solvent is removed. The gel is then suspended either directly or after being washed with 1 mM HCl in 100 mL of the chosen buffer at pH 5 to 9 (e.g., 0.1 M NaHCO$_3$, pH 8.3) containing the ligand solution (if protein is 2 to 20 mg protein/mL gel). The use of a lower pH has the advantage that hydrolysis of the active ester is minimized. Preferential reaction of α-amino groups, as opposed to ε-amino groups, can be obtained by coupling at low pH, due to the lower pK_a of α-amino groups. The reaction mixture is agitated for 10 min to 6 h at room temperature at 4°C (coupling reaction is normally very rapid).

Excess activated groups usually hydrolyze at pH > 7, but blocking can also be performed by reacting the gel with, for example, 0.1 M ethanolamine buffered to pH 7.5 to 8.5 for 1 h at 4°C.

The gel is washed and stored as described earlier.

10.6.1.8   *Immobilization with Bisepoxirane and Epichlorohydrin*   This

method is performed essentially as described by Sundberg and Porath[40] and Porath and Fornstedt.[49] A 10-mL gel is washed with 3 to 4 volumes of deionized water and is suspended in 5 to 10 mL of deionized water. One milliliter of the reagent (bisepoxirane, e.g., 1,4-butanediol-diglycidylether, or epichlorohydrin) and 3 mL of 2 M NaOH containing 20 mg of sodium borohydride are added under agitation. The activation is carried out during agitation at room temperaure for 2 h. The activated gel is then carefully washed with deionized water until the reagents are completely removed.

The activated gel is suspended in an equal volume of a buffer of pH 9 to 13 (carbonate, borate, or phosphate buffers can be used; higher pH for carbohydrate ligands) in which 0.5 to 1 mmol/mL of small ligand and 5 to 10 mg/mL gel of macromolecular ligand are dissolved. If necessary, up to 50% organic solvent (e.g., dioxane, DMF) may be used to dissolve the ligand. The mixture is agitated for 15 to 48 h at 20 to 45°C.

Oxirane groups not utilized for coupling of ligand are usually hydrolyzed at high pH. When lower pH (<10) has been used it may be necessary to block the remaining oxirane groups with ethanolamine or mercaptoethanol as described earlier. The degree of activation obtained on 4% agarose beads with a bisepoxirane such as 1,4-bisbutanediol-diglycidylether is 10 to 20 μmol of oxirane groups/ml gel and with epichlorohydrin is 30 to 40 μmol/mL gel. In most cases, a rather small number of these groups can be utilized for ligand immobilization.

The gel is then washed and stored as described earlier.

10.6.1.9   *Immobilization to Polyacrylamide with Glutaraldehyde*[41]   Ten

milliliters of gel is washed with 3 or 4 volumes of distilled water and then with 3 to 4 volumes of 0.5 M potassium phosphate buffer, pH 7.6 (other buffers with pH between 6.9 and 8.5 can also be used). The drained gel is transferred to a flask containing 100 mL of 25% aqueous glutaraldehyde, a treatment that also sterilizes the gel. Adjust the pH to 7.4. The suspension is agitated for 18 h at 37°C. The gel is then washed with 15 to 20 volumes of distilled water or with 0.5 M phosphate buffer, pH 7.7, and is then transferred to a flask containing 10 mL of 0.5 M potassium phosphate buffer, pH 7.6, in which the ligand (e.g., a protein) is dissolved. Buffers with pH from 6.9 to 8.5 containing 5 to 10 mg of protein/mL of gel should be used. After the removal of free glutaraldehyde, the activated polyacrylamide beads can also be stored at pH 7.7 and 4°C for several days.

To remove remaining reactive aldehyde groups on the gel, treat the gel with 1 volume of 0.1 M buffered ethanolamine at pH 7.5 to 8.5 for 3 h at 4°C (amino acids can also be used). Alternatively, the free aldehyde groups can be blocked by treatment with 1 volume of 0.1 M borate buffer, pH 8.5 to 9, containing 500 mg NaBH$_4$ for 15 to 20 min. (This treatment should not be used when the immobilized ligand contains disulfide bonds.) The absolute amount of protein coupled to the polyacrylamide beads depends on the nature of the protein ligand and the excess of protein used in the reaction; 0.4 to 2 mg/mL gel is typically obtained when 1 to 2 mg protein is added per milliliter of activated gel, that is, a yield of 20 to 100%. The gel is then washed and stored as described earlier.

10.6.1.10 *Immobilization to γ-Glycidoxypropylsilica (Epoxy-Silica)*[34]
Ten grams of silica, LiChrospher Si 1000, is washed briefly with 20% HNO$_3$, H$_2$O, 0.5 M NaCl, H$_2$O, acetone, and ether and put into a 500-mL three-neck flask where it is dried for 4 h at 150°C under vacuum. The reaction flask is then cooled and sodium-dried toluene (150 mL) is sucked into the flask. γ-Glycidoxypropyl trimethoxy silane (2.5 mL, Dow Corning Z6040) and trimethylamine (0.05 mL) are added and the reaction mixture is agitated by an overhead stirrer and refluxed for 16 h; a slow stream of dry nitrogen gas will ensure anhydrous conditions. The formed epoxy-silica is then washed on a glass filter with toluene, acetone, and ether is and dried under vacuum. The procedure gives an epoxy (oxirane) group content of 50 μmol/g as determined with the method described earlier for bisoxirane- and epichlorohydrine-activated agarose derivatives.

Amino- and thiol-containing ligands can be coupled directly to the obtained silica derivative according to the procedure described earlier.[34] A pH higher than 8 should not be used. This means that the coupling, at least for amino ligands, will be rather slow and, if possible, should be speeded up by using high concentrations of ligand and, if possible, an elevated temperature. The coupling is finished by converting excess epoxy groups to more hydrophilic entities. The preferred way is acid hydrolysis, provided that the ligand is stable under these conditions, which will result in a diol structure.

Heating of γ-glycidoxypropyl-silica to 50°C at pH 2 for 3 h is enough for complete hydrolysis of the epoxide groups. Excess oxirane groups can also be converted into more hydrophilic structures by treating the silica derivative with 1 M mercaptoethanol at pH 8.5 at room temperature for 2 h. This method is more suitable after the immobilization of proteins, provided these do not contain disulfides.

Other ligand coupling methods described earlier, such as the DSC and tresylchloride methods, can also be used if the epoxy-silica is first converted to diol-silica by acid hydrolysis. This is performed by mild agitation of 10 g of epoxy-silica in 1,000 mL of 0.01 M HCl at 90°C for 1 h. Diol-silica can also be obtained commerically from E. Merck AG. (Li Chrospher DIOL).

10.6.2 Affinity Chromatography

10.6.2.1 Purification of Kinases and Dehydrogenases on Blue Sepharose[95]

Affinity gels with group specificity have the potential advantage of being useful for the isolation of many compounds belonging to a particular group. This technique was nicely demonstrated by Easterday and colleagues,[95] who purified enzymes from crude yeast extract on the dye column Blue Sepharose CL-6B and eluted them by competitive elution with enzyme-specific cofactors at different pH.

Dried bakers yeast was extracted in 1 M Na_2HPO_4 for 3 h at 37°C and then centrifuged at 13,700 g for 1 h. After filtration, the supernatant was precipitated with $(NH_4)_2SO_4$ (75% saturation) at 40°C. After centrifugation, 220 mg of the precipitate was dissolved in 10 mL of starting buffer (0.02 M Tris-HCl, pH 6.4, containing 5 mM $MgCl_2$, 0.4 mM EDTA, and 2 μM 2-mercaptoethanol). Blue Sepharose CL-6B was packed in a column with the dimensions 1.6 × 5 cm (bed volume 10 mL) and equilibrated with starting buffer before application of the sample. A peak of inactive material was eluted with starting buffer (Figure 10-4); 5 mM NAD^+ and 20 mM $NADP^+$ dissolved in starting buffer were used to elute alcoholdehydrogenase (ADH) and glucose-6-phosphate dehydrogenase (Glu-6-PO4-DH), respectively. Hexokinase (HK) was eluted when the pH of the eluent was raised to 8.6 and glyceraldehyde-3-phosphate dehydrogenase (Gly-3-PO4-DH) was eluted with 10 mM NAD^+ at the same pH.

10.6.2.2 Purification of a Lectin on N-Acetyl-ᴅ-galactosamine Sepharose 6B[87]

Ground *Falcata Japonica* seeds (20 g) were suspended in 200 mL of 0.01 M PBS (pH 7.4), stirred overnight at 4°C, and then centrifuged at 2300 g for 15 min. The precipitate was extracted with a half-volume of starting PBS. Pooled supernatants were ultracentrifuged to remove the lipid materials. The crude extract (200 mL) was applied to a column of N-acetyl-ᴅ-galactosamine (GalNAc) coupled to epoxi-activated Sepharose 6B (1.6 × 5.0 cm, 10-mL bed volume). The column was washed with 0.001 M PBS (pH 7.0) until the absorbance of the elute at 280 nm was less than 0.05. The

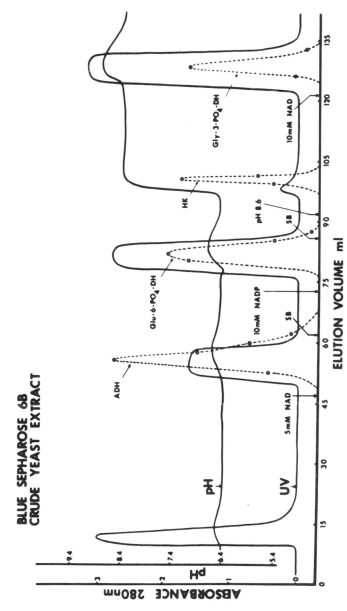

FIGURE 10-4. Purification of kinases and dehydrogenases from crude yeast extract by affinity chromatography on Blue Sepharose CL-6B. (Reprinted from Ref. 95 with permission.)

bound lectin was desorbed by a specific competitive elution adding 25 mM
GalNAc to the starting buffer (Figure 10-5). A similar result was obtained
by nonspecific elution by decreasing the pH to 2.2 by glycine–HCl buffer. A
1,000-fold purification was achieved in one step.

10 6.2.3   Binding of Immunoglobulins to Protein A Sepharose[96]   Protein
A is a group-specific ligand with affinity for many different immunoglobulins.[97]
The subclasses can be eluted separately according to binding strength by a
stepwise decrease of pH.

All subclasses of mouse IgG can be purified on Protein A Sepharose.[97,98]
Even IgG$_1$, which is known to have low affinity for protein A,[99–101] binds
efficiently under certain buffer conditions.[96] The two parameters normally
changed to affect the binding of proteins to affinity adsorbents are pH and
ionic strength. Biewenga and colleagues[102] studied the influence of these two
parameters during the binding of human myeloma IgA and human polyclonal
IgG. They found that binding of IgA decreased with decreasing pH. At salt
concentrations up to 2 M NaCl, IgA binding decreased, whereas IgG binding
was constant.

Fredriksson and co-workers[96] studied the effect of ionic strength on the
binding of mouse monoclonal antibodies of different IgG subclasses to Protein

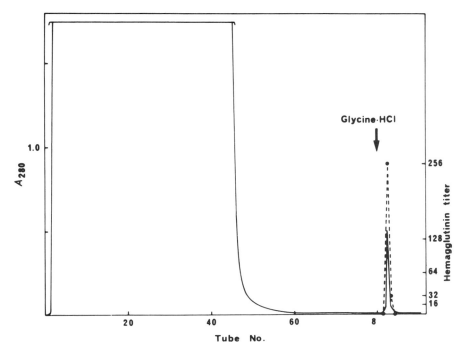

FIGURE 10-5. Specific elution of GalNAc-binding lectin from GalNAc Sepharose
6B. (Reprinted from Ref. 87 with permission.)

A Sepharose at high pH. They found that a high ionic strength (3 M NaCl) is necessary for efficient binding of IgG. This binding buffer is used (Figure 10-6) to adsorb polyclonal immunoglobulins from mouse serum to the protein A gel.

One milliliter of Protein A Sepharose CL-4B was packed in a column with the bed dimensions 1 × 1.2 cm and run at a flow rate of 0.8 mL/min. The column was equilibrated with binding buffer (1.5 M glycine, 3 M NaCl adjusted to pH 8.9 with 5 M NaOH). Five milliliters mouse serum was diluted with 5 mL of binding buffer and applied to the column. The column was then eluted using a series of buffers (0.1 M citric acid adjusted to pH 6.0, 5.0, 4.0, and 3.0). These pH values can also be used to elute the different IgG subclasses when purifying monoclonal mouse antibodies, thus eliminating acid conditions.

10.6.2.4 *Purification of Catechol-O-Methyltransferase*[103] Catechol-O-methyltransferase (COMT) is a very labile protein. Previous methods for purification of this enzyme were often laborious and time-consuming. Veser and May[103] described a rapid and specific purification method that combines ion-exchange chromatography and affinity chromatography. A low molecular

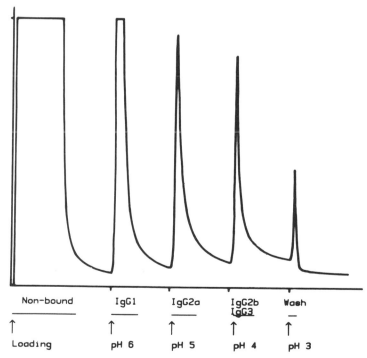

FIGURE 10-6. Elution of different IgG subclasses in mouse serum from a column of Protein A Sepharose CL-4B by stepwise decreasing pH. (Reproduced from Ref. 96 with permission.)

weight ligand, *S*-adenosyl-ʟ-homocysteine (AdoHcy) immobilized to a spacer arm-containing preactivated gel, AH Sepharose 4B, was used as the affinity adsorbent.

COMT was prepared from rat liver and partially purified by ion-exchange chromatography.[103] AdoHcy-linked AH Sepharose was packed into a column (1 × 10 cm, bed volume 8 mL) and equilibrated with 4 mM phosphate buffer, pH 6.4, containing 0.02% sodium azide, 0.2 mM magnesium chloride, 1 mM mercaptoethanol, 0.3 mM dithiothreitol, and 2% glycerol. The COMT-containing fractions from the ion-exchange chromatography were pooled and dialyzed against the equilibration buffer mentioned earlier and applied to the affinity column at a flow rate of 0.75 mL/min. The enzyme was eluted with equilibration buffer containing 0.1 mM *S*-adenosyl-ʟ-homocysteine (AdoMet) (Figure 10-7). The enzyme activity could also be eluted by a small increase in pH from 6.4 to 7.4. The column was regenerated and nonspecific bound material eluted with 0. 1 M NaCl in the equilibration buffer. The recovery of the enzyme after the affinity chromatography step was 95%.

10.6.2.5 Purification of the Angiotensin-Converting Enzyme from Human Heart Using an Immobilized Inhibitor[104]

The angiotensin-converting enzyme from human heart was isolated by a procedure containing two chromatographic steps. After extraction the enzyme was partially purified by a batch adsorption to DEAE-cellulose followed by affinity chromatography on

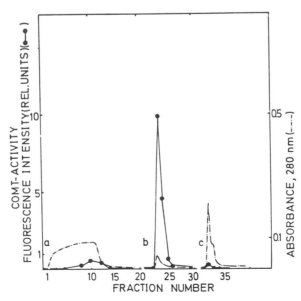

FIGURE 10-7. Affinity chromatography of catechol-*O*-methyltransferase on a column of *S*-adenosyl-ʟ-homocysteine immobilized to AH Sepharose 4B. (Reprinted from Ref. 103 with permission.)

N-[1-(S)-carboxy-5-aminopentyl]-Gly-Gly linked to Sepharose 6B via a spacer arm. The affinity adsorbent (0.9 × 8.5 cm, bed volume 5 mL) was equilibrated with 20 mM MES, pH 6.0, containing 0.5 M NaCl, 0.1 mM zinc acetate, and 0.1% Nonidet P-40 at a flow rate of 8 mLh. The enzyme was eluted by a buffer change to 50 mM sodium borate, pH 8.9 (Figure 10-8).

10.6.2.6 *High-Performance Liquid Affinity Chromatography*[105] This application demonstrates the use of HPLAC in an analytical mode for the simultaneous detection of human serum albumin (HSA) and IgG in a single sample of serum by a dual-column system (Figure 10-9). Reference 105 also presents a general scheme for the design and optimization of such a multianalyte affinity system. The ligands chosen, protein A and anti-HSA antibodies, were immobilized to activated diol-bonded LiChrospher Si-4000 and Si-500. (For a detailed description of the activation, see Refs. 34 and 106, and for the immobilization, see Ref. 105.) Quantitation of serum samples was performed on two minicolumns (6.35 mm length × 4.1 mm i.d.) connected in series and a 10-μL injection loop. The application buffer was 0.05 M phosphate and 0.05 M citrate buffer, pH 7. The elution buffer was 0.05 M phosphate and 0.05 M citrate, pH 3. The serum sample was diluted 1:5 with application buffer prior to injection. The anti-HSA column was placed before the protein

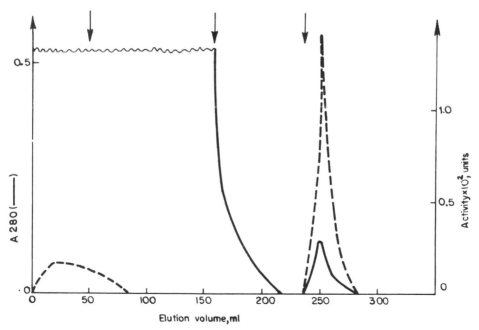

FIGURE 10-8. Affinity chromatography of human heart angiotensin-converting enzyme on immobilized N[1-(S)-carboxy-5-aminopentyl]-Gly-Gly-Sepharose 6B. Elution by increase of pH to 8.9. (Reproduced from Ref. 104 with permission.)

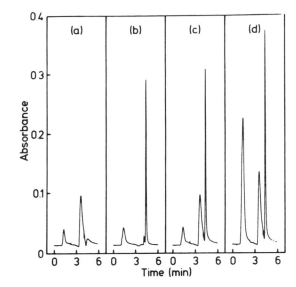

FIGURE 10-9. Chromatograms obtained after injections of (a) HSA, (b) IgG, (c) HSA plus IgG, and (d) normal serum into the dual-column system. The event sequence used was: 0.00 min, switch from pH 3 to 7 buffer; 0.50 min, sample injection; 2.25 min, protein A column switched off-line, switch to pH 3 buffer; 4.00 min, protein A column switched on-line. (Reproduced from Ref. 105 with permission.)

A column to avoid nonspecific adsorption of albumin to the protein A adsorbent (the details of this optimization experiment are given in Ref 97). Therefore, after the nonretained peak had been eluted from both columns, the protein A column was switched off-line and the anti-HSA column was eluted with pH 3 buffer. After the HSA had been eluted, the protein A column was switched on-line to elute the IgG. Standards with HSA and IgG were analyzed. This method gave results in good agreement with commerically available methods, while requiring only 2 μL of serum and 6 min per cycle.

10.6.2 7 Use of the Avidin–Biotin System for Immunosorption[107] This application describes how the avidin–biotin system can be used to prepare immunosorbents on glass beads with directed immobilization of the antibodies. The methods described by Babashak and Phillips[107] apply both conventional (IAC) and high-performance immunoaffinity chromatography (HPIAC). Monoclonal antibodies (Mabs) were biotinylated with biotin hydrazine. This reagent couples the biotin to the carbohydrate moieties of the antibodies. The carbohydrate part of most antibodies is present in the F_C region, and attachment of biotin thus ensures correct orientation of the antibody. For details of the biotinylation procedure, see Ref. 107. For silanilization and derivatization of the glass beads, see Refs. 107 and 108. The streptavidin form

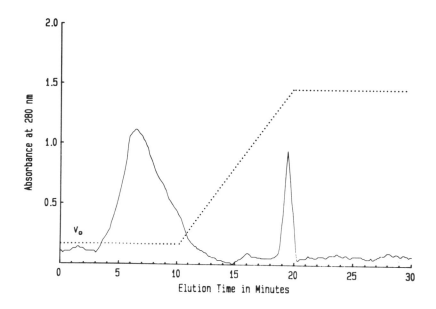

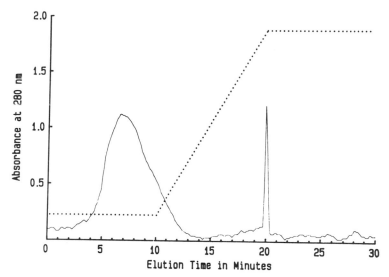

FIGURE 10-10. Chromatograms of an HPLAC isolation of the B 27 antigen isolated from detergent-solubilized membranes. The 100-μL sample was applied to 0.9% sodium chloride–0.1 M sodium acetate buffer at pH 6.5, and nonadsorbed material was washed off the column before elution started. (a) The antigen was desorbed by a pH gradient from pH 6.5 to 1.0 by the addition of 0.1 M hydrochloric acid to the initial running buffer. (b) The antigen was eluted by a chaotropic ion gradient 0 to 2.5 M sodium thiocyanate. (Reproduced from Ref. 107 with permission.)

of avidin was immobilized to the activated glass beads by incubation at pH 9 for 18 h at 4°C. The biotinylated Mabs were then attached to the streptavidin-coated beads by incubation for 1 h at 4°C in PBS on an overhead mixer. The beads were washed (PBS, pH 7) and slurry packed into 10 cm × 4.6 mm i.d. stainless-steel columns at 250 p.s.i. The immunosorbent was used to isolate the B 27 human leukocyte antigen component from detergent-solubilized human leukocyte membranes. The B 27 antigen is retained on the column, while most of the membrane components are washed through it (Figure 10-10). The bound antigen is eluted by decreased pH or by chaotropic ions. The binding between steptavidin–biotin is so strong that no elution of biotinylated Mabs occurs in these conditions. More than 20 batches of streptavidin-coated glass beads were produced and used in HPIAC. The derivatized beads bound between 1.5 and 1.85 mg of streptavidin per 2 g batch of beads, and the streptavidin beads bound between 195 and 245 μg of hydrazine biotinylated antibodies. Running the chromatograms at 4°C is preferable to room temperature (both regarding peak performance and bead lifetime). Also, chaotropic elution is preferable to acid elution, as the bead lifetime increases twofold.

10.7   REFERENCES

1. M. Wilchek, T. Miron, J. Kohn, in *Methods Enzymol. 104,* W.B. Jacoby, ed., Academic Press, New York, 1984, pp. 3–55.

2. W.H. Scouten, *Affinity Chromatography,* John Wiley & Sons, New York, 1981.

3. J.-C. Janson, T. Kristiansen, in *Packings and Stationary Phases in Chromatographic Techniques,* K.K. Unger, ed., Chromatogr. Sci., Series Vol. 47, Marcel Dekker, New York, 1990, pp. 747–782.

4. R.H. Allen, P.W. Majerus, *J. Biol. Chem., 247,* 7709 (1972).

5. D.G. Deutsch, E.T. Mertz, *Science 170,* 1095–1096 (1970).

6. E.A. Bayer, M. Wilchek, *Methods Biochem. Anal., 26,* 1 (1980).

7. See data sheet on Benzamidine Sepharose 6B from Pharmacia Biotech AB, Uppsala, Sweden, for relevant references.

8. P.D.G. Dean, F. A. Middle, C. Longstaff, A. Bannister, J. J. Dembinski, in *Affinity Chromatography and Biological Recognition,* I.M. Chaiken, M. Wilchek, I. Parikh, eds., Academic Press, Orlando, 1983, 433–443.

9. *Affinity Chromatography: Principles and Methods,* Pharmacia Biotech AB, Uppsala, Sweden.

10. K. Mosbach, in *Advances in Enzymology,* A. Meister, ed., John Wiley & Sons, New York, 1978, Vol. 46, pp. 205–278.

11. G. Kopperschläger, H.-J. Böhme, E. Hofmann, in *Advances in Biochemical Engineering,* A. Fiechter, ed., Springer-Verlag, Berlin, Heidelberg, 1982, Vol. 25, pp. 101–138.

12. R.K. Scopes, *J. Chromatogr., 376,* 131–140 (1986).

13. C.R. Lowe, S.J. Burton, J.C. Pearson, Y.D. Clonis, V. Stead, *J. Chromatogr., 376,* 121–130 (1986).

14. J. Porath, M. Belew, *Trends Biotechnol., 5,* 225–229 (1987).

15. M. Vuento, A. Vaheri, *Biochem. J., 183,* 331–337 (1979).

16. M.M. Andersson, H. Borg, L. O. Andersson, *Thromb. Res., 5,* 439–452 (1974).

17. P. Gagnon, *Purification Tools for Monoclonal Antibodies,* Validated Biosystems, Tucson, AZ, 1996.

18. Anonymous, *Monoclonal Antibody Purification,* Order No.:18-1037-46, Pharmacia Biotech AB, Uppsala, Sweden.

19. J.W. Goding, *Monoclonal Antibodies: Principles and Practice,* 2nd ed., Academic Press, London, 1986, Chapter 6.

20. H.A. Chase, B.J. Horstmann, S.L. Fowell, *J. Chem. Technol. Biotechnol., 45,* 60–67 (1989).

21. *Heparin Sepharose CL-6B for Affinity Chromatography,* Pharmacia Biotech AB, Uppsala, Sweden.

22. M.J. Desai, R.N. Zuckerman, W.H. Moos, *Drug. Dev. Res., 33,* 174–188 (1994).

23. C. Khosla, *Curr. Opin. Biotechnol. 7,* 219–222 (1996).

24. G.A. Baumbach, D.J. Hammond, *BioPharm, 5,* 25–35 (1992).

25. S.E. Cwirla, E.A. Peters, R.W. Barrett, W.J. Dower, *Proc. Natl. Acad. Sci. USA, 87,* 6378–6382 (1990).

26. J.J. Devlin, L.C. Panganiban, P.E. Devlin, *Science, 249,* 404–406 (1990).

27. R.N. Zuckerman, *Curr. Opin Struct. Biol., 3,* 580–584 (1993).

28. A. Giannis, T. Kolter, *Angew. Chem. Int. Ed. Engl., 32,* 1244–1267 (1993).

29. M. Krook, C. Lindblad, S. Birnbaum, H. Naess, J.A. Eriksen, K. Mosbach, *J. Chrom., 711,* 119–128 (1995).

30. One example of this technology is that introduced by ArQule Inc. (Boston, MA) and partially described in the patents "Oxazolone derived materials," PCT Number WO 94/00509 and "Aminimide-containing molecules and materials as molecular recognition agent," PCT Number WO 94/001102.

31. Ongoing work at Pharmacia Biotech AB, Uppsala, Sweden.

32. J. Porath, T. Låås, J.-C. Janson, *J. Chromatogr., 103,* 49–62 (1975).

33. P. Cuatrecases, M. Wilchek, C. B. Anfinsen, *Proc. Natl. Acad. Sci. USA, 61,* 636 (1968).

34. H. Schott, *Affinity Chromatography, Chromatogr. Sci., Series* Vol. 27, Marcel Dekker, New York, 1984.

35. S. Ohlson, L. Hansson, M. Glad, K. Mosbach, P.-O. Larsson, *Trends Biotechnol., 7,* 179–186, 1989.

36. K. K. Unger, in *Porous Silica, J. Chromatogr. Library,* Elsevier, Amsterdam, 1979, p. 16.

37. Superose (10 and 13 μm) and Sepharose High Performance (34 μm), Pharmacia Biotech AB, Uppsala, Sweden.

38. R.R. Harris, J.J. Rewe, *FEBS Lett., 29,* 189 (1979).

39. P. Cuatrecasas, *Nature, 228,* 1327 (1970).

40. C.R.L. Lowe, in *Affinity Chromatography.* John Wiley, New York, 1974, p. 28.

41. L. Sundberg, J. Porath, *J. Chromatogr., 90,* 87 (1974).

42. T. Ternyck, S. Avrameas, *FEBS Lett., 23,* 24 (1972).

43. M. Wilchek, T. Oka, Y.J. Topper, *Proc. Natl. Acad. Sci. USA, 72,* 1055–1058 (1975).

44. J. Porath, in *Methods Enzymol.,* B. Jakoby, M. Wilchek, eds., Academic Press, New York, 1974, Vol. 34, pp. 13–30.

45. R. Axén, J. Porath, S. Ernback, *Nature, 214,* 1302 (1967).

46. R. Axén, S. Ernback, *Eur. J. Biochem., 18,* 351–360 (1971).

47. J. Porath, K. Aspberg, H. Drevin, R. Axén, *J. Chromatogr., 86,* 53 (1973).

48. J. Kohn, M. Wilchek, *Appl. Biochem. Biotechnol., 9,* 285–304 (1984).

49. J. Porath, N. Fornstedt, *J. Chromatogr., 51,* 479 (1970).

50. K. Nilsson, K. Mosbach, in *Methods Enzymol.,* K. Mosbach, ed., Academic Press, New York, 1987, Vol. 135, Part B, pp. 65–78.

51. M. Wilchek, T. Miron, *Appl. Biochem. Biotechnol., 11,* 191–193 (1985).

52. M.T.W. Hearn, in *Methods Enzymol.,* K. Mosbach, ed., Academic Press, New York, 1987, Vol. 135, Part B, pp. 102–117.

53. G.S. Bethell, J. Ayers, W.S. Hancock, M.T.W. Hearn, *J. Biol. Chem., 254,* 2572 (1979).

54. A. Tengblad, *Biochem. J., 199,* 297 (1981).

55. S.L. Marcus, E. Balbinder, *Anal. Biochem., 48,* 448–459 (1972).

56. H. Anttinen, K.I. Kivirikko, *Biochem. Biophys. Acta, 429,* 750–758 (1976).

57. D. Robinson, N.C. Phillips, B. Winchester, *FEBS Lett., 53,* 110–112 (1975).

58. P. Cuatrecasas, I. Parikh, *Biochemistry, 11,* 2291 (1972).

59. J. Brandt, A. Svensson, J. Carlsson, H. Drevin, *J. Solid Phase Biochem., 2,* 105–109 (1977).

60. R.M.K. Dale, D.C. Ward, *Biochemistry, 14,* 2458–2469 (1975).

61. M. Lynn, in *Enzymology of Immobilized Enzymes, Antigens, Antibodies and Peptides.* H.H. Weetall, ed., 1975, 1, 1.

62. R. Axén, P. Vretblad, J. Porath, *Acta Chem. Scand, 25,* 1129–1132 (1971).

63. I. Ugi, *Anqew. Chem., 74,* 9 (1962).

64. L. Goldstein, in *Methods Enzymol.,* K. Mosbach, ed., Academic Press, New York 1987, Vol. 135, pp. 90–102.

65. J.L. Guesdon, S. Avrameas, *J. Immunol. Meth., 11,* 129 (1976).

66. J.K. Inman, in *Methods Enzymol.,* W.B. Jakoby, M. Wilchek, eds., Academic Press, New York, 1974, Vol. 34, p. 30.

67. M.B. Wilson, P.K. Nakane, *J. Immunol. Meth., 12,* 171 (1976).

68. M. Wilchek, R. Laurel, in *Methods in Enzymol.,* W.B. Jakoby, M. Wilchek, eds., Academic Press, New York, 1974, Vol. 34, p. 475.

69. I. Parikh, S. March, P. Cuatrecasas, in *Methods Enzymol.,* W.B. Jakoby, M. Wilchek, eds., Academic Press, New York, 1974, Vol. 34, pp. 77–102.

70. S. Ohlson, L. Hansson, P.-O. Larsson, K. Mosbach, *FEBS Lett., 93,* 5–9 (1978).

71. J. Woodward, H.J. Marquess, C.S. Picker, *Prep. Biochem., 16,* 337–352 (1986).

72. D.S. Secher, D. C. Burke, *Nature, 285,* 446–450 (1980).

73. W.J. Jankowski, W. von Muenchhausen, E. Sulkowski, M.A. Carter, *Biochemistry, 15,* 5182 (1976).

74. W.M. Moore, C.A. Spilburg, *Biochemistry, 25,* 5189–5195 (1986).

75. K. Miyata, Y. Yamamoto, M. Ueda, Y. Kawade, K. Matsumnoto, I. Kubota, *J. Biochem., 99,* 1681–1688 (1986).

76. L. Kiss, A. Tar, S. Gal, B.L. Toth-Martinez, F.J. Hernadi, *J. Chromatogr., 448,* 109–116 (1988).

77. G. Vlatakis, G. Skarpelis, I. Stratidaki, V. Bountis, Y. D. Clonis, *Appl. Biochem. Biotechnol., 15,* 201–212 (1987).

78. S.P.J. Brooks, V.D. Bennett, C.H. Suelter, *Anal. Biochem. 164,* 190–198 (1987).

79. L. Minbel, P. Goldschmidt-Clermont, R.M. Galbraith, P. Arnaud, *J. Chromatogr., 363,* 448–455 (1986).

80. Y. Kitagawa, E. Okuhara, E. Shikata, *J. Virol. Meth., 16,* 217–224 (1987).

81. P.T. Swoveland, *J. Virol. Meth. 13,* 333–341 (1986).

82. S.J. Busch, C.R. Duvic, J.L. Ellsworth, J. Ihm, J.A.K. Harmony, *Anal. Biochem., 153,* 178–188 (1986).

83. J.M.F.G. Aerts, W.E. Donker-Koopman, G.J. Murray, J.A. Barranger, J.M. Tager, A.W. Schram, *Anal. Biochem., 154,* 655–663 (1986).

84. T.J. Palker, M.E. Clark, M.G. Sarngadharan, T.J. Matthews, *J. Virol. Meth., 18,* 243–256 (1987).

85. P.B. Seshagiri, P.R. Adiga, *Biochim. Biophys. Acta, 916,* 474–481 (1987).

86. J.C. Zwaagstra, G.D. Armstrong, W.-C. Leung, *J. Virol. Meth., 20,* 21–32 (1988).

87. T. Nakajima, S. Yazawa, T. Kogure, K. Furukawa, *Biochim. Biophys, Acta, 964,* 207–212 (1988).

88. J. Ramwani, R.K. Mishra, *J. Biol. Chem., 261,* 8894–8898 (1986).

89. P.F. Ruhn, S. Garver, D.S. Hage, *J. Chromatogr., 669,* 9–19 (1994).

90. W.-C. Lee, C.-H. Lin, R.-C. Ruaan, K.-Y. Hsu, *J. Chromatogr., 704,* 307–314 (1995).

91. M.W. Strohsacker, M.D. Minnich, M A. Clark, R.G.L. Shorr, S.T. Crooke, *J. Chromatogr., 435,* 185–192 (1988).

92. H. Harada, M. Kamei, Y. Tokumoto, S. Yui, F. Koyama, N. Kochibe, T. Endo, A. Kobata, *Anal. Biochem., 164,* 374–381 (1987).

93. R.R. Walters, *Anal. Chem., 55,* 1395–1399 (1983).

94. K.-I. Kasai, *J. Chromatogr., 597,* 3–18 (1992).

95. R.L. Easterday, I.M. Easterday, in *Immobilized Biochemicals and Affinity Chromatography,* R.B. Dunlap, ed., Plenum Press, New York, 1974, pp. 123–133.

96. U.B. Fredriksson, L.G. Fägerstam, A.W.G. Cole, E. Walldén, Pharmacia Biotech AB, Poster presented at Sixth Int. Congr. Immunol. Toronto, Canada, 1986.

97. J.J. Langone, *J. Immunol. Meth., 55,* 277–296 (1982).

98. P.L. Ey, S.J. Prowse, C.R. Jenkin, *Immunochemistry., 15,* 429–436 (1978).

99. M.P. Chalon, R.W. Milne, J.-P., Vaerman, *Scand. J. Immunol., 9,* 359–364 (1979).

100. C.L. Villemez, M.A. Russell, P.L. Carlo, *Mol. Immunol., 21,* 993–998 (1984).

101. L. Manil, P. Motté, P. Pernas, F. Troalen, C. Bohuon, D. Bellet, *J. Immunol. Meth., 90,* 25–37 (1986).

102. J. Biewenga, F. Daus, M.L. Modderman, G.M.M.L. Bruin, *Immunol. Comm. 11(3),* 189–200 (1982).

103. J. Veser, W. May, *Chromatographia, 22,* No. 7–12, 404–406 (1986).

104. I.Y. Sahkarov, S.M. Danilov, E.A. Dukhanina, *Biochim. Biophys. Acta, 923,* 143–149 (1987).

105. D.S. Hage, R.R. Walters, *J. Chromatogr., 386,* 37–49 (1987).

106. D.S. Hage, R.R. Walters, H.W. Hethcote, *Anal. Chem., 58,* 274 (1986).

107. J.V. Babashak, T.M. Phillips, *J. Chromatogr., 444,* 21–28 (1988).

108. R.R. Walters, in *Affinity Chromatography,* P.D.G. Dean, W.S, Johnson, F.A. Middle, eds., IRL Press, Washington, DC, 1985, p. 25.

11 Affinity Partitioning of Proteins Using Aqueous Two-Phase Systems

GÖTE JOHANSSON
Department of Biochemistry
University of Lund
Chemical Centre
S-221 00 Lund, Sweden

Protein Purification: Principles, High-Resolution Methods, and Applications, Second Edition.
Edited by Jan-Christer Janson and Lars Rydén.
ISBN 0-471-18626-0. © 1998 Wiley-VCH, Inc.

11.1 INTRODUCTION

The methods for separation described in this chapter are based on partition of proteins between two water-rich phases. The two phases, in contact with each other, yield a so-called aqueous two-phase system. The water content of the phases is, in general, in the order of 70 to 95%, and the two-phase systems are obtained by dissolving two polymers in water.[1,2] Another way to achieve two water-rich phases is to mix an aqueous solution of a polymer with a suitable salt (e.g., sodium phosphate).[1] The following discussion is limited to systems based on dextran, polyethylene glycol (PEG), and water.

A protein, included in the two-phase system, is after equilibration (15 to 60 sec of careful mixing) found more or less in both phases. The relative distribution of the protein between the phases (i.e., its partition) can be described by a partition coefficient, K, defined by Eq. (11-1):

$$K = \frac{C_U}{C_L} \tag{11-1}$$

where C_U and C_L are the concentrations of the protein in the upper and lower phase, respectively. Two proteins can be resolved from a mixture by partition in a system in which their K values differ enough. Usually, several partition steps are necessary to get an acceptable resolution unless the K values are extreme (e.g., $K_1 < 0.01$ and $K_2 > 100$).

Aqueous two-phase systems can be made selective in their properties of extracting certain proteins. One way to achieve this goal is to localize an affinity ligand (for the target protein) in one of the phases. Ligands used in this way include fatty acids,[3] triazine dyes,[4,5] coenzymes,[6] and more specialized ligands.[7,8]

The aqueous two-phase systems were introduced by Albertsson as a tool for separating a vast number of cell components ranging from proteins to cell organelles.[1] It was also possible to fractionate various forms of microorganisms by partition within the two-phase system. In contrast to proteins and nucleic acids, the particles are not only recovered in bulk phases but are also found at the interface between the two phases.

11.2 GENERAL PROPERTIES

11.2.1 Composition of Two-Phase Systems

In a dextran–PEG–water two-phase system, the composition of the two phases depends on the percentage of the polymers (usually given in percent weight by weight), the molecular weights of the polymers, and the temperature.[2] The phase composition can be described by a phase diagram (Figure 11-1) where the points indicating the polymer concentrations for the upper (o) and lower

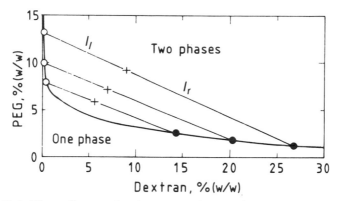

FIGURE 11-1. Phase diagram for the system dextran (M_r = 500,000), PEG (M_r = 3,000 to 3,700), and water at 0°C. The compositions of the upper phases (o) and the lower phases (•) are shown for systems of various total composition (+). (Reproduced from Ref. 1 with permission.)

(•) phase are both on the border line (the so-called binodal curve) separating one- and two-phase regions. The phase diagrams are useful tools for finding systems with the desired properties.

Systems with total compositions along one of the straight lines (tie lines) in Figure 11-1, generated by the composition points of the two phases, differ in the relative amount of upper and lower phase (but not in their composition). The volume ratio is related to the length of the sections of the tie line (l_l and l_r in Figure 11-1) when this is "divided" by the point (+) representing the total composition. This relation is given by Eq. (11-2):

$$\frac{V_U}{V_L} = \frac{l_r d_L}{l_l d_U} \tag{11-2}$$

where d_u and d_L are the densities of the upper and lower phase, respectively. Because the densities are nearly equal, the expression can, in many cases, be approximated by Eq. (11-3):

$$\frac{V_U}{V_L} = \frac{l_r}{l_l} \tag{11-3}$$

The phase diagram (Figure 11-1) also shows that if the total concentrations of dextran and PEG are both increased, the two phases will differ more and more in their compositions. The phases will consequently be increasingly pure in respect to the content of their dominating polymer. If an affinity ligand is attached to one of the polymers, it will therefore be increasingly concentrated in one phase the more the total composition is removed from the binodal curve.

11.2.2 Influence of Electrolytes

The pH value and the presence of electrolytes in the system have a pronounced effect on the partition of proteins between the two phases. Salts commonly used in biochemical procedures, such as phosphates, chlorides, or Tris-HCl, can greatly influence partition.[2,9] The salts themselves have partition coefficients close to one.

The effect of salts on the partition increases with the distance from the binodal curve (increasing percentage of the two polymers) and also depends on the temperature.[1,2] Normal salt concentrations are in the range of 25 to 250 mM.

Stronger effects than those obtainable with salt have been achieved by using charged groups covalently bound to PEG (e.g., sulfonate or trimethylamino groups).[10,11] This is an example of the use of one phase as a liquid ion exchanger. To get effective extractions in this case, the content of salt must be kept at a low level, usually well below 10 mM.

11.2.3 Theory for Effect of Salts on Partition of Proteins

The partition of a protein is influenced by the presence of salts and this effect increases with the net charge of the protein. This has led to the idea that a small portion of the ions of the salt, present in excess, form an interfacial double layer at the interface between the phases. This, in turn, gives rise to an interfacial potential that affects the partition of the charged protein. If the partition coefficient of a noncharged protein (pH = pI) is K_0, an adjustment of the pH value, which gives the protein a net charge Z, will in most cases influence the partition. The partition coefficient, K, of the protein at this pH value is given by Eq. (11-4):

$$K = K_0 \cdot (K_+/K_-)^{-Z/(m + n)} \tag{11-4}$$

The corresponding logarithmic expression,

$$\log K - \log K_0 = -Z/(m + n) (\log K_+ - \log K_-) \tag{11-5}$$

where K_+ and K_- are the (hypothetical) partition coefficients of the cation (A^{m+}) and the anion (B^{n-}) of the salt present in excess. These are the partition coefficients the ions should have had if they could partition independent of the electric field caused by the neighborhood ions. The coefficients for a two-phase system composed of 8% (w/w) dextran 500 and 8% (w/w) PEG 3400 at 20°C are given in Figure 11-2.

When K_+ is larger than K_-, the upper phase side of the interface will have a positive charge and it increases the K values for negatively charged proteins while the K values are lowered for positively charged proteins. This is the case when, for example, lithium phosphate buffer is used. If, however, $K_+ < K_-$ (e.g., with KBr), the charge dependence is the opposite. As can be seen

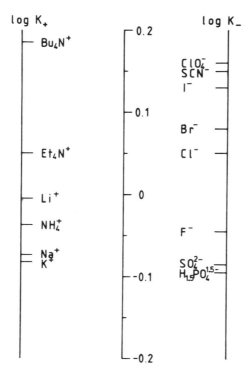

FIGURE 11-2. The (hypothetical) nonconstrainal partition coefficients for a number of cations (K_+) and anions (K_-) in a system containing 8% (w/w) dextran ($M_r =$ 500,000), 8% (w/w) PEG ($M_r =$ 3000 to 3700), and 20 to 25 mM salt at 25°C. The partition coefficient of a salt, K_s is, in the case of 1–1 electrolytes, equal to $\sqrt{K_+ \cdot K_-}$ or log $K_s = \frac{1}{2}$(log K_+ + log K_-). (Adapted from Ref. 12 with permission.)

in Eq. (11-4), the charge-dependent partition, log $K-$ log K_0 (for a given charge Z), is proportional to log K_+ − log K_-.

11.2.4   Influence of Molecular Weights of Polymers

Changes in the molecular weights of the phase-forming polymers influence both the K_0 value of a protein and the interfacial potential. A decrease in molecular weight of a polymer changes the noncharge-dependent partition toward the phase rich in this polymer.

11.3   AFFINITY PARTITIONING

An affinity ligand (for a protein) can be restricted to one of the phases by attaching it to the main polymer of this phase. Usually only a fraction of the phase-forming polymer is used as ligand carrier.

11.3.1 Preparation of Polymer-Bound Ligand Derivatives

A number of methods for the binding of ligands to water-soluble polymers have been described.[2,13,14] Ligands with reactive groups (e.g., reactive textile dyes) can be bound directly, as is described later. Otherwise, reactive groups can be introduced on the polymer (e.g., tresyl, tosyl, bromo, or carboxyl groups). The first three groups can, under mild conditions, be replaced with affinity ligands containing amino, mercapto, or phenolic groups. Alcohols and amines form ester and amide bonds, respectively, with polymers carrying carboxyl groups by using carbodiimides as a water-accepting reagent. By reacting the activated polymers with diamines (in excess), spacers with free amino groups for further binding are introduced. An increasing number of PEG derivatives are now commercially available (e.g., from Shearwater, Huntsville, AL). A number of ligands used for affinity partitioning of proteins are given in Table 11-1.

11.3.2 Partition of Ligand–Polymers

Normally not more than 5% of a polymer is used as a ligand carrier. When the two constituents of the ligand–polymer derivative are in the molar ratio 1:1, the partition of the polymer-bound ligand is comparable with that of free polymer. Higher degrees of substitution may influence the partition strongly, especially if the ligand carries ionic groups. The partition of the ligand–polymer will be more extreme at higher concentrations of the phase-forming polymers, as has been shown using Cibacron blue F3G-A bound to PEG (Table 11-2). Furthermore, the partition coefficient, K_{Cb-PEG}, is larger than the one for PEG itself. The effect of increasing the degree of substitution is illustrated by dextran-bound Procion yellow HE-3G (Table 11-3). Changes of the molecular weight of the bulk dextran also affect the partition of the ligand–dextrans.

TABLE 11-1 Examples of Ligands Used for Affinity Partitioning of Proteins

Ligand	Extracted Protein	Refs.
p-Aminobenzamidine	Trypsin	7
Palmitate	Serum albumin	3
Dinitrophenol	S-32 myeloma protein	9
Estradiol	Oxosteroid isomerase	15, 16
Lecithin	Colipase	17
Triazine dyes	Dehydrogenases and kinases	4, 5, 18–22
Nucleotides	Dehydrogenases and kinases	6
Textile dyes	Serum proteins	23
Iminodiacetate–metal complexes	Histidine-rich proteins	24

TABLE 11-2. Partition of Phosphofructokinase (PFK) From Baker's Yeast in Systems Containing Various Concentrations of the Two Phase-forming Polymers Dextran (M_r = 500,000) and PEG (M_r = 7,000 to 9,000)

System		Logarithmic Partition Coefficients		
% (w/w) Dextran	% (w/w) PEG	log K_{aff}	log $K_{Cb\text{-}PEG}$	log K_{PEG}
4.80	3.20	1.4	0.59	0.24
5.25	3.50	2.0	0.75	0.48
6.00	4.00	2.4	1.02	0.78
6.75	4.50	2.8	1.26	1.00
7.50	5.00	3.4	1.43	1.14
8.25	5.50	3.4	1.63	1.22
9.00	6.00	3.6	1.80	1.27

The systems also contained 50 mM sodium phosphate buffer, pH 7.0, 5 mM mercaptoethanol, and 0.5 mM EDTA. Temperature was 0°C. The affinity partitioning effect, log K_{aff} (= increase in log K_{PFK}), was caused by exchanging 3% of PEG for Cibracron blue F3G-A PEG (Cb-PEG). (Adapted from Ref. 25 with permission.)

11.3.3 Theory for Affinity Partitioning

The effect of the polymer-bound ligand on the partition of a protein can be described by the relative change in the partition coefficient of the protein. If the partition without ligand is given by K^*, the partition in the presence of ligand, K, is described by Eq. (11-6):

$$K = K_{aff} \cdot K^* \tag{11-6}$$

where K_{aff} is the factor quantifying the affinity partitioning effect. K_{aff} increases with the concentration of the PEG-bound ligand until it reaches a saturation value (Figure 11-3).

TABLE 11-3. Partition of Procion Yellow HE-3G Dextran (PrY-Dextran) in Systems Composed of PEG 8000 and Dextrans of Various Molecular Weights and 50 mM Sodium Phosphate Buffer, pH 7.9. Temperature, 22°C. Degree of Substitution (n) Gives the Number of Dye Molecules Per Dextran Molecule. (Reproduced by Permission From Ref. 26.)

Molecular Weight of Dextran	Composition of System		K of PrY-dextran			
	% (w/w) Dextran	% (w/w) PEG	$n = 1.3$	$n = 2.3$	$n = 5.3$	$n = 8.3$
40,000	8.00	5.00	0.15	0.30	2.0	17.5
70,000	8.00	4.50	0.23	0.50	4.1	28.2
500,000	5.00	3.80	0.97	1.62	6.2	24.3
2,000,000	5.00	4.00	1.03	1.83	7.5	22.5

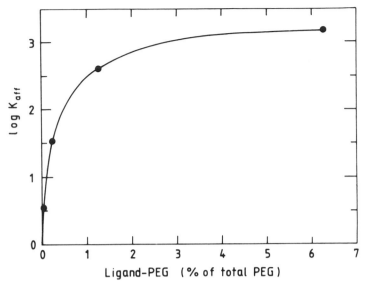

FIGURE 11-3. Log K_{aff} of phosphofructokinase when an extract of baker's yeast was partitioned in the 6% (w/w) dextran (M_r = 70,000), 4% (w/w) PEG (M_r = 35,000 to 40,000), 25 mM sodium phosphate buffer, 5 mM mercaptoethanol, and 0.25 mM EDTA system at 0°C, pH 7.0, with an increasing concentration of Cibacron blue F3G-A PEG (M_r = 6,500).

A theoretical model for affinity partitioning has been presented by Flanagan and Barondes.[8] It states that K_{aff}, in systems with an excess of polymer-bound ligand, is related to $K_{ligand\text{-PEG}}$ by Eq. (11-7):

$$K_{aff} = n \cdot K_{ligand\text{-PEG}} \, D_U/D_L \tag{11-7}$$

where n is the number of binding sites for ligand on the protein molecule, and D_U and D_L are the dissociation constants in upper and lower phase, respectively.

11.3.4   Effects of Salt, Free Ligands, and pH Value

The affinity partitioning may be affected by the presence of salt in at least two ways. First, the ligand–protein interaction might be weakened (if it is of electrostatic character) or strengthened (if hydrophobic) by increasing the salt concentration. Second, if the ligand carries ionized groups, the ligand–PEG will act, at low salt concentration, as a liquid ion exchanger and extract proteins of the opposite net charge. The salt concentration will therefore often be chosen as a compromise to minimize these two effects, with values between

10 and 150 mM. The effects of a number of salts on the partition of phosphofructokinase are shown in Figure 11-4. Salts may be used to weaken unwanted ligand–protein interaction to make the affinity extraction more selective.

Temperature also affects the affinity partitioning and this has to be checked for every ligand–protein pair. Other factors, such as pH, may play an important role. When the ligand is a trizine dye, the maximal K_{aff} value increases with decreasing pH values,[27] but at the same time the selectivity of the extraction is generally reduced.

In some cases, the molecular weights of dextran and PEG may affect the K_{aff} values[25] when systems with equal distance from the binodial curves are compared. This has been demonstrated for phosphofructokinase from baker's yeast. In several other cases, only a slight dependence on the molecular weights has been observed.

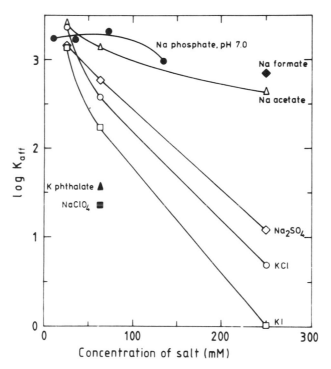

FIGURE 11-4. Effects of various salts on the affinity partitioning of phosphofructokinase from baker's yeast. The system contained 7% (w/w) dextran (M_r = 500,000), 5% (w/w) PEG (M_r 6,000 to 7,500), 10 mM sodium phosphate buffer, pH 7.0, 0.5 mM EDTA, 5 mM mercaptoethanol, and 4 nkat enzyme/g system, together with the salt, at 0°C. The affinity partitioning effect, log K_{aff}, was obtained by replacing 3% of PEG with Cibacron blue F3G-A PEG. (Adapted from Ref. 25 with permission.)

11.3.5. Batch Extraction of Proteins

When a protein is to be purified by affinity partitioning the extraction is often carried out in several steps. First, the protein mixture is included in a two-phase system (without ligand) in which most of the proteins, including the target protein, partition to the lower phase. The upper phase, containing materials with unspecific preference for this phase, is then replaced with a new upper phase supplemented with the ligand–PEG. After equilibration the target protein is found in the upper phase whereas most bulk proteins remain in the lower phase. The latter is removed and the upper phase is washed by one to three times equilibrations with fresh lower phase to remove coextracted bulk proteins. The target protein can be collected in a final lower phase by adding free ligand to the system or by adding a high concentration of phosphate to the recovered upper phase, yielding a PEG–salt two-phase system with the target protein found in the salt phase.

11.3.6 Countercurrent Distribution of Cell Extracts

The purification obtained by preparative extractions in one or a few steps is suited for isolation of one component from a mixture. Countercurrent distribution (CCD) is a procedure where the two phases stepwise are moved over each other. It often allows a number of proteins to be fractionated simultaneously. The principle of CCD is explained in Figure 11-5. The process can be carried out manually by transferring upper phases of a row of systems that are equilibrated and allowed to separate after each transfer. A large number of transfers makes this operation tedious. Instead, ordinary CCD apparatus (according to Craig) can be used or, even better, the apparatus especially constructed for aqueous two-phase systems.[28,29] After the CCD run, water can be added to get one phase. The amount necessary can be determined from the phase diagram. The measurements of protein or enzyme activities along the CCD tubes show distribution curves resembling those of a chromatogram. An example of a manual CCD with 7 transfers is shown in Figure 11-6. An automatic CCD with 55 transfers is shown in Figure 11-7.

11.4 PRACTICAL PROCEDURES

11.4.1 Strategy

The following steps are recommended:

1. Check the partition of the target protein (e.g., an enzyme) and of total protein in one chosen system (without ligand).
2. By varying parameters such as polymer concentration, salt, and pH, it is possible to see if some purification can be obtained without ligand.

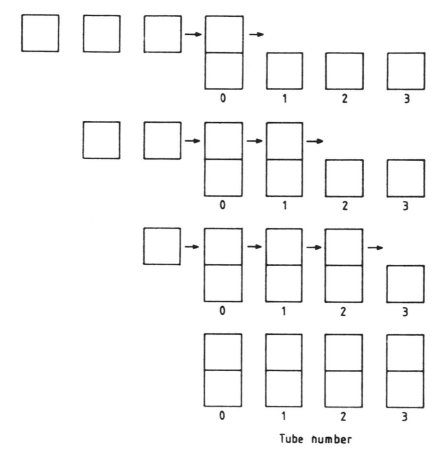

Tube number

FIGURE 11-5. Principle for countercurrent distribution. The sample to be separated is included in a two-phase system (No. 0 at the top). The system is equilibrated, and when the phases have separated, the top phase is transferred to a fresh bottom phase, yielding system No. 1. A fresh top phase is also added to the bottom phase of system No. 0. After equilibration and settling of the phases, the transfer of the top phases is repeated. A countercurrent distribution with n transfers will give rise to $n + 1$ systems (No. $0 - n$). Water can be added to the system to obtain single-phase fractions, which are analyzed.

3. Bind a ligand to the polymer occupying the phase into which target protein is to be extracted.
4. Determine the extraction curve by using increasing concentrations of the ligand–polymer.
5. Adjust the system parameters at saturated amounts of ligand–polymer to get the most effective purification. Parameters are temperature, polymer concentration, sample concentration, pH, and salt (type and concentration).

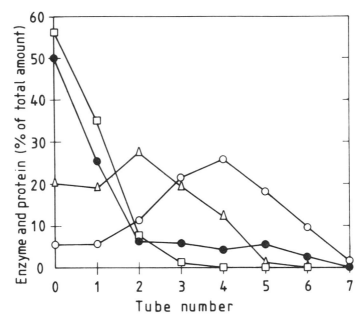

FIGURE 11-6. Countercurrent distribution of extract from baker's yeast using seven transfers. The sample was loaded in system No. 0 and the upper phases were then moved stepwise to the right. PEG-bound ligand (in the upper phase) was Cibacron blue F3G-A. System composition: 7.5% (w/w) dextran ($M_r = 70,000$), 4.2% (w/w) PEG ($M_r = 35,000$ to 40,000), 0.8% (w/w) ligand–PEG ($M_r = 6,500$), and 12.5 mM potassium phosphate buffer, pH 6.2, at 0°C. •, protein; o, 3-phosphoglycerate kinase, □, hexokinase; and △, glyceraldehydephosphate dehydrogenase.

6. Adjust the volume ratio of the system to allow a reasonable yield of the target protein in the collecting phase. Test if repeated "washings" of this phase with opposite pure phases increase the purification.

7. Carry out the extraction in the most suitable scale.

11.4.2   Preparation of Phase Systems

The phase systems are prepared from stock solutions of dextran "PEG" salts and buffer. Dextran with a molecular weight of 500,000 (dextran 500, 20% solution) and PEG with a molecular weight of 8,000 (40% solution) are normally used. Dry dextran usually contains 4 to 5% (w/w) of water. Dextran solutions, for which accurate concentrations might be of value, can be analyzed by polarimetry using the specific rotation (using sodium light) $[\alpha] = 199°$ ml dm^{-1} g^{-1}. For partitions of proteins, the following system may be tried: 8% dextran 500 and 6% PEG 8000. All the percentage values are in weight/weight. The systems also contains 25 mmol sodium phosphate buffer per kilogram (approximately equal to 25 mM).

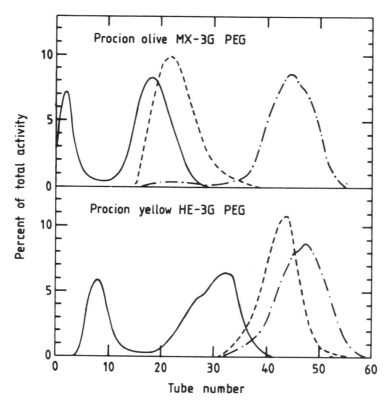

FIGURE 11-7. Countercurrent distribution with 55 transfers of an extract from baker's yeast. The graphs show the distribution of hexokinase (—), 3-phosphoglycerate kinase (---), and phosphofructokinase (— · —) when either Procion olive MX-3G or Procion yellow HE-3G is used as PEG-bound affinity ligands. System composition: 7% dextran 500, 5% PEG 8000 (including dye–PEG, 1% of total PEG), 50 mM sodium phosphate buffer, pH 7.0, 0.2 mM EDTA, and 5 mM 2-mercaptoethanol at 3°C. (Adapted from Ref. 30 with permission.)

Systems of 4- to 20-g weight can be weighed out accurately enough on a top-loaded balance. Smaller volumes can also be handled with a special technique. The preparation of a 4-g system, of the composition suggested earlier, is given here:

To a 10-ml Graduated centrifugation tube, add

- 1.6 g 20% dextran 500
- 0.6 g 40% PEG 8000
- 0.2 mL of 0.5 M sodium phosphate buffer, pH 7.5, and water to total 3.0 g
- 1.0 g protein extract

If the protein extract contains buffer or salt, this has to be taken into account. The system is thermostatted and then carefully mixed by inverting the closed tube for at least 20 sec. After centrifuging (1 min or more at 1000 g) or left standing for 10 to 20 min, the phases separate and samples are withdrawn for analysis. The partition coefficients for total protein and for the target protein are determined.

11.4.3   Preparation of Ligand–Polymers

The ligand is usually bound to PEG and the target protein will then be extracted into the upper phase.

Dextran can also be used as ligand carrier to extract proteins to the lower phase. Several detailed review articles have summarized the synthetic methods that have been used.[2,13,14] A popular choice of ligands for enzymes[4,5,18–27] and some proteins present in blood plasma[23,31] has been the triazine dyes, which are commercially available in activated form (as textile dyes). They react with both PEG and dextran in an alkaline water solution. The PEG derivative is extracted from a salt-containing water solution into chloroform and (after evaporation of solvent) purified by treatment of DEAE-cellulose. The dextran derivative is purified by precipitation. The synthetic steps of dye–polymer derivatives using Cibacron blue F3-GA are given below and apply to most other related dyes. Other polymers may also be derivatized in this way with minor modifications.[26,32]

Cibacron blue F3G-A PEG 8000: 100 g of PEG is dissolved in 150 mL of water. Three grams of Cibacron blue F3-A is added under stirring followed by 50 mL of 5 M NaOH and 10 g crystallized sodium sulfate. The mixture is mechanically stirred for 7 h, neutralized with acetic acid, and dialyzed overnight to remove salts. The polymer solution (containing unsubstituted PEG, dye–PEG, and free dye) is diluted to 1 L, and 100 g of DEAE-cellulose (Whatman DE52, neutralized and intensively washed) is added. The mixture is stirred for 1 h. Dye–PEG and free dyes both bind to the ion exchanger whereas unsubstituted PEG stays in solution. The ion exchanger is collected by suction filtration and washed with water. The dye–PEG is then eluted (without suction) with a 2 M potassium chloride solution. The free dyes remain on the ion exchanger. The dye–PEG is extracted from the salt solution by repeated treatments with chloroform (500 to 800 mL). The pooled chloroform fractions are pooled and dried with anhydrous sodium sulfate and filtered. Finally the solvent is removed by evaporation.

Procion yellow HE-3G dextran 500: 9 g of dextran 500 is dissolved in 180 mL water at 70°C, and 1 g of Procion yellow HE-3G is added. When the dye has dissolved, 18 mL of a solution containing 1 M NaOH and 0.5 M sodium sulfate is added, and the mixture is kept at the same temperature for 45 min. The solution is neutralized with acetic acid, and the dextran is precipitated by slow addition, under stirring, of 300 mL ethanol. The precipitate is dissolved in 100 mL of water and again precipitated with 300 mL ethanol. After dissolving the precipitate in 100 mL of water, the remaining free dye is removed by adding 5 mL of 2 M KCl and 30 mL wet DEAE-cellulose (Whatman DE52). The mixture is stirred for 3 h, the ion exchanger is removed by filtration, and the filter cake is washed with 10 mL of 0.1 M KCl, which is combined with the main filtrate. The treatment with DEAE-cellulose is repeated once, and the dye–dextran-containing solution is then dialyzed against water. If necessary, the solution can be concentrated by evaporation in vacuum.

A very important point for ligand–polymers, in general, is the purity. Also, minute amounts of free ligand in the system may decrease the efficiency considerably by competing in binding to the protein. The purity can be checked by thin-layer chromatography.[31] Another useful procedure is to determine the K_{aff} values obtained with the ligand–polymer after it has been further purified (e.g., by recrystallization from methanol, gel chromatography, or ion exchanger).

11.4.4 Protein Measurements

Protein determinations are assayed according to Bradford using Coomassie brilliant blue G staining in solution.[33] In contrast to other methods, the presence of polymers does not have a strong negative influence on the assay. Another possibility is to determine the absorbance at 280 nm. In both cases the phases are diluted 5 to 100 times before the assay and phases from a protein-free system are used as blanks.

11.5 APPLICATIONS

11.5.1 Isolation of Phosphofructokinase from Yeast[19]

Baker's yeast (100 g) is ground in a household mixer (with rotating knives) together with 400 g of crushed dry ice (solid CO_2) for 4 min. After evaporation of the carbon dioxide, the remains are mixed with 100 mL of 50 mM sodium phosphate buffer, pH 7.1, containing EDTA (0.5 mM), 2-mercaptoethanol (5 mM), and phenylmethylsulfonyl fluoride (PMSF, 0.5 mM). PEG 8000 is added to a concentration of 4% (w/w) at 3°C, the mixture is stirred for 15 min, and is then centrifuged at 500 g for 20 min. The supernatant is collected, and the PEG content is increased (from 4%) to 8 % (w/w), incubated, and centrifuged as just described. The phosphofructokinase (PFK) is enriched in the pellet. Purification is six to seven times relative to protein.

The pellet is dissolved in 15 mL of the buffer used earlier. The solution is included in a 50-g two-phase system containing (totally) 7.5% dextran 500, 5% PEG 8000, and 25 mM sodium phosphate buffer (with the additives mentioned earlier). The system is equilibrated at 3°C by mixing for 30 sec, and the phases are allowed to settle, which are eventually speeded up by centrifugation. The upper phase (containing some proteins with high K values) is removed and replaced with a fresh upper phase where 5% of the total PEG consists of

Cibacron blue F3G-A PEG 8000. The system is equilibrated and the phases are separated. The PFK is now in the upper (blue) phase whereas bulk proteins stay mainly in the lower phase. The upper phase is "washed" by equilibration with a new fresh lower phase. The collected upper phase is warmed to 25°C, and 12 mL of a 40% solution of potassium phosphates (KH_2PO_4 and K_2HPO_4 in molar ratio $1:1.2$) in water is added. A PEG–salt system is formed containing PFK in the lower salt-rich phase. Purification in the affinity partition step is eight to nine times.

Pure PFK is obtained from the salt phase by passing it through a Sephadex G-25 column (using 25 mM buffer with additives as described earlier) to remove salts. The protein fraction is mixed with 1 to 2 g of DEAE-cellulose for 15 min (at 3°C) and washed on a small filter (or in a column) with the buffer. The PFK is eluted with 200 mM KCl (in the same buffer). As a final step the PFK is purified on a Sepharose CL-4B column. The total purification is now 142 times. Proteolytic activity (not inhibited by PMSF) is reduced by a factor of 2,200.

11.5.2   Isolation of Glucose-6-Phosphate Dehydrogenase

Glucose-6-phosphate dehydrogenase (G6PDH) can be purified from baker's yeast in a similar way.[21] The following modifications are recommended:

1. The precipitation with PEG 8000 is carried out between 6.5 and 12.5% (w/w).
2. Two-phase partitioning is carried out in a system containing 10% dextran 500 and 7% PEG 8000. For affinity partitioning, Procion yellow HE-3G PEG (5% of total PEG) is used. For the washing with the lower phase, Cibacron blue F3G-A dextran (2% of total dextran) may be included to increase the effectivity. A more effective step than adding phosphates (to separate enzyme from dye–PEG) is to add a fresh (dextran) lower phase together with NADP (2.5 g per liter system) and sodium sulfite (0.5 g per liter system). G6PDH is then recovered in the lower phase. Total purification is 43 to 44 times relative to protein.
3. The lower (G6PDH-containing) phase is diluted with 4 volumes of a solution containing 5 mM sodium phosphate buffer, pH 7.5, 75 mM KCl, 2 mM 2-mercaptoethanol, 0.5 mM EDTA, and 2 mM magnesium sulfate. The solution is passed through a bed of DEAE-cellulose. When dextran and nonbound proteins have been washed away the enzyme is eluted by increasing the concentration of potassium chloride to 170 mM. The total purification is 330 times.

11.5.3   One-Step Isolation of Lactate Dehydrogenase from Swine Muscle[22]

Swine muscle (100 g) is cut in small pieces and homogenized in a household mixer (with rotating knives) with 330 mL of ice-cold 40 mM sodium phosphate buffer, pH 7.9, for 5 min. The mixture is centrifuged at 16000 g for 20 min. The supernatant is included in a 1-kg system with a final composition of 10% (w/w) dextran 500, 7.03% (w/w) PEG 8000, 0.07% Procion yellow HE-3G PEG 8000, and 50 mM sodium phosphate buffer, pH 7.9. The system is equilibrated at 3°C, and the phases are separated. The lactate dehydrogenase (LDH)-containing upper phase is washed twice with fresh lower phases from systems with the same composition (but not containing protein). The upper phase is mixed with half its weight of 50% phosphate solution (NaH$_2$PO$_4$ · H$_2$O + Na$_2$HPO$_4$ · 2H$_2$O is molar ratio 1:1), which has a temperature of 40°C. The obtained salt phase contains practically pure LDH.

11.6 REFERENCES

1. P.-Å. Albertsson, *Partition of Cell Particles and Macromolecules,* 2nd ed., Almqvist & Wiksell, Stockholm, l971.

2. H. Walter, D.E. Brooks, D. Fisher, *Partitioning in Aqueous Two-Phase Systems: Theory, Methods, Uses, and Applications in Biotechnology,* Academic Press, Orlando, 1985.

3. V.P. Shanbhag, G. Johansson, *Biochem. Biophys. Res. Commun., 61,* 1141–1146 (1974).

4. G. Kopperschläger, *Methods Enzymol., 228,* 121–136 (1994)

5. G. Kopperschläger, G. Lorenz, E. Usbeck, *J. Chromatogr., 259,* 97–105 (1983).

6. M.-R. Kula, G. Johansson, A.F. Buckmann, *Biochem. Soc. Transact., 7,* 1–5 (1979).

7. G. Takerkart, E. Segard, M. Monsigny, *FEBS Lett., 42,* 218–220 (1974).

8. S.D. Flanagan, S.H. Barondes, *J. Biol. Chem., 250,* 1484–1489 (1975).

9. H. Walter, G. Johansson, *Anal. Biochem., 155,* 215–242 (1986).

10. G. Johansson, *Biochim. Biophys. Acta, 222,* 381–389 (1970).

11. G. Johansson, A. Hartman, P.-Å. Albertsson, *Eur. J. Biochem., 33,* 379–386 (1973).

12. G. Johansson, *Acta Chem. Scand. B, 28,* 873–882 (1974).

13. J.M. Harris, *J. Macromol. Sci., C-25,* 325–373 (1985).

14. A.F. Bückmann, M. Morr, G. Johansson, *Makromol. Chem., 182,* 1379-1384 (1981).

15. A. Chaabouni, E. Dellacherie, *J. Chromatogr., 171,* 135–143 (1979).

16. P. Hubert, E. Dellacherie, J. Neel, E.-E. Baulieu, *FEBS Lett., 65,* 169–174 (1976).

17. C. Erlandson-Albertsson, *FEBS Lett., 117,* 295–298 (1980).

18. G. Johansson, M. Andersson, *J. Chromatogr., 303,* 39–51 (1984).

19. G. Kopperschläger, G. Johansson, *Anal. Biochem., 124,* 117–124 (1982).

20. J. Schiemann, G. Kopperschläger, *Plant Sci. Lett., 36,* 205–211 (1985).

21. G. Johansson, M. Joelsson, *Enzyme Microb. Technol., 7,* 629–634 (1985).

22. G. Johansson, M. Joelsson, *Appl. Biochem. Biotechnol., 13,* 15–27 (1986).

23. G. Birkenmeier, *Methods Enzymol., 228,* 154–167 (1994).

24. B.H. Chung, D. Bailey, F.H. Arnold, *Methods Enzymol., 228,* 167–179 (1994).

25. G. Johannsson, G. Kopperschläger, P.-Å. Albertsson, *Eur. J. Biochem.*, *131*, 589–594 (1983).

26. G. Johansson, M. Joelsson, *J. Chromatogr.*, *393*, 195–208 (1987).

27. G. Johansson, *Methods Enzymol.*, *104*, 356–364 (1984).

28. P.-Å. Albertsson, *Anal. Biochem.*, *11*, 121–125 (1965).

29. H.-E. Åkerlund, *J. Biochem. Biophys. Meth.*, *9*, 133–141 (1984).

30. G. Johansson, M. Andersson, H.-E. Åkerlund, *J. Chromatogr.*, *298*, 483–493 (1984).

31. G. Birkenmeier, G. Kopperschläger, G. Johansson, *Biomed. Chromatogr.*, *1*, 64–77 (1986).

32. G. Johansson, M. Joelsson, *J. Chromatogr.*, *411*, 161–166 (1987).

33. M.M. Bradford, *Anal. Biochem.*, *72*, 248–254 (1976).

PART III
Electrophoresis

12 Electrophoresis in Gels

TORGNY LÅÅS

Pharmacia Biotech AB
S-751–82 Uppsala, Sweden

Protein Purification: Principles, High-Resolution Methods, and Applications, Second Edition.
Edited by Jan-Christer Janson and Lars Rydén.
ISBN 0-471-18626-0. © 1998 Wiley-VCH, Inc.

12.1 INTRODUCTION

Electrophoresis in gels, primarily polyacrylamide gels, is a universally applied technique for protein analysis. Thanks to innovative developments and commercialization of equipment and precast gel media during the last decades, analytical electrophoresis has matured from an art for the specialists to an "every man's" tool. Good or even excellent results are generally obtained by any technician without special training or experience by just following the manufacturer's instructions.

This chapter therefore does not give detailed recipes of the different electrophoresis techniques; these are better obtained from more comprehensive handbooks or directly from the manufacturers of electrophoresis chemicals and gel media. Rather the purpose is to review some of the basic concepts. A better understanding of these concepts will increase the reader's possibilities in utilizing this analytical tool in the most efficient way. Another purpose is to demonstrate the power and efficiency of electrophoresis by describing a few selected applications in somewhat more detail.

The word "electrophoresis" derives from Greek and means "carried by electricity." It was initially used to describe the behavior of electrically charged colloidal particles in an electric field. The migration of true solutes was originally referred to as "ionophoresis." Eventually, electrophoresis became the recognized term for the migration of all kinds of particles in an electrical field.[1]

This chapter discusses the use of electrophoresis for the separation of proteins. The sample containing the proteins to be separated is then placed in an electric field that forces the electrically charged proteins to move. If the experiment is performed in free solution (free electrophoresis) and the proteins have different charge densities, they will move at different speeds and can thus be separated. In practice, however, the separation is normally not performed in free solution but in a supporting gel medium. The gel can either act as an "inert" support for the electrophoresis buffer or actively participate in the separation by interacting with the proteins. In the latter case, the protein–gel interaction is the actual separation factor, whereas the electrical field merely makes the proteins migrate through the gel.

Whatever the nature of the actual separation parameter, all these techniques are collectively called electrophoresis. A prefix indicating the separation principle or the separation medium used is commonly added (e.g., affinity electrophoresis, starch gel electrophoresis). Unfortunately, the nomenclature is not consistent. The prefix sometimes designates the gel media and sometimes the type of samples that are being separated. In practice, however, the meaning is normally obvious from the context.

Two electrophoretic techniques have so many special characteristics that they have independent names: *isoelectric focusing* (also called electrofocusing) and *isotachophoresis* (also called displacement electrophoresis, steady-state electrophoresis, multiphasic zone electrophoresis). Isoelectric focusing (IEF), which separates proteins according to differences in isoelectric point, is dealt with separately in Chapter 13. Isotachophooresis, however, has not gained a wide acceptance for protein separations except for the use of its principles for sample concentration by "stacking" in disc electrophoresis (Section 12.3.3.2) and in the elution of proteins and nuclei from gels after electrophoresis (Section 12.3.2 and Chapter 16).

Gel electrophoresis is primarily an analytical tool. For *preparative purposes,* the use of gel electrophoresis is hampered by several factors. First, in large-scale applications the joule heat formed causes severe problems. Second, it is difficult to recover the separated fractions without zone contamination and/ or dilution. In view of the efficiency and relative ease of scaling up high-resolution chromatographic techniques, chromatography is the natural first choice for preparative protein separations.

However, "preparative" sometimes refer to very small amounts of substance. Many techniques, such as amino acid analyses and sequencing, only need nanogram quantities of protein. In such cases the conventional analytical electrophoretic separations can readily provide the amounts necessary. The problem with preparative gel electrophoresis is then reduced to detection of the substance of interest and transferring it out of the gel and into free solution in a test tube for further treatment. With electroelution, the protein can be extracted from the gel and, at the same time, concentrated (Section 12.3.2). Recovery of proteins after gel electrophoresis is further described in Chapter 16. For pure *analytical purposes,* however, many of the electrophoretic variants have demonstrated outstanding properties. Electrophoresis is the method of choice for gathering information about the composition of a crude sample. Furthermore, approximate estimates of M_r and isoelectric point, together with a distribution profile of M_r and pI of contaminating proteins, form a good basis for a purification strategy. The technique called "titration curve analysis" provides valuable information for ion-exchange chromatography (Chapter 13). During a purification procedure it is customary to monitor the progress by electrophoresis after each purification step.

Sodium dodecyl sulfate (SDS) electrophoresis and IEF are the methods of choice for determining the molecular weight and isoelectric point of a protein.

By electrophoresis, large numbers of samples can be compared with ease and accuracy. This is used in clinical chemistry as well as in analyses of isoenzyme patterns in genetic polymorphism in, for example, forensic medicine. In cell biology the effects on protein pattern after different manipulations are often analyzed by electrophoresis.

By combining two independent separation principles, the separation power of electrophoresis can be drastically increased. Two-dimensional electrophoresis combining IEF and SDS–PAGE is one of the most powerful methods available for analyzing complex protein mixtures (Chapter 15). Immunoelectrophoresis and affinity electrophoresis (Chapter 14) are other examples. It should also be pointed out that by combining different *detection methods* after standard one-dimensional separations, additional separation dimensions may be added (Section 12.2.4).

Electrophoresis in transverse gradients of urea and/or other chaotropic agents can be used for the analysis of protein conformational changes caused by these agents.[2,3]

Electrophoresis has a number of practical advantages, including:

1. Relatively simple and inexpensive equipment.
2. High resolution results.
3. Ease of multiple sample analysis.
4. High sensitivity.
5. Specific detection easy (immunological, enzymatic, etc.).
6. Aesthetic appearance of banding pattern.

The availability of precast electrophoresis gel media from a number of manufacturers as well as semiautomatic equipment offers increased speed, convenience, reproducibility, and accuracy.[4,5]

12.2 BASIC CONCEPTS

Fundamental to electrophoretical separations is the fact that proteins are electrically charged particles. The charges are derived from amino acids with ionogenic side groups. These are the basic residues arginine, lysine, and histidine and the acidic residues glutamic acid and aspartic acid (see also Table 4-2). Tyrosine side chains also contribute to the protein's overall charge. In addition, the proteins often have associated charged components of nonprotein origin such as lipids or carbohydrates. Most of these acids and bases are relatively weak, making the overall charge of the protein strongly pH dependent. Most globular proteins are acidic, with pI values in the range of 4 to 5[6] (Figure 12-1; see also Chapter 4).

FIGURE 12-1. Proteins as charged particles. (Top) The charge dependence on pH of three imaginary proteins: a, b, and c. The pH where the net charge of the protein equals zero is called the isoelectric point (pI). Proteins with pI in the acidic region are called "acidic" proteins (a), whereas "basic" proteins have basic pI values (b and c) and "neutral" proteins have pI values close to pH 7. Note that the slope of the curve at pI can be very different. This is especially important in isoelectric focusing (Chapter 13). (Bottom) The distribution of pI values for globular proteins reported up to 1981. Although the number of reported pI values has increased since then, the overall picture, with a predominance of acidic proteins, is without doubt still the same. (Reproduced from Ref. 6 with permission of Elsevier Science Publishers B.V. Amsterdam.)

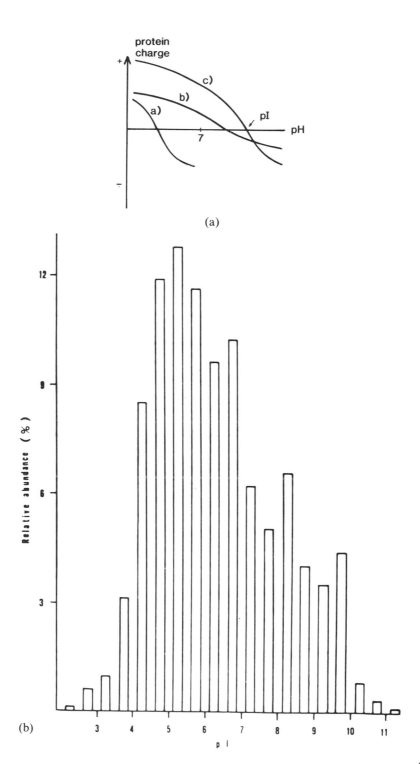

(a)

(b)

467

12.2.1 Electrophoretic Migration

When exposed to an electrical field, E (V/cm), a force, F, acts on the protein molecules. The force F depends on the field strength and the net charge, z, of the protein: $F = Ez$. The net charge for a particular protein depends not only on the pH (Figure 12-1) but also on the ionic environment. The association of ions as counterions or via hydrophobic forces (e.g., SDS, see Section 12.5) will modify the charge of the "naked" protein. As soon as the proteins start to move by the action of the electrical force, a counteracting force, F', is manifested due to the "friction" of the medium. F' is directly proportional to the migration speed and hence the protein molecule accelerates until $F = F'$. F' depends on factors such as the size of the molecule, the viscosity of the medium, and the properties of the gel medium (if there is a gel). The migration velocity of a protein in the unit electrical field ($E = 1$ V/cm) under defined conditions (temperature, buffer, pH, gel composition) is called mobility and is an intrinsic property of each individual protein. The relationship among migration velocity (v), mobility (u), and field strength (E) is therefore

$$v \text{ (cm/h)} = E \text{ (V/cm)}* u \text{ (cm}^2/\text{Vh)}$$

Note that the driving force for the migration is set up by the electrical field strength and that the current has no direct effect. However, a higher current will generate more heat, which can affect mobility and migration velocity by reducing the viscosity of the medium. (Heat evolving is proportional to the square of the current according to the formula $P = RI^2$, where P is the effect, I is the current, and R is the resistance.)

12.2.2 Gel Media

The earliest electrophoresis experiments were performed in free solution. It was soon realized, however, that the use of anticonvection media offered practical advantages, and a number of "water holding" supports such as filter paper, starch, agar, agarose, cellulose, and polyacrylamide were introduced. It was soon realized that some of these media actually contributed to the separation by sieving. Of course, the sieving depends on the relative size of the proteins and on the pores in the support.

Of all the media tried, two have proved outstanding for protein electrophoresis: Polyacrylamide (PAA) and agarose gels. In the concentrations mostly used, PAA can be considered as a sieving gel and agarose as a nonsieving gel for proteins. The sieving properties are used to advantage in the different variants of polyacrylamide gel electrophoresis (PAGE).

12.2.2.1 Polyacrylamide Gels Polyacrylamide gels are prepared by the polymerization of acrylamide in the presence of a bifunctional cross-linking agent.[7,8] Normally this is N',N'-methylenebisacrylamide. This procedure gives

a three-dimensional network. The pores of the gel can be varied within wide limits. The cross-linking reaction and the properties of the PAA gels obtained under different conditions have been extensively investigated[9-14] (see also Section 2.2).

PAA gels are conveniently described by the T and C nomenclature introduced by Hjertén[15]:

$$T = \frac{(A + B)\,100}{V}\% \text{ and } C = \frac{B\,100}{(A + B)}\%$$

where A and B are the amounts of acrylamide and bisacrylamide, respectively, expressed in grams. V was originally defined by Hjertén as the volume of *added buffer,* but for practical reasons is commonly understood as the final volume of *gelling solution.* For small values of T the difference between the two designations is small, but at higher T values it is important to clearly specify the exact meaning of V.

Typical values for T in PAGE, where the sieving effect of the gel is utilized, are 10 to 20%. C is usually in the order of 3 to 10%. Highly cross-linked gels, for example, with $C = 30$, give macroporous gels with a completely different gel structure as compared to conventional gels.[9] From C of ca. 15%, the porosity increases with C. In IEF, where the sieving is an unwanted side effect, a typical gel composition is T5 C3.

12.2.2.2 *Agar and Agarose*

Agar and Agarose Agar and agarose are both composed of a gel-forming polysaccharide of alternating galactose and 3,6-anhydrogalactose.[16] Various amounts of methyl ether, sulfate, and pyruvate groups are then attached to this gel-forming backbone. No principal difference exists between agar and agarose. The name "agarose" is reserved for the more charged-free and well-defined fraction of agar (see Section 2.2).

Agarose with different specifications suitable for different applications can be obtained from a number of suppliers.

Because agarose is a nonsieving matrix for proteins, the separations in agarose gels do not benefit from sieving and suffer from increased diffusion in the large agarose pores. The separations are thus in general not as good as in PAGE unless the pure electrophoresis is combined with some other separation parameter, such as in immunoelectrophoresis or affinity electrophoresis.

The only really high-resolution electrophoretic technique where agarose may be used with advantage is isoelectric focusing. However, for this purpose, only the purest forms of agarose, with an extremely low content of charged groups, are suitable (Chapter 13). For optimal performance, the small number of unavoidable negative charges can be counterbalanced by the introduction of an exactly balanced number of positively charged groups. Such agarose is available from Pharmacia Biotech as "Agarose IEF."

12.2.3 Electro(endo)osmosis

When applying a voltage over a gel it is frequently observed that the gel shrinks close to one of the electrodes, mostly the anode, and swells at the other end. Sometimes swelling is so large that buffer is extruded from the gel, causing the gel to "sweat." Occasionally the gel shrinkage may cause the gel to completely dry out, which of cause totally ruins the experiment. These effects are all due to the phenomenon called electroosmosis.[17,18] All gel media contain charged, mostly acidic, groups. Polyacrylamide is inevitably deamidated to some extent, and agar and agarose always contain sulfate and carboxylic groups. These immobilized, negatively charged groups attract positively charged ions from the buffer as counterions to keep the electroneutrality of the system. These counterions are not completely immobilized but will occasionally dissociate into solution, where they will be carried away in the voltage gradient until they are trapped by the next charged group on the gel matrix. Because the counterions are normally small, highly hydrated, positively charged cations, the macroscopic effect is a transportation of liquid from anode to cathode. The magnitude of the electroosmotic flow depends primarily on the charge density of the gel, the voltage, and the buffer concentration. More immobilized charges mean more counterions carry the electroosmotic flow, a higher voltage increases the migration velocity of the hydrated ions, and a lower buffer concentration forces a higher proportion of the current to be transported by the counterions. A consequence of this is that the effects of electroosmosis are especially pronounced in isoelectric focusing with its extremely high voltage and low ionic strength. In IEF, therefore, only extremely low charged gel media such as PAA or charged-balanced agarose will work properly.

However, in some systems, as for example in immunoelectrophoresis, a controlled electroosmotic flow may be an advantage (Chapter 15).

In agar and agarose gels, electroosmosis can be expressed quantitatively as a relative mobility index, m_r, where $m_r = -D/(D + A)$ when D and A are the migration distances of uncharged dextran and serum albumin when the experiment is performed under standardized conditions at pH 8.2 with an ionic strength of 0.05 M.[19]

12.2.4 Detection

Mostly the electrophoretic separation is interrupted with the separated protein zones still in the gel and the separation pattern visualized by some protein-staining procedure.

The most frequently used stain is Coomassie (brilliant) blue (CBB) R-250 marketed under trade names such as Serva Blue R, PhastGel Blue R tablets, and so on. Normally the proteins are first precipitated by agents such as trichloroacetic acid (ca. 10%) and/or formaldehyde. The gel is then exposed to a solution of stain dissolved in a dilute acetic acid/ethanol mixture. Finally, excess stain is removed by soaking the gel in a destaining solution of acetic acid/ethanol until the background is clear.[20] Among the numerous investigations on staining with CBB can be mentioned the use of CBB for quantitative protein determinations on electrophoresis gels,[21] description of an extremely fast staining method (6 min!) using phosphoric acid,[22] and an extremely sensitive procedure based on prolonging the staining (up to 72 h) at low pH in the presence of ammonium sulfate.[23] With this method, less than 1 ng/mm[2] of protein can be detected. The most sensitive protein-staining methods are based on the deposition of silver, which are often claimed to be up to 100 times as sensitive as conventional CBB-based methods, with detection limits of less than 1 ng of protein.[24] The first proposed methods were extremely tedious and expensive due to a large number of experimental steps and a high consumption of silver nitrate. Modern methods are much faster and more economical. Rabilloud and colleagues made a comprehensive review of the chemistry of the different silver-staining variants described[25] and also a summary of recommended staining protocols for different electrophoretic applications.[26] Commercially available ready-to-use silver-staining kits that greatly facilitate the otherwise quite complicated and tedious procedures are available from several suppliers of electrophoresis chemicals. The semiautomatic PhastSystem is provided with a staining unit that can automatically perform the commonest staining procedures.[27–29] By combining alcian blue with the standard PhastGel silver kit, Möller and co-workers have devised a very sensitive and selective staining method for proteoglycans and glycosaminoglycans, which otherwise stain poorly with standard Coomassie and silver-staining methods.[30] Using a minor modification of a standard protein protocol, lipoproteins could also be successfully detected with silver staining.[31]

As already mentioned, another dimension of information can be obtained by combining general protein staining with *specific detection methods*. Such methods include staining for lipids,[32] carbohydrates,[33,34] and enzyme activity.[35–37]

Proteins with antigenic binding properties are often analyzed by transferring the separated proteins to a membrane ("Western blotting") and then exposing the membrane to a suitable, usually labeled, antiserum (Chapter 16). When used in conjunction with total protein staining, the component(s) of interest can be localized in the whole protein spectrum.

After detection, PAA gels can be stored either wet or dry. Wet gels without plastic backing are preferably soaked in acetic acid (10%) and stored in a sealed plastic bag. Special equipment for drying slab electrophoresis gels is commercially available.

Gels used in horizontal electrophoresis are usually supported by a plastic backing foil. Such gels are usually dried by just being left in open air. To avoid gel cracking, it is customary to include 10% glycerol in the last washing solution. To protect the gel surface during long-term storage, it is recommended to cover the gel with a thin foil, a gel protective sheet, or something similar.

12.3   ELECTROPHORETIC TECHNIQUES

As pointed out in Section 12.2.2, the highest resolution is obtained in sieving media and polyacrylamide has proved to be especially outstanding. The following discussion is therefore restricted to polyacrylamide gel electrophoresis, PAGE. The space allows only an overview of the most important techniques and some selected hints of the possibilities available. For more detailed information about casting and running electrophoresis gels, the cited references or more comprehensive handbooks should be consulted.[14,38]

12.3.1   Equipment and Mode of Operation

PAGE can be performed in equipment with the gel positioned either vertically or horizontally. Vertical electrophoresis, with the sample migration downward, is the commonest way to run PAGE. In vertical electrophoresis the gel is usually contained in a cassette of glass or plastic, with the upper and lower part of the cassette open and in contact with the electrolyte solutions. The sample is loaded on top of a gel, usually in special sample application wells. In horizontal electrophoresis, the gel, usually supported by a plastic foil, is layered on a horizontal cooling plate. The sample is applied on top of the gel, either in preformed sample wells, as droplets directly on the surface, or using special sample application masks. The electrical contact can be accomplished in different ways (Figure 12-2).

The earliest PAGE experiments were performed in cylindrical gel rods in vertically positioned glass tubes, but vertical electrophoresis is now performed in slab gels in cassettes of glass or plastic. The major advantage with slab gels is that a number of samples can be run simultaneously on the same gel, facilitating comparison of the electrophoresis patterns and minimizing the risk of sample confusion. Versatile equipment is available from a number of commercial sources. Commercially available, precast, ready-to-use, gel slabs, both for horizontal and vertical equipment, are becoming more and more common (Table 12-1).

A particular advantage with horizontal electrophoresis systems is the lack of liquid electrode buffer. Not only is the use of buffer strips in gel form very convenient, they also greatly reduce the amount of radioactive waste when labeled proteins are analyzed.[39]

The following fundamental principles are equally applicable to both the horizontal and the vertical mode (Figure 12-3).

In nucleic acid electrophoresis, a special form of horizontal electrophoresis, "submarine," is used where the whole gel is immersed in the electrophoresis buffer.[40] This technique has not been applied to proteins so far and is therefore not dealt with here.

12.3.2   Sample Application and Recovery

Obviously only protein molecules in solution can be analyzed by electrophoresis. If the sample is difficult to dissolve, agents such as detergents or urea may

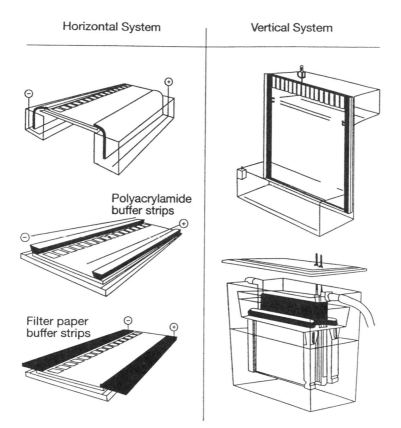

Horizontal System   Vertical System

Polyacrylamide
buffer strips

Filter paper
buffer strips

FIGURE 12-2. Schematic figures illustrating the essential parts of setups for horizontal and vertical electrophoresis. In horizontal electrophoresis the electrical contact between the electrodes and the electrophoresis gel can be mediated in three different ways: (1) filter paper wicks soaked in buffer and dipped into buffer tanks; (2) gel buffer strips (polyacrylamide or agarose, commercially available), making liquid buffer obsolete; and (3) filter paper wicks soaked in concentrated buffer; no buffer tanks are required.

TABLE 12-1. Comparison between Vertical and Horizontal Electrophoresis Systems: Specific Advantages

Vertical Systems	Horizontal Systems
Inexpensive, simple equipment	Minimal buffer consumption
Can take large sample volumes	No leakage problems
Even, efficient cooling	Gels supported by plastic backing are
Well suited for blotting	Easy to handle
	Stable, do not shrink or swell
	Thin, i.e., efficiently cooled allowing fast
	separations and sharp bands
	Versatile: Same equipment can be used for all
	techniques (agarose, PAGE, IEF, etc.)

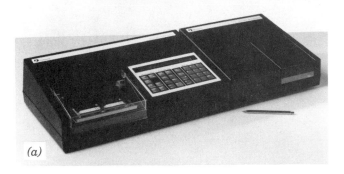

FIGURE 12-3. (a) With the microprocessor-controlled PhastSystem, the electrophoresis conditions, including running temperature, sample application, and staining, can be preprogrammed and accurately controlled during processing. For speed, economy, and convenience, the PhastSystem operates with ca. 5×5 cm minisized gels. Despite the small format, perfectly satisfactory results are mostly obtained. (b) Multiphor II is a horizontal system designed for gel sizes up to ca. 20×26 cm, which is suitable for most gel PAA-based techniques, as well as agarose and immunoelectrophoresis. (c) Conventional equipment for vertical electrophoresis suitable for PAGE and SDS–PAGE.

prove necessary. Unlike chromatography, where insoluble material in the sample may easily ruin the column, the presence of particles or even fibrous matter in the electrophoresis sample is generally not critical. In fact, with a concentrating technique such as isoelectric focusing, even whole tissue samples can be directly applied and the extraction performed during the IEF process itself. In some cases, such as in the zone-sharpening techniques discussed in the following section, the sample must be transferred into a special buffering

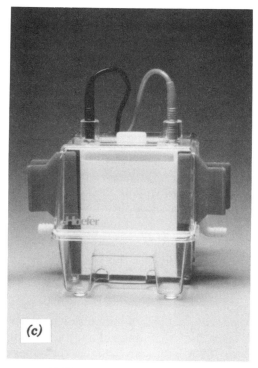

FIGURE 12-3. *(Continued)*

environment. This can be done by dialysis or by gel filtration on a small column packed with Sephadex G-25 or G-50, such as the NAP-5, -10 or -25, or PD-10 columns from Pharmacia Biotech.

The purified protein band may be recovered easily by cutting out the band from the gel, after having localized it by any of the detection methods available, and then simply soaking the crushed gel in buffer. More efficient methods, however, utilize the principles of isotachophoresis to accomplish a simultaneous extraction and concentration, which is further described in Chapter 16.

12.3.3 Zone-Sharpening Techniques

The separated protein zones will inevitably broaden by time due to diffusion. Fortunately, this effect can be counteracted by different methods. The importance of these techniques cannot be overestimated, as they are fundamental to the high-quality results that are regularly obtained today! Zone broadening can be minimized by concentrating the protein zone *before* it enters the gel (sample ionic strength and protein "stacking") as well as *during* and *after* separation (pore gradient PAGE). Various combinations are common.

12.3.3.1 Sample Ionic Strength   A very simple and straightforward zone-sharpening technique is to make sure that the electrical conductance of the sample buffer is lower than that of the running buffer. This will increase the field strength in the sample zone. Proteins will then migrate fast under the influence of the high field strength within the sample zone and arrive at the separating gel in a narrower zone than originally applied.[41] A 2 to 10 times dilution of the buffer will decrease the conductance proportionally with a concomitant proportional increase in field strength. Because migration velocity is directly proportional to field strength (Section 12.2.2), this results in a 2 to 10 times compression of the protein zone as it leaves the original sample zone.

12.3.3.2 Protein Stacking and Moving Boundary Electrophoresis   As mentioned in Section 12.1, this is an application of the principles of isotachophoresis. Typical for this very efficient technique for sample concentration is a special concentrating or "stacking" gel on top of the separation gel. Because the technique utilizes *discontinuities* in buffers as well as in PAA concentration, it is frequently referred to as "disc electrophoresis."[42,43] The purpose of the stacking gel is to simply hold the buffer required for protein concentration before the protein zone enters the separating gel. The stacking gel is therefore prepared as a low porosity gel with minimal sieving properties.

In the simplest form, the separating gel and the stacking gel both contain a fast-moving ion (e.g., Cl^-) at the start, whereas the upper electrolyte solution contains a slow-moving ion (e.g., glycine). When the voltage is applied, the fast Cl^- ions leaving the upper part of the stacking gel are replaced with the slower-moving ions entering from the upper buffer reservoir. Because the faster ions migrate first, they tend to leave the slower ions behind, creating a zone of "ionic vacuum." The low ionic strength in this region causes a high field strength, and when the slower ions enter this region, their migration speed increases so that they catch up with the fast ions. Thus, immediately behind the fast ions, the slower ions are diluted to a concentration with a field strength that is just right to give both the slower and the faster ions the same migration speed. A self-regulating moving boundary is set up. The slower "trailing" ions cannot pass the faster "leading" ions from the buffer. Should they happen to enter the zone with the fast ions, they will immediately be surpassed by the faster ions and be caught up by the zone with the other ions of the same kind. Should the faster ions happen to lag behind the boundary into the zone with the slower ions and the higher field strength, they will immediately overtake these ions and migrate to the right side of the boundary again.

Suppose now that the mobility of the fast ion is higher, and that of the slow is lower, than the mobilities of the proteins in the sample. The proteins will then be trapped between the two zones just described and, furthermore, they will be sorted in mobility order according to the same principles as for the buffer ions.

The concentration of the buffer ions and the sample proteins will be adjusted so that the electrophoretic migration in all zones is the same as in the zone with the leading fast electrolyte. The migration velocity and electrolyte concentration in each zone are determined by the "Kohlrausch regulating function."[44] Mathematical details will not be given here, suffice it to mention that since the protein concentration expressed as charge equivalents/volume must be in the same order of magnitude as that of the leading buffer ion, the concentration effect can be enormous and the sample concentrated to a very narrow zone. Finally, when the proteins enter the separating gel, their mobilities are reduced. The much smaller trailing ions continue with unchanged speed and pass the proteins. Because of the sieving effect of the gel the proteins will now separate (Figure 12-4).

It is evident that the stacking gel must be long enough for protein stacking to be finished before the moving boundary reaches the separating gel; 5 to 10 mm is usually sufficient.

Proteins with mobilities outside those of the leading and trailing ions are not concentrated. Under nondenaturing conditions and with standard buffer systems, no protein will ever move faster than the leading ion. Many proteins will, however, in their native state move slower than the trailing ion, with some even in the opposite direction, depending on the pI of the protein and the pH of the electrophoresis buffer. In SDS electrophoresis, however, all proteins have a strong negative charge and will move faster than the trailing ion. Small proteins (M_r less than ca. 10,000) may move to the front immediately behind the leading ion (see Section 12.5.2). By choosing suitable buffers and buffer concentrations, the principle of protein stacking can be applied to different pH values for the separation of different types of proteins.[45]

12.3.3.3   Pore Gradient Gels

In this technique it is not the sample, but the protein zones that are concentrated during the separation. This is accomplished by using a gel with a continuously changing acrylamide concentration.[46–48] The sample is applied in the low concentration region. When the proteins migrate into a denser gel network, the "friction" or sieving effect from the gel increases. The electrophoretic migration speed thus decreases and asymptotically reaches zero as the protein approaches the gel concentration of it's "pore limit," primarily dependent on the Stoke's radius. Properly used, gradient gel PAGE is a versatile technique for estimating native protein molecular weights.[49–52] For the most accurate M_r determinations of native proteins, gradient gel PAGE should be performed at a pH where the protein is highly charged, which for the majority of proteins means a basic pH. Buffers with pH values of 8.2 to 8.6 are standard.

Very often the gradient gels are not run to the "pore limit." The zone-sharpening effect of the gel gradient will still improve the band sharpness significantly, but no reliable conclusions about molecular weights can be drawn from such experiments.

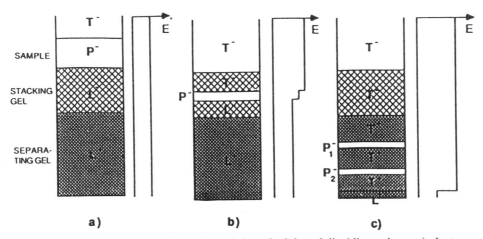

FIGURE 12-4. Schematic illustration of the principles of disc(discontinuous) electrophoresis. (a) *Starting conditions.* The system is composed of a relatively dense sieving gel as the separating gel, with a highly porous, low density stacking gel on top. Both these gels contain the fast, leading ion L^- as the anion. The catholyte anion, T^-, has a lower mobility than both the proteins, P^-, and L^-. The protein sample, which can be very dilute, is layered on top of the stacking gel. The cation can be considered to be the same throughout the system and can be neglected for the sake of the argument. (b) *Concentrating phase.* Under the influence of the voltage, all ions start to migrate. Given the same field strength in the whole system, the L^- ion would leave the other slower ions behind, creating an "ionic vacuum." This will not happen, as in this imaginary "ionic vacuum," the field strength would be very high and the slower ions would catch up. In reality, the concentrations of the P^- and T^- ions will automatically be adjusted to the concentrations that give exactly the field strength that makes all ionic species migrate at the same speed as the L^- ion. Because the mass/charge ratio is much higher for proteins than for buffer salts, this means that the proteins have to be very concentrated to generate the necessary ionic strength. The concentration effect can be enormous. In addition to being concentrated, the different proteins will also be separated into adjacent zones in order of electrophoretic mobility. (c) *Separating phase.* Eventually the proteins enter the denser, separating gel where their migration velocity decreases because of the sieving interaction with the gel network. The much smaller T^- ion will not be affected by the gel and will now pass the proteins, and a new self-regulating moving boundary is set up between the T^- and the L^- ions (the "front"). Marker dyes (bromophenol blue, etc.) will normally migrate in this front as they have a mobility intermediate between most L^- and T^- ions used. The proteins will be separated by sieving in the gel matrix.

The casting of gradient gels is most conveniently done by filling the cassette from a gradient mixer of the type used in, for example, low-pressure ion-exchange chromatography. With suitable equipment, several gels can be cast simultaneously. The acrylamide concentration can be varied within wide limits, but concentrations below 4% and above 30 to 40% are not recommended for practical reasons. Detailed instructions for gel casting are provided by companies supplying gel casting equipment. Precast gels with gradients suitable for most applications are available from several manufacturers, both for conventional vertical electrophoresis and for horizontal electrophoresis.

A consequence of the successively decreasing protein mobilities in gradient gels is that the whole separation pattern is more compressed than on a homogeneous PAA gel. Gradient gels are thus very useful for screening purposes, with all proteins from the very smallest to the largest ones separated on the same gel. However, the best resolution between two closely spaced proteins is obtained on a homogeneous gel of suitable concentration.

12.4 OPTIMIZING THE SEPARATION

12.4.1 Choice of Buffer pH and Gel Porosity; The Ferguson Plot Analysis

The electrophoretic mobility of a protein in a gel is determined by the balance between the electrical force and the retarding, "frictional" force. The electrical force is a function of the voltage applied (which is the same for all proteins in the sample) and the protein charge. The retardation force is a function of gel concentration and protein size. The parameters to adjust in optimization are the buffer concentration and pH and the gel porosity. In most cases, standard procedures with a buffer pH of 8 to 9 and a gel concentration of 7 to 10% in the separating gel will work for neutral and acidic proteins, whereas reversed polarity and a buffer pH of 4 to 5 will separate basic proteins. If, however, the separation problem is difficult and needs to be optimized, the so-called Ferguson plot analysis can be used as a powerful tool for establishing the optimal conditions.[53]

A conventional Ferguson plot analysis is done as follows: Run the separation at different gel concentrations (T values) while keeping C and the buffer fixed. Measure the migration distance of the protein(s) of interest and divide this distance with the migration distance of the marker stain (usually bromophenol blue) to obtain the relative migration value, R_f. Plot the R_f values of the protein to be purified and the major contaminants as a function of T. This is called a "Ferguson plot." Repeat the procedure for different buffers. By analyzing the plots, the gel concentration and the buffer that gives the maximal distance between the proteins to be separated can be deduced[54–57] (Figure 12-5). For careful optimizations, the optimal value of T can be mathematically calculated from data obtained by the Ferguson plot analysis.[56] To simplify the optimization procedure, a system of 19 buffers with different operational pH values has been described.[45]

The easiest way to perform a Ferguson plot analysis is to run the separation on precast gels of different concentrations in a fast operating system such as PhastSystem.[58]

12.4.2 Choice of Buffer Composition and Detergent

Electrophoresis under denaturing conditions in SDS is so efficient, straightforward, and well established that it is sometimes adopted too uncritically. Before

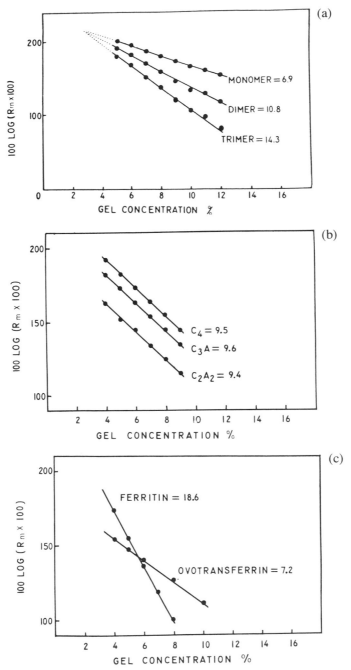

FIGURE 12-5. Ferguson plot analysis. (a) A Ferguson plot of monomer, dimer, and trimer of bovine serum albumin. Because all molecules have the same charge density, they will asymptotically approach the same mobility at T = 0, with better separation at higher gel concentrations. (b) Charge isomers: A Ferguson plot of aldolase isoenzymes. These proteins all have the same size and are therefore affected by changes in gel concentration in the same way. Separation is, in principle, not dependent on gel concentration.

setting up the electrophoretic analysis it may prove worthwhile to contemplate for a moment which method is best apt to achieve the desired goals. In many cases where the biological function of the protein can be utilized to gain further relevant information, electrophoresis (or isoelectric focusing) of the native protein, followed by specific detection of enzymatic or other activity, will normally be a better alternative or a powerful complement to SDS electrophoresis of denatured proteins. Ordinary globular proteins with good solubility in standard buffers can easily be separated in native conformation and with full activity. The separation conditions can be chosen primarily from a pure electrophoretic separation point of view.

Many proteins, however, such as hydrophobic membrane proteins and filamentous proteins with poor solubility in ordinary buffers, pose special problems in electrophoresis. To get these proteins in solution for electrophoresis, detergents such as SDS and reducing agents [mercaptoethanol, dithiothreitol (DTT), etc.] for the cleavage of S-S bridges must be used. Mixtures of phenol, acetic acid, and urea have been used successfully for the solubilization of membrane proteins.[59]

All types of detergents—nonionic, zwitterionic, cationic, and anionic—can be used.[60] A good strategy to keep the protein in its native configuration is to first try the mildest possible detergent. For the solubilization of membrane proteins without denaturation, Hjelmeland and Chrambach[60] recommend starting with CHAPS (3-[(cholamidopropyl)dimethylammonio]-1-propanesulfonate) and systematically exploring the optimal solubilization conditions.

As a last resort for proteins that are extremely difficult to solubilize, stronger detergents such as SDS (see later) can be tried. Some unusually stable proteins retain their activity in SDS, and in other cases proteins can regain activity by a careful exchange of the stronger detergent with a weaker one.[61] Once the solubilization conditions are found, the electrophoresis can then be performed according to standard procedures described later. Renaturation and the use of different detergents in PAGE have been reviewed.[62]

12.5. SDS GEL ELECTROPHORESIS AND M_r DETERMINATIONS

Proteins can be denatured by a variety of agents such as urea and guanidinium hydrochloride as well as strong positive or negative ionic detergents. Among

---

(c) Different charge and molecular weight: A Ferguson plot of ferritin ($M_r > 10^6$) and ovotransferrin (M_r 87,000). Ovotransferrin has a low charge and is overrun by ferritin at low gel concentrations. The mobility of the larger ferritin is more dependent on the gel concentration than on the mobility of the ovotransferrin. At higher gel concentrations the transferrin is highly retarded because of its size. Note that at $T = 6\%$ and the pH used, the two proteins cannot be separated by PAGE. (Reproduced from Ref. 57 with permission of Academic Press).

these, the negatively charged detergent dodecyl sulfate, normally in the form of its sodium salt, has proved extremely useful in connection with electrophoretical separations in acrylamide gels (SDS–PAGE).[63,64] Occasionally, as described below, SDS is used in conjunction with other agents such as urea.

Reasons for the popularity of SDS–PAGE can be summarized as follows:

- It separates proteins strictly according to a single molecular parameter, molecular size, and can therefore be used for molecular characterization if the system is calibrated with molecular standards.
- The resolution is generally very high.
- The technique is inexpensive and easy to set up and perform.
- The separation is fast.
- The method is applicable to most separation problem, as SDS solubilizes most existing proteins.

The major drawback with the technique is that SDS-denatured proteins have lost their biological activities, which decreases the possibilities to gain extra information by specific detection methods. Immunogenic detection methods can, however, be used as sufficient immunogenic identity is mostly inherent in the peptide sequence. It is also sometimes possible, as already mentioned, to renature protein activity.[62]

12.5.1 Principles

Fundamental to SDS–PAGE is the very strong interaction between the dodecyl sulfate detergent and the protein peptide chain causing the SDS–protein complex to migrate as one well-defined entity. The exact nature of this interaction is not fully elucidated, but it appears that a flexible rod containing about 1.4 g SDS/g protein is formed.[65,66]

Some facts about SDS–protein interaction can be summarized:

- The strong solubilizing effect of SDS make essentially all proteins accessible for electrophoretic analysis, including normally insoluble ones such as filamentous proteins and hydrophobic membrane proteins.
- The SDS–protein interaction is strong enough to make the composition of the SDS–protein complex essentially pH independent.
- All protein–SDS complexes acquire the same conformation and differ only in size.
- All protein–SDS complexes have the same, very high, charge/mass ratio.

Because the charge density and conformation are the same, all SDS–protein complexes migrate with essentially the same velocity in free solution. The separation in SDS–PAGE is thus due entirely to the sieving effect from the gel. The high charge content of the SDS–protein complexes gives them a high electrophoretic mobility and short separation times. The various theories describing the sieving effect have been reviewed.[67] For practical use it is sufficient to know that there is a direct relationship between log M_r and R_f, making the determination of protein molecular weight by SDS electrophoresis a both simple and reliable method.[68–70] R_f is defined as the migration distance of the protein relative to that of the tracking dye. By calibrating the system with proteins of known molecular weight, this relationship can be used to determine the molecular weight of unknown proteins.

The condition most commonly used for SDS–PAGE is that described by Laemmli[71] or variants thereof where discontinuous buffers are used for zone sharpening by proteins stacking as described earlier.

The merits of three commonly used buffer systems, described by Wyckoff and co-workers,[72] Neville,[69] and Laemmli,[71] have been compared.[73] The system of Wyckoff and colleagues was found to be particularly suitable for smaller proteins/peptides, down to ca. 6,000 in molecular weight, if the SDS concentration was increased from 0.03 to 0.1%.

Generally, SDS is added to all gels and buffers to concentrations of 0.05 to 0.1% (w/v). However, in principle, it is only necessary to add SDS to the cathodic buffer.[74] The dodecyl sulfate molecules enter the gel before the proteins. Certainly, SDS in the anodic buffer fulfills no purpose unless the buffer is circulated between the two compartments.

Other buffer systems suggested for SDS electrophoresis include a bis Tris–bicine/Tris–sulfate system claimed to separate polypeptides in the 100,000 to 1,000 range[75] and a Laemmli buffer variant with higher pH that extends the separation range to smaller peptides.[76]

The quality of the SDS used, particularly the presence of alkyl analogs, has been demonstrated to significantly influence the separation.[77,78] The SDS quality also affects the renaturation of proteins.[79] Different qualities of SDS can undoubtedly explain the many discrepancies found in SDS electrophoresis.[80] For reproducible results, only high-quality, well-defined SDS should be used, and other detergents or alkyl-sulfate analogs should only be used in an controlled way in the rare cases where this is called for.

12.5.2 Practical Procedures in SDS–PAGE

To assure complete denaturation, the sample should be heated to 95 to 100°C for at least 3 min in the presence of excess SDS (1 to 2%).[64] If desired, a reducing agent may be included to cleave S-S bridges.[69] For releasing individual polypeptide chains, 5% mercaptoethanol or 1.5% DTT is commonly used. To protect the sample from any tendencies to reoxidate the S-S bridges, the sample can be alkylated by treatment with iodoacetamide after reduction. This is accomplished by adding iodoacetamide to a concentration of 2% immediately after having heated the sample in SDS and the reducing agent. Alkylation very often results in sharper bands and significantly improved results.

The sample buffer is preferably the same buffer as that in the stacking gel diluted to half the concentration. Although a low ionic strength in the sample in principle always is preferable, it was found that the negative effects of the ionic strength of the sample was not always as strong as expected. In one investigation, it was found that up to 0.8 M NaCl was well tolerated.[81] Although the best results are generally obtained with reduced and alkylated samples, good molecular weight estimates can be obtained even without the reduction of intramolecular S-S bonds. Unreduced protein migrated only ca. 14% slower.[82] Normally, in vertical electrophoresis, sucrose or glycerol is added to the sample to increase its density for easier sample application. A small amount of bromophenol blue (from a 1% stock solution) further aids in sample application on horizontal as well as on vertical gels and also serves as a tracking dye.

The amount of sample that is required depends primarily on the sensitivity of the detection method used. With the comparatively low sensitivity of Coomassie, at least ca. 500 ng protein per band or 5 to 10 μg per sample well is adequate, whereas with highly sensitive immunological, enzymatic, or silver staining methods 10- to 25-ng quantities (silver stain) or less are sufficient. Optimal conditions have to be worked out for each application. For complete control of sample concentration and maximal reproducibility, the total protein concentration in SDS containing electrophoresis buffers can be determined by densitometry of sample spots on nitrocellulose stained with silver[83] or amido black.[84]

12.5.3 M_r Determinations

If molecular weight calibration proteins are run on the same gel as the sample, the protein pattern obtained after SDS electrophoresis will give a good survey of the molecular weight distribution of proteins. Molecular weights of individual proteins can be calculated with good accuracy from log M_r versus R_f plots. Kits with selected calibration proteins can be obtained from several manufacturers. If the sample is run in both reduced and unreduced forms, the number of covalently linked subunits can be derived.

Molecular weights down to about 10,000 to 14,000 can be estimated fairly accurately using standard procedures. Smaller proteins and peptides will not migrate strictly according to molecular size or migrate in the buffer front and not separate at all. Also, these small peptide–SDS complexes are of similar size as the SDS micelles,[65] and all peptides below a certain size (ca. 10,000) will concentrate in the SDS–micelle front. By adding 8 M urea to the system (0.1% SDS, 0.1 M trisphosphate, pH 6.8, T = 12.5%), Swank and Munkres[85] succeeded in separating peptides down to ca. 1,000, essentially according to molecular weight. By running the buffer system of Wyckoff and co-workers[72] in a PAA gradient (10 to 30.2%), proteins and peptides in the range of 1,500 to 100,000 in molecular weight can be separated in the same gel.[86] Increasing the pH of the classical Laemmli buffer also extends the separation range to smaller polypeptides.

Some proteins do not behave normally in SDS–PAGE. Glycoproteins do not bind SDS to the same extent as "normal" proteins and M_r estimations must be made with much care. Molecular weights may easily be overestimated by as much as 30%.[87,88] With extremely positively charged proteins such as histones, it seems that the charges of SDS to a significant degree are balanced by the proteins' own charges, which makes M_r determinations difficult.[89] Very acidic proteins such as pepsin, papain, and glucose oxidase were found to bind very small amounts of SDS[90] (Figure 12-6).

12.6 APPLICATIONS

12.6.1 Characterization of Wheat Grain Allergens Involved in Bakers' Asthma by SDS Gradient PAGE

Proteins from seven different cultivars of whole-meal flour were extracted into three fractions of sequentially decreasing solubility (albumin/globulin, gliadin, and glutenin).[91] The fractions were separated by SDS gradient PAGE and blotted onto a polyvinylidene difluoride (PVDF) blotting membrane. By exposing the membrane to pooled sera from allergic bakers it was found that IgE antibodies bound to numerous polypeptides of all three fractions. The highest IgE binding occurred to certain 27,000 M_r polypeptides in the albumin/globulin fraction with less binding to gliadin and glutenin peptides (Figure 12-7). The allergens were further characterized by two-dimensional electrophoresis (not shown).

Experimental Procedure

Briefly, the albumin/globulin contains proteins soluble in ordinary Tris-HCl buffer, with gliadin proteins soluble in 75% ethanol. The remaining part of the proteins that are soluble in SDS/DTT constitute the glutenin.

Gels for horizontal SDS–PAGE were cast on GelBond plastic backing foils (Pharmacia Biotech). The gels contained a 4% stacking gel and a linear polyacrylamide gradient in the separating gel from 10 to 15% using the Laemmli buffer system (C was kept constant at 4% in the whole system). Sample volumes were 3 μL (silver staining) and 6 μL (blotting).

The gels were run on Multiphor II electrophoresis apparatus (Pharmacia Biotech). For silver staining the gels were soaked in methanol/acetic acid/water (40/10/50) for 30 min for fixation of the proteins.

Before blotting the gels were removed from the plastic backing with Gel-Remover. ^{125}I-labeled antihuman IgE (Phadebas RAST) was from Pharmacia AB.

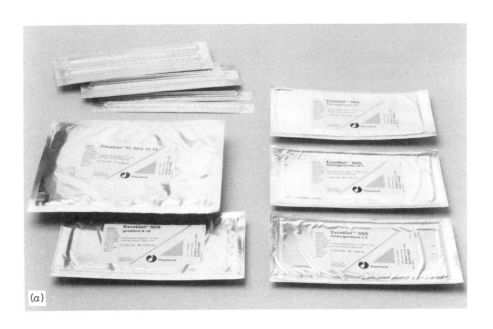

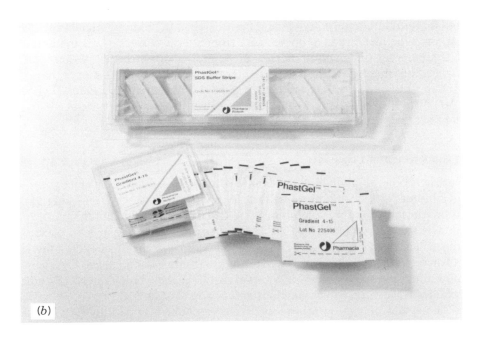

FIGURE 12-6. The wide range of precast gel media for conventional vertical equipment as well as for horizontal in both standard format (a) and miniature format PhastGel media (b) is now available for most standard electrophoretic techniques: native and SDS–PAGE, both gradient and homogeneous; IEF; and media for two-dimensional electrophoresis.

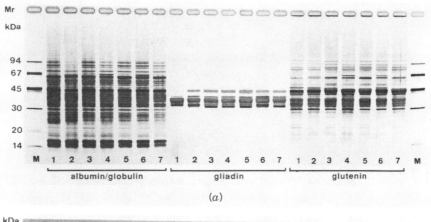

(a)

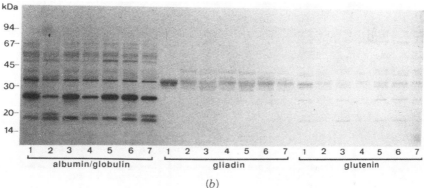

(b)

FIGURE 12-7. (a) Horizontal SDS–PAGE (10 to 15%) of albumin/gliadin, gliadin, and glutenin proteins from seven wheat cultivars. Silver stain. Wheat cultivars: (1) Apollo, (2) Ares, (3) Avis, (4) Boheme, (5) Bussard, (6) Fregatt, and (7) Futur; M_r marker proteins. (b) Same separation, but analyzed for IgE-binding properties (i.e., allergens) after being blotted to a PVDF membrane and exposed to pooled serum from four asthmatic bakers. The bound IgE was detected by exposing to ^{125}I-labeled antihuman IgE and autoradiography (a 4-week exposure time). (Reproduced from Ref. 91 with permission of VCH Verlagsgesellschaft mbH.)

12.6.2 Monitoring Chemical Modifications of Human Interleukin-3 by Native Electrophoresis on the PhastSystem

Human interleukin-3 (hIL-3) is a protein with a M_r of 13,000 that plays a central role in the generation of blood cells from bone marrow cells. Its action mechanisms on an molecular level can be explored by studying the effects of chemical modifications of recombinant hIL-3. In this context, there was a need for a rapid, convenient, and sensitive method to analyze the different chemical derivatives of hIL-3.[92]

Standard PhastGel homogeneous 20% gel media and PhastGel native buffer strips gave reasonably but not quite acceptable results for this application

(Figure 12-8). hIL-3 has eight amino groups susceptible to derivatization. All derivatives obtained by succinid anhydride and acetic acid anhydride modifications could not be analyzed in the same experiments on the standard system. However, by devising a special buffer composition for the buffer strips, this was made possible. (An instructive example on how standard techniques and precast gel media can be optimized to meet even very specific requirements when needed!)

Experimental Procedure

The electrophoretical separation was performed as described in PhastSystem Technique File 121. The special buffer strips used in Figure 12-9 (BTH system) were prepared by dissolving 3% agarose in 0.25 M BTH (bis–Tris/HEPES), pH 7.5, by boiling. After cooling to 70°C, the solution was poured into an empty PhastGel buffer strip package and left to gel before being wrapped in Saran wrap and left overnight at 4°C.

Staining with Coomassie was done as described in Table 2 in PhastSystem Technique File 200.

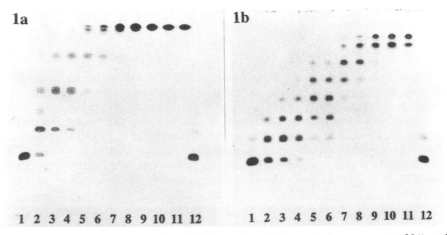

FIGURE 12-8. Electrophoresis of hIL-3 on PhastGel media homogeneous 20% and PhastGel native buffer strips at manufacturers' conditions. hIL-3 was modified with (a) 3 mM succinic anhydride and (b) 3 mM acetic anhydride. Lane 1, hIL-3 negative control, at pH 9.5, no modifying agent added; lanes 2–11, hIL-3 modified at pH 5.0 to pH 9.5 in steps of 0.5 pH units, respectively; lane 12, hIL-3 negative control with 3 mM modifying agent and 90 mM lysine at pH 9.5. Succinic anhydride modification is a charge reversal reaction that converts a positive charge to a negative while the acetic anhydride reactions just neutralize the positive charge, hence the double distance between subsequent bands in (a) compared to (b). (Reproduced from Ref. 92 with permission from VCH Verlagsgesesslschaft mbH.)

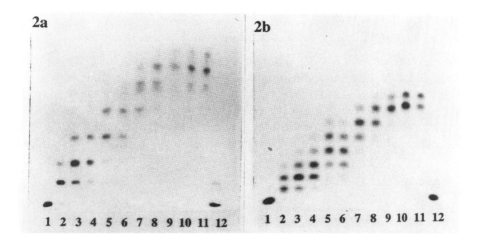

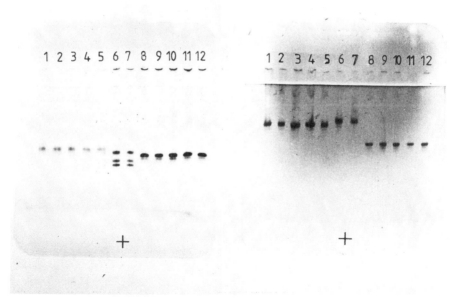

FIGURE 12-10. Isoenzyme analysis of proteins from different species of *Meloidogyne* separated by native PAGE on PhastGel 8–25. Lanes 1–5 are *M. chitwoodi;* lanes 8–12 are *M. hapa;* and lanes 6 and 7 are *M. javanica.* (a) Esterase staining and (b) malate dehydrogenase staining. (Reproduced from Ref. 93 with permission from VCH Verlagsgesellschaft mbH.)

12.6.3   Identification of Root-Knot Nematode Species by Electrophoresis on Polyacrylamide Gradient Gel in PhastSystem and Isozyme Staining

The parasitic nematodes *Meloidogyne hapa, M. chitwoodi,* and *M. javanica* are of economic importance in temperate zones and tropical highlands. Species identification by classical light microscopic characteristics is difficult because of morphological similarities. By taking previously described electrophoresis methods and adapting them for PhastSystem, it was possible to design a fast and efficient method for routine diagnosis (Figure 12-10).[93]

Experimental Procedure

Sample preparation: Briefly, young, egg-laying nematode females were isolated under the microscope and transferred to special sample wells after a desalting step. One female was macerated in 0.6 µL extracting buffer (20% sucrose, 2% Triton X-100, and 0.01% bromophenol blue) and 0.3 µL was transferred directly to the special PhastSystem sample applicator comb (Type 12/0.3) for electrophoresis.

Electrophoresis

The following modified running program was used:

Sample applicator down at:					3.2 Vh
Sample applicator up at:					3.3 Vh
Sep 3.1	400 V	10 mA	2.5 W	10°C	10 Vh
Sep 3.2	400 V	1 mA	2.5 W	10°C	2 Vh
Sep 3.3	400 V	10 mA	2.5 W	10°C	125 Vh

Staining:

After electrophoresis, the gels were incubated in petri dishes at 37°C.

Esterase activity (incubation time 60 min):	
0.1 M phosphate buffer, pH 7.3	100 mL
Fast Blue RR salt	0.06 g
EDTA	0.03 g
α-Naphtyl acetate (dissolved in 2 ml acetone)	0.04 g
Malate dehydrogenase (incubation time 5 min):	
β-NAD	0.05 g
Nitro blue tetrazolium	0.03 g
Phenazine methosulfate	0.02 g
0.5 M Tris, pH 7.1	5.0 mL
10.6 g Na_2CO_3 + 1.34 g L-malic acid in 100 ml water	7.5 mL
Water	70 mL

12.7 REFERENCES

1. J.Th.G. Overbeek, B.H. Bijsterbosch, in *Electrokinetic Separation Methods*, P.G. Righetti, C.J. van Oss, J.W. Vanderhoff, eds., Elsevier/North Holland, 1979, pp. 1–32.

2. D.P. Goldenberg, T.E. Creighton, *Anal. Biochem.*, *138*, 1–18 (1984).

3. T.E. Creighton, D. Shortle, *J. Mol. Biol.* *242*, 670–682 (1994).

4. J. Brewer, E. Grund, P. Hagerlid, I. Olsson, J. Lizana, in *Electrophoresis '86*, M.J. Dunn, ed., VCH Verlagsgesellschaft Meinheim BRD, 1986, pp. 226–229.

5. I. Olsson, U.-B. Axiö-Fredriksson, M. Degerman, B. Olsson, *Electrophoresis*, *9*, *1*, 16–22 (1988).

6. E. Gianazza, P.G. Righetti, *J. Chromatogr.*, 193, 1–8 (1980).

7. S. Raymond, L.S. Weintraub, *Science*, *130*, 711–713 (1959).

8. L. Ornstein, B.J. Davies, *Disc Electrophoresis:* Distillation Products Div., Eastman Kodak Co., Rochester, New York, 1962.

9. C. Gelfi, P.G. Righetti, *Electrophoresis*, *2*, 213–219 (1981).

10. C. Gelfi, P.G. Righetti, *Electrophoresis*, *2*, 220–238 (1981).

11. P.G. Righetti, C. Gelfi, A.B. Bosisio, *Electrophoresis*, *2*, 291–295 (1981).

12. T. Lyubimova, S. Caglio, C. Gelfi, P.G. Righetti, T. Rabilloud, *Electrophoresis*, *14*, 40–50 (1993).

13. S. Caglio, M. Chiari, P.G. Righetti, *Electrophoresis*, *15*, 209–214 (1994).

14. *Acrylamide Gel Casting Handbook*, Pharmacia Biotech, Uppsala, Sweden, 1994.

15. S. Hjertén, *Arch. Biochem. Biophys. Suppl. 1*, 147–151 (1962).

16. P. Serwer, *Electrophoresis*, *4*, 375–382 (1983).

17. T.W. Nee, *J. Chromatogr.*, *105*, 231–249 (1975).

18. P. Serwer, S.J. Hayes, *Electrophoresis 3*, 80–85 (1982).

19. R.J. Wieme, *Agar Gel Electrophoresis*, Elsevier, Amsterdam, 1965, pp. 110–113.

20. O. Vesterberg, L. Hansen, A. Sjösten, *Biochim. Biophys. Acta*, 160–166 (1977).

21. V. Neuhoff, R. Stamm, I. Pardowitz, N. Arold, W. Ehrhardt, D. Taube, *Electrophoresis*, *11*, 101–117 (1990).

22. P. Mitra, A.Kr. Pal, D. Basu, R. Hati, *Anal.Biochem.*, *223*, 327–329 (1994).

23. V. Neuhoff, R. Stamm, H. Eibl, *Electrophoresis*, *6*, 427–448 (1985).

24. R.C. Switzer, C.R. Merril, S. Shifrin, *Anal. Biochem.*, *98*, 231–237 (1979).

25. T. Rabilloud, *Electrophoresis*, *11*, 785–794 (1990).

26. T. Rabilloud, L. Vuillard, C. Gilly, J.-L. Lawrence, *Cell. Mol. Biol. Noisy-le-Grand* *40*, 57–75 (1994).

27. I. Olsson, R. Wheeler, C. Johansson, B. Ekström, N. Stafström, R. Bikhabhai, G. Jacobson, *Electrophoresis*, *9*, *1*, 22–27 (1988).

28. J. Heukeshoven, R. Dernick, *Electrophoresis*, *9*, *1*, 28–32 (1988).

29. H. Kierdorf, H. Melzer, H. Mann, H.-G. Sieberth, *Electrophoresis*, *14*, 820–822 (1993).

30. H.J. Møller, D. Heinegård, J.H. Poulsen, *Anal.Biochem.*, *209*, 169–175 (1993).

31. A. Fomsgaard, M.A. Freudenberg, C. Galanos, *J.Clin.Microbiol.* 2627–2631 (1990).

32. Ö. Gaal, G.A. Medgyesi, L. Vereczkey, in *Electrophoresis in the Separation of Biological Macromolecules,* John Wiley & Sons, 1980, pp.327–335.

33. A.H. Wardi, G.A. Michos, *Anal. Biochem., 49,* 607–609 (1972).

34. G. Dubray, G. Bezard, *Anal.Biochem., 119,* 325–329 (1982).

35. *Detection of Enzymes on Electrophoresis Gels: A Handbook,* CRC Press Inc., 1994.

36. *Electrophoresis of Enzymes: Laboratory Methods,* G.M. Rothe, ed., Springer Verlag, New York, 1994.

37. *Practical Protein Electrophoresis for Genetic Research,* Timber Press Inc., 1992.

38. R. Westermeier *Electrophoresis in Practice: A Guide to Theory and Practice,* VCH Verlagsgesellschaft, Weinheim, 1997.

39. B. Kleine, G. Löffler, H. Kaufmann, P. Scheipers, H. Schickle, R. Westermeier, W.G. Bessler, *Electrophoresis, 13,* 73–75 (1992).

40. P.G. Sealy, E.M. Southern, in *Gel Electrophoresis of Nucleic Acids: A Practical Approach,* D. Rickwood, B.D. Hames, eds., 1982, pp. 39–76.

41. S. Hjertén, S. Jerstedt, A. Tiselius, *Anal.Biochem., 11,* 219–223 (1965).

42. L. Ornstein, *Ann. N.Y. Acad. Sci., 121,* 321–349 (1964).

43. B.J. Davies, *Ann. N.Y. Acad. Sci., 121,* 404–427 (1964).

44. F. Kohlrausch, *Ann. Phys. Chem., 62,* 209–239 (1897).

45. A. Chrambach, T.M. Jovin, *Electrophoresis, 4,* 190–204 (1983).

46. G.G. Slater, *Fed. Proc., 24,* 225 (1965).

47. J. Margolis, K.G. Kenrick, *Biochem. Biophys. Res. Commun., 27,* 1, 68–73 (1967).

48. G.G. Slater, *Anal.Biochem., 24,* 215–217 (1968).

49. P. Lambin, J.M. Fine, *Anal. Biochem., 98,* 160–168 (1979).

50. M. Lasky, in *Electrophoresis '78,* Catsimpoolas, ed., Elsevier North Holland, 1978, pp. 195–209.

51. G.M. Rothe, H. Purkhanbaba, *Electrophoresis, 3,* 33–42 (1982).

52. D. Riebe, W. Thorn, *Electrophoresis, 12,* 287–293 (1991).

53. K.A. Ferguson, *Metabolism 13,* No 10, Part 2, 985–1002 (1964).

54. A. Chrambach, *Molecular and Cellular Biochemistry, 29, 1,* 23–46 (1980).

55. A. Chrambach, *J. Chromatogr., 320,* 1–14 (1985).

56. D. Rodbard, A. Chrambach, G.H. Weiss, in *Electrophoresis and Isoelectric Focusing in Polyacrylamide Gel,* R.C. Allen, H.R. Maurer, eds., W. deGruyter, Berlin/New York, 1974, pp. 62–104.

57. J.L. Hedrick, A.J. Smith, *Arch. Biochem. Biophys., 126,* 155–164 (1968).

58. M.C. Diemert, L. Musset, O. Gaillard, S. Escolano, A. Baumelou, F. Rousselet, J. Galli, *J. Clin. Pathol., 47,* 1090–1097 (1994).

59. S.R. Gallagher. R.T. Leonard, *Anal. Biochem., 162,* 350–357 (1987).

60. L.M. Hjelmeland, A. Chrambach, *Meth. Enzymol.,* B. Jacoby, ed., pp. 305–318 (1984).

61. S. Hjertén, *Biochim. Biophys. Acta, 736,* 130–136 (1983).

62. G.M. Rothe, W.D. Maurer, in *Gel Electrophoresis of Proteins,* M.J. Dunn, ed., Wright Bristol, 1986, pp. 108–112.

63. J.V. Maizel, Jr., *Science, 151,* 988–990 (1966).

64. A.L. Shapiro, E. Vinuela, J.V. Maizel, Jr., *Biochem. Biophys. Res. Commun., 28,* 5, 815–820 (1967).

65. J.A. Reynolds, C. Tanford, *J. Biol. Chem., 245, 19,* 5161–5165 (1970).

66. T. Takagi, K. Tsuji, K. Shirahama, *J. Biochem., 77,* 939–947 (1975).

67. G.M. Rothe, W.D. Maurer, in *Gel Electrophoresis of Proteins,* M.J. Dunn, ed., Wright Bristol, 1986, pp. 37–140.

68. K. Weber, M. Osborn, *J. Biol. Chem., 244,* 4406–4412 (1969).

69. D.M. Neville, Jr. *J. Biol. Chem., 246, 20.* 6328–6334 (1971).

70. A.K. Dunker, R.R. Rueckert, *J. Biol. Chem., 244, 18,* 5074–5080 (1969).

71. U.K. Laemmli, *Nature, 227,* 680–685 (1970).

72. M. Wyckoff, D. Rodbard, A. Chrambach, *Anal. Biochem., 78,* 459–482 (1977).

73. A.F. Bury, *J. Chromatogr., 213,* 491–500 (1981).

74. J.T. Stoklosa, H.W. Latz, *Biochem. Biophys. Res. Commun., 58,* 74–79 (1974).

75. J. Wiltfang, N. Arold, V. Neuhoff, *Electrophoresis, 12,* 352–366 (1991).

76. G.S. Makowski, M.L. Ramsby, *Anal. Biochem., 212,* 283–285 (1993).

77. M.M. Margulies, H.L. Tiffany, *Anal. Biochem., 136,* 309–313 (1984).

78. H.D. Matheka, P.-J. Enzmann, H.L. Bachrach, B. Migi, *Anal. Biochem., 81,* 9–17 (1977).

79. D. Best, P.J. Warr, K. Gull, *Anal. Biochem., 114,* 281–284 (1981).

80. H. Anwar, P.A. Lambert, M.R.W. Brown, *Biochim. Biophys. Acta, 761,* 119–125 (1983).

81. Y.P. See, P.M. Olley, G. Jackowski, *Electrophoresis, 6,* 382–387 (1985).

82. A. Chrambach, *Anal. Biochem., 20,* 150–154 (1967).

83. P. Dráber, *Electrophoresis, 12,* 453–456 (1991).

84. A.W. Henkel, S.C. Bieger, *Anal. Biochem., 223,* 329–331 (1994).

85. R.T. Swank, K.D. Munkres, *Anal. Biochem., 39,* 462–477 (1971).

86. D. Bothe, M. Simonis, H. von Döhren, *Anal. Biochem. 151,* 49–54 (1985).

87. J.P. Segrest, R.L. Jackson, E.P. Andrews, V.T. Marchesi, *Biochem. Biophys. Res. Commun. 44,* 390–395 (1971).

88. H. Glossman, D.M. Neville, Jr., *J. Biol. Chem., 246, 20,* 6339–6346 (1971).

89. S. Panyim, R. Chalkley, *Arch. Biochem. Biophys. 130,* 854–857 (1969).

90. C.A. Nelson, *J. Biol. Chem., 246,* 3895–3901 (1971).

91. W. Weiss, C. Vogelmeier, A. Görg, *Electrophoresis, 14,* 805–816 (1993).

92. V. Smit, *Electrophoresis, 15,* 251–254 (1994).

93. G. Karssen, T.V. Hoenselaar, B. Verkerk-Bakker, R. Janssen, *Electrophoresis, 16,* 105–109, (1995).

13 Isoelectric Focusing

TORGNY LÅÅS

Pharmacia Biotech AB
S-751 82 Uppsala, Sweden

Protein Purification: Principles, High-Resolution Methods, and Applications, Second Edition.
Edited by Jan-Christer Janson and Lars Rydén.
ISBN 0-471-18626-0. © 1998 Wiley-VCH, Inc.

13.1   INTRODUCTION

Isoelectric focusing (IEF) can be described as electrophoresis in a pH gradient
set up between a cathode and an anode with the cathode at higher pH than
the anode. Proteins, being amphoteric species, are positively charged at pH
values below their pI and negatively charged above their pI. This means that
whenever a protein is in the gradient, it will migrate toward its pI (Figure 13-1).

Ikeda and Suzuki[1] were the first to apply the principles of isoelectric focusing
in their isolation of glutamic acid from plant hydrolysates, whereas Williams
and Waterman[2] were the first to describe the basic principles behind today's
IEF clearly. The first separations resembling IEF as it is known today were
performed by Kolin[3] in the 1950s in sucrose gradients composed by careful
layering of buffers with different pH values on top of each other.

The major obstacle at that time was the stabilization of the pH gradient.
It was not until Svensson (later named Rilbe) and Vesterberg managed to
synthesize "carrier ampholytes" with properties to create and stabilize the
pH gradient that IEF could be developed into the common tool that is known
today.[4] The most useful property of good carrier ampholytes is buffer capacity
at the isoelectric point. Svensson and Vesterberg called their pH gradients
"natural" as they developed automatically during electrophoresis, in contrast
to "artificial" pH gradients created in advance.

The use of artificial pH gradients got a revival with the introduction of
Immobiline™ in 1982.[5] With Immobiline, the pH gradient is literally immobilized as part of the gel matrix.

Many of the problems connected with conventional IEF are circumvented
with Immobiline gradients, which in turn cause some new problems, as is
discussed later. The two systems therefore complement each other.

13.2   THEORY OF ISOELECTRIC FOCUSING

Under the influence of the electrical force, the pH gradient is established by
the carrier ampholytes, and the protein species are focused at their isoelectric
points, as described in Figure 13-1. The focusing effect of the electrical force
is counteracted by diffusion, which is directly proportional to the protein

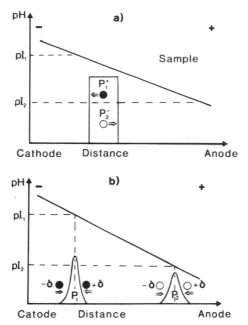

FIGURE 13-1. Principle of isoelectric focusing. (a) Schematic illustration of a sample with two proteins, P_1 and P_2, placed in the center of a pH gradient. P_1, with the more alkaline pI, is positively charged at the application spot whereas the acidic P_2 is negatively charged. Under the influence of the electric field, the two proteins start to migrate as indicated by the arrows. (b) As the proteins approach their pI, they gradually become less and less charged and migrate slower and slower. The proteins will thus concentrate at the position where pH = pI. However, the proteins cannot concentrate in an indefinitely concentrated zone. Zone widening by diffusion is inevitable. Any protein molecule diffusing away from pI will acquire a net charge and be transferred back to pI again by electrophoresis. A balance will be reached between electrophoretical accumulation at pI and diffusion.

concentration gradient in the zone. Eventually, a steady state is established where the electrokinetic transport of protein into the zone is exactly balanced by the diffusion out of the zone (Figure 13-1). From the factors that regulate the widths of the protein zones and the distance between the zones, Svensson and Vesterberg derived an equation for the resolution of two similar proteins[6] based on the following assumptions:

- Straight and continuous pH gradients, dpH/dx.
- Even field strength, E.
- The two different proteins have the same diffusion coefficient, D.
- The electrophoretic mobility change with pH, $d\mu/dpH$, is constant and the same for both proteins.

- Two closely spaced proteins are considered separated when the position of their peak maxima differs by a minimum of 3 standard deviations.

The minimum difference in pI, (ΔpI), for two proteins to be resolved is then

$$\Delta pI = -3\sqrt{\frac{D(dpH/dx)}{E(d\mu/dpH)}}$$

The equation shows that reducing the *diffusion, D,* would increase resolution. With a given separation, the only way to accomplish this is to increase the viscosity of the medium. Inert noncharged substances such as sucrose and glycerol may be added or the experiment can be performed in a sieving media such as a high-concentration polyacrylamide (PAA) gel. Increased viscosity will, however, also affect the mobility of the proteins. Not only will this make the whole experiment slower, but it also counteracts the resolution by decreasing the factor dμ/dpH in the equation.

Therefore, increasing the viscosity is not generally a successful way to improve the resolution, although it may explain why there is a clear tendency for better resolution in sieving PAA gels than in the more porous agarose gels.

Because diffusion is inversely related to molecular size, it follows that larger proteins will tend to focus better than smaller ones, with other things being equal. Isoelectric focusing (IEF) of peptides is particularly difficult, although with suitable precautions, good results can often be obtained.[7]

The shallower the *gradient, dpH/dx* (low value of dpH/dx), the further apart will two proteins be and hence better separated. Note, however, that the factor only applies as the square root (as in fact all other factors). There are some drawbacks with the use of extremely shallow gradients: Long focusing times as proteins must migrate a relatively long distance close to pI with a very low charge; only a limited number of proteins with pI values within the narrow pH interval can be analyzed simultaneously; and the carrier ampholytes may not manage to maintain a completely smooth pH gradient.

With Immobiline IEF, this situation is different. Narrow pH range gradients are in fact easier to cast than broad range pH gradients. The possibility of obtaining high-resolution separation in very shallow gradients is relatively easy and is in fact the greatest advantage with Immobiline IEF.

Higher *field strength, E,* not only increases the resolution, but also reduces the experimental time. Too high a field strength may give heat problems if the cooling is inefficient, especially when focusing in the very basic or acidic pH regions or if carrier ampholytes with unevenly distributed conductance are used. Recommendations by manufacturers are generally safe.

The approach should be to always work with the highest "safe" field strength.[8] The higher the *pH dependence of the mobility (dμ/dpH),* the better the focusing. A high electrophoretic mobility close to the pI will efficiently transfer diffused protein molecules back to pI. This is essentially an intrinsic factor of the protein that cannot be manipulated. The effect of modifying mobility by affecting the viscosity is counteracted by the effect of viscosity on diffusion, as already discussed, and the overall effect is difficult to predict. A high value of $d\mu/dpH$ results from the presence of many groups with pK values close to the pI. Statistically, this is more likely to be the case for a larger protein than for a smaller protein. Both $d\mu/dpH$ and diffusion thus favor the focusing of large proteins, and the influence of these factors explains the difficulties in focusing small proteins and peptides to sharp zones.

Good indications of the prospect for IEF are obtained by titration curve analysis (see Section 13.4.8). Proteins with shallow profiles close to their pI are less likely to focus well than ones that cross the pI at a steep angle.

13.3 NATURAL pH GRADIENTS

13.3.1 Carrier Ampholytes and pH Gradient Formation

The formation of a pH gradient is schematically illustrated in Figure 13-2.

Hydrogen ions form at the anode and hydroxyl ions at the cathode in electrode reactions. This results in regions of low and high pH near the anode and cathode, respectively, and steep gradients as one moves into the bulk solution. An amphoteric species with a pI lower than the average pH in the system will concentrate in the steep gradient close to the anode. A substance with good buffering capacity at its pI creates a pH plateau around its pI. Given a sufficient number of such substances with evenly distributed pI values, their corresponding plateaus overlap, resulting in a continuous pH gradient. The amphoteric substances that form and stabilize the pH gradient are collectively called "carrier ampholytes".

The most essential property for a good carrier ampholyte molecule is a good buffering capacity at its isoelectric point.[9,10] This requires many pK values close to the isoelectric point for each molecular species, making most naturally occurring ampholytic substances, especially most naturally occurring amino acids, useless as carrier ampholytes. After numerous trials with limited success using mixtures of different amphoteric compounds, Svensson came to the conclusion that the only way to obtain useful carrier ampholytes was to synthesize substances with the required properties, a task performed by Vesterberg. The established pH gradient is maintained by hundreds or thousands of carrier ampholyte molecules lined up in the order of pI with partially overlapping distributions. Because there are no other ionic species in the system, each carrier ampholyte must act as a counterion to other carrier ampholytes. Consequently, each position in the pH gradient has a unique chemical composition. Electrical conductance and buffer capacity therefore vary over the pH gradient. Regions with low buffer capacity are more prone to distortions. In preparative experiments with high protein loads, buffering capacity from the proteins may affect the pH gradient.

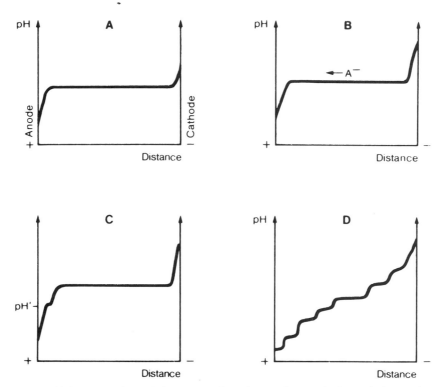

FIGURE 13-2. Formation of the pH gradient by carrier ampholytes. (A) Imagine a system with electrolysis of water. Electrode reactions will produce H^+ at the anode and OH^- at the cathode, causing the pH to change close to the electrodes. (B) Now introduce the amphoteric substance A in the system. If A has an acidic pH it is negatively charged in the center of the system. Under the influence of the electric field, A will migrate toward the anode. (C) A will accumulate at its pI in the (steep) pH gradient close to the anode as described for proteins in Figure 13-1. Suppose further that A has a good buffering capacity at its pI. A will then tend to create a pH plateau at that position. (D) Given a number of species similar to A but with different pI values, a stepwise pH gradient will be formed. If the number of different "A molecules" is large enough (thousands as in commercial carrier ampholyte mixtures), the different steps are smoothed out to a continuous even pH gradient, with the different carrier ampholytes extending through the whole system, each focused at its pI.

Local heating occurs in the regions with the highest field strength (lowest conductance), and these regions determine how high an overall voltage can be used.[11] Consequently, other regions with a lower field strength are not focused at optimal conditions. Optimal conditions over the whole pH gradient thus require even field strength (conductance) and buffering capacity across the entire gradient (Figure 13-3).

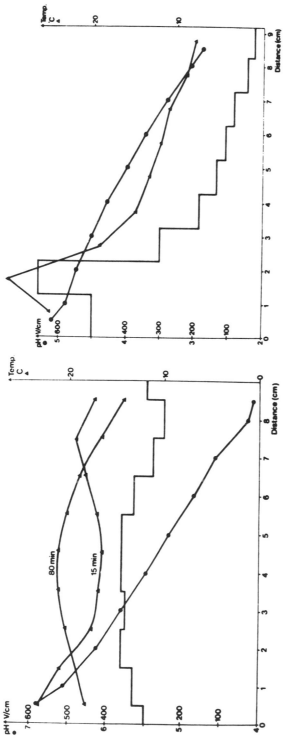

FIGURE 13-3. An example of distribution of buffer capacity and field strength over the pH gradient. The ideal carrier ampholyte mixture has a completely even distribution of buffer capacity and field strength over the pH gradient to assure even focusing and even heat evolution. Pharmalyte 4–6.5 comes close to this ideal, whereas Pharmalyte 2.5–5 has much more uneven properties. At the low pH of this pH interval, it is inevitable that the high conductance contribution from the protons at the acid end of the pH gradient reduces the field strength in this region. The consequence of this is poor focusing in the acidic region because of the low field strength and high temperature in the other end of the gradient with risk of overheating. (Reproduced from Ref. 11 with permission from Pergamon Press.)

13.3.2 pH Gradient Engineering

A large number of carrier ampholyte mixtures covering a wide spectra of broad and narrow pH range intervals are commercially available, as well as precast ready-to-use gels covering different pH ranges. The gradient of choice depends on the purpose of the analysis. For screening purposes, a broad range interval (pH 3 to 10 or similar) should be used. A narrow pH range interval is useful for careful pI determinations or when analyzing proteins with very similar pI points. Generally, one should not use a narrower gradient than necessary because the shallower gradient will lead to longer focusing times and more diffuse bands. When chosing a pH gradient, one should be aware that the interval stated by the manufacturer can never be more than approximate. The exact gradient obtained depends on many factors, such as choice of electrolyte solutions, gradient media (PAA or agarose), and focusing time.

Despite the large number of pH intervals and precast IEF gels available, there may be occasions where none fit perfectly. In such cases, one can work either with Immobiline or use "pH gradient engineering" in any of the following variants:

- Extend a given pH interval by adding carrier ampholytes covering the adjacent or a partly overlapping region. Extension into the extreme pH ends can be accomplished by adding acidic[12] or basic compounds.[13]
- Extend a certain pH area by adding an amphoteric substance "spacer," such as an amino acid. The spacer should be a "bad" ampholyte so that it does not focus too well but instead exerts its effect over a wider distance.[14–17]
- Extend a certain pH range by manipulating the thickness of the gel. The gradient will be shallower in areas with thinner gel.[18–20]
- Manipulating the carrier ampholyte concentration also affects the steepness of the final gradient. Areas with lower concentration will give shallower gradients.[18]

The different methods can be combined, as was demonstrated by Gill,[20] and can also be applied to modify the pH gradients of PhastGel IEF media.[21]

Generally, IEF will give a true representation of the isoelectric spectrum of the sample. However, IEF of immunoglobulins in standard carrier ampholyte mixtures results in distinct bands in the otherwise continuous smear of immunoglobulin molecules. This was shown to depend on heterogeneous carrier ampholyte distribution.[22] For a more truthful representation of the distribution of pI points in an immunoglobulin sample, the best results were obtained in mixtures of different carrier ampholyte preparations. A mixture of three different Pharmalyte intervals to maximize the number of carrier ampholytes in the interesting region was found to give the best results.[22]

13.4 EXPERIMENTAL TECHNIQUES

13.4.1 Equipment for Preparative and Analytical IEF

Traditionally, *preparative IEF* was performed in a vertical, cooled glass column stabilized toward heat convection by a density gradient of sucrose, glycerol, or similar substance.[23] The many drawbacks of this technique can be circumvented by performing the separation in a horizontal bed of Sephadex gel.[24,25] Isoelectrically precipitated protein will not move in the Sephadex bed and disturb other regions. More efficient cooling allows higher voltages to be used. Although easy in principle, the technique requires some practice to get the right slurry density when preparing the gel bed. In addition, a new separation problem is created, the elution of the protein out of the Sephadex gel. Furthermore, carrier ampholyte consumption is high.

Since both techniques have severe drawbacks, the first choice for preparative protein separation according to the pI should be chromatography by chromatofocusing (see Chapter 5). Some proteins tend, however, to form complexes or aggregates at the low ionic strength used in chromatofocusing; in such cases, preparative IEF may still be the best alternative.

Analytical IEF is exclusively performed in gels of either PAA or agarose. Polyacrylamide gel isoelectric focusing (PAGIF) can be performed vertically in gel rods or horizontally in thin gels. PAGIF in gel rods is rarely used except as for the first-dimension separation in two-dimensional electrophoresis, but here again the use of gel rods is declining in favor of IEF in Immobiline DryStrip (see Chapter 15). IEF in gel rods is therefore not discussed further.

Horizontal slab gel IEF is the technique of choice when high-resolution, high-reproducibility, and high-quality results are required. It is ideal for screening and comparing many samples run side by side. In comparison with agarose, PAA gels tend to give sharper bands, have less problems with electroosmosis and gradient drift (see later), and generally have more reproducible results. Ready-to-use, precast PAGIF slabs are commercially available for both conventional and Immobiline horizontal IEF.

As already mentioned, IEF in agarose requires special quality agarose, and even then, focusing does not always proceed as trouble-free as with PAA, especially at extreme pH values. Even high-quality agaroses contain some sulfate and carboxylic groups.[26] At the extremely low ionic strength conditions used in IEF, these groups generate electroosmosis with concomitant problems. Only the purest forms of agarose are suitable for IEF.[27,28] By "charge balancing" the few residual negatively charged groups with the careful addition of positively charged groups, the properties of the agarose can be further improved, as in agarose IEF from Pharmacia Biotech. In comparison with PAA, agarose offers the following advantages:

1. Easy gel casting with nontoxic substances.
2. Suitable for specific detection methods, such as immunoprinting, immunoblotting, and certain enzymatic staining procedures.

3. Better focusing of large proteins due to the larger pores in the agarose gel.

Standard format horizontal electrophoresis equipment such as Multiphor II (Figure 12-3b) is suitable for both preparative and analytical IEF, whereas PhastSystem (Figure 12-3a) with its smaller dimensions is primarily an analytical tool.

13.4.2   IEF Gel Media

Precast gel media, in standard format, ready to use, are available from a number of suppliers in different pH ranges. Precast gels are safe, convenient, reliable, and highly recommended when the pH range available is suitable and the economy allows.

In addition to wet precast gels, *dry* gels are also available, both in standard size and for PhastSystem (CleanGel Dry IEF and PhastGel Dry IEF, respectively, both from Pharmacia Biotech). These gels offer the flexibility of laboratory-cast gels in that they can be reswelled in any carrier ampholyte/additive combination required for the specific application without the labor of gel casting and the hazards of handling toxic acrylamide.

Despite the attractiveness of precast gel media, sometimes gels have to be cast in the laboratory.

PAA gels down to ca. 0.5 to 1 mm in thickness are most easily cast in gel cassettes composed of two glass plates sealed with a rubber gasket. The gelling mixture of acrylamide monomers and carrier ampholytes is prepared and carefully deaerated by vacuum. Usually, the casting mixture also contains 10% of sorbitol, glycerol, or another low molecular weight uncharged substance. This increases the background osmotic pressure of the system, thereby counteracting uneven osmotic pressure due to different osmotic properties of the focused carrier ampholytes, which tend to form ripples on the gel surface. Immediately before filling the cassette, the polymerization reaction is initiated by the addition of a freshly prepared solution of ammonium persulfate. The amount of persulfate should be minimized, as excess ionic components from the polymerization mixture tend to give disturbances (wavy bands, pH gradient modifications). A low concentration of persulfate makes extensive air removal critical for gelling to occur, as oxygen is an effective inhibitor of polymerization. It should also be realized that the persulfate decomposes in solution.

It is customary to prepare a stock solution of acrylamide and to store it over some ion exchange material (mixed bed in H^+ and OH^- form). Detailed procedures are provided by carrier ampholyte manufacturers. For casting *ultrathin PAA gels,* either the "flap technique"[29] or the "sliding technique"[30] is recommended.

A common formula for a PAGIF gel is T5 C3 (see Section 12.2.2.1 for nomenclature), 3% Ampholine or Pharmalyte diluted 1 : 15, 10% glycerol, and 20 to 50 mg/100 ml ammonium persulfate. Normally TEMED is not used in

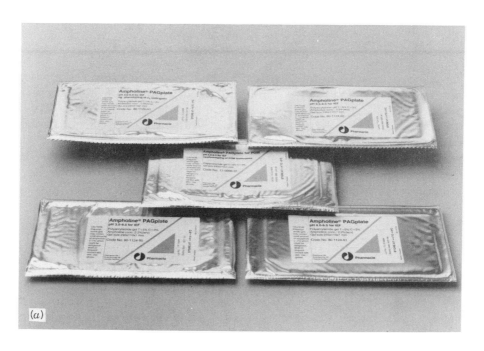

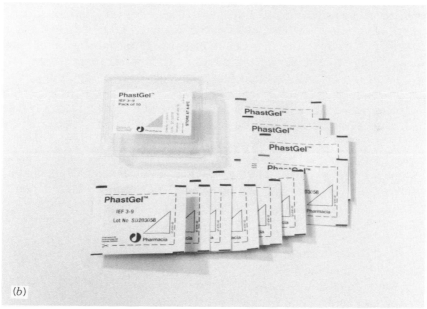

FIGURE 13-4. Gel media for analytical IEF are available in a range of wide and narrow pH intervals in the form of ready-to-use wet gels as well as precast *dry* gels. Dry gels have to be reswelled before use, thus offering complete flexibility as to the use of carrier ampholytes and additives at the expense of some of the convenience of the wet precast gels. (a) Gels in standard format (125 × 250 × 1 mm) for Multiphor II and similar equipment for horizontal IEF, and (b) gels in a miniature format for PhastSystem.

casting IEF gels. When preparing *agarose gels,* the agarose is first dissolved in boiling water to the desired concentration (commonly 1 to 2%). The addition of a neutral substance such as sorbitol to ca. 10% will improve the physical properties of the gel, giving less ripple formation during IEF and less risk of water extrusion. After cooling of the solution to 70 to 80°C, the carrier ampholytes are added and carefully mixed. The gel is then cast by simply pouring the viscous solution on the gel support. The surface tension will prohibit the solution from overflowing. As gel support, a glass plate can be used or preferably a specially treated plastic foil (GelBond™). A few drops of water between the plastic foil and the flat support (glass plate) pressed to a thin film will ensure that the plastic foil stays flat during gel casting.

13.4.3 Sample

The sample should have as low an ionic strength as possible. Too high a salt concentration in the sample will result in curved protein bands. It is seldom necessary to concentrate the sample, due both to the concentrating effect of the IEF process itself and to the extremely sensitive detection procedures available. If necessary, concentration is most conveniently done in an Amicon or equivalent concentration cell. Crude extracts may be very viscous due to the presence of nucleic acids. Treatment of the sample with DNase usually helps. An even simpler alternative that sometimes work is to apply the sample close to the anode. The negatively charged nucleic acids will then quickly separate from the proteins, which can then focus normally. The sample is most often applied with the use of a plastic mask with cut-out holes or with the use of filter paper pieces soaked in the sample. Plastic masks work excellently on PAA gels with their sticky surface. On agarose gels, the samples have a tendency to penetrate under the plastic. The use of filter paper pieces soaked in the sample is therefore generally a safer method for sample application on agarose gels. If the sample is very crude, this method also has the advantage that precipitated proteins or other solid material in the sample will stay in the filter paper and can be removed from the gel. High-quality paper without unspecific adsorption of protein or other undesired side effects must be used. Such papers are available from companies providing electrophoresis equipment.

A good guide to the best sample application spot can be obtained by titration curve analysis (Section 13.4.8). If possible, apply the protein from the side with the highest electrophoretic mobility. Any regions with pH instability will also be revealed by this analysis.

If possible, do not apply the sample directly over the area of greatest interest as there will always be some disturbances from precipitated protein and so on on the spot of sample application.

One advantage of IEF is the possibility of applying very crude samples. Soluble proteins may be analyzed by applying pieces of tissue directly on the gel surface.[31-34] Proteins will diffuse out of the tissue and into the gel where they will focus.

In preparative IEF in Sephadex,® the sample is usually mixed with the carrier ampholytes in the gel slurry and applied as a sample zone with the aid of a special sample application tray.[25]

Within reasonable limits, the sample concentration is of less importance for the IEF process than the total amount of each protein. The optimal amount depends on a number of factors. With Coomassie staining of total protein after focusing in gels of ca. 10 cm length, 10 to 50 μg protein or 1 to 5 μg per protein band will generally give a good result. In shorter gels with steeper gradients the bands are narrower and smaller amounts give a suitable staining intensity.[36] With silver staining the amounts should be reduced about 50 to 100 times.

The optimal amount of sample depends on the purpose of the experiment. If maximal sharpness is required for accurate pI determination, the smallest detectable amount should be used. If, however, the whole protein pattern is to be screened or small amounts of impurities are to be looked for, larger amounts must be used. When specific detection methods such as immunochemical or enzymatic detections are being used, much smaller amounts can be detected. With radioactively labeled antibodies, picogram amounts of protein can be detected.[36]

13.4.4 Controlling Experimental Parameters; The "Steady State"

At the beginning of the experiment when the voltage is applied, the majority of the carrier ampholytes find themselves in a pH far away from their pI values. They are therefore highly charged and the electrical conductivity of the system is good. As the pH gradient forms and the carrier ampholytes concentrate at their pI points, their electrical charges, and consequently also the electrical conductivity of the whole system, approach zero. The establishment of the pH gradient can therefore be followed by monitoring the change of the electrical properties of the system.

Generally, IEF experiments are performed at constant power (wattage). The pH gradient development can then be followed by monitoring either the increase in voltage toward an asymptotic steady-state value or the decreasing current (Figure 13-5).

However, because the carrier ampholytes move much faster than the proteins, focusing must continue for some time after the establishment of the pH gradient. How long this continues depends on the electrophoretic mobilities of the proteins in the sample. The time required to reach the pI is a function of the protein mobility over the pH values it has to pass and the voltage. For a given experimental setup (gel size, type, and concentration and sample application spot) the necessary extent of the experiment is most accurately expressed in volt-hours (Vh). It was found that, with the same experimental setup, the same degree of focusing was reached after the same number of volt-hours independent of the size of the voltage used and the brand of carrier ampholytes used.[37] With a focusing distance of ca. 10 cm, the fastest proteins

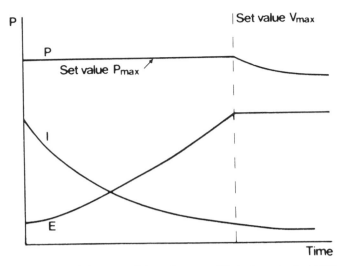

FIGURE 13-5. Voltage (E), power (P), and current (I) during a typical IEF experiment run initially at constant power and changed to constant voltage close to the end of the experiment.

were focused after ca. 1,500 Vh, whereas slower ones needed at least ca. 3,000 Vh.

Once the pH gradient is established and the proteins are focused at their isoelectric points, a "steady state" is reached that should ideally persist indefinitely. However, in practice, a true steady state is never established, but the system gradually changes, mainly through electroosmosis-induced gradient drift, and eventually collapses. Consequently, for each experiment there is an optimal extension of the focusing. The carrier ampholyte manufacturers give guidelines, but for best results the optimal focusing time has to be experimentally determined for each application. Particular care has to be taken if the goal of the experiment is to determine an unknown isoelectric point. Because the proteins have different mobilities, they will need different times to reach their isoelectric points, and it is necessary to ascertain that the protein of interest actually is focused and not still migrating when the experiment is interrupted. A safe way to ensure this is to apply the sample at both the acidic and the basic side of the gel and to make sure that the band stops at the same position. If this approach is not possible, the sample can be applied at intervals or else focused for different times. Finding the protein band at the same pH after focusing for different Vh is also a good indication that the steady state has been reached.

The Vh are monitored most accurately and conveniently with a volt-hour registering device. Most power supplies carry this possibility. Other parameters that must also be recorded for the Vh integration to be meaningful and the experiment to be reproducibly described are gel type and composition, temperature (preferably that in the gel, but at least the cooling conditions should be given), carrier ampholyte used, electrode distance, point of sample application, and additives, if any. Even though gel width and thickness and carrier ampholyte concentration are theoretically not needed in defining the steady-state conditions as these factors will be compensated for in the Vh integration, it is recommended that they are given anyway. The carrier ampholyte concentration can especially affect other aspects of the IEF experiment as, for example, pH gradient stability or protein solubility.

13.4.5 Measuring pH and Determination of pI

To obtain reproducible and relevant results for the comparison of different samples or for an approximate estimation of the pI spectra of the sample, an approximate pH gradient determination may be adequate. To determine a proper pI value, one must first ensure that the protein really is focused at its pI. Focusing for 20 to 50% longer than the minimal number of Vh will ensure this (see earlier discussion). Then the pH must be measured accurately, which is not a trivial task. The easiest and, in most cases, the most accurate way is to use commercially available pI marker proteins. Only two factors are needed to pay special attention to: (1) that the focusing is complete (see earlier) and (2) that conditions are chosen so that the final temperature in the gel at the protein position is the same as the temperature for which the pI values of the marker proteins are given, usually 24 to 25°C.[8]

If pI markers are not used, but the pH is measured directly in the gel, there are two especially important factors that have to be considered: *temperature* and *solvent composition*. Glass electrodes used for pH measurements are normally calibrated with standard buffers with carefully specified compositions. These can either be prepared in the laboratory according to formulas given by the National Bureau of Standards[38] or bought premade ready to use. The theory of pH determination is quite complicated. This chapter only discusses the factors of special relevance for IEF.

The phenomenon called *liquid junction potential* (the potential drop from glass to liquid) enters into the output of any pH reading. Any factor that affects the chemical potential of the liquid also affects the liquid junction potential. Consequently, for accurate pH readings, the conditions must be as similar as possible to those used at the calibration. Factors of particular importance are stirring, temperature, and solvent composition. pH readings in solutions with a significantly different composition to the standard buffers must be interpreted with great care.[39–43] For example, pH readings of solutions containing high concentrations of urea should be referred to as "apparent" pH or properly corrected.[42]

Measuring high pH values accurately is particularly difficult. At high pH, carbon dioxide in the air is eagerly absorbed, lowering the pH of the high pH end of the gel. The pH measurement should therefore be taken as soon as possible and the standard buffers used should be prepared fresh from CO_2-free water and discarded after use.[44]

pK_a values, and consequently pI values, are temperature dependent. This is especially prominent at basic pH.[45] The temperature of interest is obviously that in the gel where the focusing takes place, which will necessarily be significantly higher than that of the coolant. If carrier ampholytes with an uneven conductivity profile are used, the temperature will be different at different places in the gradient[8,11] (Figure 13-3). Using, for example, a thermocouple thermometer it is relatively easy to measure the temperature of the gel at steady state toward the end of the experiment. To avoid complicated and not very exact temperature compensations, it is advisable to calibrate the pH electrode at the temperature at which the reading is to be taken.

13.4.6   IEF at Acidic and Basic pH Values

The electrical conductance of the system should be even over the whole pH gradient for optimal performance. At low pH (below pH 3 to 3.5) this condition cannot be fulfilled due to the high concentration of protons. The electrical field strength at these pH values is thus very low, making focusing of acid proteins at their pI very slow and uncertain.

Another problem with focusing acid proteins is related to the nature of the proteins themselves. Acid proteins usually contain large amounts of acidic carbohydrates, making them very difficult to precipitate with standard procedures. Also, the binding of common stains such as Coomassie depends to a large extent on ionic forces between the negatively charged stain and the proteins. At the acid pH used for staining, "normal" proteins are highly positively charged in opposition to acidic proteins. Very acidic proteins therefore tend to stain poorly with standard procedures.

The extreme alkaline end of the pH scale also offers special problems that one must be aware of:

1. The stability of the PAA gel is not very high at basic pH values. The amido bond is sensitive to alkaline hydrolysis, giving rise to negative charges on the PAA gel matrix, causing electroosmosis and gradient drift. PAA gels with basic carrier ampholytes should therefore not be stored for more than a few days in the refrigerator.

2. Carbon dioxide in the system will dissolve at the alkaline end of the gradient, lowering the pH in that region. It will then migrate as bicarbonate ions toward the acid region where it will evaporate as CO_2. The result is a continuous acidification of the alkaline end of the pH gradient. In addition, the bands tend to be tilted, as the gel surface is more affected by the CO_2 then the bottom of the gel. The migration

of the pH gradient also makes protein bands fuzzier and broader than normal.

3. Agarose gels tend to cause problems with gradient drift at high pH values. Generally, PAA gels give better performance. In a closed system, such as a gel rod for use in two-dimensional electrophoresis, carbon dioxide in the system cannot evaporate but will displace carrier ampholytes and concentrate in a pH plateau around pH 5 to 7. This is often seen as a region with apparently very little proteins due to the flat pH gradient. The lack of carrier ampholytes in this zone makes pH measuring in this region very unreliable. Sometimes inverse pH gradients have been reported. Such gradients are theoretically impossible but are measuring artifacts due to this phenomenon.

The remedy for CO_2 problems is to exclude CO_2 (and HCO_3^- ions) from the system as far as possible. Deaerated or nitrogen-saturated water should be used when preparing the buffers. This is especially important with basic buffers and electrode solutions, as HCO_3^- in these buffers cannot be removed by deaerating. In flat bed focusing, the equipment can be flushed with nitrogen or argon, whereas in rod focusing, the cathodic buffer should be prepared as free as possible of carbon dioxide. Carbon dioxide traps in the form of 1 M NaOH in a vessel or sponge close to the gel will also improve the result. As already mentioned, the pH readings should be taken as soon as possible to minimize the risk from pH being reduced by CO_2. The sometimes used procedure of soaking gel pieces in water before pH measurement is a very unreliable method, especially at alkaline pH values and in the carbon dioxide pK region at pH 5 to 7.

If required, the pH span may be extended by adding basic amino acids such as lysine and arginine.[13]

13.4.7 Additives; IEF in Urea

In principle, it should be possible to use all noncharged, or zero net charged, additives without problem. This is generally the case. The most commonly used additives are detergents and urea. Zwitterionic and nonionic detergents do not interfere with the focusing process, provided they are pure. Unfortunately, this is not always the case. In such cases the charged contaminants must be removed from the detergents by, for example, passage over a zwitterionic ion exchanger in H^+ and OH^- form.

The use of urea is accompanied by other problems as urea decomposes to ammonium isocyanate, which is very reactive to primary amines in the proteins (carbamylation).[46] This reaction transfers the positive amino group into a negative group and lowers the pI.

Fortunately, the rearrangement of urea to isocyanate and ammonia is a slow reaction at temperatures at and below room temperature. In the cold, urea solutions can be kept safely for at least a week, but for critical experiments it is wise to use freshly prepared solutions of high-quality urea and/or purify the solution over a zwitterionic ion exchanger as described for detergents. Also note that by prefocusing before sample application the isocyanate will migrate out of the gel. By using commercially available dry IEF gel media the problems and hazards connected with gel casting are avoided, simply reswell the gel in any carrier ampholyte/additive mixture!

The use of high concentrations of urea in agarose IEF is particularly difficult. The high temperatures needed to dissolve the agarose will enhance the isocyanate formation of the urea. In addition, high concentrations of urea impair the formation of agarose gels. These problems can be handled by suitable modification of the gel casting procedure and by using a 2% agarose gel.[47] Special precautions must be taken when measuring pI in gels containing urea.[43]

13.4.8 Titration Curve Analysis

By using a combination of IEF and electrophoresis it is possible to prepare a physically visible titration curve of a protein. A square slab gel for horizontal PAGIF with broad range carrier ampholytes (pH 3 to 10 or similar) is prepared. Normally the gel casting mold is prepared so that a sample application well, extending almost across the whole gel, is formed. An IEF experiment is run without sample just enough to ensure that a stable pH gradient is formed. The sample well should be along the pH gradient so that pH is acid at one end and basic at the other. The gel is then rotated 90° and the well filled with sample. The sample is now electrophoresed across the pH gradient. Because a gel with low porosity is used, the electrophoretic mobility is essentially a function of protein charge, and the pattern for each protein will correspond to its titration curve.[48,49]

Titration curve analysis can be very useful in the prediction of purification strategies[50,51] (see Chapter 4) and can also be used for Ferguson plot analysis (Section 12.4.1)

13.5   IEF IN IMMOBILIZED pH GRADIENTS; IMMOBILINE™

The general theory and most of the factors discussed above for conventional IEF are applicable also to IEF in Immobiline gradients. The presentation below therefore concentrates on factors of special relevance to IEF in Immobiline.

13.5.1   Principles and pH Gradient Generation

The only principal difference between conventional IEF in carrier ampholytes as described earlier, and IEF in Immobiline is the way the pH gradient is formed and maintained. In 1982, a new technique for creating and maintaining the pH gradient was made available by the introduction of Immobiline.[5] Immobiline is the trademark for special derivatives of acrylamide: CH = CH-CO-NH-R, where "R" is an amino or carboxylic group with a well-defined pK value. Immobiline is used to form immobilized artificial pH gradients in PAA gels. At present there are seven different types commercially available: four basic (pK values 6.2, 7.0, 8.5, and 9.3) and three acidic ones (pK values 3.6, 4.4, and 4.6).

When casting an Immobiline gel, two solutions of gel monomers are made: Immobiline titrated to the acidic end pH and Immobiline titrated to the alkaline end pH. These solutions are then mixed in a gradient mixer to form a continuous gradient during the filling of the gel casting cassette containing a GelBond PAG plastic support. The practical procedure is quite analogous to the casting of PAA gradient gels. The two Immobiline solutions titrate each other and, if properly chosen, form the desired pH gradient. To stabilize the gradient during casting, the bottom solution (usually the most acidic) is made denser with glycerol (25%, w/v). Since the Immobiline molecules contain the acrylic double bond, they will be covalently incorporated in the gel network during the polymerization process and hence the pH gradient will be covalently immobilized. When polymerization is complete, the cassette is opened and the gel on the plastic support is used for flat bed IEF essentially as in the conventional case.[52] The introduction of charges in the gel does not give problems with electroosmosis during IEF since the gel *net charge* is zero at all pH values. (Compare charge-balanced agarose.)

Before use, the gel must be carefully washed in water to remove residual reagents (at least 1 h for a 0.5-mm-thick gel). Care must also be taken to ensure that the gel is swollen to the same degree after washing as during casting.[53,54] Narrow-range pH gradients are prepared by chosing an Immobiline with a pK within the same desired pH range as the buffering Immobiline. This Immobiline is then titrated with nonbuffering Immobiline types having pK values outside the pH range. In this way gradients spanning up to about 1 pH unit can be prepared.[5] Detailed instructions on how to prepare Immobiline gradients are obtainable from the manufacturer as well as from the literature.[55,56] To make longer pH gradients, more complicated procedures must be used.[57] Computer programs have been developed to facilitate the calculation of gradient mixing.[58]

Urea and neutral detergents are well tolerated in Immobiline systems.[59] This, together with the high loading capacity and absolutely stable, reproducible pH gradients has made Immobiline IEF the method of choice for the first dimension in two-dimensional electrophoresis (see Chapter 15).

A range of precast Immobiline gels are commercially available. The gels are supplied dry and must be reswelled before use.

13.5.2 Characteristics of Immobiline Gels

Immobiline gels have many attractive advantages over conventional IEF gels:

- By definition, gradient drift cannot take place.
- Well-defined chemicals give better control of pH gradients.
- Ultraflat pH gradients can easily be prepared (down to 0.01 pH unit/cm) with a concomitant increase in resolution.
- Problems such as wavy bands related to inadequate carrier ampholyte quality are eliminated.
- They have high loading capacity.
- They are less sensitive to salts and buffers in the sample.

The separation power was clearly demonstrated already in the very first paper.[5] It was estimated that bands with pI values differing as little as 0.001 pH unit could be resolved. However, limitations in the technique as compared to conventional IEF soon became obvious: The gel casting is quite cumbersome (this has been solved, at least partly, with the advent of commercially available precast Immobiline gel plates); long preruns[60] and long separation times are needed (overnight runs are standard); it is very difficult to measure pH in the gels, in fact it seems to be impossible without the addition of carrier ampholytes to the system; and lower detection limits as required for protein than in conventional IEF due to loss of sample material.[61] The reason for the longer separation times is probably related to interactions between the proteins and the gel matrix. Strong evidence suggests that this is primarily due to a hydrophobic interaction between the proteins and basic Immobiline molecules.[62] Ionic interactions may also contribute. Consequently, a fraction of the proteins is more or less tightly "bound" to the matrix, and only the proteins in solution can actually migrate. Adding carrier ampholytes to Immobiline systems (hybrid IEF) circumvents problems such as precipitation at sample application and lateral band spreading as well as smearing and background.[61,63–656] However, it has also been shown that by using proper conditions for gel reswelling and sample preparation and a sample application position adapted to sample pH, together with sufficiently low voltage during the initial focusing, good results can also be obtained in carrier ampholyte-free Immobiline gels.[66] Another approach to minimize lateral band spreading and other problems related to the application of the sample on the surface is to make sample application wells by simply punching holes in the gel.[67]

Using PhastSystem with its small dimensions, silver-stained two-dimensional Immobiline gels can be produced in only 3.5 h.[68]

13.6 OPTIMIZING THE IEF SYSTEM

13.6.1 Choosing the IEF Type; Conventional/Immobiline

The optimal system and experimental conditions will of course depend on the separation problem, the type of sample, and the purpose of the separation. For preparative separations, flat bed IEF in Ultrodex is generally recommended. Choose the narrowest possible pH gradient and follow the recommendations of the carrier ampholyte manufacturer. To separate proteins differing by less than ca. 0.05 to 0.1 pH units, IEF in Immobiline gradients should be tried. Proteins can be extracted from the gel by isotachophoresis by any of the suggested techniques.[69-73]

In analytical IEF there are three basic techniques, of which two can be performed on commercial precast gels:

1. Conventional IEF in agarose: Lab-cast gels only.
2. Conventional IEF in PAA: Lab-cast or precast gels.
3. IEF in Immobiline (in PAA gel): Lab-cast or precast gels.

If the protein to be analyzed has a very high molecular weight (higher than 150,000 to 200,000), agarose should be the first choice. In the majority of cases, however, IEF in PAA is preferred.

In PAGIF the choice stands between conventional IEF with carrier ampholytes and IEF in Immobiline gels. In most cases, conventional IEF is the natural method of choice because of its higher speed and convenience. Staining and pH measurements are also more straightforward in conventional IEF, which is important when pI values are recorded.

For more demanding applications, Immobiline gels are recommended due to a higher loading capacity and resolving power and more accurate performance. In general, it is advisable to use precast gels as much as possible to save time and effort and ensure a high degree of reproducibility.

13.6.2 Gel Length and Thickness

Intuitively it seems advantageous to use long gels in order for the proteins to be spaced apart as much as possible. In practice, however, the use of very long gels is accompanied with a number of drawbacks:

1. The longer the gel, the longer the path for the proteins to migrate. Eventually, some proteins may not be able to reach their pI during the lifetime of the pH gradient. At the very least, the duration of the experiment will be inconveniently long.
2. At equal field strength, the resolution is inversely proportional to the square root of the gradient slope (Section 13.2). Since for practi-

cal reasons lower field strengths normally must be used on longer gels, the increase in resolution with long gels is only marginally higher at best.

After investigating the effect of gel length on overall performance, it was concluded that there is generally little to gain by using gels longer than ca. 10 cm and, in fact, considerably shorter gels can be used advantageously in many cases.[35] This is proven in practice by the excellent results routinely obtained on PhastSystem with its small format gels.[74,75]

Thin gels offers advantages such as more efficient cooling, apparently sharper bands (the effect of tilted bends is less severe), fast solvent penetration in staining and destaining, and lower consumption of carrier ampholytes. Too thin a gels will, however, be accompanied by a number of practical problems: Small nominal differences in gel thickness will become relatively large with concomitant effects on field strength, heat production, and gradient distortions. Because of the small amounts of carrier ampholytes, the whole system becomes more sensitive to all kinds of disturbances, for example, from carbon dioxide and evaporation. The optimal compromise for most applications seems to be a thickness of 0.4 to 1 mm.

13.6.3 Running Conditions; Voltage and Power

As discussed earlier, the sharpest bands and best resolution are obtained with maximum voltage. The limits are set only by the ability of the system to control the heat produced, which is directly proportional to the voltage. The following points should be considered for optimal performance:

1. Use an experimental setup with good cooling capacity. For analytical purposes, this is thin or ultrathin (0.2 to 0.5 mm) layer PAGIF on an apparatus with an efficient cooling plate.

2. Use high-quality carrier ampholytes with the most even conductivity profile. This may sometimes require blending. In regions with lower conductivity, the voltage and heat production are higher. Those areas will dictate the overall voltage that can be applied. Focusing in the rest of the gradient will suffer from a focusing voltage less than optimum.

3. Do not use more carrier ampholytes and/or Immobiline than needed to stabilize the pH gradient. Excess carrier ampholytes will increase the current and the heat produced, making it impossible to use high voltage.

4. Run the experiment at constant power. This will assure maximum voltage during the whole separation, which minimizes separation time and maximizes resolution at the final stage of the experiment.

13.7 TROUBLESHOOTING

A good result in slab IEF requires that the iso-pH lines extend straight over the plate. Uneven gradients give rise to wavy protein bands. With conventional IEF using carrier ampholytes, a number of factors affect the straightness of the iso-pH lines. The most important is the chemical composition of the system, especially the *carrier ampholytes* used. Some carrier ampholyte mixtures are more easily disturbed than others. This depends mostly on the buffer capacity distribution during the pH gradient development. Most disturbances appear during the development of the gradient. Straighter iso-pH lines are therefore generally achieved if the pH gradient is allowed to form before the sample is applied, called "prefocusing." Most often the disturbances appear from the anode as "acid fingers." Due to the great mobility of the protons, they sometimes manage to break through the buffering barrier of the carrier ampholytes.

Uneven *contact between the electrodes* and the gel will affect the gradient. Make sure that the electrodes are straight and in good condition and that the strips are evenly wet and in good contact with the gel. The electrodes must be applied to the strips or gel with a certain pressure.

Sample composition is also very important. Bow-shaped bands generally occur because of excess salt in the sample. The effect of buffering salts is especially strong. Neutral salts (e.g., NaCl) or buffering salts applied at a nonbuffering pH can generally be tolerated up to ca. 0.5 M. With reasonable amounts of buffering salts, an improved result may be obtained by changing the sample application site to a pH where the buffering capacity of the salt is smaller. If neither this nor prefocusing is sufficient, the salts have to be removed before sample application. Standard methods such as gel filtration or dialysis can be used. The *ionic strength of the gel casting mixture* should be kept as low as possible. The persulfate concentration should be kept to a minimum as sulfate emanating from the persulfate is covalently attached to the ends of the PAA chains during the initiation of the polymerization reaction. A low concentration of persulfate makes exhaustive deaerating a necessity, as oxygen inhibits the polymerization. The choice of *electrode solutions,* when used, also strongly affect the pH gradient. A suitable electrode solution compensates for disturbances due, for example, to uneven electrode contact. The composition, concentration, and pH of the electrode solutions also affect the exact span of the pH gradient and can be used for gradient fine-tuning.

The phenomenon of *gradient drift* is the major reason why conventional IEF is not quite the steady-state technique it theoretically should be. The major cause of gradient drift is electroosmosis, although other factors such as the absorption of carbon dioxide at high pH and evaporation at acidic pH or the different mobilities of the positive and negative forms of the carrier ampholyte molecules may play a part.[76] Gradient drift is mostly toward the cathode ("cathodic drift"), reflecting a negatively charged matrix. Sometimes the opposite can be seen, and sometimes both directions of drift may be manifested in the same experiment, reflecting a pH dependent net charge of the gel matrix. The physical manifestation of gradient drift in flat bed IEF is liquid transportation from one electrode toward the other. In severe cases, liquid droplets are extruded from the gel and/or the gel dries out at the other side, eventually to complete dryness. To eliminate the problem, a gel matrix with a lower charge content must be used. Change to higher quality, IEF-grade chemicals or try to deionize the monomers as with a mixed bed ion exchanger as described earlier. Be sure that high pH gels are used fresh as the acrylamide deamidates at a high pH. It seems that gradient drift can never be completely eliminated in conventional IEF, but is seldom a problem in practice when using PAA gels or IEF grades of agarose.

Proteins are generally poorly soluble at their isoelectric point. Isoelectric precipitation was often a serious problem with focusing in vertical gradient-stabilized systems. In horizontal systems, both preparative and analytical, isoelectric precipitation is seldom a problem. Should there be precipitation, urea or nonionic detergents can be added to increase the solubility.

13.8   APPLICATIONS

13.8.1   Phenotyping of Apolipoprotein E from Plasma by IEF in Immobiline DryPlate 4-7 and Silver Staining

Determination of the Apo-E phenotype is of great clinical importance in the diagnosis of lipid metabolism disorders. Because of the high resolution of Immobiline IEF as compared to conventional IEF in carrier ampholyte systems, the discrimination among the six different Apo-E isoforms could be made more conclusive. The increased sensitivity of silver staining reduced the necessary amount of starting material to delipidated very low density lipoproteins from 1 mL serum. The procedure reported here, which is based on prefabricated Immobiline DryPlate 4-7, is convenient, reproducible, less expensive, and less time-consuming than alternative techniques. In addition, silver staining gave an increased resolution compared to immunoblotting.[77]

Experimental Procedures

Very low density lipoproteins (VLDL) were isolated from 1 mL serum by ultracentrifugation (2.5 h, 10°C, 100,000 rpm). The top fraction (200 to 300 μl) containing the VLDL was recovered. Delipidation was accomplished by treatment with 4 mL ethanol/diethyl ether (3/1, v/v) in a rocker for 1 h at 4°C. The delipidated VLDL was pelleted and washed once with 1 mL diethyl ether and dried under a stream of nitrogen.

Finally the protein pellet was solubilized in 0.01 *M* Tris-HCl, 0.01 *M* DTT, 30% glycerol, 4% Ampholine pH 5-7, and 6 *M* urea by 24-h incubation at 4°C. Both this sample and the VLDL fraction after ultracentrifugation can be stored at −20°C for several weeks.

Immobiline DryPlate 4-7 (Pharmacia Biotech) was reswelled in the same 6 *M* urea buffer as just described, but without carrier ampholytes. Catholyte and anolyte solutions were 10 m*M* NaOH and 10 m*M* glutamic acid, respectively, both containing 4% Ampholine 5-7. The gel was placed on a Multiphor II electrophoresis unit (Pharmacia Biotech) thermostatted at 13°C.

After prefocusing for 90 min, first at 200 V, increasing to 3000 V, 15 to 25 μL of sample was applied on the gel. The sample application was ca. 1 cm from the cathode, and 20 samples were normally analyzed simultaneously. Focusing was then started at 1000 V for 1 h to allow a smooth sample penetration and was continued overnight with the following settings: 3000 V, 5 mA, 5 W as maximum.

After completion, the gels were soaked in a mixture of 11% trichloroacetic acid and 35% sulfosalicylic acid for 1 h to precipitate the proteins before being washed three times with methanol/acetic acid (10/5, v/v). After incubation for 30 min in DTT (5 mg/L) the gels were stained for 30 min in silver nitrate (2 g/L). Sodium carbonate (3%) containing 0.5 mL/L formaldehyde developed the protein pattern. Finally the gels were rinsed with water and allowed to dry in open air (Figure 13-6).

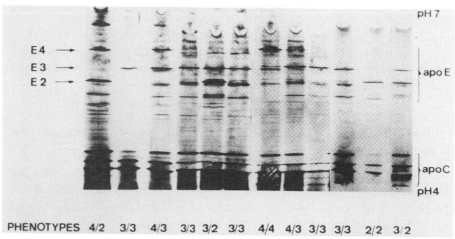

FIGURE 13-6. Typical patterns after separation of delipidated VLDL proteins on Immobiline DryPlate 4–7 and silver staining. (Reproduced from Ref. 77 with permission from VCH Verlagsgesellschaft mbH and the authors.)

13.8.2   Using IEF in Immobiline pH Gradients for the Analysis of Recombinant Leech Hirudin, Human α_1-Antitrypsin, and *Schistosoma mansoni* Parasite Antigen, p28

The superior resolving power of Immobiline IEF was used to detect impurities in these three recombinant proteins that were undetected by other methods such as HPLC by reversed phase, anion exchange, and gel filtration as well as SDS–PAGE under reducing and denaturing conditions.[78]

Hirudin is potentially interesting in human therapy because of its anticoagulant properties. By IEF in IPG gels, pH 3.8 to 4.8, the purity and alkaline stability of recombinant hirudin were assessed. Deamidation upon prolonged storage at pH 9 was efficiently monitored.

Analysis by IEF-IPG, pH 5 to 6, of α_1-antitrypsin from six different production runs using different procedures showed the same type of inhomogeneity in all cases. Based on theoretical calculations of the pI values of possible deamidation products and comparison with the observed patterns, it was concluded that the inhomogeneities were due to deamidations that obviously occurred to about the same extent, despite the production procedure.

Recombinant p28 from *S. mansoni* is a 210 amino acid protein that is a candidate for a vaccine toward schistosomiasis, a chronic debilitating disease affecting about 200 million people in tropical and subtropical countries. Analysis of recombinant p28, expressed in *Saccharomyces cerevisiae* and purified by ion-exchange chromatography and affinity chromatography on glutathione agarose, revealed considerable inhomogeneity and a pI of the major component of pH 7.81 instead of the expected 7.17 (Figure 13-7).

For further analysis by sequencing, the protein bands were electroblotted onto PVDF membranes after having removed the gel from the GelBond plastic backing. A CAPS buffer, pH 11 without SDS, proved to give a blotting yield of 75%. For the sequencing analysis, the high loading capacity of the IPG gels without a loss in resolution was found to be particularly useful. As much as 24 g was focused into an area of ca. 12 mm^2 (= ca. 80 pmol/mm^2) without any special optimizations.

13.8.3   Identification of Whale Species by Thin-Layer IEF on Ampholine PAGplate 3.5–9.5

Chemotaxonomical methods for identifying whale species is both of a scientific interest in clarifying some uncertainties still remaining in the classification of toothed whales (*Odontoceti*) and of interest as a practical control method of meat because of regulations of commercial whaling.

Species-specific bands were obtained for all 8 of the 12 species investigated. Four dolphin species could only be identified as a group but showed no species-specific banding among themselves.[79]

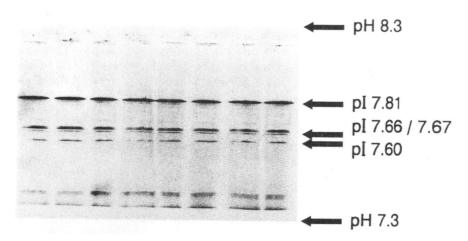

← pH 8.3

← pI 7.81

⇐ pI 7.66 / 7.67

⇐ pI 7.60

← pH 7.3

FIGURE 13-7. IEF-IPG pH 7.3 to 8.3 of recombinant p28 from *Schistosoma mansoni.* The sample applied had been purified by ion-exchange chromatography and affinity chromatography on glutathione–agarose. Protein bands, indicated by arrows, were further analyzed after electroblotting to PVDF membrane. (Reproduced from Ref. 78 with permission from VCH Verlagsgesellschaft and the authors.)

Experimental Procedures

Sarcoplasmic proteins from 1 g of whale muscle preserved at −30°C were isolated by homogenizing in 20 mL of cold distilled water. The extracts were clarified by centrifugation (1,600 g for 20 min at 4°C) and the supernatants were used for analysis. Ampholine PAGplate 3.5-9.5 were positioned on the cooling plate of a Multiphor II unit (Pharmacia Biotech) on a film of kerosene, assuring a good cooling contact. Electrolyte strips soaked in 1 M NaOH (catholyte) and 1 M H_3PO_4 (anolyte) were placed at the edges of the gels. Samples (15 μL) were applied on the center of the plate by using sample application pieces. Isoelectric focusing separation was conducted for 1.5 h with the following maximum settings: power = 30 W, voltage = 1,500 V, and current = 50 mA.

After finishing the focusing, the gels were placed in a fixing solution of 57.5 g trichloroacetic acid and 17.25 g sulfosalicylic acid in 500 mL water for 5 min. The gels were then placed in staining solution (0.46 g Coomassie brilliant blue R-250 in 400 mL destaining solution) for 10 min at 60°C. Background staining was removed by placing the gels in destaining solution (500 mL ethanol and 160 mL acetic acid in 2 L water) overnight. Finally the gels were preserved by first soaking for

1 h in preserving solution (40 mL glycerol and 400 mL destaining solution) to prevent cracking during drying, then drying at 50°C for 40 min, followed by covering with cellophane preserving sheets. Figure 13-8 shows a typical result of the isoelectrophoretic patterns obtained. By densitometry and computer analysis of the protein patterns, the precision of species identification could be enhanced further.

13.8.4   Identification of Potato Cyst Nematode Species by IEF on PhastSystem and Silver Staining

The parasitic nematodes *Globodera pallida* and *G. rostochiensis* are of economic importance in temperate zones and tropical highlands. Species identification by classical light microscopic characteristics is difficult because of morphological similarities. By taking previously described electrophoresis methods and adapting them for PhastSystem, it was possible to design a fast and efficient method for routine diagnosis (Figure 13-9).[80]

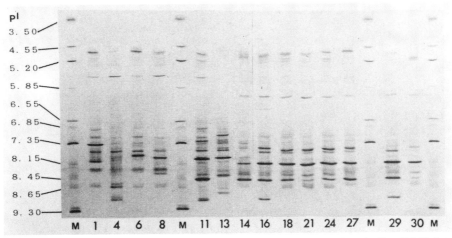

FIGURE 13-8. Protein patterns after isoelectric focusing on Ampholine PAGplate 3.5–9.5 of sarcoplasmic proteins from 12 authentic (known) whale species and 2 unknown species. Samples 1,4,6, and 8 are known baleen species (fin whale, Sei whale, Bryde's whale, and Minke whale, respectively). Samples 11, 13, 14, 16, 18, 21, 24, and 27 are toothed whale species (sperm whale, Baird's beaked whale, short-finned pilot whale, Dall's porpoise, Northern right whale dolphin, Pacific white-sided dolphin, common dolphin, and striped dolphin, respectively). M are pI marker proteins, and samples 29 and 30 are the unknown samples. Sample 29 was sold as meat from Doll's dolphin, which was confirmed by the IEF protein pattern (compare samples 29 and 16). Sample 30, however, marketed as meat from Sei whale, was obviously wrongly identified (compare samples 30 and 4). The pattern resembles more the patterns of the toothed whales 18, 21, 24,and 27. (Reproduced from Ref. 79 with permission.)

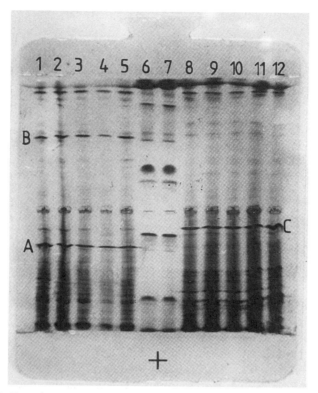

FIGURE 13-9. Protein patterns after isoelectric focusing of *Globodera pallida* (lanes 1–5) and *G. rostochiensis* (lanes 8–12) and broad pI marker kit proteins (lanes 6 and 7) on PhastGel 5–8 and silver staining. The primary bands that differentiate the patterns of the two species are marked (A,B, and C). (Reproduced from Ref. 80 with permission from VCH Verlagsgesellschaft mbH.)

Experimental Procedures

Sample preparation: Protein extraction was performed in the cavities of the PhastGel sample well stamp, chilled on ice. Briefly, one dry cyst, presoaked in 1% glycerol for at least 12 h, was placed in each cavity together with 0.7 μL of 1% glycerol, except for the middle wells, which were used as references and filled with 0.7 μL of marker kit protein solution. To release the egg content, each cyst was crushed with a glass rod and macerated for 15 s. Finally, the samples were transferred to the 12/0.3 PhastSystem sample applicator, which was placed in the middle position of the applicator arm.

Isoelectric focusing: The following running program was used (see PhastSystem Technique File 100 for more detailed information):

Sample applicator down at 2.2:					0 Vh
Sample applicator up at 2.3:					0 Vh

Sep 2.1	2,000 V	2.0 mA	3.5 W	15°C	75 Vh
Sep 2.2	200 V	2.0 mA	2.5 W	15°C	15 Vh
Sep 2.3	2,000 V	5.0 mA	3.5 W	15°C	510 Vh

Total running time was ca. 30 min.

Staining: After electrophoresis the gels were stained with the PhastGel silver kit in the PhastSystem developing unit.

13.8.5   Detection of Oligoclonal IgG in Serum and Cerebrospinal Fluid by IEF in pH Gradient-Modified PhastGel IEF 3–9, Immunofixation, and Silver Staining

The presence of oligoclonal IgG in cerebrospinal fluid (CSF) is one of the major laboratory signs of multiple sclerosis and other immunostimulating central nervous system diseases, such as meningitus caused by *Borrelia spirochetes.*

The method described is based on prefabricated PhastGel IEF 3–9 extended in the alkaline end by the application of Pharmalyte 8–10.5 at the alkaline side during the prefocusing step. It further utilizes polyethylene glycol (PEG)-enhanced immunofixation and silver staining.

The new method was compared with previous IEF methods and was found to be faster and more convenient (Figure 13-10).[81]

Experimental Procedures

The samples (serum and CSF) were diluted with 6.5 mM phosphate-buffered saline (PBS), pH 7.2, to an IgG concentration of 10 to 15 mg/ L. The next step was to pipette 3.6 μL of each sample onto the sample applicator, which was placed in the anodic position in the apparatus. To expand the pH gradient in the alkaline region, a 2 × 43-mm filter paper strip (Pharmacia Biotech) was evenly wetted with 15 μL Pharmalyte 8–10.5 and placed ca. 2 mm from the anode. After a prefocusing step of 50 Vh the paper strips were removed.

The separation was performed for 510 Vh using the following running program:

Extra alarm at 2.1					50 Vh
Sample applicator down at 2.2:					0 Vh
Sample applicator up at 2.3:					100 Vh

Sep 2.1	1,000 V	3.0 mA	3.5 W	15°C	60 Vh
Sep 2.2	300 V	3.0 mA	3.5 W	20°C	50 Vh
Sep 2.3	2,000 V	3.0 mA	3.5 W	15°C	400 Vh

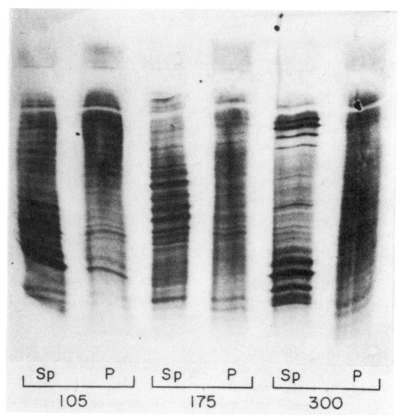

FIGURE 13-10. Examples of IgG patterns obtained from patients with multiple sclerosis when analyzing serum and CSF on PhastGel IEF 3–9 according to the procedure described in the text. (Reprinted from Ref. 81 with permission from Scandinavian University Press.)

During prefocusing step 2.1, the sample was loaded onto the sample applicator. At the alarm after 50 Vh, a pause in the program was selected, the filter paper strip was removed, and the sample applicator was applied in the apparatus in the anodal position.

After separation, the IgG proteins were precipitated by immunofixation: Rabbit antihuman IgG Fc (Dakopatts A/S, Copenhagen, Denmark) were diluted 10 times with 3% (w/v) PEG 6000 in PBS. A droplet (300 μL) of this was placed onto a hydrophobic plastic foil, and the IEF gel was placed upside down on this droplet and left for about 30 min at room temperature.

Before automatic silver staining in the PhastSystem development unit, excess proteins were removed by washing the gel with PBS, four to six times for 30 min or overnight.

13.9 REFERENCES

1. K. Ikeda, S. Suzuki, *US Patent 10, 5891* (1912).
2. R.R. Williams, R.E. Waterman, *Proc. Soc. Exp. Biol. N.Y., 27,* 56–59 (1929).
3. A. Kolin, *J. Chem. Phys., 22,* 1628–1629 (1959).
4. O. Vesterberg, *Acta Chem. Scand., 23,* 2653–2665 (1969).
5. B. Bjellqvist, K. Ek, P.G. Righetti, E. Gianazza, A. Görg, R. Westermeier, W. Postel, *J. Biochem. Biophys. Meth., 6,* 317–339 (1982).
6. O. Vesterberg, H. Svensson, *Acta Chem. Scand., 20,* 820–834 (1966).
7. T. Bibring, J. Baxandall, *Anal. Biochem., 85,* 1, 1–14 (1978).
8. T. Låås, I. Olsson, L. Söderberg, *Anal. Biochem., 101* 449–461 (1980).
9. H. Svensson, *Acta Chem. Scand., 16,* 456–466 (1962).
10. H. Svensson, in *Protides of Biological Fluids,* H. Peeters, ed., Pergamon Press, 1967, Vol. 15, pp. 515–522.
11. T. Låås, I. Olsson, L. Söderberg, in *Protides of Biological Fluids,* H. Peeters, ed., Pergamon Press, 1979, Vol. 27, pp. 683–686.
12. U.-H. Stenman, R. Gräsbeck, *Scand. J. Clin. Lab. Invest. Suppl., 116.* 26 (1971).
13. G. Yao-Jun, R. Bishop, *J. Chromatogr., 234,* 459–462.
14. M.L. Caspers, Y. Posey, R.K. Brown, *Anal. Biochem., 79,* 166–180 (1977).
15. P. Gill, J.G. Sutton, *Electrophoresis, 6,* 23–26 (1985).
16. P. Gill, *Electrophoresis 6,* 282–286 (1985).
17. J. Lizana, I. Olsson, *J. Clin. Chem. Clin. Biochem., 22,* 545–549 (1984).
18. T. Låås, I. Olsson, *Anal. Biochem., 114,* 167–172 (1981).
19. K. Altland, M. Kaempfer, *Electrophoresis, 1,* 57–62 (1980).
20. P. Gill, *Electrophoresis, 6,* 552–555 (1985).
21. R. Hackler, T.O. Kleine, *Electrophoresis, 9,* 262–267 (1988).
22. I. Olsson, T. Låås, in *Electrophoresis '82,* D. Stathakos, ed., Walter de Gruyter, 1982, pp. 157–167.
23. H. Svensson, *Arch. Biochem. Biophys. Suppl., 1,* 132–136 (1962).
24. B.J. Radola, *Biochim. Biophys. Acta,* 181–195 (1974).
25. L. Wahlström, R. Björkman, in *Protides of Biological Fluids,* H. Peeters, ed., Pergamon Press, Oxford, 1979, Vol. 27, pp. 703–706.
26. M. Duckworth, W. Yaphe, *Carbohyd. Res., 16,* 189–197 (1971).
27. C.A. Saravis, N. Zambeck, *J. Immunol. Methods, 29,* 91–96 (1979).
28. A. Rosén, K. Ek, P. Åman, *J. Immunol. Methods, 28,* 1–11 (1979).
29. B.J. Radola, *Electrophoresis, 1,* 43–56 (1980).
30. H. Garoff, W. Ansorge, *Anal. Biochem., 115,* 450–457 (1981).
31. C.A. Saravis, M. O'Brien, N. Zamcheck, *J. Immunol. Meth., 29,* 97–100 (1979).
32. B.J. Thompson, A.H.M. Burghes, M.J. Dunn, V.Dubowitz, *Electrophoresis, 2,* 251–258 (1981).
33. B.J. Thompson, M.J. Dunn, A.H.M. Burghes. V. Dubowitz, *Electrophoresis, 3,* 307–314 (1982).

34. A.L. D'Andrea, V.M.T. Leedham, *Electrophoresis, 6,* 468–469 (1985).

35. T. Låås, I. Olsson, in *Electrophoresis '81,* R.C. Allen, P. Arnaud, eds., Walter de Gruyter, New York, 1981, pp. 191–203.

36. J.M. Gershoni, G.E. Palade, *Anal. Biochem., 131,* 1–15 (1983).

37. T. Låås, I. Olsson, *Electrophoresis, 2,* 235–238 (1981).

38. R.G. Bates, *Determination of pH,* 2nd ed. Wiley-Interscience, New York, 1973.

39. W.J. Gelsema, C.L. de Ligny, *J. Chromatogr., 130,* 41–50 (1977).

40. W.J. Gelsema, C.L. de Ligny, N.G. van der Veen, *J. Chromatogr., 140,* 149–155 (1977).

41. W.J. Gelsema, C.L. de Ligny, N.G. van der Veen, *J. Chromatogr., 154,* 161–174 (1978).

42. W.J. Gelsema, C.L. de Ligny, N.G. van der Veen, *J. Chromatogr., 171,* 171–181 (1979).

43. J.A. Illingworth, *Biochem. J., 195,* 259–262 (1981).

44. H. Delincee, B.J. Radola, *Anal. Biochem., 90,* 609–623 (1978).

45. S. Fredriksson, *J. Chromatogr., 151,* 347–355 (1978).

46. G.R. Stark, W.H. Stein, S. Moore, *J. Biol. Chem., 235,* 11, 3177–3181 (1960).

47. I. Olsson, T. Låås, *J. Chromatogr., 215,* 373–378 (1981).

48. A. Rosengren, B. Bjellqvist, V. Gasparic, in *Electrophocusing and Isotachophoresis,* B.J. Radola, D. Graesslin, eds., Walter de Gruyter, New York, 1977, pp. 165–171.

49. P.G. Righetti, E. Gianazza, in *Electrophoresis '81* R.C. Allen, P. Arnaud, eds., Walter de Gruyter, New York, 1981, pp. 655–665.

50. L.A. Haff, L. Fägerstam, A.R. Barry, *J. Chromatogr., 266,* 409–425 (1983).

51. U.-B. Axiö-Fredriksson, R. Bikhabai, J. Brewer, L.G. Fägerstam, H. Lindblom, P. Steffner, in *Protides of Biological Fluids,* H. Peeters, ed., Pergamon Press, 1986, Vol. 34, pp. 735–739.

52. P.G. Righetti, *J. Chromatogr., 300,* 165–223 (1984).

53. C. Gelfi, P.G. Righetti, *Electrophoresis, 5,* 257–2262 (1984).

54. K. Altland, A. Banzhoff, R. Hackler, U. Rossman, *Electrophoresis, 6,* 379–381 (1984).

55. P.G. Righetti, K. Ek, B. Bjellqvist, *J. Chromatogr., 291,* 31–42 (1984).

56. LKB Application Notes 321, 324.

57. E. Gianazza, F. Celentano, G. Dossi, B. Bjellqvist, P.G. Righetti, *Electrophoresis, 5,* 88–97 (1984).

58. K. Altland, *Electrophoresis, 11,* 140–147 (1990).

59. E. Gianazza, G. Artoni, P.G. Righetti, *Electrophoresis, 4,* 321–326 (1983).

60. C. Gelfi, A. Morelli, E. Rovidaq, P.G. Righetti, *J. Biochem. Biophys. Meth.,* 113–124 (1986).

61. K. Altland, U. Rossman, *Electrophoresis, 6,* 314–325 (1985).

62. T. Rabilloud, C. Gelfi, M.L. Bissi, P.G. Righetti, *Electrophoresis, 8,* 305–312 (1987).

63. M.A. Rimpiläinen, P.G. Righetti, *Electrophoresis, 6,* 419–422 (1985).

64. J. Fawcett, A. Chrambach, in *Protides of Biological Fluids,* H. Peeters, ed., Pergamon Press, 1985, Vol. 33, pp. 439–442.

65. A. Görg, W. Postel, J. Weser, S. Gunther, J.R. Strahler, S.M. Hanash, L. Somerelot, *Electrophoresis, 8,* 45–51 (1987).

66. B. Bjellqvist, M. Linderholm, K. Östergren, J. Strahler, *Electrophoresis, 9,* 453–463 (1988).

67. M.J. Loessner, S.Scherer, *Electrophoresis, 13,* 461–463 (1992).

68. A. Görg, W. Postel, S. Gunther, C. Friedrich, *Electrophoresis, 9,* 57–59 (1988).

69. L.G. Öfverstedt, K. Hammarström, N. Balgobin, S. Hjertén, U. Pettersson, J. Chattopadyaya, *Biochim. Biophys. Acta, 782,* 120–126 (1984).

70. S. Hjertén, Q.Z. Liu, S.L. Zhao, *J. Biochem. Biophys. Meth., 7,* 101–113 (1983).

71. B. An der Lan, R. Horuk, J.V. Sullivan, A. Chrambach, *Electrophoresis, 4,* 335–337 (1983).

72. P.G. Righetti, A. Morelli, C. Gelfi, R. Westermeier, *J. Biochem. Biophys. Meth., 13,* 151–159 (1986).

73. R. Charlionet, C. Morcamp, R. Sesboue, J.P. Martin, *J. Chromatogr., 205,* 355–366 (1981).

74. J. Brewer, E. Grund, P. Hagerlid, J. Lizana, in *Electrophoresis '86,* M.J. Dunn, ed., VCH Verlags-gesellschaft BRD, 1986, pp. 226–229.

75. I. Olsson, U.-B. Axiö-Fredriksson, M. Degerman, B. Olsson, *Electrophoresis, 9,* 16–22 (1988).

76. H. Rilbe, in *Electrofocusing and Isotachophoresis* B.J. Radola, D. Graesslin, eds., 1977, pp. 35–40.

77. R.Cartieer, A. Sassolas, *Electrophoresis, 13,* 252–257 (1992).

78. R. Bischoff, D. Roecklin, C. Roitsch, *Electrophoresis, 13,* 214–219 (1992).

79. Y. Ukishima, M. Kino, H. Kubota, S. Wada, S. Okada, *J. Assoc. Off. Anal. Chem., 74,* 6, 943–950 (1991).

80. G. Karssen, T.V. Hoenselaar, B. Verkerk-Bakker, R. Janssen, *Electrophoresis, 16,* 105–109 (1995).

81. L.-O. Hansson, H. Link, L. Sandlund, R. Einarsson, *Scand. J. Clin. Lab. Invest., 53,* 487–492 (1993).

14 Immunoelectrophoresis

JORGE A. LIZANA

Pharmacia Biotech AB
S-751 82 Uppsala, Sweden

Protein Purification: Principles, High-Resolution Methods, and Applications, Second Edition.
Edited by Jan-Christer Janson and Lars Rydén.
ISBN 0-471-18626-0. © 1998 Wiley-VCH, Inc.

Abbreviations

CA, cellulose acetate
Tricine, N-tris(hydroxymethyl)methylglycine
I, ionic strength
IE, immunoelectrophoresis
IEF, isoelectric focusing
Tris, 2-amino-2-(hydroxymethyl)propane-1,3-diol
CIE, crossed immunoelectrophoresis
CRIE, crossed radioimmunoelectrophoresis
SDS, sodium dodecyl sulfate
DOC, deoxycholate
PEG, polyethylene glycol 6000
FITC, fluorescein isothiocyanate
FPLC, fast protein liquid chromatography
RAST, radioimmunosorbent test
M_r, molecular weight
ELP, electrophoresis

14.1   INTRODUCTION

The use of methods based on the specific reaction between an antigen (protein) and its antibody is more than 80 years old. The formation of specific immunoprecipitates allows several visualization assays not possible with ordinary protein electrophoresis methods. Proteins that migrate very close to each other or even overlap in gel electrophoresis or isoelectric focusing will normally give different precipitation lines and thus can be distinguished from each other. The possible similarities or identities of one or several proteins (e.g., different forms of aggregates) can be discovered, as the immunoprecipitin lines will fuse or cross. The pioneer work of Oudin in the 1940s, extended and modified by Ouchterlony and Elek, set the basis for the use of methods employing free diffusion of the antigens and antibodies into an agar/agarose matrix. However, these immunodiffusion methods were very slow, prompting several groups of researchers to develop a new generation of assays using an electric field as the motive force, giving better separations and faster results.

The introduction of immunoelectrophoresis by Grabar and Williams[1] in 1953 was a great technical advance and opened the way for the study of complex mixtures of proteins (antigens) in biological fluids. In this technique, the proteins are first separated by electrophoresis in agar gels. Subsequently, the immunoprecipitin reaction is performed by diffusion of the antigens and antibodies on the same gel. Two serious disadvantages limit the value of this technique: the impairment of electrophoretic resolution due to the long immunodiffusion step (usually overnight migration) and the nonquantitative nature of the technique. Laurell and colleagues,[2] later developed several qualitative and quantitative methods based on the electrophoresis of proteins through agarose gels containing antibodies. Extension and modifications of these methods have resulted in other types of closely related immunoelectrophoretic techniques (Figure 14-1) that are being used by an increasing number of laboratories in many areas of the biomedical sciences. Most of these immunoelectrophoretic methods have been compiled in some very useful publications.[2–7]

Parallel to the development of immunoelectrophoresis, the associated instrumentation, such as the electrophoresis apparatus, power supplies, and related accessories, has also been developed and improved to give reliable long-term operation, proper cooling, high power output, and low running costs. Agarose gels and antibodies have also been improved. Agaroses with specifications tailored to the various immunoelectrophoretic methods are currently available, and a growing range of antibody preparations with different qualities are available from commercial sources.[8]

This chapter describes immunoelectrophoretic methods commonly used as tools to follow the progress of protein (antigens) purification schemes, to check the purity of the fractioned materials, and to quantitate specific proteins.

14.2 IMMUNOCHEMICAL PRINCIPLES

14.2.1 The Precipitin Reaction

The specific reaction between an antigen and its corresponding antibody, known as the precipitin reaction, constitutes the basis for the development of the array of methods available for researchers in biomedical sciences.

The addition of increasing amounts of an antigen to a fixed amount of antibody produces antigen–antibody complexes that grow in size until they form a precipitate. This region of increasing precipitation is known as the "antibody excess" zone. Above this zone, an equivalence between antigen and antibody concentrations is obtained, called "the equivalence zone," and the amount of precipitate shows no further increase. A further addition of antigen gives rise to a solubilization of the immunoprecipitates and formation of soluble complexes, the "antigen excess" zone.[5,9,10]

The precipitin reaction can be performed in free solution or in agarose gels where the antigen–antibody complexes can be seen as opaque precipitates. The precipitin reaction is temperature and pH dependent. Immunoprecipitation is usually performed at a pH between 7 and 9; the immunoprecipitates

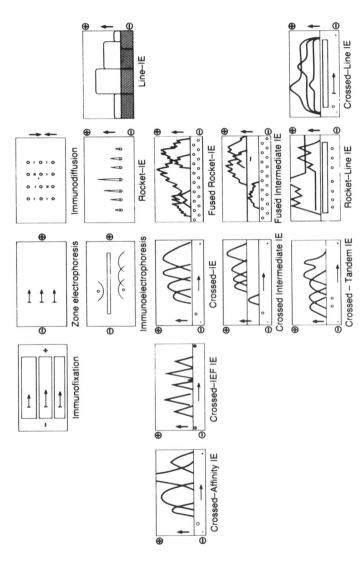

FIGURE 14-1. Gel arrangements of the various immunoelectrophoretic and immunodiffusion techniques. Zone electrophoresis can be combined with immunoprecipitation by placing a CA membrane containing specific antibodies onto the gel (immunofixation) or by diffusing the antibodies into the gel (immunofixation)—CIE. The resolution of CIE can be sped up by using electrophoresis in both dimensions—CIE. The resolution of CIE can be improved by the addition of carrier ampholytes (IEF), lectins, or affinity media into the first-dimension gel. The incorporation of an intermediate gel containing antibodies or affinity media increases the flexibility of CIE. Immunodiffusion can be sped up by electrophoresis of the samples into an antibody-containing gel: rocket IE. If a trench (line) is used instead of sample wells, larger samples can be applied—crossed line IE. When the sample wells are closely placed, and the samples are allowed to diffuse before electrophoresis, fused rockets are obtained. The incorporation of an intermediate gel with specific antibodies will trap the corresponding proteins and fused rockets will be obtained in the intermediate gel. The combination of line IE with rocket IE increases the discriminating power of rocket IE.

are soluble below pH 4.5 and above pH 10. Temperatures between 4 and 37°C can be used.[4]

14.2.2 Antigens

An antigen is any foreign substance that, when introduced parenterally into an animal, will elicit an immune response with the production of antibodies. The antibodies formed are generally found in plasma and should react specifically with the antigen used to induce its formation. Actually the whole antigen is not recognized as foreign or "nonself" but only parts of the antigen molecule. These parts are called "antigenic determinants" or epitopes and are defined as the elementary structure that can combine with only one antibody molecule (or one antibody site). Ovalbumin (M_r 43,000), when injected into a rabbit, has at least 10 antigenic determinants whereas thyroglobulin (M_r 669,000) has as many as 40.

Of great importance for a successful immune response is the accessibility of the antigenic determinants on the surface of the antigen molecule. This is closely related to the molecular size, complexity, and rigidity of the antigen.[5,9,10] Other important aspects in the elicitation of the immune response by the antigen are the dosage, immunization site, and use of adjuvants to increase the persistence of the antigen in the host.[4,9,11]

14.2.3 Antibodies

An antibody is a plasma protein belonging to the family immunoglobulins. All immunoglobulins, independently of their specificity, can be described on the basis of a common structure of four polypeptides: Two identical heavy chains and two identical light chains with constant and variable regions. In humans the immunoglobulins can be divided into five major classes with the indicated name of the heavy chain: IgG (γ), IgA (α), IgM (μ), IgD (δ), and IgE (ε). There are two types of light chains: κ and λ. Each immunoglobulin molecule contains two light chains. Individual molecules may possess either κ or λ chains, but never both. Each immunoglobulin molecule has two antigen-binding sites, or paratopes, which are identical. The most abundant immunoglobulin in serum is IgG; its concentration (~7 mg/mL) is 10 to several hundred times higher than the other immunoglobulins. IgG is the immunoglobulin of choice in immunoelectrophoresis.

It is now well established that the antigenic determinants of an antigen are recognized by B lymphocytes carrying specific receptors. This results in a division and differentiation process, and the lymphoid cells start to produce an antibody with specificity for the antigenic determinant recognized by the receptor-carrying B lymphocyte first encountered by the antigen. All antibodies produced by this cell line thus have the same chemical structure. This is called a monoclonal antibody.

When an antigen, which usually contains a number of antigenic determinants, is introduced into a host, the host responds with a number of antibodies against the various determinants. This is called a polyclonal antibody response. The antibodies made against one antigen are thus a population of immunoglobulins with the same specificity toward that antigen, "monospecific," but directed to the various antigenic determinants of that antigen. In addition, two antibodies binding to the same determinant may also have different affinities (i.e., they bind with different affinity constants). Monospecific, polyclonal antibodies with high affinity are preferred for the immunoelectrophoresis of specific proteins.

When two (or more) antigens are injected into an animal, "polyspecific" antibodies are obtained (i.e., a population of immunoglobulins composed of subpopulations with specificity for one antigen or the other). Again, within each subpopulation there are immunoglobulin molecules against the various determinants and with different affinities.

Polyclonal antibodies are required for immunoprecipitation as the antigen–antibody network can only be formed through the binding of several antibody molecules to different binding sites on the antigen. Such a three-dimensional complex is impossible between antigens and monoclonal antibodies unless the antigen itself is an aggregate of identical subunits.

The lymphoid cell lines producing antibodies specific for antigenic determinants (monoclonal antibodies) have a very short life when cultured *in vitro*. In 1975, Kohler and Milstein[12] reported that these cells can be immortalized by fusion with myeloma cell lines. The resulting hybrid cells, "hybridomas," can be cloned and maintained indefinitely. At any time hybridoma samples can be grown in culture or injected into animals for the large-scale production of monoclonal antibodies.[13]

The choice of polyclonal antibodies for the various immunochemical techniques is discussed in Section 14.4.3.

14.3 CHOICE OF IMMUNOELECTROPHORETIC METHOD

There is a growing trend to replace the traditional immunoelectrophoresis methods with methods based on the electrophoresis of proteins in agarose gels containing antibodies.[2] Figure 14-1 shows the gel arrangements used most often in immunoelectrophoretic techniques. For researchers engaged in protein fractionation and isolation, the choice of method is influenced by the capacity, ease of use, and speed, as well as by the availability of antibodies. When dealing with eluates from chromatographic separations, the first choice is usually a semiquantitative screening method, such as fused rocket IE. For further peak characterization, crossed IE or one of its modifications is often used. Quantitative determinations are best obtained by rocket IE.

14.3.1 Rocket Immunoelectrophoresis

Rocket IE, also called electroimmunoassay,[2] has become a very popular technique for the quantification of specific proteins and is widely used in routine analysis.[6]

A protein migrating electrophoretically through a gel containing specific antibodies to that protein will form antigen–antibody complexes. These complexes will migrate further into the gel until they reach the point of equivalence between the antigen and antibody concentration. Here the antigen–antibody complexes precipitate and no further migration occurs. The result is a rocket-shaped precipitate, the area of which is proportional to the concentration of antigen (protein) in the sample. It is important to run the experiment at a pH such that the antibodies are not migrating *too much* (see Section 14.4.1). Recommendations for improving the curve fitting in rocket IE[14] and a computer program to better describe the antigen concentration/peak height relationship have been proposed.[15] A very useful compilation of artifacts encountered in the practise of immunoelectrophoresis has been reported by Bjerrum.[16]

Rocket IE can be performed at high voltage (10 V/cm, "fast rockets") for about 2 to 4 h or at low voltage (2 V/cm, "slow rockets") for about 12 to 18 h. The choice is often a question of available time and sufficient cooling capacity. With the modern instrumentation available there is a trend to run fast rocket IE. Added benefits include reducing lateral diffusion and improving precision (Figure 14-2).

Proteins that migrate very slowly (e.g., IgG, IgA) form cigar-shaped anodic-cathodic precipitates around the application well, as the total area of these precipitates must be used for quantitation. In such cases the precision of the assay is impaired. Carbamylation of the samples lowers the pI of the protein and increases its anodic mobility. Another possibility is to carbamylate the antibodies[17] (see Section 14.4.7). True rockets are thus obtained and the precision of the assay is improved.

14.3.2 Fused Rocket IE

This technique is a modification of rocket IE and is a very useful tool to follow the purification of a protein during the various fractionation steps.[2–4] Peaks that appear to be symmetrical in a chromatogram are often shown to be composed of several components by this technique (Figure 14-3).

Samples of the fractions from column chromatography are applied in a row of closely placed wells and allowed to diffuse into the surrounding antibody-free agarose gel for 30 to 60 min. Thereafter samples are electrophoresed into an antibody-containing agarose gel. The result of the first diffusion step is that the rocket-shaped precipitates of the different samples, formed during the second dimension, are "fused" together, and elution profiles for each individual protein are thus obtained in a single run.[18]

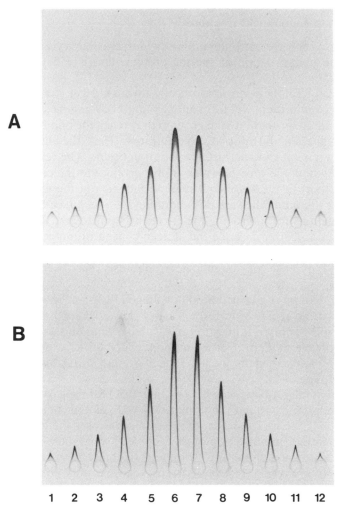

FIGURE 14-2. Rocket immunoelectrophoresis of human serum albumin. Agarose A (1%) gel containing antialbumin (7 μL/cm^2 of gel). Tris–barbital buffer (pH 8.6, I = 0.020). (A) Low-voltage rockets (2 V/cm) run for 18 h and (B) high-voltage rockets (10 V/cm) run for 3 h. Wells 1 to 6: albumin calibration solutions 5.6 to 120 mg/L; wells 7 to 12 are duplicates of wells 1 to 6. Sample volume: 5 μL.

In addition to its analytical discrimination, fused rocket IE also gives a semiquantitative estimation of all the proteins present in the different peaks, provided that there are enough high-titer antibodies against each of them. A modification called fused rocket IE with an intermediate gel[19] (Figure 14-1) solves the identification problem of the pattern obtained with common fused rockets, but is limited by the availability of monospecific antibodies (which are incorporated into the intermediate gel) to the protein under purification.

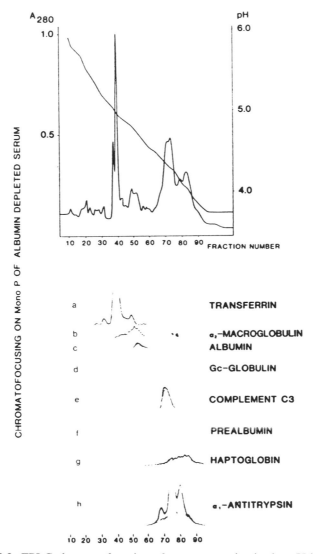

FIGURE 14-3. FPLC chromatofocusing of serum proteins in the pH interval 6.0 to 3.8, followed by fused rocket IE. Aliquots (6 μL) of the fractions were analyzed by fused rocket IE using monospecific antisera against transferrin (a), α_2-macroglobulin (b), albumin (c), Gc-globulin (d), complement C3 (e), prealbumin (f), haptoglobin (g), and α_1-antitrypsin (h). (Reproduced from Ref. 18 with permission of the authors and editors.)

14.3.3   Crossed IE

In crossed immunoelectrophoresis (CIE) the proteins are first separated by
zone electrophoresis into an agarose gel followed by a second electrophoretic
run, perpendicular to the first dimension, into an antibody-containing agarose
gel.[2] During the second dimension, antigen–antibody complexes for each
antigen–antibody system present are formed. Because of antigen excess, these
complexes migrate further into the antibody-containing gel until all the antigen
is used. Precipitation zones are thus formed (Figure 14-4). For a given concentration of an antibody component, the area of the individual precipitation
zone (or peak) is proportional to the concentration of the corresponding
antigen. CIE is therefore a powerful combination of qualitative electrophoresis
and semiquantitative immunoelectrophoresis.

With this technique it is possible to analyze complex mixtures of proteins
(e.g., proteins in human plasma and other body fluids). The resolution is very

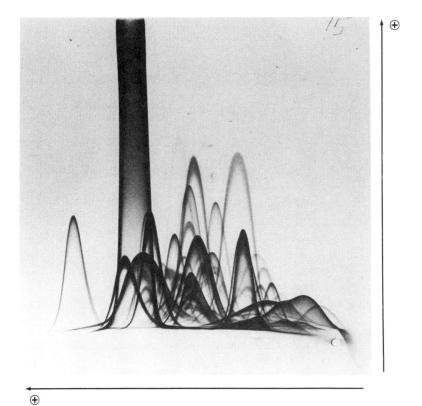

FIGURE 14-4. Crossed immunoelectrophoresis of human serum. First dimension:
Agarose A (1%) gel run for 90 min at 10 V/cm. Barbital/barbiturate buffer (pH 8.6,
I = 0.075). Sample: 3 μL of fresh serum. Second dimension: Agarose C (1%) containing
antibodies against whole human serum (9 μL/cm^2 of gel). Run overnight at 3 V/cm.
Tris–barbital, pH 8.6, I = 0.020.

high compared to traditional IE. The sensitivity is also high and can be increased by increasing the sample size to 20–30 μL in the first-dimension gel and by decreasing the amount of antiserum used in the second-dimension gel.[4] The information obtained is mainly semiquantitative, but the Clarke and Freeman modification[20] should give quantitative results. CIE is very common in biomedical research, but is too complicated for routine diagnosis as it requires experience in the evaluation of the results. In some areas it is being replaced by immunofixation[21] or immunoblotting (see Chapter 16). Other areas of use are the estimation of subpopulations of heterogeneous antigenic compounds (e.g., demonstration of polyclonal, oligoclonal, and monoclonal IgG mixtures), studies of association and dissociation phenomena (e.g., complex formation between protein A and IgG), studies of genetic variants of proteins; and classification of antigens as immunologically identical (sharing epitopes), nonidentical, or partially identical.[4,9,22]

A number of modifications of this method and hybrids with other techniques have been developed. These include CIE with intermediate gel, crossed IEF IE, crossed affinity IE, crossed line IE and tandem CIE (Figure 14-1).[2,4,9] Some of these techniques are described in the following sections.

14.3.4   Crossed Isoelectric Focusing IE

In crossed IEF IE the proteins are first separated by isoelectric focusing in polyacrylamide or agarose IEF gels containing carrier ampholytes, followed by electrophoresis in antibody-containing agarose gels.[4,22,23] The resolution obtained by IEF is much superior to that obtained with agarose zone electrophoresis and, similarly, higher resolution can be expected with crossed IEF IE (compare Figures 14-4 and 14-5).

A wider acceptance of the technique has been delayed by technical problems caused by the electroosmosis of the different supporting media in the IEF (polyacrylamide) and electrophoresis (agarose) steps. The "laying-on" technique[22] was proposed to overcome such problems and involves simply applying a lane of the IEF gel, containing the focused proteins, on top of the second antibody–agarose gel. With the introduction of agarose IEF, a charge-balanced agarose specially designed for isoelectric focusing, agarose can now be used as the supporting gel in both dimensions, eliminating the gel junction problems caused by electroosmosis.[23-25]

The technique is very straightforward. IEF gels should be cast on plastic films. After running IEF for 1 to 2 h at 15 W, 1500 V (2,000 to 3,000 Vh), the portions of the IEF gel that are under the electrode strips are removed as they contain very acid or basic electrode solutions and will produce disturbances in the second-dimension antibody-containing agarose gel. A lane containing the focused proteins is excised and placed on the second-dimension gel. Tris–barbital buffer, pH 8.6 (I = 0.050), is recommended for the second dimension. The high ionic strength is needed to buffer the carrier ampholytes contained in the lane of the IEF gel. Fast electrophoresis at 10 V/cm for 3 to 4 h is recommended if the resolution of the sharply focused protein band is to be retained during the second-dimension run.[23] A typical crossed IEF IE experiment using agarose IEF is shown in Figure 14-5.

Crossed IEF IE is very well suited for checking the purity of isolated proteins and for studying hereditary polymorphism, microheterogeneity, fragmentation of proteins and demonstration of abnormal fractions.[22-27]

14.3.5 Crossed Affinity IE

The combination of affinity chromatography and immunoelectrophoresis constitutes a general technique for the study of ligand–protein interactions during electrophoresis.[28,29] Successful applications of this principle include the use of free and immobilized lectins to study glycoproteins[30,31] and hydrophobic interaction media combined with CIE for the study of amphiphilic proteins.[32]

Characterization and purification by interaction with lectins are common procedures for many glycoproteins. Two general approaches can be used to study lectin–glycoprotein interactions during electrophoresis: (1) The free or immobilized lectin can be included in the first-dimension gel of CIE or

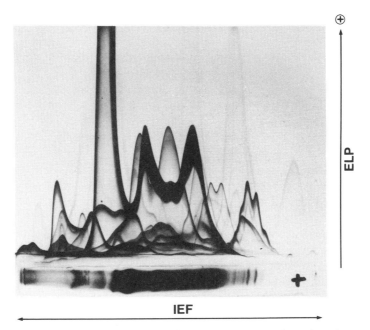

FIGURE 14-5. Crossed Isoelectric Focusing immunoelectrophoresis of human serum. First dimension: Agarose IEF (1%) gel containing Pharmalyte pH 2.5–5 + 4–6.5 (1:1), run at 15 W, 1500 V for 2,500 Vh at 10°C. Sample: 3 μL of fresh serum. Second dimension: Agarose C (1%) containing antibodies against whole human serum (9 μL/cm²) run for 4 h at 10 V/cm. Tris–barbital buffer, pH 8.6, I = 0.050. (Reproduced from Ref. 23 with the permission of the authors and editors.)

(2) the free or immobilized lectin can be included in an intermediate gel between the first and the second dimension of CIE.[28-31] The presence of a lectin (free or immobilized) in the first-dimension gel or in the intermediate gel does not inhibit or disturb the formation of immunoprecipitates of other proteins present in the sample.

Concanavalin A (Con A) has been the lectin most widely used, although other lectins (e.g., lentil lectin, wheat germ lectin, soybean lectin, *Ulex europeus* lectin, and peanut agglutinin) have also been tried.[28,29] Finding a minor form of α-fetoprotein that does not react with Con A and may be helpful in the laboratory diagnosis of neutral tube defects has increased the interest in this technique in clinical situations.[32,61,62]

When lectins are included in the first-dimension gel of crossed immunoelectrophoresis, the biospecific interactions between the lectin and the carbohydrate moiety of the glycoproteins present in the sample produce a retardation, partial or total, of the electrophoretic mobility of the glycoproteins. The second-dimension run into an agarose gel containing antibodies will reveal the heterogeneity of the carbohydrate part by the appearance of several peaks for a single protein. Control runs without the lectin in the first-dimension gel are required for comparison of the electrophoretic pattern.

Introduction of the lectin in an intermediate gel, after the sample proteins have been separated in the first-dimension agarose gel, produces disappearance, diminishing, or even splitting of immunoprecipitates in the second-dimension gel. Again, control runs without the lectin are required for comparison of the patterns.[28,30,31]

Crossed hydrophobic interaction immunoelectrophoresis is a related technique in which Phenyl- or Octyl-Sepharose is incorporated into the first-dimension gel or in an intermediate gel. The amphiphilic proteins present in the sample are trapped or retarded by the gel media, which is seen by the shape and characteristics of the immunoprecipitates formed in the second dimension.[4,32]

14.3.6 Crossed Radioimmunoelectrophoresis

Crossed radioimmunoelectrophoresis permits direct identification of individual allergens in complex allergen extracts without previous separation.[7] In this respect the technique is unique. CRIE is also a very valuable tool for the control of allergen extracts. The technique is based on binding of specific IgE to immunoprecipitates in a crossed immunoelectrophoresis plate after incubation with serum samples from patients with an IgE-mediated allergy. The IgE bound to the allergen is demonstrated using anti-IgE labeled with ^{125}I and subsequent autoradiography.[34] The high resolving power of the CRIE method is useful for establishing the allergenic composition of various allergen extracts. Furthermore, the amount of specific IgE binding to the precipitates may be semiquantified using a CRIE reference system.[35] The often used CRIE method, however, requires a very long exposure time.

A procedure for performing CRIE with significantly reduced exposure times has been developed.[36] The rapid method has been affected by using a highly sensitive X-ray film and X-ray intensifying screen and by exposing at a low temperature ($-70°C$) in a Kodak X-Omatic cassette. Exposure at $-70°C$ appears to be critical for obtaining optimal radiostaining. Furthermore, the isotope concentration has been doubled without any loss of specificity or reproducibility. The rapid procedure makes it possible to obtain a CRIE screening within a day and a complete CRIE classification in 5 days. A CRIE experiment of *Aspergillus fumigatus* antigens[33] is shown in Figure 14-6.

14.4 GENERAL TECHNIQUES

14.4.1 Choice of Buffer

Although an optimal buffer can be found for each system, most workers use buffers based on barbituric acid (barbital) because the resolution obtained is sufficient for most of the immunoelectrophoretic techniques.

To make the Tris–barbital buffer, pH 8.6, ionic strength 0.100, add 224 g 5,5-diethylbarbituric acid, 443 g Tris, 20 g calcium lactate, 10 g sodium azide and distilled water to 10 L. This stock solution can be stored at 4°C for several weeks.

The pH of the buffer should be above the pI of the proteins under study. At pH 8.6 most of the serum proteins have a net negative charge and migrate toward the anode. The antibodies should not migrate *too much* during the electrophoretic run (i.e., the pH of the buffer should be similar to their pI).

The majority of the immunoelectrophoretic techniques use an ionic strength of 0.020. Dilute 200 ml of the stock solution to 1 L with distilled water. If other ionic strengths are required, dilute the stock solution accordingly. High ionic strengths give sharper bands and improved resolution, but a lot of heat develops, making it necessary to use an apparatus with a high cooling capacity to avoid drying/burning of the gel. The aforementioned buffer is not recommended for agarose zone electrophoresis of plasma proteins. The barbital/Na-barbital buffer (pH 8.6) of ionic strength 0.075, as suggested by Jeppsson and colleagues,[37] should be used as it gives sharper bands and better resolution. Barbiturate has potential as a drug of abuse, and government regulations have made the handling of this substance difficult. A nonbarbital buffer using Tris–tricine has been proposed and should give similar results to Tris–barbiturate.[38]

To make the Tris–tricine buffer, pH 8.6, ionic strength 0.100, add 226 g tricine, 443 g Tris, 20 g calcium lactate, 10 g sodium azide, and distilled water to 10 L. This stock solution can be stored at 4°C for several weeks. To obtain an ionic strength of 0.020, dilute accordingly with distilled water.

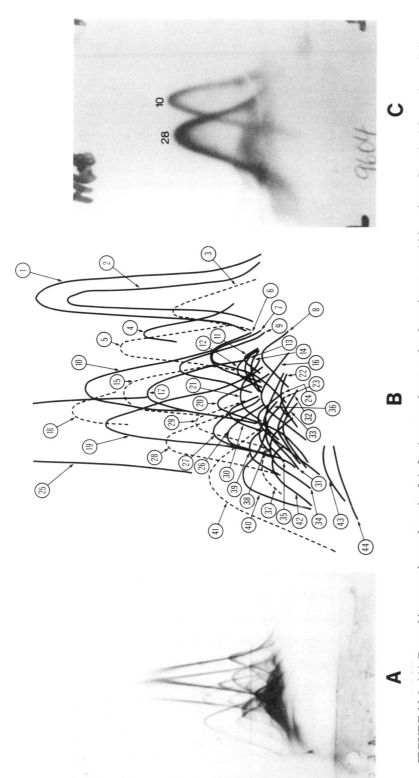

FIGURE 14-6. (A) Crossed immunoelectrophoresis of *A. fumigatus* antigens against hyperimmune rabbit antisera. Sample: 0.2 mg of mycelial extract. First dimension: Agarose A (1%) gel run for 35 min at 10 V/cm (15°C). Second dimension: Agarose A (1%) containing rabbit antiserum (10 µL/cm²) run at 2V/cm for 18 h. (B) Schematic drawing of the immunoprecipitates arbitrarily numbered 1–44 from the anode: ——— denotes sharp precipitates seen after CBB staining and -------- denotes precipitates only visible after radiostaining. (C) X-ray film from CRIE showing an allergenic patient serum containing high amounts of immunoprecipitates 10 and 28. Immunoprecipitates 24, 30, and 40 are also present. (Reproduced from Ref. 33 with the kind permissions of the authors and editors.)

543

14.4.2 Choice of Gel Matrix

Immunoelectrophoresis can be performed in polyacrylamide, starch, cellulose acetate, fibrin/agar, and agar/agarose gels (reviewed in Ref. 9). Agarose gel and cellulose acetate membranes are the two matrices most used. Cellulose membranes are mainly used in immunofixation following zone electrophoresis or isoelectric focusing for the identifying and phenotyping of proteins.[21,39,40] Over the years agarose has emerged as the matrix of choice for immunoelectrophoretic techniques. In the majority of applications it is important to use an agarose that has a low ionized group content in order to maintain a suitable degree of electroosmosis (see Chapter 12) and to avoid any risk of adsorption of materials by the gel.

For most electrophoretic techniques, agarose with an electroosmosis of 0.13 (relative mobility, M_r −0.13) should give good results. Electroosmosis can even be beneficial as a mechanism that may increase the separation between particular proteins during electrophoresis and is essential for successful counterflow immunoelectrophoresis (see Ref. 9 for review). Recommendations concerning which type of agarose to use for the various immunoelectrophoretic techniques are given in Table 14-1. Of great importance is the batch-to-batch reproducibility of the electroosmosis of the selected agarose in order to avoid shifts in the migration of the proteins and confusion in the interpretation of the results.

TABLE 14-1. Selection of Agarose for Different Techniques

Technique	Agarose Type[a]		
	A	B	C
Zone electrophoresis	●	o	o
Counter immunoelectrophoresis		●	
Zone immunoelectrophoresis	●		
Immunoelectrophoresis	●	●	
Rocket immunoelectrophoresis	●	o	o
Fused rockets	●	o	o
Line immunoelectrophoresis	●	o	
Crossed immunoelectrophoresis	●		●
Crossed electrofocusing IE	.o		●
Crossed affinity IE	●		o
Double immunodiffusion	●	o	
Radial immunodiffusion	●	o	

[a] ●, recommended type; o, used by some authors in modified procedures. Agarose A, medium electroosmosis ($M_r = −0.13 ± 0.01$); agarose B, high electroosmosis ($M_r = −0.25 ± 0.01$); agarose C, low electroosmosis ($M_r = −0.02 ± 0.01$); and agarose IEF, charge balanced for isoelectric focusing. These agaroses are from Pharmacia Biotech AB, Uppsala, Sweden. Other suppliers offer similar agaroses (e.g., FMC Bioproducts, Portland, Maine, USA).

Agarose will form a gel in a 0.1% solution, but concentrations of around 1% are commonly used to give a much stronger gel that is easier to handle. A 1% gel has a molecular exclusion limit for spherical particles well above 50 million M_r and allows free diffusion of antigens, antibodies, and immune complexes. A 1% solution of agarose A in Tris–barbital buffer, pH 8.6, ionic strength 0.020, is prepared as follows: Weigh out the appropriate amount of agarose and dissolve in the buffer by gentle heating. Ensure that all the agarose has dissolved by boiling the solution a few moments. A 1% solution can be stored at 4°C for several weeks and melted repeatedly by heat. If antibodies are to be included in the gel, cool the solution to about 55 to 57°C, add the antisera, mix, and pour on the plate. If lectins or other proteins are to be included, proceed as with antibodies. Agarose gels can be cast on glass plates or on the hydrophilic side of plastic polyester sheets (GelBond™). The agarose solution can be poured directly onto a plate fitted with a gel casting frame (Figure 14-7). Casting between glass plates with a U-shaped frame is done only when an absolutely even gel thickness is desired.

The sizes vary greatly depending on the particular technique and on different authors, but the trend is to use small gels with a recommended thickness of 1 mm. Thicker gels will tend to develop a temperature gradient during the running and give skewed bands.

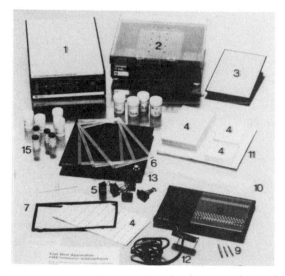

FIGURE 14-7. Instruments and accessories for immunoelectrophoresis. (1) Power supply, (2) electrophoresis apparatus, (3) cooling plate with printed pattern for easy centering of gels, (4) glass plates and plastic film, (5) leveling table and clamps, (6) gel casting frame, (7) gasket for casting gels between glasses, (8) punching template and holder, (9) punches, (10) sample application foil for agarose electrophoresis, (11) electrode wicks, (12) voltage probe, (13) spirit level, (14) agaroses, and (15) antibodies.

14.4.3 Choice of Antibodies

Although whole unfractionated antisera can always be used in immunoelectro-phoretic techniques, these preparations have many disadvantages: the high level of non-antibody proteins present in a crude preparation is such that a longer washing time will be required to obtain a clear background. Crude preparations usually become turbid on storage due to the precipitation of denatured lipoproteins. They also contain excess antigen and soluble com-plexes of antibody and antigen used to absorb the antiserum. In addition, there is always the uncertainty of an interaction(s) between the non-antibody proteins of the antiserum and proteins in the sample, which can cause artifacts.

It is recommended to use antibody preparations that contain only the immunoglobulin fraction as they are more stable than crude preparations and the titer is better controlled. In addition, the protein content is much lower and the tedious washing steps can be considerably shortened or even elimi-nated. Crude antisera preparations can be fractionated by salting out with ammonium sulfate (250 g/L rabbit serum), buffer exchange with a small column of Sephadex G-25 (PD-10), and chromatography on DEAE Sephadex A-50 or DEAE Sepharose Fast Flow at pH 5.0 in sodium acetate/acetic acid buffer (I = 0.05). After another buffer exchange with 0.1 M NaCl/15 mM sodium azide, the γ-globulin fraction obtained can be stored for years at 4°C. The loss in activity is around 5% per year. Storage at -20°C reduces the loss in activity even further.[11]

In most instances, the choice of species in which antibodies are raised may reasonably be made on the basis of what is available and the volume of antiserum required. Rabbits have been shown to produce antibodies with optimal precipitating activity and are widely used in immunoelectrophoretic techniques. Rabbits are easy to handle, and by pooling the serum from several animals the individual immune reaction, which can give atypical antibodies, is eliminated. Goat, sheep, and swine antisera are also equally useful. An extensive list of suppliers of crude and fractionated antibody preparations can be found in Ref. 8.

14.4.4 Choice of Electrophoretic Conditions

A suitable DC power supply for immunoelectrophoresis should deliver stabi-lized voltage up to 500 V (400 mA). Most of the experiments are run at constant voltage. Slow electrophoresis at 2 to 5 V/cm requires 80 to 130 V and an overnight run. Fast electrophoresis at 10 to 20 V/cm requires 350 to 400 V and 2 to 3 h to run; a good cooling capacity is critical. Ultrarapid electrophoresis with potential gradients of up to 62 V/cm and running times down to 9 to 25 min have been reported (see Chapter 23 in Ref. 4).

The electrophoresis apparatus (Figure 14-7) consists of a box with two removable buffer vessels and a cooling plate. Each buffer vessel contains a built-in electrode. A printed pattern on the cooling plate facilitates the centering of the gel plates and application of the samples. A black background allows easy visualization of the white immunoprecipitates.

For most applications, tap water (10 to 16°C) circulating at a flow rate of 1.5 to 2 L/min is sufficient. Ethanol or ethylene glycol (20%, v/v) solutions may also be used to cool the system. The recommended maximum flow of coolant is 4 L/min. Too intensive cooling may cause water condensation on the gel surface and flooding of the immunoprecipitates. Insufficient cooling will increase evaporation from the gel and result in gel drying and/or burning. The lid seals the apparatus from the surrounding air and protects the operator from any electric hazard. The pairs of holes on either side of the center line of the lid are for insertion of the voltage probe for measuring the field strength (V/cm) across the gel during the course of an experiment.

The best electrophoresis wicks are those made of highly absorbent paper protected on one side by a film of polyethylene, which prevents the evaporation of buffer from the wicks. The wicks should be cut to the exact width of the gel. The plastic surface should be uppermost. If wicks with plastic film are not available, two layers of Whatman No. 3 filter paper (or one layer of Whatman No. 17) can be used. Ensure that there is good contact between the wicks and the gel, as artifacts due to uneven distribution of the electric field may be obtained.

The necessary instrumentation for running immunoelectrophoresis and accessories for gel casting and punching are shown in Figure 14-7.

14.4.5 Staining and Destaining

14.4.5.1 Coomassie Brilliant Blue After the electrophoretic step the immunoelectrophoresis plates do not need fixation as the antigens are already fixed in the immunoprecipitates. The plates are covered with several layers of filter paper and a clean glass plate to which a 1-kg weight is placed. Press for 15 min. Carefully remove the wet papers. The plate is now reduced to a thin film that is easy to handle and allows easy washing, staining, and destaining.

Place the thin film in a tray containing NaCl (9 g/L) to wash away unreacted proteins (from sample and antiserum). Depending on the purity of the antiserum used, the washing time can be shortened from overnight to less than 2 h. Rinse the plate with distilled water (30 sec). Press again for 15 min and dry the gel in hot air.

Staining solutions consist of a stock solution of 5 g of Coomassie brilliant blue R-250 (CBB) or Page Blue 83 in 1 L of 96% ethanol. The alcoholic stock solution is stable for weeks. Add an equal volume of 20% acetic acid to an aliquot of this solution immediately before use. Filtering is not necessary. The alcoholic stain with acetic acid results in precipitation of the dye and should be discarded after a few days of use. Staining of the immunoprecipitates is completed in 15 to 30 min. Staining for longer times (overnight) will not increase the uptake of the dye by the immunoprecipitates, it will only increase the background staining of the gel.[42] CBB is three to five times more sensitive than amido black 10 B and should be the dye of choice for immunoprecipitates.

After staining, the immunoplates are rinsed in water (10 to 20 sec) to remove excess dye and placed in a tray containing the destaining solution: 450 mL ethanol (96%), 100 mL acetic acid, and 450 mL distilled water. Destain until a clear background is obtained (10 to 15 min). The destaining solution can be reused after filtering through activated charcoal.

Staining and destaining can be accelerated by a factor of two or more if all the steps are carried out at 40 to 50°C. Working under a fume hood is recommended.

14.4.5.2 Silver Staining Various authors have described silver staining of proteins and immunoprecipitates in agarose gels.[42–45] In general, all these silver staining methods are 20 to 40 times more sensitive than those using CBB and are therefore very useful for the detection of small amounts of proteins. For instance, McLachlan and Burns[44] have reported a sensitivity of 0.9 ng/mm^2 in human IgG–anti-IgG precipitates, about 20 times higher than immunoprecipitates stained with CBB.

After electrophoresis the immunoplate is washed with saline and twice with distilled water for 5 min. This is then dried to a thin film as described in Section 14.4.5.1. Staining is carried out as follows: Prepare two solutions. Solution A contains 5 g of Na_2CO_3 in 100 mL of distilled water. Solution B contains 0.4 g NH_4NO_3, 0.4 g $AgNO_3$, 1.0 g silicotungstic acid, and 2.8 mL of a 370-g/L solution of formaldehyde in 200 ml of distilled water. Add solution B to solution A with constant stirring in a volume ratio of 2:1, just before staining and pour the mixture into the staining tray containing the dried immunoplate. Agitate the tray. When the color of the mixture changes from white to gray after about 3 min, remove the gel and immerse it in water. Meanwhile, freshly prepare another mixture of stain and again stain the gel with it. When the gel pattern shows the desired intensity (usually after 2 to 3 min), stop the reaction by immersing the gel in a 5% acetic acid solution for 15 min. Finally, rinse the gel in water for 5 min and dry it.

14.4.6 Enhancing the Sensitivity of Immunoelectrophoresis

14.4.6.1 Nonionic Polymer Solutions Neutral polymers (dextrans, polyethylene glycols, polyvinyl pyrrolidones) have been shown to enhance the antigen–antibody reaction. This enhancement has been explained in terms of a steric exclusion of the antigen–antibody complexes from the domain of the polysaccharide (for review see Ref. 46). The use of this effect in immunoelectrophoresis was first reported in 1968.[47] Dextran T-10 and PEG, in concentrations of 2 to 4%, are at present the most used polymers for enhancement of the immune reaction.

The polymers are usually included in the gel by addition to the agarose solution before casting. Immunoelectrophoresis is then performed as usual.

The result is a dramatic sharpening of the immunoprecipitin lines; weak immunoprecipitates become more apparent. An alternative immersion technique has also been reported.[48] The agarose plate with the immunoprecipitates is immersed in 4% Dextran T-40 or PEG 6000 solutions for 60 min. Thereafter the gels are washed, pressed, and stained as usual. Somewhat better results are obtained with the inclusion technique as compared to the immersion technique.

14.4.6.2 Enzyme-Labeled Antibodies

The visualization of weak immunoprecipitates can be enhanced with enzyme-labeled antibodies. The increase in sensitivity is about 50 times compared to Coomassie Brilliant Blue (CBB) staining.

After electrophoresis the immunoplates are incubated with a second antibody to which an enzyme has been coupled. The second antibody is specific for the antibody fixed on the immunoprecipitate. Development of the enzyme color reaction will yield stained immunoprecipitates.[49,50] Do not use labeled antibodies as the first antibody, as the introduction of an enzyme or any other group to the antibody molecule will cause a shift in the isoelectric point. New pH conditions must then be found to prevent the migration of the labeled antibody contained in the agarose plates.

Several enzymes can be used to label the antibody: Peroxidase, β-galactosidase, glucose oxidase, alkaline phosphatase, and so on. The detailed methodology of coupling enzymes to antibodies has been reviewed by Avrameas and colleagues.[51] For labeling, either the purified IgG fraction or the affinity-purified antibody can be used. Better results were obtained when labeled Fab antibody fragments, rather than the whole antibody molecule, were used.[51] An extensive list of suppliers of enzyme-labeled antibodies can be found in Ref. 8.

14.4.6.3 Fluorescent-Labeled Antibodies

A fluorescent label can be used in the same way as an enzyme label for the amplification of immunoprecipitates (see earlier discussion). In this case the immunoprecipitates are visualized by UV illumination. The most frequently used fluorochrome is fluorescein isothiocyanate. Conjugation of proteins by FITC is a well-standardized technique that does not require too much bench work.[22] Low FITC concentrations (around 10 μg/mg of protein) are used with antibodies and fragments of antibodies to avoid hyperconjugation and loss of antibody activity. Conjugation is carried out at pH 9 for 18 h. Unreacted FITC is separated from FITC antibodies by gel filtration on Sephadex G-25. Conjugated and unconjugated antibody molecules can be further separated by ion exchange chromatography.[9] A list of suppliers of fluorescent antibodies can be found in Ref. 8.

An alternative is to incubate the immunoprecipitates with FITC-protein A. Protein A will bind to the F_c region of the IgG in the immunoprecipitate.[52]

14.4.6.4 Radiolabeled Antibodies

The immunoprecipitates can be made visible by radiolabeling the antigen or antibody. Radioiodine (^{125}I) is the most frequently used tracer to label antibodies; the procedure is well known and standardized.[22] The gel arrangements for radioimmunoelectrophoresis are similar to those in Figure 14-1. After electrophoresis, the plates are washed, dried, and subjected to autoradiography. In one application the detection limit for α-fetoprotein, as measured by rocket radio IE, was as low as 0.3 μg, an increase in sensitivity of 50 to 100 times as compared to CBB staining.[53]

Another approach is to incubate the "cold" immunoprecipitates with a second radiolabeled antibody. The main application of this approach is CRIE for identifying major and minor allergens by reaction of the immunoprecipitate with a patient's specific IgE followed by reaction with ^{125}I-anti-IgE and autoradiography (for review see Ref. 7).

14.4.7   Carbamylation of Proteins

Carbamylation is used in immunoelectrophoresis to increase the mobility of proteins that migrate very slowly (see Section 14.3.1).

Mix equal volumes of the protein sample (antigens) and standards with freshly prepared KCNO (2 M). Incubate the mixture at 45°C for 30 min (alternatively, leave overnight at room temperature). Cool to 10 to 15°C in a water bath. Apply the samples and run the rocket IE as usual. Carbamylation of the *antibodies* contained in the agarose antibody plate can be performed in the same way as described earlier. In this case the electrophoresis must be performed at pH 5.0 instead of pH 8.6. The recommended buffer is 0.01 M TEMED and 0.02 M acetic acid (pH 5.0). The carbamylated antibodies will not migrate during electrophoresis because carbamylation has lowered their pIs to about 5. The sample antigens will migrate toward the cathode (the reverse side compared to using pH 8.6).

14.4.8   Use of Detergents in Immunoelectrophoresis

Nonionic detergents solubilize the proteins, usually without denaturation. This means that the antigenic determinants of the protein remain available for interaction with the antibodies. A number of nonionic detergents (Table 14-2), at concentrations of 0.05 to 1.0% (w/v), have been shown to have no effect on the antigen–antibody reaction.[54–56] The detergent is used at a concentration below the critical micellar concentration and is incorporated into the agarose gel before casting. Thereafter, IE is carried out as normal. It is not necessary to add the nonionic detergent to the buffer.

Ionic detergents (anionic and cationic) unfold the tertiary structure of proteins and very often denature the protein they solubilize. Therefore, it is not surprising that in most cases the antigen–antibody reaction is inhibited when these detergents are present in the medium[55,57] (Table 14-2). However, in some cases the direct immunochemical identification of proteins is possible in gels containing SDS.[10,58]

TABLE 14-2. Effect of Detergents on the Immune Reaction

Detergent Type	Concentration (%, w/v)	Inhibition (Extent %)
Ionic		
SDS	0.05	Yes (60%)
	0.10	Yes (80%)
	0.20	Yes (90%)
DOC	0.10	Yes (60%)
	0.25	Yes (80%)
Nonionic		
Triton X-100	0.05–1.0	No
Nonidet NP-40		No
Tween 20		No
Tween 80		No
Brij 58		No
Berol EMU-043		No
Lubrol WX		No

Data from Ref. 52.

If there is a combination of SDS or DOC and another nonionic detergent in the reaction medium, the inhibition by SDS or DOC is reduced. Thus, membrane proteins can be solubilized in SDS or DOC and run in agarose gels containing nonionic detergents.

Sodium dodecyl sulfate is often used to solubilize membrane proteins, and its presence during electrophoresis is necessary to avoid aggregation of the solubilized proteins. However, when SDS is present in concentrations above 0.05% (w/v), it binds antigens or antibodies and charges their electrophoretic mobility, spoiling the resolution of the separation. In addition, the antigen–antibody reaction is inhibited. Distorted immunoprecipitates can be obtained.[55,59] If there is a mixture of SDS and a nonionic detergent in the reaction medium, the inhibition of the antigen–antibody reaction by SDS is reduced. This is possibly due to the formation of mixed micelles between SDS and the nonionic detergent resulting in a decrease of the free SDS monomers and/or to a competitive binding of SDS and the nonionic detergent for the proteins.[55,56] Anderson and Thorpe[58] have reviewed the methods available for removing SDS from polyacrylamide gels prior to immunochemical analysis.

Zwitterionic detergents (e.g., Empigen BB and the sulfobetaine series) can be used directly in connection with immunochemical analysis as they do not interfere with the immune reactions.[10,59]

Chaotrophic agents such as urea can be used to solubilize membrane antigens. Urea inhibits the immune reaction.[15] However, concentrations of 1.5 M in conjunction with nonionic detergents have been used successfully in immunoelectrophoresis.[54]

14.5 APPLICATIONS

14.5.1 FPLC™ Chromatofocusing Fractionation of Human Serum Proteins Assessed by Fused Rocket IE

Human serum (1 mL) was first depleted of albumin by affinity chromatography on a column (115 × 10 mm) packed with Blue Sepharose CL-6B and equilibrated in 50 mM Tris-HCl with 100 mM NaCl, pH 7.0 (Figure 14-3, top). The albumin-depleted material was then equilibrated in the chromatofocusing start buffer (bis/Tris 25 mM, pH 6.3) and applied to a Mono P column mounted in an FPLC system. The eluent was polybuffer 74 (diluted 1 : 10 in HCl), pH 4.0. The flow rate 1 mL/min. Fractions of 0.4 mL were collected. Aliquots (6 μL) of the fractions were analyzed by fused rocket IE (Figure 14-3, bottom). Gels (120 × 240 × 1 mm) were cast with 26 mL of 1% agarose A in Tris–barbital buffer (pH 8.6, I = 0.020) and 300 μL of the monospecific antiserum toward (1) transferrin, (2) α_2-macroglobulin, (3) albumin, (4) Gc-globulin, (5) complement C3, (6) prealbumin, (7) haptoglobin, and (8) α_1-antitrypsin. In each gel, 80 samples were run (fraction Nos. 10 to 90). The gels were run at 5 V/cm overnight and later dried, stained, and destained as described in Section 14.4.5.1. Fractions with the selected proteins were pooled and submitted to further fractionation by chromatofocusing in other pH ranges.[18]

14.5.2 Microheterogeneity of α_2-Macroglobulin Studied by Crossed Electrofocusing IE

Serum α_2-macroglobulin (M_r 820,000) is an acute-phase reactant and shows proteinase inhibitor activity that controls the clotting and fibrinolytic system. Evidence also exists that this protein shows genetic polymorphism.[60] The use of crossed IEF IE allows the microheterogeneity of this serum protein to be studied (Figure 14-8).

An aliquot of fresh human serum (10 μL) was run on an agarose IEF gel containing Pharmalyte pH 4–6.5 and 2.5–5 (1 : 1 ratio) for 2500 Vh. IEF was run at constant power (15 W) with a maximum set voltage and a current of 1500 V and 25 mA. Water cooled to 10°C was circulated through the apparatus. After IEF, a lane containing the focused proteins was cut out and placed on top of a second-dimension gel (1% agarose C) containing anti-α_2-macroglobulin (2 μL/cm^2). Electrophoresis was run for 4 h at 10 V/cm in Tris–barbital buffer, pH 8.6, I = 0.05. Four components with pIs between 4.9 and 5.3 were found. This approach can be further extended to study the variations and interrelations of these components of α_2-macroglobulin in normal subjects and patients with various diseases.

14.5.3 Antigenic Variability of Different Strains of *Aspergillus fumigatus* Studied by Crossed IE and Crossed Radioimmunoelectrophoresis

The antigenic and allergenic activity of different strains of fungi *A. fumigatus* can be examined by a combination of CIE, CRIE, and RAST inhibition

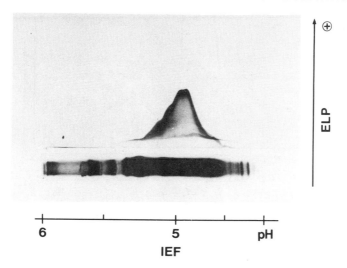

FIGURE 14-8. Crossed IEF IE of α_2-macroglobulin. First dimension: Agarose IEF (1%) gel containing Pharmalyte pH 2.5–5 + 4–6.5 (1:1) run at 15 W, 1,500 V for 2,500 Vh (10°C). A lane (5 mm width) containing the focused proteins was cut out and laid on the second-dimension gel. Sample: 10 μL of fresh serum. Second dimensions: Agarose C (1%) containing anti-α_2-macroglobulin (90 μL/cm^2 of gel) run as described in Figure 14-5. (Reproduced from Ref. 23 with permission of the authors and editors.)

(Figure 14-6). CIE was performed as described in Section 14.3.3. Samples of 10 μL of crude *A. fumigatus* extract (20 mg dry weight/mL) were run in the first dimension at 10 V/cm for 35 min. The second-dimension electrophoresis was performed at 2 V/cm for 18 h on an agarose A (1%) gel containing rabbit antiserum against *A. fumigatus* (10 μL/cm^2). The immunoplate was then stained with CBB (see Section 14.4.5.1). Radiostaining was performed by incubation of the immunoplate with the patient's serum for 18 h at 22°C followed by incubation with ^{125}I-labeled anti-IgE solution (72,000 cpm/mL) for 18 h at 22°C. After washing and drying, the plate was placed on X-ray photographic film for rapid autoradiographic exposure in a cassette at low temperature (−70°C) as described in Ref. 36.

Forty-four different antigens were demonstrated in mycelial extracts of *A. fumigatus* by crossed IE stained with CBB and radiostaining (Figures 14-6A and 14-6B, respectively, see p. 543). Eighteen different allergens were identified using CRIE: 2 major allergents (antigens No. 10 and 40 in Figure 14-6B), 10 intermediate allergens, and 6 minor allergens.[33] In the next step, individual serum samples of patients with fungal allergy were screened for reactivity to *A. fumigatus* antigens. In Figure 14-6C the X-ray film from a CRIE experiment shows a patient with high levels of IgE antibodies against antigens No. 10 and No. 28. Three other antigens also showed a weak IgE response.[33]

Using this approach, it is possible to identify individual allergens in extracts and to obtain the specific pattern of IgE response for each patient.

14.6 REFERENCES

1. P. Grabar, C.A. Williams, *Biochim. Biophys. Acta, 10,* 193–202 (1953).
2. C.-B. Laurell, *Scand. J. Clin. Lab. Invest.,* 29, Suppl. (1972).
3. N.H. Axelsen, *Scand. J. Immunol.,* 4, Suppl. 2 (1975).
4. N.H. Axelsen, *Scand. J. Immunol., 17,* Suppl. 10 (1983).
5. Ö. Ouchterlony, L.-A. Nilsson, In *Handbook of Experimental Immunology,* D.M. Wier, ed., Blackwell Scientific Publications, Oxford, 1978, Vol. 1, Chapter 19.
6. R. Verbruggen, *Clin. Chem., 21,* 5–43 (1975).
7. H. Løwenstein, *Prog. Allergy, 25,* 1–62 (1978).
8. W.D. Linscott, *Linscott's Directory of Immunological and Biological Reagents.* 9th ed. Mill Valley, California, 1996.
9. J. Clausen, *Immunochemical Techniques for the Identification and Estimation of Macromolecules,* Elsevier/North Holland Biomedical Press, Amsterdam, 1981.
10. K. Burridge, In *Methods in Enzymology.* S.P. Collowick, N.O. Kaplan, eds., Academic Press, New York, 1978, Vol. 50, pp. 54–64.
11. R.J. Mayer, J.H. Walker, *Immunochemical Methods in the Biological Sciences: Enzymes and Proteins.* Academic Press, New York, 1980.
12. G. Köhler, C. Milstein, *Nature, 256,* 495–497 (1975).
13. C. Milstein, In *Antibodies as a Tool: The Applications of Immunochemistry.* J.J. Marchalonis, G.W. Warr, eds., John Wiley & Sons, Chichester, 1982.
14. U.W. Mueller, J.M. Potter, *Anal. Biochem., 100,* 51–53 (1979).
15. H.P. Prince, D. Burnett, D.B. Ramsden, *J. Chromatogr., 143,* 321–323 (1977).
16. O.J. Bjerrum, *Electrophoresis, 6,* 209–226 (1985).
17. B. Weeke, *Scand J. Clin. Lab. Invest., 21,* 351–354 (1968).
18. L.G. Fägerstam, J. Lizana, U.-B. Axiö-Fredriksson, L. Wahlström, *J. Chromatogr., 226,* 523–526 (1983).
19. P.J. Svendsen, N.H. Axelsen, *J. Immunol. Meth., 1,* 169–173 (1972).
20. H.G.M. Clarke, T. Freeman, *Protides Biol. Fluids, 14,* 503–509 (1966).
21. R.J. Ritchie, R.M. Smith, *Clin. Chem., 22,* 497–499 (1976).
22. J. Söderholm, C.J. Smyth, T. Wadstrom, *Scand. J. Immunol., 4,* Suppl. 2, 107–113 (1975).
23. J. Lizana, I. Olsson, In *XI International Congress of Clincal Chemistry,* E. Kaiser, F. Gabl, M.M. Muller, M. Bayer, eds., Walter de Gruyter, Berlin, 1982, pp. 1225–1233.
24. J. Lizana, I. Olsson, A. Savill, *Clin. Chem., 28,* 1569 (1982).
25. M. Thyman, *Hum. Hered., 31,* 214–221 (1981).
26. P.D. Eckersall, J.A. Beeley, *Electrophoresis, 1,* 62–67 (1980).
27. A. Siden, *J. Neurol., 217,* 103–109 (1977).
28. T.C. Bøg-Hansen, ed., *Lectins-Biology, Biochemistry, Clinical Biochemistry,* Vol. 1. Walter de Gruyter, Berlin, 1980.
29. V. Horejsi, *Anal. Biochem., 112,* 1–8 (1981).
30. T.C. Bøg-Hansen, *Anal. Biochem., 56,* 480–488 (1973).
31. P. Owen, J.D. Oppnheim, M.S. Nachbar, R.E. Kessler, *Anal. Biochem., 80,* 446–457 (1977).

32. O.J. Bjerrum, *Anal. Biochem.*, *90*, 331–348 (1978).

33. I. Wallenbeck, L. Aukrust, R. Einarsson, *Int. Arch. Allergy. Appl. Immunol.*, *73*, 166–172 (1984).

34. B. Weeke, H. Lowenstein, *Scand. J. Immunol.*, *2*, Suppl. 1, 149–154 (1973).

35. J. Breborowicz, A. Mackiewicz, D. Breborowicz, *Scand. J. Immunol.*, *14*, 15–20 (1981).

36. T. Uhlin, R. Einarsson, *Int. Arch. Allergy. Appl. Immunol.*, *70*, 213–219 (1983).

37. J.O. Jeppsson, C.-B. Laurell, B. Franzén, *Clin. Chem.*, *25*, 629–638 (1979).

38. J.F. Monthony, E.G. Wallace, D.M. Allen, *Clin. Chem.*, *24*, 1825–1828 (1978).

39. A.M. Johnson, *Annals Clin. Lab. Sci.*, *8*, 195–216 (1978).

40. J. Lizana, A. Savill, I. Olsson, In *Electrophoresis '81*, R.C. Allen, P. Arnaud, eds., Walter de Gruyter, Berlin, 1981, pp. 550–561.

41. V. Neuhoff, R. Stamm, H. Eibl, *Electrophoresis*, *6*, 427–448 (1985).

42. L. Kerenyi, F. Gallyas, *Clin. Chim. Acta*, *38*, 465–467 (1972).

43. E.W. Willoughby, A. Lambert *Anal. Biochem.*, *130*, 353–358 (1984).

44. R. McLachlan, D. Burns, In *Electrophoresis '84*, V. Neuhoff, ed., Verlag-Chemie, Wienheim, 1984, pp. 324–327.

45. O. Vesterberg, B. Gramstrup-Christensen, *Electrophoresis*, *5*, 282–285 (1984).

46. K. Hellsing, In *Automated Immunoanalysis*, R.F. Ritchie, ed., Marcel Dekker, New York, 1977, pp. 67–112.

47. M. Céska, F. Grossmüller, *Experientia*, *24*, 391–392 (1968).

48. N. St. G. Hyslop, D.G. Cochrane, *J. Immunol. Meth.*, *6*, 99–107 (1974).

49. J.-L. Guesdon, S. Avrameas, *Immunochemistry*, *11*, 595–598 (1974).

50. S. Avrameas, J. Uriel, *Comp. Rend. Acad. Sci. Paris Ser. D*, *262*, 2543–2545 (1966).

51. S. Avrameas, T. Ternynck, J.-L. Guesdon, *Scand. J. Immunol*, *8*, Suppl. 7, 7–23 (1978).

52. J.W. Goding, *J. Immunol. Meth.*, *20*, 241–253 (1978).

53. B. Nørgaard-Pedersen, K. Toftager-Larsen, N. H. Axelsen, *Scand. J. Immunol.* *17*, Suppl. 10, 259–263 (1983).

54. O.J. Bjerrum, P. Lundahl, *Scand. J. Immunol.*, *2*, Suppl. 1, 139–146 (1973).

55. G. Dimitriadis, *Anal. Biochem.*, *98*, 445–451 (1979).

56. O.J. Bjerrum, J.H. Gerlach, T.C. Bøøg-Hansen, J.B. Hertz, *Electrophoresis*, *3*, 89–92 (1982).

57. L.D. Lee, H.P. Baden, *J. Immunol. Meth.*, *18*, 381–385 (1977).

58. B.H. Anderson, R.C. Thorpe, *Immunol. Today*, *1*, 122–127 (1980).

59. L.M. Hjelmeland, A. Chrambach, *Electrophoresis*, *2*, 1–11 (1981).

60. A. Rosén, K. Ek, P. Åman, *J. Immunol. Meth.*, *28*, 1–11 (1979).

61. K. Toftager-Larsen, E. Kjaersgaard, J.-Chr. Jacobsen, B. Nørgaard-Pedersen, *Clin. Chem.*, *26*, 1656–1659 (1980).

62. L. Aukrust, K. Aas, *Scand. J. Immunol.*, *6*, 1093–1099 (1977).

15 Protein Mapping by Two-Dimensional Polyacrylamide Electrophoresis

REINER WESTERMEIER

Pharmacia Biotech Europe
D-79111 Freiburg, Germany

ANGELIKA GÖRG

Lehrstuhl für Allgemeine Lebensmitteltechnologie
Technische Universität München
D-85350 Freising-Weihenstephan, Germany

Protein Purification: Principles, High-Resolution Methods, and Applications, Second Edition.
Edited by Jan-Christer Janson and Lars Rydén.
ISBN 0-471-18626-0.   © 1998 Wiley-VCH, Inc.

15.1   INTRODUCTION

One-dimensional separations have a limited resolving power. When two different polyacrylamide gel electrophoresis techniques are combined, resolution can be enhanced by a magnitude. Protein mapping is mostly applied for resolving complex samples with more than 100 proteins. The first method was introduced by Margolis and Kenrick[1]: disc electrophoresis followed by a second-dimension porosity gradient electrophoresis. The combination of isoelectric focusing and polyacrylamide gel electrophoresis was published by Macko and Stegemann,[2] as well as by Dale and Latner,[3] both in 1969. Several groups improved the methodology: Barrett and Gould,[4] MacGillivray and Rickwood,[5] and Scheele.[6]

In 1975, Klose[7] and O'Farrell[8] defined "high resolution two-dimensional electrophoresis" for the complete separation of highly complex protein mixtures such as body fluids, cell lysates, tissue extracts, or entire organisms. Both separations must be performed under denaturing conditions: sample solubilization with highly concentrated urea (9.5 mol/L) and nonionic detergent, isoelectric focusing in the presence of highly concentrated urea and nonionic detergent in thin, nonsieving gel rods, followed by vertical slab gel SDS–polyacrylamide electrophoresis (see Figure 15-1). Both dimensions must provide a minimum of separation distance (>16 cm). The description of the method by O'Farrell covers the full procedure from the lysis buffer, gel production, and running parameters to the detection of the protein spots by autoradiography.

The most important feature of this method is that the protein mixture is separated by two different physicochemical parameters: the charge in the first dimension and the molecular mass in the second dimension.

The highest number of spots that could be detected in a two-dimensional gel were 1,500 to 2,000. It is, of course, not easy to evaluate such a complex

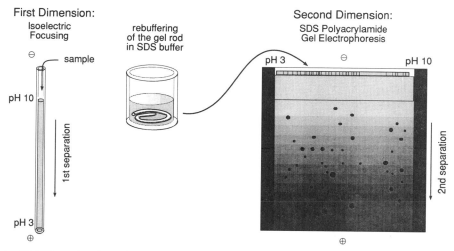

FIGURE 15-1. The principle of high-resolution two-dimensional electrophoresis according to O'Farrell.[8]

protein map, and it is even more difficult to compare different gels with each other. This has to be done with the help of a computer. An important preposition for a computer-aided analysis and comparison is a high reproducibility of the spot positions, and thus a high reproducibility of the method. Therefore Anderson and Anderson[9,10] introduced a methodology based on the O'Farrell technique, which simplified the procedures and made it possible to perform multiple gel casting and runs: the ISO-DALT system.

Because several technical problems are caused on the vertical rod and the vertical slab gel system, Görg and co-workers[11] have modified the entire procedure to perform both dimensions in the horizontal plane. With this method it is possible to use gels polymerized on plastic films and the handling of the horizontal technique and apparatus is much easier.

The main problems of unsatisfactory reproducibility, inadequate resolution, and loss of proteins with acidic and basic isoelectric points are, however, connected with the isoelectric focusing methodology. These drawbacks have led to the development of immobilized pH gradients[12] in order to achieve stable pH gradients and to abolish the gradient drift. Immobilized pH gradients are built by acrylamido buffers (Immobiline™), which are covalently attached to a polyacrylamide gel. Detailed information on the theory and methodology of immobilized pH gradients can be found in Ref. 13.

The combination of isoelectric focusing in immobilized pH gradients with a horizontal SDS gel on a film support has been optimized by Görg and colleagues[14] and gives the most reliable and reproducible protein maps. This "IPG-Dalt" method (isoelectric focusing in immobilized pH gradients followed by SDS electrophoresis) can be easily established in every protein laboratory.

The following chapter describes the methodologies of the most advanced two-dimensional methods areas: important detection methods, evaluation possibilities, computer-aided image analyses, and applications.

15.2   SAMPLE PREPARATION

The following points are important:

- Complete solubilization, also of hydrophobic proteins
- Disaggregation of complexes and complete unfolding of the polypeptide chains
- Low contents of salts
- Absence of lipids and phenols
- Heating of the urea solution and urea-containing samples must be avoided because of the danger of carbamylation.

15.2.1   General Procedure

The following solubilizing mixture has been optimized for many different applications[15]: 9 mol/L urea, 2% (v/v) Triton X-100 or, preferably, 2% CHAPS [3-(3-cholamidopropyl)dimethylammonio-1-propane sulfonate], 0.8% (w/v) Pharmalyte pH 3 to 10, 2% (v/v) 2-mercaptoethanol, and 8 mmol/L PMSF (phenylmethylsulfonyl fluoride). To make the *"gel slurry,"* add 30 mg of Sephadex G-25 Superfine to 1 mL of solubilizing mixture and swell overnight in the refrigerator. The sample application is sample solution + gel slurry (1 + 1) partial volumes.

The functions of the compounds are:

1. High urea concentration: Cleaves hydrogen bonds, unfolds the secondary and tertiary structure, avoids aggregates and hydrophobic interactions, and keeps hydrophobic proteins in solution.
2. Triton X-100 (nonionic detergent) or CHAPS (zwitterionic detergent): Solubilizes hydrophobic proteins.
3. Pharmalyte pH 3 to 10: The combination of a heterogeneous carrier ampholyte mixture and a nonionic detergent form a mixed zwitterionic detergent, which solubilizes very hydrophobic proteins.[16]
4. 2-Mercaptoethanol: Cleaves the disulfide bonds for complete denaturation of tertiary and quartery structures.
5. PMSF: Protease inhibitor.
6. *"Gel slurry":* The granulated gel in the sample decreases the proteins' mobility in the start phase in order to prevent them from aggregating when they enter the gel. Note: The dextran gel must be prehydrated, as dry dextran particles adsorbs proteins.

An optimal protein concentration for the sample application is about 2 mg/mL of sample solution, which corresponds to 40 μg of protein/20 μL of applied sample (inclusive dextran gel) per run (silver staining concentration).

To separate protein solutions such as serum and plasma, the solubilizing mixture is diluted accordingly after a quantitative protein determination.

Tissue and cell proteins must be solubilized with the solubilizing mixture by mechanical disintegration. Examples include:

- Yeast cell lysate: Mix 300 mg of lyophilized yeast (*Saccharomyces cerevisiae*) with 2.5 mL of solubilizing mixture, sonicate for 10 min at 0°C, centrifuge for 10 min at 15°C at 42,000 g, mix 20 μL of the supernatant with 20 μL of gel slurry, and apply 20 μL to the *anode* of the gel (silver staining concentration).
- Myeloblast cell lysate: Mix 5×10^8 cells with 1 mL solubilization mix and continue as described for yeast.
- Plant seeds: Mix 20 to 100 mg (dependent on protein content) of ground material with 1 mL solubilization mix and continue as described for yeast.

15.2.2 Problematic Samples

Delipidation: High lipid contents are incompatible with the method. The lipids are removed by protein precipitation with 80% acetone at −20 °C.

Removing phenols: Plant material can contain a high amount of phenols, which would form irreversible complexes with proteins. Grind sample material (e.g., leaves) in a mortar, cooled with liquid nitrogen. Suspend the powder in 10% trichloroacetic acid/0.07% 2-mercaptoethanol/90% acetone; precipitate proteins at −18°C for ca. 45 min; centrifuge and wash the pellet with 0.07% 2-mercaptoethanol/acetone at −18°C; and decant the supernatant, dry the protein pellet under vacuum, dissolve it in solubilization mix, mix 20 μL of this solution with 20 μL of gel slurry ($\rightarrow$ 40 μL), and apply 20 μL to the *anode* of the gel (silver staining concentration).

If *DNA-* and *RNA*-rich samples cause problems, they should first be treated with DNase and RNase.[8,17]

If a sample contains some very hydrophobic and large molecules such as cytoskeletal proteins, tubulin and big membrane proteins, adding low concentrations of SDS (ca. 0.1%) to the solubilization buffer can help.[8,18] There is an extensive discussion on sample preparation methods with many references in the first part of the review by Dunn and Burghes.[19]

15.3 ISOELECTRIC FOCUSING

The principle of isoelectric focusing and its features is described in Chapter 13. However, there are special demands on this method when it is employed as the first dimension in high-resolution two-dimensional electrophoresis:

- A high concentration of urea (8 to 9 M) and a certain content of a nonionic detergent are required, thus the migration of proteins is very slow due to high viscosity. This results in long focusing times.
- The pore size of the gels should be as large as possible to include high molecular proteins in the map. Thus the gel matrix has low mechanical stability.
- Preferably each sample should be run in an individual gel, in order to have full quantitative control over the sample to be analyzed.
- A high reproducibility of the pH gradient profile and gel length (= separation distance) is required in order to allow intergel comparisons; preferably even from laboratory to laboratory.

15.3.1 Carrier Ampholyte-Generated pH Gradients

Dunn and Burghes[19] have compiled a list of the features and disadvantages of the existing two-dimensional electrophoresis techniques and their modifications, most of them based on the O'Farrell method. The main technical problems are connected with the isoelectric focusing step, performed in vertical gel rods:

- The slight anodic drift and the strong cathodic drift of the carrier ampholyte pH gradient. The final pH gradient does not extend below pH 4.5 and above pH 7.
- That true equilibrium of the pH gradient cannot be reached and the pattern is time dependent.
- Diffusion and loss of proteins during the equilibration and transfer of the first-dimension gel.
- Pattern matching can be difficult because of varying spot positions.

The very severe problem of losing the complete fraction of basic proteins due to the cathodic drift was identified by the O'Farrell group.[17] An additional type of two-dimensional electrophoresis method was introduced: NEPHGE (nonequilibrium pH gradient electrophoresis). Here, the sample was loaded on the anodal side of the isoelectric focusing gel, the run was performed for a relatively short time, and no steady state was attained. This means that the sample components were separated in a pH gradient while it was moving toward the cathode. The resulting pattern was even more time dependent compared to the original method, but other experimental conditions also influenced the result.

The reasons for the problems with isoelectric focusing are:

- Electroendosmosis in the basic part of the gel, which is enhanced by the glass material of the vertical tubes.

- Because the carrier ampholytes are in solution, the gradient profiles are modified by the buffering power of the proteins themselves and by other sample components.
- Batch-to-batch variations of carrier ampholytes.
- Length variations of the very soft gel rods.

15.3.2 Immobilized pH Gradients

Isoelectric focusing in immobilized pH gradients (IPG) is a true equilibrium method because the gradient is fixed to the matrix and the matrix is polymerized on a polyester film support. Also, instead of employing a mixture of 600 to 700 different individuals of amphoteric buffers, only 6 to 8 different substances of Immobiline are needed. These compounds are acrylamido derivatives with the following general structure: $CH_2=CH-CO-NH-R$, where R contains either a carboxyl or a tertiary amino group.

The immobilized pH gradients are prepared like casting a conventional linear gradient gel by using a density gradient to stabilize the Immobiline concentration gradient. As these Immobiline molecules contain double bonds, they copolymerize with the acrylamide and methylenebisacrylamide network (see Figure 15-2). Exact instructions and recipes for casting these gels are provided by the manufacturer and can be found in the literature.[13,20] With tailor-made pH gradients the resolution and loading capacity can be optimized, particularly for preparative uses. When extremely wide immobilized gradients

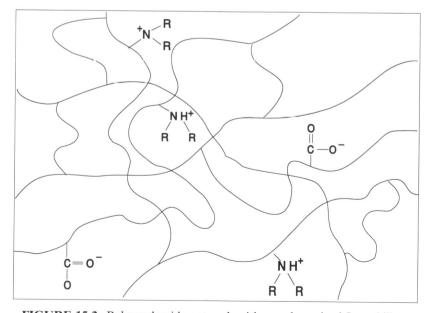

FIGURE 15-2. Polyacrylamide network with copolymerized Immobilines.

(e.g., pH 2.5 to 11) are needed to include nearly all possible cellular products in a protein map, the gels are prepared according to Sinha and co-workers.[21]

In a conventional electrophoresis or isoelectric focusing experiment, the conductivity of the buffer or the free carrier ampholytes is high enough to transport the ionic polymerization catalysts TEMED and ammonium persulfate out of the gel. Immobiline gels exhibit a relatively low conductivity because the buffering groups are fixed to the matrix. Therefore Immobiline gels must be washed with double distilled water after polymerization.

Because polyacrylamide gels swell during washing and shrink in all directions during drying, Immobiline gels are polymerized as slab gels on polyester film supports. Several attempts of using gel rods with immobilized pH gradients have had limited success because of the necessity of washing the gels.[22,23]

However, a much easier and better method has been developed by Görg and co-workers[14] for running individual separations in immobilized pH gradients. The washed and dried Immobiline gel is cut into 3- to 5-mm-wide strips with a paper cutter (see Figure 15-3). As these strips are cast on plastic backings, the gels cannot stretch or shrink and are much easier to handle than gel rods. There are no edge effects: the iso-pH lines are absolutely straight and fixed to the matrix. The width of the strips should be less than 5 mm as this minimizes the electroendosmotic effects between the first and the second dimension[14,24] and reduces the amount of nonionic detergent that is transferred from the first to the second dimension to a minimum.[14,15] The average loading capacity is 60 to 100 μg of a complex protein mixture per strip.

The name or number of the sample can be easily engraved or marked on the back of the film with water-resistant ink for identification.

dried gel with immobilized pH gradient

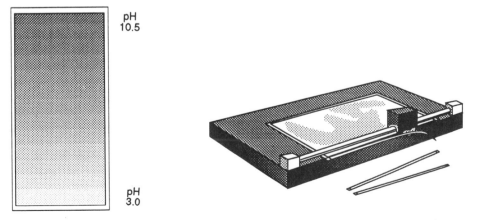

FIGURE 15-3. Cutting individual IPG strips from a dried gel slab (stabilized by a polyester film support).

Three-millimeter-wide strips are readily available in different lengths (11 and 18 cm) and with different immobilized pH gradients: 4.0 to 7.0, 3.0 to 10.5 with linear and nonlinear gradient profiles, respectively (Immobiline™ DryStrips). In order to identify the polarity of the pH gradient, the anodal side of a strip is cut like an arrow.

The nonlinear gradient 3 to 10.5 is flattened between pH 4 and 6.5. It is mainly applied for the separation of protein mixtures that contain acidic proteins in high relative abundance, such as serum. In this way, streaking of these proteins due to local overloading of the second dimension is abolished.

15.3.3 Procedure of Isoelectric Focusing in Individual IPG Strips

15.3.3.1 Rehydration of Strips   The strips are rehydrated for a minimum of 6 h or overnight before use to their original thickness (0.5 mm). This is performed in a mold with a solution containing the necessary additives for the first dimension [e.g., 8 M urea, 0.5% (v/v) Triton X-100 as nonionic detergent (or 0.5% CHAPS as zwitterionic detergent), 10 mmol/L dithiothreitol as reducing reagent, and 2 mmol/L acetic acid[25] for improved sample entry (anodal sample application)]. When a sample must be loaded at the cathodal end, the addition of 2 mmol/L Tris is recommended instead of acetic acid. The addition of 0.1% (w/v) orange G makes the strips faintly yellow and helps in strip alignment. Note that the solution has to be prepared fresh because urea degrades to isocyanate with time, causing carbamylation of proteins during isoelectric focusing.

Usually, a low amount of carrier ampholytes is added to the rehydration solution because (1) the presence of carrier ampholytes improves the solubility of hydrophobic proteins, and (2) they raise the conductivity in the gel. However, adding carrier ampholytes to the immobilized pH gradient gel has some disadvantages with respect to equilibrium isoelectric focusing, particularly when their concentration exceeds 0.5% (w/v).[14]

One possible procedure: the strips are inserted into the mold, which has already been filled with the rehydration solution. The manufacturer of the strips suggests an alternative technique: Apply the strips on a moistened glass plate with the gel surface facing upward. Then assemble the reswelling cassette and fill through the little tube from the bottom (see Figure 15-4).

The surfaces of rehydrated IPG gel strips are blotted with wet filter paper before use.

15.3.3.2 Loading of the Sample   In most cases, anodal sample application is preferred.[26] The sample solution is pipetted directly on the gel surface. Paper or cellulose applicators must not be used.

To avoid leakage, it is very important that the surface is dry before sample holders (see Figure 15-5) are applied.

The addition of the dextran gel to the sample is not mandatory, but it is very helpful in avoiding the development of protein aggregates and the blocking of the gel surface by protein precipitates.

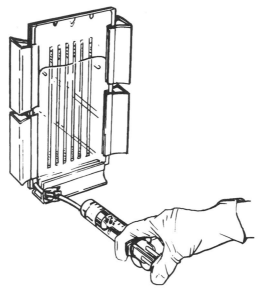

FIGURE 15-4. Rehydrating Immobiline DryStrips in a mold.

The amount of protein loaded is dependent on the heterogeneity of the complex protein mixture, the type of pH gradient, and the desired detection method. In some cases, the optimal amount has to be determined by a prerun. Here are some hints for a start:

- For an analytical, silver-stained, large-size protein map with 18×20 cm and a wide gradient pH 3 to 10, apply up to 50 to 100 μg of total protein.

- For smaller gels, less protein must be applied, as some spots will not separate from each other: Areas with protein agglomerates are produced. Overloading is the most frequent cause for horizontal and vertical streaks in the pherogram.

- When "preparative" two-dimensional separations for sequencing of protein spots are performed, apply 100 to 500 μg of total protein to the strips with a relatively narrow pH gradient interval.[27] Between 1 mg[28] and even up to 10 mg[29] of crude protein preparations have been applied onto single strips for subsequent two-dimensional separations.

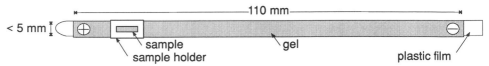

FIGURE 15-5. Sample application to an IPG strip with a silicone rubber frame.

The volume should not be less than 20 μL: With lower volumes the sample entrance would be worse because of protein aggregation in highly concentrated solutions. As there is slight water transport during an IPG experiment, the sample could also dry out if a smaller volume is applied. Volumes higher than 100 μL can lead to an increase of water accumulation on the gel surface.

There are two possible methods for loading the sample and running the strips. In the first method, the rehydrated strips (up to 40) are placed directly on the cooling plate of a horizontal electrophoresis/isoelectric focusing apparatus with the gel surface facing upward.

For optimal cooling contact, the cooling plate must be coated with a thin layer of kerosene or *n*-decan before applying the strips. Figure 15-5 shows the assembly of a rehydrated IPG strip and a sample holder. It can be prepared by carving a little frame from a 2-mm-thick silicone rubber plate or by cutting off a piece from a sample applicator strip. If a sample is highly diluted, up to 100 μL can be applied by repeated loading.

There is dedicated equipment on the market, which is placed on the cooling plate of a horizontal electrophoresis/isoelectric focusing apparatus: The Immobiline DryStrip kit. A strip aligner plate with parallel grooves is laid into the tray. In this way, up to 12 strips can be arranged absolutely parallel and with defined distances. The walls of the tray contain conducting rails for the electrode bars, which are placed on the respective ends of the gel strips (see Figure 15-6). The special feature of this equipment is the sample application

FIGURE 15-6. Placing the electrodes into the Immobiline DryStrip kit (Pharmacia Biotech AB).

system: An additional bar is placed into the tray, which holds sets of sample cups with open bottoms (see Figure 15-7). These funnel-shaped cups are positioned in the same distances like the strips. Because they are sitting on plastic clips like springs, they can be slightly pressed onto the surface of the gel strips to ensure good contact without leakage. Up to 100 μL of sample solution can be pipetted into each cup. The amount of sample can be increased by repeated loading.

The Immobiline DryStrip kit is also designed for running the strips completely under silicone or paraffin oil. This procedure is recommended when samples should have no air contact during the separation (e.g., native separation without reducing agent) or when the technique is performed in a climate area with very high or extremely low humidity.

15.3.3.3 Sample Entry

Isoelectric focusing has mostly been performed at a 15°C cooling temperature in the thermostatic circulator. Lower temperatures must not be used because the urea in the sample and in the gel would precipitate. The present state of the art is that the temperature be set to 20°C.[30,31]

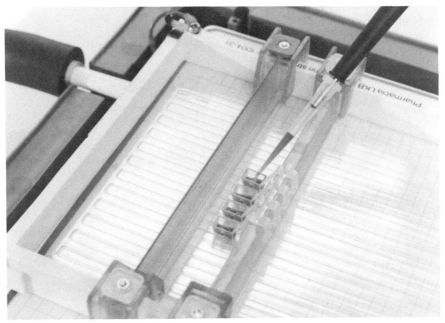

FIGURE 15-7. Loading the samples into the sample cups of the Immobiline DryStrip kit (Pharmacia Biotech AB).

No prefocusing step is performed because the pH gradient exists already in the gel, and the higher conductivity, existing at the beginning, improves the protein transport into the matrix.

The initial field strength must be limited to 10 to 30 V/cm for 1 to 2 h in order to avoid local concentration of proteins, which can result in the development of aggregates.

Paper strips soaked in 5 mol/L sodium hydroxide are added to the chamber to diminish the carbon dioxide content inside the chamber.

15.3.3.4 *First-Dimension Run* After 1 to 2 h, the field strength (E) applied is raised to a maximum of 300 V/cm for IPG gels with 0.5% carrier ampholytes. The maximal current is set to 1 mA, the maximal power to 5 W for running 12 strips. The electrical conditions are controlled with the set maximal voltage because the IPG strips exhibit a very low conductivity.

The focusing time has to be long enough to reach a constant pattern. The time is dependent on the separation distance, the pH interval, the field strength applied, the amount of protein loaded, and the amount of urea and detergent. The focusing time must be longer for:

- Longer gels.
- Narrow pH gradients.
- Lower field strength.
- Higher protein load.
- Higher urea and detergent concentrations.

Isoelectric focusing in immobilized pH gradients produces protein patterns, which are stationary over several days. However, the focusing time should not be extended too long after reaching equilibrium because some proteins have a limited stability at their isoelectric points.[32] In addition, it should be noted that water exudation may occur with a prolonged focusing time. Too much liquid on the surface destroys the focusing pattern.

An important parameter for reproducibility in isoelectric focusing in IPG is the value of the volt/hour integral (Vh). Typical separation conditions are listed in Ref.[30]: from 11,000 Vh for 11 cm strips at pH 3 to 10.5 to 42,000 Vh for 18-cm-long strips with a narrow pH of 4 to 7.

After the completed run, the strips are rebuffered in SDS sample buffer and loaded onto a SDS electrophoresis gel for the second-dimension run. This has to be done immediately after the electric field has been switched off from the focusing gels.

Before equilibration the strips can be stored in liquid nitrogen or deep frozen with at least $-60°C$ between two sheets of plastic film for several weeks without changing the pattern.[33] Storage around $-20°C$ is not recommended because ice crystals can develop and the stability of proteins is limited.

Intermediate fixing and/or staining of the proteins before equilibration has resulted in severe losses of proteins spots in most cases[14] and is therefore not recommended.

15.4 REBUFFERING THE FIRST-DIMENSION GEL FOR SDS ELECTROPHORESIS

The separated proteins have to be converted to SDS–polypeptide micelles after isoelectric focusing. This is performed in a SDS sample buffer at room temperature. Heating is not necessary because the urea in the first dimension has already unfolded the polypeptide chains.[18] In conventional (carrier ampholyte gel) systems, equilibration is done only for a few minutes to keep band diffusion and protein loss to a minimum.[8] Proteins diffuse much slower in IPG gels because the IPG gels behave as weak ion exchangers. After isoelectric focusing, the proteins form something similar to an "Immobilinate–protein" complex.[21] Thus the equilibration time can be prolonged.

In general, immobilized pH gradient gels have to be equilibrated in SDS sample buffer for at least 30 min[14,23,27] in test tubes on a shaker. Shorter equilibration can cause higher background staining, reduced protein transfer from the first to the second dimension, and vertical streaking.

Because Immobiline gels become negatively charged in alkaline milieu, which is the case in an SDS buffer, there is an electroendosmotic water transport toward the cathode occuring in an electric field. This can cause some problems in two-dimensional electrophoresis. Görg and co-workers[34] have found an easy way to overcome this phenomenon:

- Add 30% glycerol and 6 mol/L urea to the equilibration buffer to reduce the water transport.
- When a *horizontal* SDS gel is employed for the second dimension, remove the IPG strip after the proteins have left the gel, then move the cathodal buffer wick or strip forward to cover the area of the removed IPG strip by 1 to 2 mm.

The SDS sample buffer for IPG strips contains 0.05 mol/L Tris–HCl, pH 8.5, with 2% (w/v) SDS and 30% glycerol, 6 mol/L urea, and a trace of bromophenol blue. For cleaving the disulfide bonds, 65 mmol/L dithiothreitol is added. However, staining artifacts unrelated to protein, such as horizontal lines and vertical point streaking, were observed in silver-stained protein maps. With the following protocol, these problems are abolished[34]: Equilibrate for 15 min in 10 mL sample buffer plus 65 mmol/L dithiothreitol, and subsequently equilibrate another 15 min with 10 mL sample buffer plus 65 mmol/L dithiothreitol and 260 mmol/L iodoacetamide. In this way, the excess reducing agent reacts with the iodoacetamide without alkylating the proteins. When very slim IPG strips (<3 mm) are used, the equilibration steps can be reduced to 2 × 10 min.

15.5 SDS ELECTROPHORESIS

15.5.1 General Aspects

The SDS electrophoresis step is less critical than isoelectric focusing. Whereas the classical O'Farrell method[8] employs a discontinuous gel according to Laemmli[35] with a porosity gradient, it has been shown that a stacking gel is not always necessary.[17] In the Laemmli system, the resolving gel contains 0.375 mol/L Tris–chloride, pH 8.8, and 0.1% SDS; in the cathodal running buffer glycine is used. Because the pH value in these gels is relatively high, they have a limited shelf life due to a hydrolyzation of the polyacrylamide matrix. Consequently, the gels containing this buffer system have to be prepared not longer than 10 days before use.

However, the more complex the methodology, the more influences can come from the "human" factor of making solutions, casting the gels, and operating the system. The homemade gels can be replaced by ready-made SDS polyacrylamide gels, which employ a Tris–acetate buffer, pH 6.7, in the gel in combination with tricine in the cathodal buffer.[36]

15.5.2 Vertical Systems

For multiple separations, vertical setups such as the IsoDalt system according to Anderson and colleagues[10] are employed because up to 20 gels can be run at the same time. For an optimal contact from the first- to the second-dimension gel, the IPG strip is cemented on the upper edge with 0.5% agarose containing electrode buffer (25 mmol/L Tris, 192 mmol/L glycine, and 0.1% SDS).[34,37–39]

Independently from this, the procedures of the first dimension and the rebuffering step are identical for vertical and horizontal SDS electrophoresis.

When two-dimensional electrophoresis is employed for preparative techniques (e.g., for subsequent protein sequencing or amino acid analysis), it must be remembered that isoelectric focusing gels have a higher loading capacity than SDS gels.[18] Therefore, isoelectric focusing in IPG strips, which have even a higher capacity than carrier ampholyte gels, is combined with relatively thick (>1 mm) vertical slab gels.[27–29,40]

15.5.3 Horizontal Systems

Using horizontal gels polymerized on film supports is much easier because the IPG strip does not have to be cemented onto it with agarose. The dissipation of

Joule heat is very efficient because the gels are placed onto the cooling block of the horizontal apparatus.

Thin-layer (0.5 mm) SDS porosity gradient gels are readily available with the Tris–acetate/tricine buffer system under the name ExcelGel SDS. There is a choice between two types: a gradient 8 to 18 in a short separation distance gel (11 × 25 cm) and a gradient 12 to 14 in a large gel (18 × 25 cm). Both gels contain a stacking gel zone with 5% T. The running buffers are supplied in precast polyacrylamide strips, which are laid along the anodal and the cathodal edges of the gel layer.

After equilibration, the IPG strip is blotted with filter paper and applied with the gel side facing down onto the stacking area of the SDS gel (see Figure 15-8).

As shown in Figure 15-9, the platinum wire electrodes are put directly onto the surface of the electrode strips.

Total separation time is 1 h and 40 min at maximal 600 V, 50 mA, and 30 W for the short distance gel, and 3 h and 30 min at maximal 1000 V, 40 mA, and 40 W for the large gel with temperature set to 15°C.

A very important feature of these horizontal gels on carrier films is that there is no breaking, swelling, or deformation during staining, drying, and storage, which makes computer-aided evaluation much easier and faster.

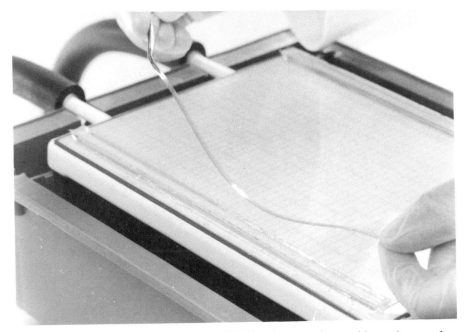

FIGURE 15-8. Applying the rebuffered IPG strip onto the stacking gel area of an ExcelGel SDS XL. Note the cathodal and the anodal buffer strips at the edges.

FIGURE 15-9. Adjusting the electrode wires at their positions on the electrode strips. Note the faint IPG strip on the gel surface close to the cathodal electrode strip on the left-hand side.

15.6 PROTEIN DETECTION

Upon completed separation, the spots can be detected inside the gel or on the surface of an immobilizing membrane after an intermediate transfer step (blotting). A comprehensive review on the analysis of protein spots can be found in Dunn and Burghes.[41]

When the spots have to be analyzed inside the gel, the proteins are, in most cases, fixed in a 30% ethanol/10% acetic acid for 30 min to 1 h, depending on the gel thickness.[42] For an efficient fixation of low molecular mass peptides (<10 kDa), the proteins are cross-linked in the gel with 0.2% glutardialdehyde in the presence of 0.2 mol/L sodium acetate and 30% ethanol.[43]

15.6.1 Autoradiography and Fluorography

Autoradiography is highly sensitive: Down to 0.1 pg of protein is detected per spot.[8] Protein can be labeled by growing a cell or tissue culture in the presence of [^{35}S]-methionine or [^{14}C]-amino acids or sugars by labeling specific groups of proteins such as $^{32}PO_4$ for phosphoproteins and sugar precursors for glycoproteins. Gels can be stained with ^{59}Fe. The gels are dried and then exposed to X-ray film for an appropriate time, the magnitude of one or several days.

Horizontal 0.5-mm-thin gels on carrier films are easily dried and thus very practical for this detection method.

Fluorography is a scintillation autoradiography for low-energy β particles such as ^3H[44]: The gel is first impregnated with dimethyl sulfoxide (DMSO) and then with the organic scintillator 2,5-phenyloxazole (PPO). After this the gel is dried and exposed to a blue-sensitive X-ray film at $-70°C$. Because DMSO and PPO are toxic and expensive, they have been replaced by the water-soluble fluoro sodium salicinate.[45]

15.6.2 Staining Methods

15.6.2.1 Coomassie Brilliant Blue Staining After SDS electrophoresis, the proteins are fixed and stained in one step[42,43]: the gel is directly laid into a 50 to 60°C warm solution of 0.02% Coomassie brilliant blue R-250 or 350 in 10% acetic acid for ca. 15 min. Destaining is performed at room temperature in 10% acetic acid. The sensitivity is dependent on the protein; 20 ng of bovine serum albumin can be detected per spot. If the dye-binding property of a protein is known, this method can be employed for quantitation of a protein.

15.6.2.2 Silver Staining Silver staining is 50 to 100 times more sensitive than Coomassie staining. However, quantitation can only be performed in a very limited concentration area. The modifications, which are applied most frequently for two-dimensional gels, can be found in Merrill and co-workers,[46] in Heukeshoven and Dernick,[42] and in Blum and colleagues.[47] When horizontal gels on carrier films are stained, the ethanol content in Blum's staining protocol has to be reduced from 50 to 30% to avoid the separation of gel and carrier film. In a comprehensive general overview, Rabilloud and co-workers,[48] have tested and discussed all of the different silver staining modifications and optimizations and published an overview.

15.6.3 Blotting

Blotting techniques are employed for protein identification with immunological or lectin-glycoprotein affinity methods. Even amino acid analysis, protein sequencing, and mass spectrometry analyses are performed on blots.[49] Details on blotting methods can be found in Chapter 16.

Figure 15-10 shows IPG-Dalt protein maps of *Dactylis glomerata* pollen-soluble extracts that have been treated with different detection methods: Silver staining, blotting, and autoradiography of the blot.[50]

15.7 EVALUATION

15.7.1 Physicochemical Parameters

The isoelectric points (pI) and the molecular weights can be read directly from the gel.

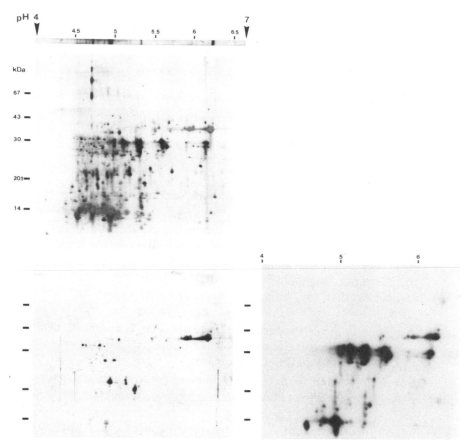

FIGURE 15-10. IPG-Dalt (pH 4 to 7, SDS gradient 12 to 18%) protein maps of a *Dactylis glomerata* pollen-soluble extract. (Top) Silver staining according to Blum.[47] (Bottom left) Blot on nitrocellulose, stained with Indian ink. (Bottom right) Autoradiography with 5 days exposure: Blot incubated with an allergic patient serum and with radiolabeled anti-IgE. (Courtesy of Virginie Leduc-Brodard and Gabriel Peltre, Institute Pasteur, Paris.)

The definition of the pI in the presence of 8 mol/L urea is a matter of discussion. There is still no standardization that is respected by all researchers. The pH value shifts in this environment and the conformations of the proteins differ from that under native conditions. When immobilized pH gradients are used, the pH gradient is part of the gel and is physicochemically defined in the absence of urea. It would be very easy and reliable to use this fact for a basis of definition.

For the estimation of the molecular weights of the polypeptide chains in the SDS electrophoresis step, molecular weight standards are run as one-dimensional lanes parallel to the two-dimensional sample. Interpolation of the spot isoelectric point and molecular weight can also be made with help of the appropriate software in computer-aided analysis.

15.7.2   Recording of the Pattern

As already mentioned in Section 15.1, the complex patterns are not easy to evaluate with just the eye. Highly sophisticated softwares have been developed to take over this task. However, the information has to be digitized first.

Several possibilities are available with the latest developments of electronic devices:

- *Video cameras* are easy to use; however, problems of unproper recording can be caused by uneven illumination.
- *Desk top scanners* can record a two-dimensional map within several minutes with high resolution in the area as well as in the optical density scale. For reliable scanning, the use of the transmission is preferable to reflectance. It should be noted that there are big differences in performances that are mostly directly proportional to the investment costs.
- *Laser densitometer:* Proper quantification demands a scanning laser densitometer because it is the only instrument that has a long linear range up to 4 OD (optical densities) extinctions.
- For radiolabeling or fluorescent-labeling or staining, *phosphor-storage plate readers* or *direct fluorescent imagers* are employed.

At present, no instrument is available that manages every task absolutely perfect, but because technology develops very fast, further improvements can be expected in the near future.

15.7.3   Computer-Aided Analysis

The hardware possibilities have been improved so fast and so far that soon there will be no limits at all concerning performance, speed, and memory, all at an affordable cost. Almost all necessary operations will be feasible with a personal computer.

The most important features of a two-dimensional gel analysis software are:

- Scanning parameters management.
- Spot detection.
- Background definition.
- Spot integration (quantitation).
- Spot matching.
- Image comparison.

- Reference image.
- Interface to data bases.

More details on this subject can be found in Refs. 41, 51, and 52.

Many problems with evaluating two-dimensional maps are abolished using immobilized pH gradients on plastic support in first and carrier film-supported SDS gels in a horizontal system.

Figure 15-11 shows a computer screen displaying a PDQUEST image of barley proteins.

15.7.4 Data Bases and Data Communication

The first big project of a protein data base was started by Anderson and Anderson[53] in their molecular anatomy program: the human protein index. However, more data bases have been developed, a few are described in a

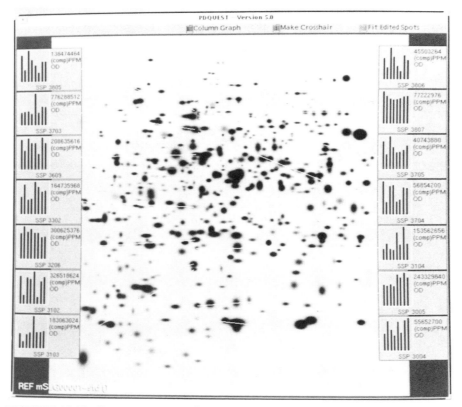

FIGURE 15-11. Computer screen displaying PDQUEST image of a protein map of barley proteins. (Courtesy of A. Posch, Technical University Munich, Weihenstephan.)

series of paper symposia that are now updated annually: Protein data bases in two-dimensional electrophoresis.[54–56] Moreover, data bases have been set up on servers that may be assessed from any computer connected to the internet[57] combined with a mailbox for fast communication.

15.8 APPLICATIONS

Figure 15-12 shows a two-dimensional electrophoresis of mouse liver proteins. Figure 15-13 shows a protein map of carrot proteins that was entirely run on ready-made gels. The first dimension was an Immobiline DryStrip pH 3.0–10.5, the second dimension was an ExcelGel SDS XL gradient 12–15.

The following lists have tried to find a certain order to summarize the many different applications of protein mapping.

15.8.1 Clinical Chemistry

"High resolution two-dimensional electrophoresis is the only technique currently available which can resolve and map the very complex mixtures of

FIGURE 15-12. Two-dimensional electrophoresis of mouse liver proteins. IPG-Dalt (pH 4 to 9, SDS gel gradient 12 to 15% T). Silver Staining according to Merrill.[46] (From Ref. 30 (Görg, A.) with permission of the publisher.)

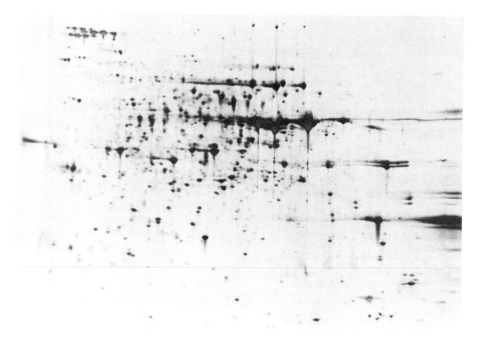

FIGURE 15-13. Two-dimensional electrophoresis of carrot seed proteins on ready-made gels: Immobiline DryStrip pH 3.0–10.5, the second dimension an ExcelGel SDS XL gradient 12–14.

proteins found in cells and body fluids" (Anderson and Anderson[58]). Accordingly, the applications in this area are widely spread[59]:

- Human genetics.[60]
- Diagnostics.[61]
- Detecting known or discovering new disease markers or patterns.
- Detecting allergens.[50,62,63]
- Detecting mono-, oligo-, or polyclonal IgGs.[64]
- Following disease process and protein expression.
- Monitoring of therapies.[65]
- Identifying the origin of body fluid.
- Analyzing protein phenotypes.
- Cancer research.[65]

15.8.2 Biological Research

- Analysis of cell differentiation.
- Microbiology research.

- Genetic analysis and genetic engineering.
- Cell biology.
- Plant research and breeding.[66,67]
- Zoological research and breeding.

15.8.3   Protein Chemistry

- Proteins characterized by subsequent blotting and sequencing and amino acid analysis.[39,40,49,68]

15.8.4   Pharmaceutical

- Mutagenity tests via cell cultures.
- Purity and identity tests.
- Development of diagnostics.
- Quality test of diagnostics.
- Genetic engineering.

15.9   METHODICAL PERSPECTIVES

The highest resolution and the highest number of spots have been revealed in very large gels such as 32×40 cm[69] or 42×33 cm[70] (IEF $\times$ SDS–PAGE), with subsequent fluorography, preferably with sequential exposures for different times.

However, the long separation times, the handling of such big plates, and the time-consuming and costly detection procedures are limiting the introduction of this optimized procedure into but a few research laboratories. Thus, the method of "high-resolution two-dimensional electrophoresis," which is practically applied in most laboratories, is always a compromise between resolution and detection limit on the one hand and handling, time, and cost efforts on the other hand.

For several applications, "low resolution two-dimensional procedures"[71] are preferred, not only because of the complexity of the pattern, but also because of time consumption and handling. Görg and co-workers[72] have optimized the application of immobilized pH gradients for two-dimensional electrophoresis for 3×3-cm protein maps in PhastSystem (see Figure 15-14). Total time, including automated silver staining, is 3.5 h.

In two-dimensional electrophoresis, influences on the pattern and problems with reproducibility can also be caused by the "human" factor of handling the many different steps.

An automat has been developed that operates the entire two-dimensional electrophoresis procedure[73] in order to minimize this factor; however, it runs only one separation at a time and it uses the classical O'Farrell approach.

Horizontal IPG–Dalt   PhastSystem

+ ←IPG–CA 4–7→ –

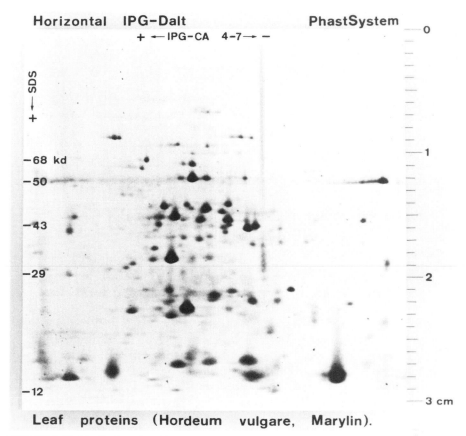

Leaf   proteins   (Hordeum   vulgare,   Marylin).

FIGURE 15-14. IPG-Dalt of leaf proteins from barley. First dimension: IPG pH 4–7, 8 mol/L urea, 0.5% (w/v) carrier ampholytes, 15% (v/v) glycerol, and 10 mmol/L dithiothreitol. Second dimension: PhastGel gradient 10–25 with SDS buffer strips. (From Görg and colleagues.[72])

A different approach to a "fast screening two-dimensional electrophoresis" with a fewer number of operational steps has been achieved by Schickle[74]: Both dimensions are performed in one single gel matrix (see Figure 15-15). The gel employed in this technique is a discontinuous polyacrylamide gel on carrier film, which is divided into a stacking and a resolving zone. Further, the gel has been washed with distilled water after polymerization and dried down on the carrier film, as described in Ref. 20.

Here, the separation of the first dimension is performed inside the low sieving stacking zone. Before sample application, a narrow strip of this area is selectively rehydrated with carrier ampholytes, urea, and nonionic detergent. A strip of cotton is used to keep the rehydration restricted to a defined area. After isoelectric focusing, the entire gel is flooded with SDS gel buffer for a

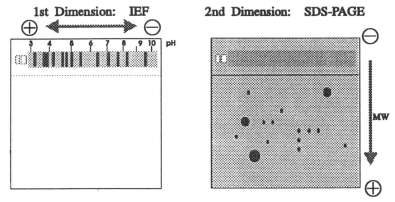

FIGURE 15-15. Schematic representaion of a two-dimensional electrophoresis in a single gel.

few minutes. In this way, the dried polyacrylamide layer swells to a SDS gel and the focused proteins are converted into SDS–protein micelles. Then SDS electrophoresis is performed.

A series of experiments have shown that the protein loss between the two steps is not higher compared to existing two-dimensional techniques.

Unfortunately, it seems to be very difficult to perform this method with immobilized pH gradients. Thus, it has to be performed with carrier ampholytes. However, the benefits of this method are:

- The handling of the system is easy: After the sample has been applied, there are no possibilities of manipulations. It is particularly suitable for small gels, similar to PhastSystem. It can be automated easily.
- The first dimension gel is not lost, thus the stained protein map also includes those proteins that did not migrate out of the first dimension.
- The lateral diffusion of proteins is low because the proteins do not have to migrate through the surface of another gel matrix.

Figures 15-16 and 15-17 show separations of bovine meat juice and bovine oviduct liquid.

Without a doubt, high-resolution protein mapping has taken a big step forward:

- Most of the technical problems associated with the isoelectric focusing step have been abolished by the proper use of immobilized pH gradients in the first dimension.
- The SDS electrophoresis step has been simplified and higher reproducibility can be obtained from laboratory to laboratory by employing ready-made and standardized gels.

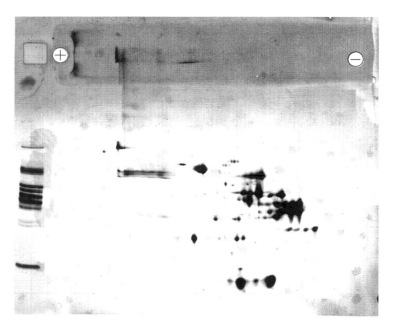

FIGURE 15-16. Protein map of bovine meat juice. Both separations are performed in a single discontinuous polyacrylamide gel with a 5% T stacking gel and a 10% T resolving gel. Silver staining according to Heukeshoven.[42] Reference one-dimensional sample. (Courtesy of H.P. Schickle, ETC GmbH Leonberg, FRG.)

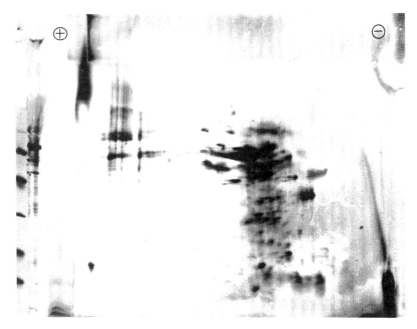

FIGURE 15-17. Protein map of bovine oviduct liquid: 50-μL protein load. Silver staining according to Heukeshoven.[42] (Courtesy of R. Einspanier, Technical University Munich, unpublished data.)

- Computer-aided evaluation has become faster, more reliable, and easier due to the development of high-performance hardware and easy-to-use and intelligent software.

15.10 REFERENCES

1. J. Margolis, K.G. Kenrick, *Nature (London)*, *221*, 1056–1057 (1969).
2. V. Macko, H. Stegemann, *Hoppe-Seyler's Z. Physiol. Chem.*, *350*, 917–919 (1969).
3. G. Dale, A.L. Latner, *Clin. Chim. Acta*, *24*, 61–68 (1969).
4. T. Barrett, H.J. Gould, *Biochim. Biophys. Acta*, *294*, 165–170 (1973).
5. J. MacGillivray, D. Rickwood, *J. Biochem.*, *41*, 181–190 (1974).
6. G.A. Scheele, *J. Biol. Chem.* *250*, 5375–5385 (1975).
7. J. Klose, *Humangenetik*, *26*, 231–243 (1975).
8. P.H. O'Farrell, *J. Biol. Chem.* *250*, 4007–4021 (1975).
9. N.G. Anderson, N.L. Anderson, *Anal. Biochem.* *85*, 331–340 (1978).
10. N.L. Anderson, N.G. Anderson, *Anal. Biochem.* *85*, 341–354 (1978).
11. A. Görg, W. Postel, R. Westermeier, E. Gianazza, P.G. Righetti, *J. Biochem. Biophys. Meth.* *3*, 273–284 (1980).
12. B. Bjellqvist, K. Ek, P.G. Righetti, E. Gianazza, A. Görg, R. Westermeier, W. Postel, *J. Biochem. Biophys. Meth.* *6*, 317–339 (1982).
13. P.G. Righetti, *Immobilized pH Gradients: Theory and Methodology*, R.H. Burdon, P.H. van Knippenberg, eds., Elsevier, Amsterdam, 1989.
14. A. Görg, W. Postel, S. Günther, *Electrophoresis*, *9*, 531–546 (1988).
15. A. Görg, W. Postel, J. Weser, S. Günter, J.R. Strahler, S.M. Hanash, L. Somerlot, *Electrophoresis*, *8*, 45–51 (1987).
16. M. Rimpilainen, P.G. Righetti, *Electrophoresis*, *6*, 419–422 (1985).
17. P.Z. O'Farrell, H.M. Goodman, P.H. O'Farrell, *Cell*, *12*, 1133–1142 (1977).
18. J.I. Garrels, *Dev. Biol.*, *73*, 134–152 (1979).
19. M.J. Dunn, A.H.M. Burghes, *Electrophoresis*, *4*, 97–116 (1983).
20. R. Westermeier, *Electrophoresis in Practice: A Guide to Theory and Practice.* VCH, (1993).
21. P. Sinha, E. Köttgen, R. Westermeier, P.G. Righetti, *Electrophoresis*, *13*, 210–214 (1992).
22. J. Asakawa, *Electrophoresis*, *9*, 562–568 (1988).
23. D. Hochstrasser, V. Augsburger, M. Funk, R. Appel, C. Pellegrini, A.F. Muller, In *Electrophoresis '86*, M.J. Dunn, ed. VCH, Weinheim, 1986, pp. 566–568.
24. R. Westermeier, W. Postel, J. Weser, A. Görg, *J. Biochem. Biophys. Meth.* *8*, 321–330 (1983).
25. B. Bjellqvist, M. Linderholm, K. Östergren, J.R. Strahler, *Electrophoresis*, *9*, 453–462 (1988).
26. A. Görg, W. Postel, J. Weser, S. Günther, JR. Strahler, S.M. Hanash, L. Somerlot, R. Kuick, *Electrophoresis*, *9*, 37–46 (1988).

27. B. Werner, K. Andersson, F. Lottspeich, M. Kehl, Sonderdruck Pharmacia Freiburg SD-097, 1990.

28. S.M. Hanash, J.R. Strahler, J.V. Neel, N. Hailat, R. Melham, D. Keim, X.X. Zhu, D. Wagner, D.A. Gage, J.T. Watson, *Proc. Natl. Acad. Sci. USA, 88,* 5709–5713 (1991).

29. B. Bjellqvist, J.-C. Sanchez, C. Pasquali, F. Ravier, N Paquet, S. Frutiger, G.J. Hughes, D. Hochstrasser, *Electrophoresis, 14,* 1375–1378 (1993).

30. A. Görg, *Biochem. Soc. Trans., 21,* 130–132 (1993).

31. A. Görg, W. Postel, C. Friedrich, R. Kuick, J.R. Strahler, S.H. Hanash, *Electrophoresis, 12,* 653–658 (1991).

32. A. Görg, W. Postel, S. Günther, J. Weser, *Electrophoresis, 6,* 599–604 (1985).

33. A. Görg, In *Cell Biology: A Laboratory Handbook,* J.E. Celis, ed., Academic Press Inc., 1994.

34. A. Görg, W. Postel, J. Weser, S. Günther, J.R. Strahler, S.M. Hanash, L. Somerlot, *Electrophoresis, 8,* 122–124 (1987).

35. U.K. Laemmli, *Nature, 227,* 680–685 (1970).

36. A. Görg, *Nature, 349,* 545–546 (1991).

37. J.R. Strahler, S.M. Hanash, L. Somerlot, J. Weser, W. Postel, A. Görg, *Electrophoresis, 8,* 165–173 (1987).

38. S.M. Hanash, J.R. Strahler, L. Somerlot, W. Postel, A. Görg, *Electrophoresis, 8,* 229–234 (1987).

39. D.F. Hochstrasser, S. Frutiger, N. Paquet, A. Bairoch, F. Ravier, C. Pasquali, J.-C. Sanchez, J.-D. Tissot, B. Bjellqvist, R. Vargas, R.D. Appel, G.J. Hughes, *Electrophoresis, 13,* 992–1001 (1992).

40. C. Eckerskorn, P. Jungblut, W. Mewes, J. Klose, F. Lottspeich, *Electrophoresis, 9,* 830–838 (1988).

41. M.J. Dunn, A.H.M. Burghes, *Electrophoresis, 4,* 173–189 (1983).

42. J. Heukeshoven, R. Dernick, *Electrophoresis, 6,* 103–112 (1985).

43. J. Heukeshoven, R. Dernick, *Electrophoresis, 9,* 60–61 (1988).

44. W.M. Bonner, R.A. Laskey, *Eur. L. Biochem., 46,* 83–88 (1974).

45. J.P. Chamberlain, *Anal. Biochem., 98,* 132–135 (1979).

46. C.M. Merrill, D. Goldman, S.A. Sedman, M.H. Ebert, *Science, 211,* 1437–1438 (1981).

47. H. Blum, H. Beier, H.J. Gross, *Electrophoresis, 8,* 93–99 (1987).

48. T. Rabilloud, L. Vuillard, C. Gilly, J.-J. Lawrence, *Cell. Mol. Biol., 40,* 57–75 (1994).

49. R. Kellner, F. Lottspeich, H.E. Meyer, *Microcharacterization of Proteins,* VCH Weinheim, 1994.

50. V. Leduc-Brodard, B. David, G. Peltre, *Cell. Mol. Biol., 40,* 1–8 (1994).

51. M.J. Dunn, ed. Paper symposium: Quantitative evaluation and densitometry. *Electrophoresis, 11,* 355–424 (1990).

52. P.J. Collins, C. Juhl, J.-L. Lognonné, *Cell. Mol. Biol., 40,* 77–83 (1994).

53. N.G. Anderson, N.L. Anderson, *Clin. Chem., 28,* (1982).

54. J.E. Celis, P. Madsen, B. Gesser, S. Kwee, H.V. Nielsen, H.H. Rasmussen, B. Honoré, H. Leffers, G.P. Ratz, B. Basse, J.B. Lauridsen, A. Celis, In *Advances in Electrophoresis* A. Chrambach, M.J. Dunn, B.J. Radola, eds., VCH Weinheim, 1989, Vol. 3, pp. 3–181.

55. J.E. Celis, ed. Paper symposium: Protein databases in two-dimensional electrophoresis. *Electrophoresis, 10,* 71–164 (1989).

56. J.E. Celis, ed. Paper symposium: Protein databases in two-dimensional electrophoresis. *Electrophoresis, 14,* (1993).

57. R.D. Appel, J.-C. Sanchez, O. Golaz, M. Miu, J.R. Vargas, D.F. Hochstrasser, *Electrophoresis, 14,* 1232–1238 (1993).

58. N.L. Anderson, N.G. Anderson, In *Clinical Chemistry. Special Issue: Two-Dimensional Electrophoresis,* 1984, Vol. 30, pp. 1897–2108.

59. D.F. Hochstrasser, J.-D. Tissot, In *Advances in Electrophoresis,* A. Chrambach, M.J. Dunn, B.J. Radola, eds., VCH Weinheim, 1993, Vol. 6, pp. 270–376.

60. B.B. Rosenblum, J,V. Neel, S.M. Hanash, *Proc. Natl. Acad. Sci. USA, 80,* 5002–5006 (1983).

61. D.S. Young, R.P. Tracy, *Electrophoresis, 4,* 117–121 (1983).

62. V. Brodard, B. David, A. Görg, G. Peltre, *Int. Arch. Allergy Immunol., 102,* 72–80 (1993).

63. W. Weiss, C. Vogelmeier, A. Görg, *Electrophoresis, 14,* 805–816 (1993).

64. K.E. Willard-Gallo, In *Advances in Electrophoresis* A. Chrambach, M.J. Dunn, B.J. Radola, eds., VCH Weinheim, 1989, Vol. 3, pp. 221–272.

65. S.M. Hanash, In *Advances in Electrophoresis* A. Chrambach, M.J. Dunn, B.J. Radola, eds., VCH Weinheim, 1993, Vol. 2, pp. 341–384.

66. C. Damerval, M. Zivy, F. Granier, D. de Vienne, In *Advances in Electrophoresis* A. Chrambach, M.J. Dunn, B.J. Radola, eds., VCH Weinheim, 1993, Vol. 2, pp. 263–340.

67. C. Damerval, D. de Vienne, eds., Paper symposium: Two-dimensional electrophoresis of plant proteins. *Electrophoresis, 9,* 679–796 (1988).

68. J. Klose, ed. Paper symposium: Blotting and sequencing. *Electrophoresis, 11,* 515–594 (1990).

69. B.P. Voris, O.A. Young, *Anal. Biochem., 104,* 478–484 (1980).

70. J. Klose, In *Modern Methods in Protein Chemistry: Review Articles,* H. Tschesche, ed., Walter de Gruyter, Berlin, 1983, pp. 49–78.

71. A.T. Andrews, *Electrophoresis: Theory, Techniques, and Biochemical and Clinical Applications,* Clarendon Press, Oxford, 1986.

72. A. Görg, W. Postel, S. Günther, C. Friedrich, *Electrophoresis, 9,* 57–59 (1988).

73. K. Nokihara, N. Morita, T. Kuriki, *Electrophoresis, 13,* 701–707 (1992).

74. H.P. Schickle, German patent 42 44 082, 1992, U.S. patent pending, Japanese patent pending.

16 Protein Elution and Blotting Techniques

REINER WESTERMEIER

Pharmacia Biotech Europe
D-79111 Freiburg, Germany

16.1 INTRODUCTION

After electrophoretic separation, further characterization of the protein bands or spots can be performed outside the gel matrix. Either the desired fraction is eluted directly into a liquid phase or the complete separation pattern is transferred onto immobilizing membranes. The latter method is called blotting.

Direct elution either needs a previous detection of the fraction or a series of gel segments have to be eluted in parallel. Because the protein is often diluted during the procedure, it is mostly followed by a concentrating step.

Protein Purification: Principles, High-Resolution Methods, and Applications, Second Edition.
Edited by Jan-Christer Janson and Lars Rydén.
ISBN 0-471-18626-0. © 1998 Wiley-VCH, Inc.

Blotting techniques have been more preferable than direct elution, as many new and sophisticated analytical micromethods have become available. Gentle detection methods can be performed on the blotting membrane and no concentrating steps are necessary.

16.2 PROTEIN ELUTION

When a protein should be purified by an electrophoretic method and is needed in a liquid phase, the sample is preferably separated under native conditions either by native basic or acidic electrophoresis or by isoelectric focusing.

16.2.1 Elution by Diffusion

Elution by diffusion is only recommended for agarose and dextran gels because polyacrylamide gels are very restrictive.

Because isoelectric focusing has a high loading capacity and an inbuilt concentration effect, it is the most suitable electrophoretic method for the preparative purification of proteins.[1] Furthermore, this method can be performed in a granulated gel (e.g., in Sephadex) as it does not need the sieving properties of a gel. The pH gradient can be measured directly on the flat bed surface or paper prints can be taken from the surface to detect the positions of the desired fractions.[2] With a fractionating grid and a spatula, the fractions of interest are removed from the flat bed, placed in a column, and eluted with buffer. It is not always necessary to separate the carrier ampholytes (Ampholine or Pharmalyte) and protein before performing further steps.

16.2.2 Electrophoretic Elution

Elution from polyacrylamide has to be performed in an electric field.

Nguyen and co-workers[3] used isotachophoresis (displacement electrophoresis) to move the proteins out of a gel slice in a compact zone and collected them inside a dialysis membrane. Because of the danger of adsorption of protein to the membrane, Öfverstedt and colleagues[4] electrophoretically drove the proteins into Sephadex G-25, from where further elution can be performed in an easy manner.

Very effective methods for electroelution from polyacrylamide gels combine the electrophoretic transport with a concentrating step. This can be done by employing a discontinuous conductivity gradient[5,6] Figure 16-1a shows the principle of this technique: A polyacrylamide supporting gel is polymerized into a glass tube. This gel contains 10% T and 25 mmol/L Tris/75 mmol/L glycine, pH 8.8. The gel slices to be eluted are imbedded in 1.5% agarose containing this buffer. The recovery solution contains this buffer and 40% (w/v) glycerol. The tube is carefully filled with 2 mol/L sodium chloride. The running

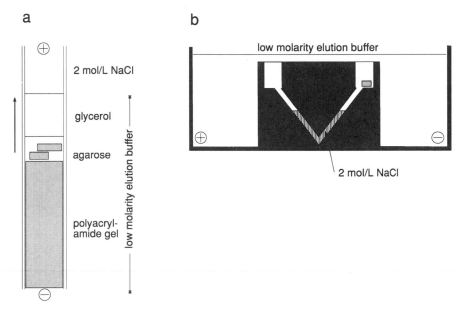

FIGURE 16-1. Electroelution of proteins from polyacrylamide gel slices into a discontinuous conductivity gradient. (a) See Refs. 5 and 6. (b) See Ref. 7.

buffer consists of 50 mmol/L Tris/150 mmol/L glycine, pH 8.8. The eluted sample collects in the interface between the liquid buffer and the salt layer due to the steep conductivity step.

This principle is also used in the constuction of a widely used elution device:[7] Here the salt layer is introduced in a V-boring in the divider of a horizontal buffer chamber (see Figure 16-1b).

16.3 PROTEIN BLOTTING

16.3.1 Principle

Blotting is the transfer of large molecules onto the surface of an immobilizing membrane. The proteins adsorbed on the membrane surface are freely available for macromolecular ligands, for example, antigens, antibodies, lectins, or nucleic acids. Before the specific detection, the free binding sites of the membrane must be blocked with substrates that do not take part in the ensuing reaction (Figure 16-2). Several detection steps can be applied one after the other because the proteins are fixed on the membrane surface.

In addition, blotting is an intermediate step in protein sequencing and an elution method for subsequent amino acid analysis. A very comprehensive review on protein blotting can be found in Beisiegel.[8]

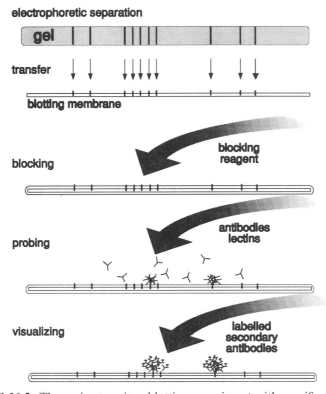

FIGURE 16-2. The main steps in a blotting experiment with specific detection.

The main features of protein blotting are:

- Native fixing of proteins, they do not lose their biological activities.
- Accessibility for specific detection methods, (e.g., antigen–antibody reactions, glycoprotein–lectin affinity).
- Immunological detection is also possible with monoclonal antibodies, which do not precipitate.
- Low consumption of antibodies.
- Short incubation times.
- Easy long-term storage of proteins (up to 1 yr).
- Highly sensitive nonradioactive detection methods possible (e.g., chemiluminescence).
- Purification method for monospecific antibodies.

16.3.2   Transfer Methods

Diffusion blotting: The blotting membrane is just laid on the gel surface. The proteins are transferred by diffusion. It is mainly used after isoelectric focusing in gels with large pores (e.g., agarose).[8,9] For polyacrylamide isoelectric focusing, it is recommended to place a stack of filter paper on top of the blot sandwich, followed by a glass plate and a weight. Thus, a high portion of the proteins (up to 90%) are pressed out of the gel onto the surface of the membrane.

Capillary blotting: Braun and Abraham[10] have introduced a simple, but effective technique for transferring proteins out of PhastGel media gradient 10–15 (Figure 16-3): The gel is laid on the blotting membrane, which is laid on a strip of filter paper soaked with phosphate-buffered saline, pH 7.2. The sandwich is clamped between two small glass plates. One end of the filter paper strip is immersed in the buffer, with the long end hanging down freely, preferably over the edge of a table. Alternatively, for thermolabile proteins, the setup is placed into a refrigerator. The constant flow of the buffer below the membrane causes a more efficient transfer of proteins than diffusion alone. Although the gels are not removed from the plastic support, an almost complete protein elution can be achieved within 2 h.

Vacuum blotting: This technique is used only for agarose gels because polyacrylamide gels can stick indefinitely to the membrane after this procedure.

Electrophoretic blotting: Electrophoretic transfers are the most effective for proteins after electrophoresis in polyacrylamide.[11,12]

Originally, vertical **buffer tanks** with platinum wire electrodes on two side walls were used. For this technique, the gel and blotting membrane were clamped in grids between filter papers and sponge pads, and suspended in the tank filled with cooled buffer. These transfers usually were run overnight.

Semidry blotting between two horizontal graphite plate electrodes is simpler, less expensive, and faster and a discontinuous buffer system can be used.[13,14] Graphite is the best material for electrodes in semidry blotting because it conducts well, does not have to be cooled, and does not catalyze

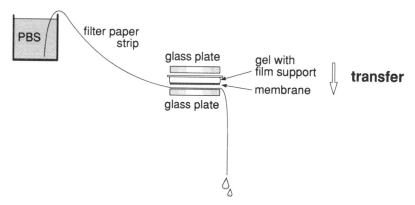

FIGURE 16-3. Assembly of a so-called modified diffusion blot. The effectiveness and speed of the transfer are supported by the capillary transport of buffer through the paper strip. (From Ref. 10.)

reactions with anodal oxidation products. A current no higher than 0.8 per cm^2 of blotting area is recommended. If higher currents are used, the gel can warm up, resulting in the precipitation of some proteins.

The transfer time is approximately 1 h and depends on the thickness and the concentration of the gel. When longer transfer times are required, as for thick (>1 mm) or highly concentrated gels, a weight is placed on the upper plate so that the electrolyte gas is expelled out of the sides.

Figure 16-4 shows a diagram of a semidry blot. The stacks of filter paper are soaked in the respective buffer before they are laid onto each other on the graphite blotter.

When gels supported by films, such as ready-made horizontal gels, are blotted electrophoretically, the film has to be removed without damage of the gel. Figure 16-5 shows an apparatus for pulling a taut, thin, stainless-steel wire between gel and film support.

Figure 16-6 shows the graphite electrode system NovaBlot, which is placed into the electrophoresis chamber instead of the cooling plate. A detailed methodological description of the practical steps and a troubleshooting guide can be found in Ref. 15.

It is also possible to perform electrophoretic transfers on two membranes simultaneously using *double replica blotting*.[16] An alternating electric field is applied on a sandwich with a membrane on each side of the gel with an increasing pulse time, resulting in two symmetrical blots.

16.3.3   Blotting Membranes

Nitrocellulose is the most commonly used membrane. It is available in pore sizes from 0.05 to 0.45 μ. The pore size is a measure of the specific surface: the smaller the pores, the higher the binding capacity. Disadvantages include limited binding capacity and poor mechanical stability. A better adsorption of glycoproteins, lipids, and carbohydrates is obtained by precoating the nitrocellulose with ligand.[17]

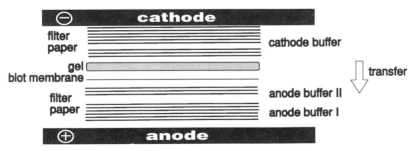

FIGURE 16-4. Schematic representation of semidry blotting with a discontinuous buffer system.[13]

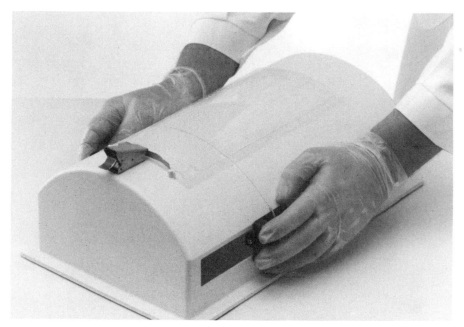

FIGURE 16-5. Removing the carrier film from the gel with a taut, thin wire (Pharmacia Biotech AB).

Polyvinylidenedifluoride (PVDF) membranes on a Teflon base possess a high binding capacity and a high mechanical stability similar to nylon membranes. PVDF membranes can also be used for direct protein sequencing.[18,19]

Polyamide (nylon) membranes possess a high mechanical stability and a high binding capacity, usually due to electrostatic interactions. This means that staining can be a problem because small molecules are also strongly bound. Positively and negatively charged, as well as neutral, nylon membranes are available. Fixing with glutaraldehyde after transfer is recommended to increase the binding of low molecular weight peptides to nylon membranes.[20]

Ion-exchange membranes, diethylaminoethyl (DEAE) or carboxymethyl (CM), are used for preparative purposes because of the reversibility of the ionic bonds.

Activated glass fiber membranes are used when blotted proteins are directly sequenced. Several methods can be used to activate the surface: bromocyanide treatment, derivatization with positively charged silanes,[21] or hydrophobization by siliconizing.[22]

During electroblotting a portion of the proteins—mostly smaller individuals—migrate through the film whereas larger proteins do not completely leave the gel. To trap low molecular weight proteins, sometimes a second blotting membrane is placed behind the first membrane. It would be desirable to obtain as regular transfers as possible. Unfortunately, a blotting membrane that binds 100% of the molecules does not yet exist.

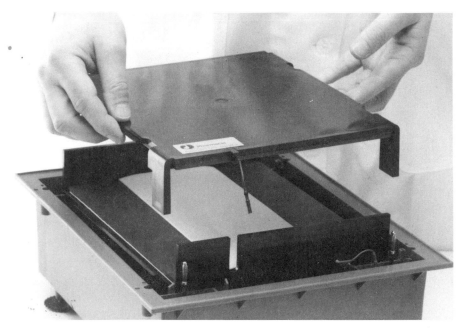

FIGURE 16-6. Assembling the blot sandwich in the graphite blotter NovaBlot (Pharmacia Biotech AB).

Thus, blotting conditions have to be optimized for a simultaneous transfer of high and low molecular weight proteins. When SDS electrophoresis is used for the separation, a *porosity gradient* gel should be employed in order to provide high retardation for small proteins and low retardation for high molecular weight proteins. Choosing the correct buffer is also important for optimal transfer.

16.3.4   Buffers for Electrophoretic Transfers

A continuous Tris–glycine–SDS buffer is often recommended for semidry blotting. However, experience has shown that *discontinuous buffer system* is preferable as it yields sharper bands and more regular and efficient transfers.

The continuous buffer (e.g., for double replica electroblotting) consists of 48 mmol/L Tris, 39 mmol/L glycine, 0.0375% (w/v) SDS, and 20% methanol. The discontinuous buffer system (according to Ref. 13) consists of

- Anode I: 0.3 mol/L Tris, 20% methanol
- Anode II: 25 mmol/L Tris, 20% methanol
- Cathode: 40 mmol/L 6-aminohexanoic acid (same as ε-aminocaproic acid), 20% methanol, and 0.01% SDS

This buffer system can be used for SDS as well as for native and IEF gels. Methanol serves two functions: (1) it avoids swelling of gel during the transfer and (2) it improves the binding capacity of nitrocellulose, which can be necessary, particularly in the presence of SDS.

If the transfer efficiency of high molecular weight proteins (<80,000 Da) is not satisfactory, the gel can be equilibrated in the cathode buffer for 5 to 10 min before blotting. For enzyme detection the buffer must not contain methanol, or else biological activity is lost. A brief contact with a small amount of SDS does not denature the proteins.

Figure 16-7 shows a stained nitrocellulose membrane after semidry blotting with the discontinuous buffer system after SDS electrophoresis in a ready-made porosity gradient gel: an ExcelGel SDS gradient 8–18.

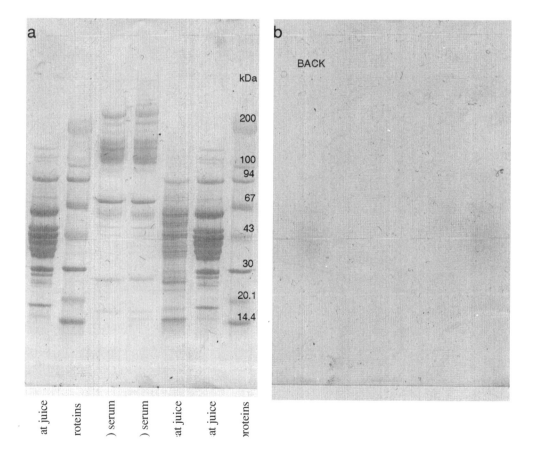

FIGURE 16-7. Blot of muscle and human serum proteins (* desalted and albumin is removed with blue dextran) with the discontinuous buffer after SDS–electrophoresis in an ExcelGel SDS gradient 8–18. General staining with Fast Green FCF. Also note that high molecular weight proteins (>200,000) have been efficiently transferred. (a) Front side. (b) Back: no through blotting of low molecular weight proteins.

16.3.5   General Staining

Proteins adsorbed on nitrocellulose can be reversibly stained so that the total protein can be estimated before specific detection.[23] In addition to staining with amido black or Coomassie brilliant blue, mild staining methods such as the very sensitive Indian ink method[24] exist as well as reversible ones with Ponceau S[23] or Fast Green FCF (see Figure 16-7). The sensitivity of Indian ink staining and the antibody reactivity of the proteins can be enhanced by alkaline treatment of the blotting membrane.[25] Unfortunately, indian ink is at present not produced any more, but it is hoped that a replacement for this nontoxic, inexpensive, and sensitive dye can be found as soon as possible.

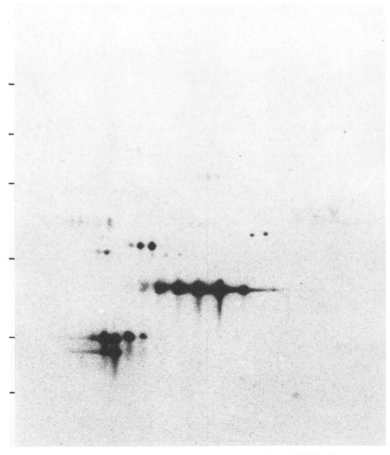

FIGURE 16-8. Two-dimensional separation of specific IgE-binding proteins (allergens), visualization with ^{125}I-labeled antihuman IgE followed by autoradiography (2-wk exposure time).[33] (Courtesy of W. Weiss, Technische Universität München, Weihenstephan.)

Blotting is often applied because very sensitive general detection methods are possible on a membrane: a general immunostain,[26] colloidal gold,[27] fluorography,[11] and chemiluminescence[28] with the highest sensitivity currently available.

Because nylon membranes bind anionic dyes very strongly, normal staining is not possible; however, nylon membranes can be stained with cacodylate iron colloid (FerriDye).[29]

16.3.6 Blocking

Macromolecular substances that do not take part in the visualization reaction are used to block the free binding sites on the membrane.

- A number of possibilities exist: 2 to 10% bovine albumin[11]
- Skim milk or 5% skim milk powder[30]

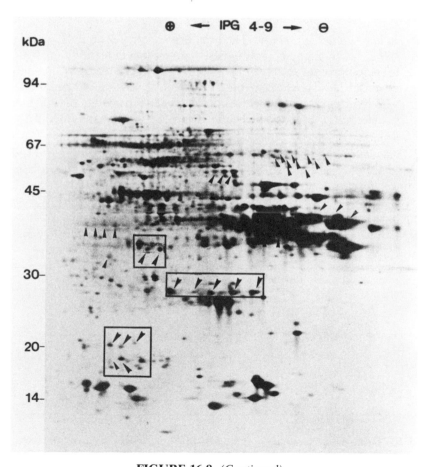

FIGURE 16-8. *(Continued).*

- 3% fish gelatine
- 0.05% Tween 20
- Casein preparations with a wide spectrum of different molecular sizes block membranes very effectively

16.3.7 Specific Detection

Enzyme blotting: The transfer of native separated enzymes onto blotting membranes has the advantage that the proteins are fixed without denaturation and thus do not diffuse during slow enzyme–substrate reactions and the coupled staining reactions.[31]

Immunoblotting: Specific binding of immunoglobulins (IgG) or monoclonal antibodies are used to probe for individual protein zones after blocking. An additional marked protein is then used to visualize the zones. Several possibilities exist:

1. Radioiodinated protein A. The use of radioactive protein A, which attaches itself to specific binding antibodies, enables high detection sensitivities,[32] but ^{125}I-labeled protein A only binds to particular IgG subclasses. In addition, radioactive isotopes should be avoided as much as possible in the laboratory. Figure 16-8 shows a blot of a two-dimensional separation of IgE-binding proteins (allergens), visualization with ^{125}I-labeled antihuman IgE, and autoradiography (2-wk exposure time).

2. Secondary antibodies with an enzyme label. An antibody to the specific binding antibody is used, which is conjugated to an enzyme. Peroxidase[33] or alkaline phosphatase[34] is usually employed as the conjugated reagent. The ensuing enzyme-substrate reactions have a high sensitivity. Figure 16-9 shows wheat gliadins separated by SDS–electrophoresis and detected by a rabbit antigliadin horseradish peroxidase conjugate on a PVDF membrane.[33]

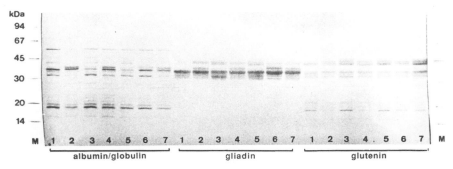

FIGURE 16-9. Wheat gliadins separated by SDS–electrophoresis and detected by a rabbit antigliadin horseradish peroxidase conjugate on a PVDF membrane.[33] (Courtesy of W. Weiss, Technische Universität München, Weihenstephan.)

3. Gold coupled secondary antibody. Detection by coupling the antibody to *colloidal gold* is very sensitive[35]; in addition, the sensitivity can be increased by subsequent *silver enhancement:* the lower limit of detection lies around 100 pg.[36]

4. Avidin–biotin system. Higher sensitivity is achieved with an amplifying enzyme detection system, with peroxidase complexes,[37] or complexes with alkaline phosphatase.

5. Chemiluminescence. The highest sensitivity without using radioactivity can be achieved with enhanced chemiluminescent detection methods.[28]

Lectin blotting: The detection of glycoproteins and specific carbohydrate moieties is performed with lectins. The visualization methods are carried out by aldehyde detection or, in analogy to immunoblotting, with the avidin–biotin method.[38]

16.3.8 Protein Sequencing

The use of blotting for direct protein sequencing[18,21,22,39,40] and amino acid composition analysis[41] has been a big step forward for protein chemistry and molecular biology. One-dimensional SDS gels or two-dimensional gels are usually blotted. If the proteins were separated by isoelectric focusing, an immobilized pH gradient should be used because carrier ampholytes would interfere with the sequencing signals.[42] References 43 and 44 present comprehensive reviews of different methods. The different membranes on the market have been checked for their sequencing properties in Ref. 45. The techniques are thoroughly described in Ref. 46.

16.3.9 Mass Spectrometry

Since the introduction of matrix-assisted laser desorption ionization mass spectrometry,[47] the molecular mass of proteins can be measured. This is performed on polymer membranes after SDS electrophoresis or two-dimensional electrophoresis and subsequent blotting with high precision.[48,49]

16.4 REFERENCES

1. J.C. Janson, Ph.D. thesis, Acta Universitatis Upsaliensis No. 5 (1972).
2. B.J. Radola, *Biochim. Biophys. Acta,* 412–428 (1973).
3. N.Y. Nguyen, J. diFonzo, A. Chrambach, *Anal. Biochem., 106,* 78–91 (1980).
4. L.-G. Öfverstedt, Johansson, G. Fröman, S. Hjertén, *Electrophoresis, 2,* 168–173 (1981).
5. M. Otto, M. Sneidárková, *Anal. Biochem., 111,* 111–114 (1983).
6. P. Strålfors, P. Belfrage, *Anal. Biochem., 128,* 7–10 (1983).

7. S. Diekmann, In *Electrophoresis '84,* (V. Neuhoff ed., VCH Weinheim, 1984, pp. 154–155.

8. U. Beisiegel, *Electrophoresis, 7,* 1–18 (1986).

9. G. Peltre, J. Lapeyre, B. David, *Immunol. Lett., 5,* 127–131 (1982).

10. W. Braun, R. Abraham, *Electrophoresis, 10,* 249–253 (1989).

11. H. Towbin, T. Staehelin, J. Gordon, *Proc. Natl. Acad. Sci. USA, 76,* 4350–4354 (1979).

12. W.N. Burnette, *Anal. Biochem., 112,* 195–203 (1981).

13. J. Kyhse-Andersen, *Biochem. Biophys. Meth., 10,* 203–209 (1984).

14. E.R. Tovey, B.A. Baldo, *Electrophoresis, 8,* 384–387 (1987).

15. R. Westermeier, *Electrophoresis in Practice: A Guide to Theory and Practice,* VCH 1993.

16. K.-E. Johansson, *Electrophoresis, 8,* 379–383 (1987).

17. E. Handmann, H.M. Jarvis, *J. Immunol. Meth., 83,* 113–123 (1985).

18. P. Matsudaira, *J. Biol. Chem., 262,* 10035–10038 (1987).

19. H.D. Kratzin, J. Wiltfang, M. Karas, V. Neuhoff, N. Hilschmann, *Anal. Biochem., 183,* 1–8 (1989).

20. K.P. Karey, D.A. Sirbasku, *Anal. Biochem., 178,* 255–259 (1989).

21. R.H. Aebersold, D. Teplow, L.E. Hood, S.B.H. Kent, *J. Biol. Chem., 261,* 4229–4238 (1986).

22. C. Eckerskorn, W. Mewes, H. Goretzki, F. Lottspeich, *Eur. J. Biochem., 176,* 509–519, (1988).

23. O. Salinovich, R.C. Montelaro, *Anal. Biochem., 156,* 341–347 (1986).

24. K. Hancock, V.C.W. Tsang, *Anal. Biochem., 133,* 157–162 (1983).

25. M.W. Sutherland, J.H. Skerritt, *Electrophoresis, 7,* 401–406 (1986).

26. J.M. Kittler, N.T. Meisler, D. Viceps-Madore, *Anal. Biochem., 137,* 210–216 (1984).

27. M. Moeremans, G. Daneels, J. De Mey, *Anal. Biochem., 145* 315–321 (1985).

28. P. Laing, *J. Immunol. Meth., 92,* 161–165 (1986).

29. M. Moeremans, M. De Raeymaeker, G. Daneels, J. De Mey, *Anal. Biochem., 153,* 18–22 (1986).

30. D.A. Johnson, J.W. Gautsch, J.R. Sportsman, *Gene Anal. Technol., 1,* 3–8 (1984).

31. B.G. Olsson, B.R. Weström, B.W. Karlsson, *Electrophoresis, 8,* 377–464, (1987).

32. J. Renart, J. Reiser, G.R. Stark, *Proc. Natl. Acad. Sci. USA., 76,* 3116–3120 (1979).

33. W. Weiss, C. Vogelmeier, A. Görg, *Electrophoresis, 14,* 805–816 (1993).

34. M.S. Blake, K.H. Johnston, G.J. Russell-Jones, *Anal. Biochem., 136,* 175–179.

35. D. Brada, J. Roth, *Anal. Biochem., 142,* 79–83 (1984).

36. M. Moeremans, G. Daneels, A. Van Dijck, G. Langanger, J. De Mey, *J. Immunol. Meth., 74,* 353–360 (1984).

37. D.-M. Hsu, L. Raine, H. Fanger, *J. Histochem. Cytochem., 29,* 577–580 (1981).

38. E.A. Bayer, H. Ben-Hur, M. Wilchek, *Anal. Biochem., 161,* 123–131 (1987).

39. J. Vandekerckhove, G. Bauw, M. Puype, J. Van Damme, M. Van Montegu, *Eur. J. Biochem., 152,* 9–19 (1985).

40. C. Eckerskorn, F. Lottspeich, *Chromatographia, 28,* 92–94 (1989).

41. C. Eckerskorn, P. Jungblut, W. Mewes, J. Klose, F. Lottspeich, *Electrophoresis, 9,* 830–838, (1988).

42. R.H. Aebersold, G. Pipes, L.H. Hood, S.B.H. Kent, *Electrophoresis, 9,* 520–530, (1988).

43. R.J. Simpson, R.L. Moritz, G.S. Begg, M.R. Rubira, E.C. Nice, *Anal. Biochem., 177,* 221–236 (1989).

44. J. Klose, *Electrophoresis, 11,* 515–594 (1990).

45. C. Eckerskorn, F. Lottspeich, *Electrophoresis, 14,* 831–838. (1993).

46. R. Kellner, F. Lottspeich, H.E. Meyer, *Microcharacterization of Proteins,* VCH Weinheim, 1994.

47. M. Karas, D. Bachmann, U. Bahr, F. Hillenkamp, *Int. J. Mass Spectrom. Ion Proc., 78,* (1987).

48. C. Eckerskorn, K. Strupat, M. Karas, F. Hillenkamp, F. Lottspeich, *Electrophoresis, 13,* 664–665 (1992).

49. K. Strupat, M. Karas, F. Hillenkamp, C. Eckerskorn, F. Lottspeich, *Anal. Chem., 66,* 464–470 (1994).

17 Capillary Electrophoretic Separations

WOLFGANG THORMANN

Department of Clinical Pharmacology
University of Bern
CH-3010 Bern, Switzerland

Abbreviations

ACE, affinity capillary electrophoresis

BLB, β-lactoglobulin B

Brij 35, polyoxyethylene 23 lauryl ether

CAL, conalbumin

CE, capillary electrophoresis

CGE, capillary gel electrophoresis

CHAPSO, 3 [(3-cholamidopropyl)-dimethylammonio]-2-hydroxy-1-propanesulfonate

Protein Purification: Principles, High-Resolution Methods, and Applications, Second Edition.
Edited by Jan-Christer Janson and Lars Rydén.
ISBN 0-471-18626-0.   © 1998 Wiley-VCH, Inc.

CREAT, creatinine

CYTC, cytochrome c

CZE, capillary zone electrophoresis

CIEF, capillary isoelectric focusing

CITP, capillary isotachophoresis

CFE, continuous flow electrophoresis

DZE, discontinuous zone electrophoresis

FER, ferritin

FEP, fluorinated ethylenepropylene polymer

FFF, field flow fractionation

EFFF, electrical field flow fractionation

EHFFF, electrical hyperlayer field flow fractionation

EMMA, electrophoretically mediated microanalysis

EPC, electropolarization chromatography

GC-MS, gas chromatography–mass spectrometry

Hb, hemoglobin

HEC, high-performance electrochromatography

HGM, human growth hormone

HPLC, high-performance liquid chromatography

HPMC, hydroxypropylmethyl cellulose

IEF, isoelectric focusing

ITP, isotachophoresis

LC, liquid chromatography

LIF, laser-induced fluorescence

LPA, linear polyacrylamide

LYSO, lysozyme

MC, methylcellulose

MECC, micellar electrokinetic capillary chromatography

MS, mass spectrometry

MYO, myoglobin

OVA, ovalbumin

PAGE, polyacrylamide gel electrophoresis

PEG, polyethylene glycol

PTFE, polytetrafluoroethylene (Teflon)

PVA, polyvinyl alcohol

RNase, ribonuclease A

SDS, sodium dodecyl sulfate

SEC, size exclusion chromatography

Tris, tris(hydroxymethyl)aminomethane
ZE, zone electrophoresis

SUMMARY

Electrophoretic separations in capillaries of small inner diameters and thin fluid films have received considerable attention since the mid-1960s. In this period a number of instrumental approaches for capillary zone electrophoresis, capillary isotachophoresis, capillary isoelectric focusing, capillary gel electrophoresis, electrokinetic capillary chromatography, continuous flow electrophoresis, and electrical field flow fractionation or electropolarization chromatography have emerged. High separation efficiencies have been demonstrated in conjunction with high-resolution on-column sample detection. For the separation and analysis of proteins, capillary electrophoresis complements other existing techniques, particularly high-performance liquid chromatography (HPLC) and slab gel electrophoresis. The principles of capillary electrophoresis and instrumentation for the various techniques and their use for protein separation and analysis are reviewed. Special applications dealing with affinity electrophoresis, immunoassays, reaction based chemical analysis, hyphenation with mass spectrometry, and two-dimensional separations combining liquid chromatograpy and capillary electrophoresis are also described.

17.1 INTRODUCTION: CAPILLARY ELECTROPHORETIC TECHNIQUES

Electrophoresis is the premier analytical separation method for biological compounds such as proteins and polynucleotides despite recent applications of HPLC to these biopolymers. Unfortunately, unlike chromatographic methods, classical electrophoresis, that is, electrophoresis conducted in solid support media such as slab gels, is rather slow, laborious, difficult to quantify, and not really adaptable to complete automation. For example, although an actual gel electrophoretic separation can be achieved within minutes and multiple samples can be processed simultaneously, completion of the entire experiment typically requires several hours. Obviously, a more instrumental approach that is capable of effectively handling the complex interaction of mass transport by electromigration, diffusion, flow, chemical equilibria, and partitioning between phases is required. One possible solution has been the exploration and development of electrophoretic separations and analyses in the capillary format, generally referred to as capillary electrophoresis (CE). CE techniques have received considerable attention since the mid-1960s, having demonstrated highly efficient separations.[1-9] However, their widespread use in bio-polymer separations and analyses and other fields of applications has only emerged in the past few years.

CE methods should be regarded as complementary or as attractive alternatives to other capillary separation techniques, such as gas chromatography, liquid chromatography, supercritical fluid chromatography, and field flow fractionation. The advantages of CE are high resolution, efficiency, mass sensitivity and speed, full automation, minute sample size, rapid method development, and the use of small amounts of inexpensive and nonpolluting chemicals, as well as simple adaptation for micropreparative work. More important, CE techniques can exploit numerous separation principles, making them flexible and easily applied to a variety of separation problems. For example, CE has been used to separate a broad spectrum of species and compounds ranging from small molecules (inorganic and organic ions) to large molecules and particles (proteins, oligonucleotides, and cells). However, the concentration sensitivity is somewhat lower than in many other techniques, including HPLC, often calling for effective on-line or off-line preconcentration of analytes prior to analysis. Fortunately, electrophoretic techniques feature unique concentration effects (which are inherent to electrophoretic mass transport and very rarely seen in other separation techniques), providing some compensation for the low concentration sensitivity.[10,11]

The various CE techniques can be categorized according to their mode of operation (Figure 17-1). In the conventional configuration the electric field is applied parallel to the capillary axis (Figure 17-1A). Electrophoretic separations of this nature have been conducted successfully in quiescent or flowing solutions with narrow-bore plastic tubes, glass or fused-silica capillaries and in rectangular troughs (for an overview refer to Refs. 8 and 9). The specific techniques comprising this group are capillary zone electrophoresis (CZE), capillary isotachophoresis (CITP), capillary isoelectric focusing (CIEF), and a range of electrokinetic capillary chromatography techniques, including micellar electrokinetic capillary chromatography (MECC). The differentiating features of these methods lie in the initial and boundary conditions applied, which determine the character of the migrating sample zones (vide infra). Furthermore, sieving according to molecular size is affected using capillaries filled with gels or entangled polymers. These approaches are typically referred to as capillary gel electrophoresis (CGE).

There is a fundamental unity that underlies all electrophoretic processes. A single mathematical model can describe the characteristic behavior of all basic modes of electrophoresis.[12,13] The superimposition of additional constraints, such as specific affinities, molecular sieving, cross- or counterflow, magnetic fields, and fixed charges, yield all the electrophoretic methods in use. Simulation data presented in Figure 17-2 depict sample zone dynamics in CZE, CITP, and CIEF. CZE is conducted in a continuous buffer where the samples are the only discontinuities present. Under the influence of the

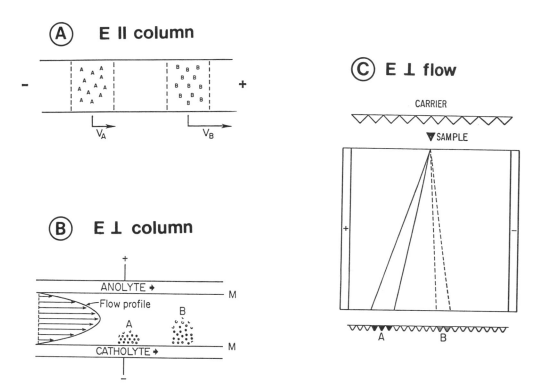

FIGURE 17-1. Schematic representation of the separation of two components by electrokinetic capillary methods with the electric field (A) parallel to column axis and flow (conventional CE), (B) perpendicular to column axis and flow (EFFF), and (C) perpendicular to buffer flow (CFE). M, membrane.

electric field, sample zones migrate without exhibiting any steady-state behavior and thus their shape and position continuously change with time (Figure 17-2, left). In this technique, separation is based on differences in net mobility. Conversely, CITP[1-3] is performed in a discontinuous buffer system, the so-called leading and terminating electrolytes. The sample components are introduced in small quantities at the interface between the two electrolytes. The establishment of isotachophoretic zones requires the net mobility of samples to be intermediate to the mobilities of the buffers. Under the influence of an applied electric field, the sample components separate according to their net mobilities by forming a pattern of consecutive zones between the leader and the terminator (Figure 17-2, center). The system attains a migrating steady state in which all components have the same velocity (hence the prefix isotacho). Ideally, enough sample is applied to produce zones with constant composition, whose lengths are proportional to the amount present. In CIEF, sample components are sorted according to their isoelectric point in an equilibrium

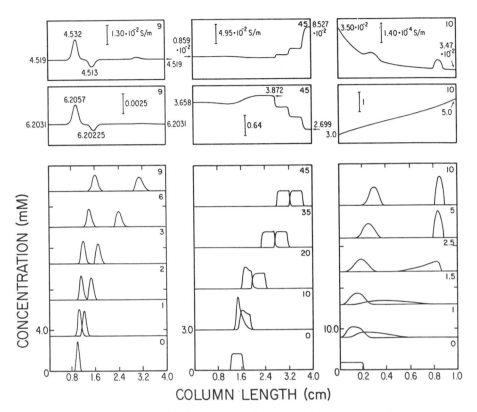

FIGURE 17-2. Computer-simulated separation dynamics of two sample components in quiescent solution for, from left to right, zone electrophoresis, isotachophoresis, and isoelectric focusing. For presentation purposes, concentration profiles at indicated time points (min) after power application are depicted with an offset in the y axis. Concentrations of buffer constituents are not shown. The graphs on the top represent conductivity (uppermost) and pH distributions for each final time point. (Left) Cationic separation at 6,000 A/m^2 of equal quantities of histidine (slower compound) and lysine at pH 6.2 in the background buffer composed of 2 M cacodylic acid and 1 M Tris. (Center) Anionic separation at 9 A/m^2 of equal quantities of chloroacetic acid (first zone) and glycolic acid having a leader composed of 2.5 mM HCl and acetic acid as terminator. (Right) Separation at 20 A/m^2 of equal quantities of two ampholytes (pIs of 3.6 and 4.6; $\Delta pK = 1$ each; anode is to the left) on an immobilized background pH gradient of 3.0 to 5.0. For details on computer simulations of electrophoretic processes, refer to Ref. 13.

gradient. In its most common form, ampholytes are focused in a pH gradient. Typically, a mixture of carrier ampholytes, having different pI values, is exposed to an electric field in a convection-free medium. Eventually, a pH gradient is formed where the most acidic component has condensed in the most anodal position of the capillary, the most basic at the cathode and constituents of intermediate pIs in between. Proteins and other amphoteric compounds can be separated in this gradient provided their isoelectric points are sufficiently different (Figure 17-2, right). In an absence of any flow along the column, a stationary steady state is attained. Many other electrophoretic processes can be classified as combinations of these methods (e.g., DZE[14] is conceptually a composite of CZE and CITP). As in CITP the sample is introduced at the interface between two buffers. However, the terminating constituent's mobility is slightly larger than the mobilities of the sample components. It therefore penetrates the sample zones from the rear side upon current flow. After an initial sharpening at the buffer transition the sample components separate zone electrophoretically within the terminating electrolyte. Similar efficient sample stacking can be achieved by having one or more major components of like charge in the sample. Also, stacking at the initial sample buffer interfaces occurs when the conductivity of the sample solution is lower compared to that of the buffer.[10,11,14]

MECC[15] is an electrokinetic method allowing the separation of neutral and charged small molecular mass molecules. In this technique, two distinct phases are used: an aqueous and a micellar, or pseudostationary, phase. These two phases are established by employing buffers containing surfactants, which are added above their critical micellar concentration. Electrophoresis takes place in open tubular capillaries with electroosmotic flow. Nonionic solutes partition between the two phases and elute with zone velocities between those of the two phases. Separation is of chromatographic nature. For ionic solutes that are also differentially distributed between the two phases, separation is of chromatographic and electrophoretic nature.

Without inclusion of solute partitioning between two phases, a fundamental requirement for the application of electrophoresis is the presence of fully or partly charged solutes. CE has displayed considerable versatility, having been applied successfully to numerous analyses and separations. CE, unlike chromatography, has proved to be an excellent method for very small molecules such as metal ions, amino acids, drugs, and particles like eukaryotic cells. CE is also well suited to the analysis and isolation of molecules found in complex matrices, such as those encountered in the biotechnology industry, particularly proteins, peptides, enzymes, and oligonucleotides, as well as small biomolecules such as antibiotics. Although CE is primarily conducted in aqueous solutions, there have been reports of the successful use of nonaqueous and mixed media.

In the second category, the electric field is applied perpendicular to the capillary axis, and the separation of charged solutes is based on their differential alignment across the mobile-phase profile of a field flow fractionation (FFF) device (Figure 17-1B). Such separations are performed in thin, ribbon-like channels,[16–18] in hollow ultrafiltration fibers,[19–21] or in channels of trapezoidal cross section.[22,23] The techniques representing this class are electrical field flow fractionation (EFFF), electropolarization chromatography (EPC), and electrical hyper-layer field flow fractionation (EHFFF) or isoelectric focusing FFF, respectively. In EFFF and EPC the electric field, applied perpendicular to the flow direction, causes the solute molecules to accumulate in a layer of distinct thickness near one channel wall. The solute is then transported by flow along the channel at a rate determined by the mean thickness of the layer. In EHFFF,[22–24] separation is based on partitioning by the concomitant presence of a pH equilibrium gradient and an electric field, as in isoelectric focusing (IEF). Solute layers having different depths from the walls and which travel at different velocities along the column are thereby established. Although protein separations with EFFF and EPC have been documented quite some time ago,[16–21] successful EHFFF separations have been described relatively recently.[22,23]

In the third category, the electric field is also oriented perpendicular to the direction of flow, but parallel to the flow profile of a thin fluid film flowing between two parallel plates (Figure 17-1C). This represents the configuration of continuous flow electrophoresis (CFE), which can be operated in various modes, including zone electrophoresis (ZE), isotachophoresis (ITP), and IEF,[25,26] or recycling electrophoresis, in which outlet and inlet ports of the separation cells are connected by closed circuitry loops for process fluid recirculation by means of a multichannel peristaltic pump (for details, see Chapter 18). In CFE the buffers and the sample are continuously introduced at one end of the electrophoresis chamber and are fractionated by an outlet array at the other end (Figure 17-1C). CFE is considered a CE technique as fluid stabilization is achieved by a small gap (typically 0.5 mm or less) between the two parallel plates. Although CFE is mainly a preparative methodology for simultaneous separation and fractionation under mild operating conditions, it is also used for analytical purposes, particularly sample preparation. Thus far, applications of analyticel CFE are mainly concerned with the separation and characterization of cells, cell organelles, and cell membrane systems. CFE permits quick determination of electrophoretic mobility distribution of cells or proteins. It can be used as an aid in understanding the phenomena that take place on the cell surface, such as the effect of drugs, antigens, and mitogens.[25]

Finally, high-performance electrochromatography (HEC) can be considered as a fourth category of capillary electrophoretic separations. HEC is a new, developing technique that comprises a capillary column packed with a chromatographic stationary phase and having an electric field applied along the column. It can be divided into (1) HEC with and without pressurized flow and (2) HEC with electroosmosis.[27] HEC involves both, partitioning between two phases (chromatography) and electrophoresis, thus uncharged (neutral) and charged solutes can be separated. HEC provides a mean for separating compounds that have similar electrophoretic mobilities and partition coefficients (i.e., molecules that are difficult to separate by CZE or open-tubular liquid chromatography).[28] HEC can be used for continuous focusing of a minor compound in a complex mixture and for sample preparation by solid-phase extraction. Thus far, no analytical applications dealing with proteins have been reported. Hence, this technique is not discussed further in this chapter. However, electrochromatography on a preparative scale has been described (for details, refer to Chapter 18).

17.2 CONVENTIONAL CAPILLARY-TYPE ELECTROPHORETIC INSTRUMENTATION

The design variables for electrokinetic capillary analyzers are presented in Table 17-1, and the basic arrangement used in conventional CE is depicted schematically in Figure 17-3. It features a capillary tube mounted between two electrode compartments (E). Typically, the sample is injected in small quantities into the beginning of the separation column (S) and on-column detected toward the capillary end (D). The power supply and a method for buffer introduction (e.g., with a pump or through application of vacuum) complete the basic arrangement (Figure 17-3A). Introduction of the buffers and the sample, data manipulation (i.e., collection, storage, evaluation), and fraction collection (Figure 17-3B) are ideally controlled by a computer. The computer system also allows the establishment of data dictionaries as in other analytical procedures, such as in gas chromatography–mass spectrometry (GC-MS). Pulse-free counter or a coflow of electrolyte is useful for (1) changing the column volume without changing the physical capillary length in CZE and CITP, thereby manipulating loading capacity and analysis time; (2) mobilization of the zone spectrum in CIEF; (3) sample fractionation; and (4) sample detection outside of the capillary. The last two features (Figure 17-3b) do not require a pump for the open tubular configuration with electroosmotic flow. The capillary cross sections used in conventional CE with the applied electric field parallel to the capillary axis are depicted in Figure 17-4. Cell configurations used in particle (cell, micro) electrophoresis[29] in which the mobilities of the particles are measured without achieving their separation are included in this graph for purposes of completeness.

In CE, a high-voltage DC electric field is applied along the column that not only induces electrophoretic transport and separations of charged compounds, but in case of a charged inner wall and no fluid entrapment via use of a gel, also a movement of the entire liquid along the capillary. This latter process is termed electroosmosis. In open tubes and having a negative (positive) surface charge, an electroosmotic flow toward the cathode (anode) is

TABLE 17-1. Design Variables for an Electrokinetic Capillary Analyzer

1. Modes
   Zone electrophoresis, discontinuous zone electrophoresis
   Isotachophoresis
   Isoelectric focusing
   Micellar electrokinetic chromatography
   Electrophoresis in sieving media
2. Separation capillary
   Material (fused silica, glass, Teflon, or other plastic)
   Length (1 to 200 cm depending on mode, material, and instrument)
   Cross section (circular, rectangular)
   Cross-sectional area (capillary i.d.)
   Gel filled/free solution
   Surface modification of capillary wall (permanent/dynamic coating)
3. Sample injection
   Electrokinetic
   Hydrodynamic flow (vacuum/pressure/gravity)
   Syringe/sample splitting
4. Sample detection
   On-column/off-column
   Single/multiple/array/scan sensors
   Pre-/postcolumn derivatization
5. PC-based data analysis/automation
   Software for data collection/evaluation
   Autosampler
   Fraction collector
6. Peripheral components
   Electrode compartments
   Power supply (voltage/current/power regulated)
   Electrolyte pumps (optional)
   Sample preparation unit (optional)

generated. In contrast to pressure-driven hydrodynamic flow, electroosmosis is characterized by a plug flow profile that has little impact on zone boundary dispersion. Thus, high-resolution separations can take place in the presence or absence of electroosmosis along the separation axis.

Depending on application and instrumental setup, capillaries of about 1 to 120 cm length are connected to buffer reservoirs containing the driving electrodes. Small amounts of samples (nL volumes or pmol to subfmol solute quantities) are introduced by electrokinetic or hydrodynamic techniques. Upon application of power (about 2 to 30 kV, 1 to 150 μA), samples are transported through the capillary by the combined action of electrophoresis and electroosmosis. In principle, many of the detection methods developed for HPLC can also be employed for CE (Table 17-2), with spectroscopic techniques (on-column direct and indirect absorbance, direct and indirect

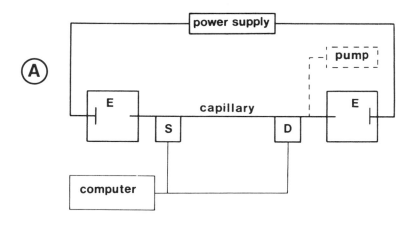

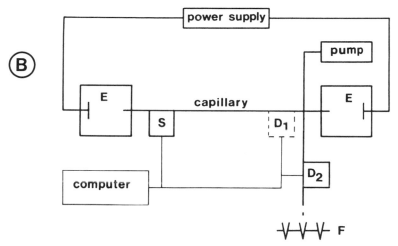

FIGURE 17-3. Schematic representation of a conventional CE setup with (A) on-column solute detection and (B) off-column sample monitoring and fractionation. E, electrode compartments; S, sample inlet system; D, D_1, on-column detectors; D_2, off-column detector; F, fraction collector.

fluorescence detection) being the most popular. Conductivity and radiometry, as well as off-column monitoring employing mass spectrometry (MS) or amperometry, are enjoying increased attention. In addition, unique principles can be used that rely on the electric current flow through the separation capillary, such as the monitoring of potential gradient, conductivity, and temperature in CITP.[2,3] Most electrical sensors read with microelectrodes that have direct contact with the solution in the capillary, whereas most optical measurements are taken through the capillary walls. The advantage of electrical sensing is

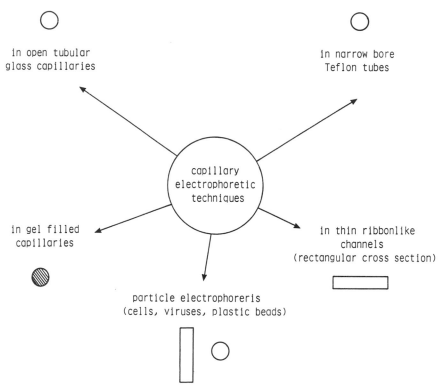

FIGURE 17-4. Column cross sections employed in conventional capillary electrophoretic techniques.

TABLE 17-2. Detection Methods for Electrokinetic Capillary Analyses

Detection Method	On Column	Off Column
UV-VIS absorbance	x	x
Fluorescence	x	x
Polarimetry–interferometry	x	x
Circular dichroism	x	x
Chemiluminescence		x
Temperature	x	x
Conductivity	x	x
Electric field	x	x
Amperometry/voltammetry	x	x
Potentiometry/pH		x
Mass Spectrometry		x
Radiometry	x	x

high resolution and sensitivity, whereas the disadvantage is the possibility of troublesome alterations of the tiny electrode surfaces. Spectroscopic measurements through fused-silica capillaries are usually trouble free, even for UV absorbance detection at 195 nm. In the case of plastic tubes (Teflon, FEP), however, absorption of the capillary walls can be significant, especially at wavelengths below 254 nm. Therefore a molded detection block with optical fibers conducting the light directly into and out of the liquid has been constructed (LKB 2127–140, Bromma, Sweden; see also Ref. 30).

In CE, tiny zone volumes must be measured accurately. In any case, the detection limit is determined by both operational and equipment parameters. The technology for the design of detector cells with picoliter to microliter volumes is now well established. In most arrangements, one or two on-column sensors placed toward the capillary end are used and monitoring occurs while the electrophoretic zone structures migrate across their location. The advent of versatile and inexpensive microprocessor systems also permits simple control of sensor arrays, thereby providing increased analytical information through the repeated detection of the whole separation pattern along the capillary or the comparison of electropherograms from at least three completely different locations of the separation trough.[31,32] The advantages of using imaging detection (e.g., with a CCD camera) have been demonstrated.[33,34]

First, capillary electrophoresis comprised electrophoretic separations in quiescent solutions via the use of coated glass capillaries of 0.1 to 1 mm i.d., narrow-bore plastic tubes of 0.2 to 0.8 mm i.d., or thin ribbon-like channels of rectangular cross sections (0.1 to 0.5 mm channel height and 0.5 to 2 mm channel width) having various plastic and glass walls (for a review, refer to Refs. 2, 3, 8, and 9). With this kind of instrumentation, microliter sample quantities were applied with a conventional microliter syringe. In the seventies and early eighties, a number of commercial instruments equipped with plastic capillaries emerged. These are the Tachophor 2127 of LKB (Bromma, Sweden), the Tachophor Delta of Itaba (Jarfalla, Sweden), the IP-1A, IP-2A, and IP-3A analyzers of Shimadzu (Tokyo, Japan), and the CS ITP analyzer of VVZ PJT and the CS Capillary Electrophoresis Analyzer EA 100 (Spisska Nova Ves, Czechoslovakia; distributed by Salus-Braumapharm, Vienna, Austria). Only the IP-3A and the CS analyzers are still available. A few reports described procedures for fractionation employing microsyringes or elution onto a strip of cellulose acetate. The second approach was commercialized by LKB, Sweden, under the tradename Tachofrac.

More recently, electrokinetic capillary instrumentation featuring open-tubular fused-silica capillaries of very small inside diameters (10 to 100 μm) became the focus of both active research and introduction to routine analysis.[4–7,35] Fused-silica capillaries permit rapid removal of ohmic heat, application of high voltages (300 V/cm), and hence realization of short runs, and are transparent to UV light. Furthermore, they exhibit a pH-dependent, strong electroosmotic flow toward the cathode if no special wall coatings are applied. Modifications of the surface properties that result in reduced, no, or even

TABLE 17-3. Specifications of Selected Commercial CE Instruments Featuring Fused-Silica Capillaries

Instrument/Supplier	Sample Detection	Sample Injection	Autosampler	Capillary Temperature Control	System Control	Special Features
270A-HT/Applied Biosystems (Perkin Elmer)	UV-VIS, 190–800 nm	Electrokinetic, vacuum	50 positions, temp. control	Forced air	Front panel, PC	Fraction collector, Z detection cell
P/ACE 5000/Beckman Instruments	UV (filter), diode array, LIF	Electrokinetic, pressure	34 positions, temp. control	Liquid	Front panel, PC	MS interface
BioFocus 3000/Bio-Rad Laboratories	UV-VIS, 190–800 nm, fast scanning	Electrokinetic, pressure	32 positions, temp. control	Liquid	PC	MS interface
SpectraPhoresis 1000/Thermo Separation Products	UV-VIS, 190–800 nm, fast scanning	Electrokinetic, vacuum	80 positions	Forced air	Front panel, PC	
HP^{3D}/Hewlett Packard	UV-VIS, 190–600 nm, diode array	Electrokinetic, pressure	48 positions, temp. control	Forced air	PC	Fraction collector, bubble cell

Instrument/Manufacturer	Detector	Injection	Autosampler	Cooling	Control	Additional
Eureka 2100/Kontron Instruments	UV-VIS 190–800 nm, diode array	Electrokinetic, pressure, gravity	20 positions	Forced air	PC	
CES 1/Dionex	UV-VIS, 190–800 nm, fluorescence	Electrokinetic, pressure, gravity	39 positions	Forced air	Front panel, PC	Fraction collector
Quanta 4000E/Waters	UV (185 nm)	Electrokinetic, gravity	20 positions	Forced air	Front panel, PC	Fraction collector
Model 3850/Isco	UV 190–360 nm	Electrokinetic, split flow	No autosampler	Fan	Front panel	Syringe flush
Crystal CE 300(310)/ ATI Unicam (Prince, Lauerlabs)	Modular[a]	Electrokinetic, pressure	4 positions (48 positions)	Forced air	Front panel (PC)	
SpectraPhoresis 100/ Thermo Sep. Products (Prime Vision I, Europhor)	Modular[a]	Electrokinetic, vacuum	12 positions (manual sampler)	Fan	Front panel	
Model 100/Stagroma	Modular[a]	Electrokinetic, gravity	No autosampler	Fan	Front panel	HPLC-CE coupling unit

[a] For these sampling units there is a choice of detectors, including UV-VIS (also fast scanning), fluorescence, conductivity, and MS.

reversed electroosmosis have been developed. Fused-silica capillaries are manufactured with a thin outside polymer coating that makes them extremely flexible, permitting easy manipulation. Using these capillaries makes the customary microliter syringe for sample injection obsolete. The sample is typically introduced either by hydrodynamic flow [i.e., a displacement technique based on (1) application of a vacuum to the capillary outlet, (2) application of pressure to the sample vial with an inert gas, (3) changing the relative heights of the sample and outlet buffer vials (gravity injection), or (4) a split-flow injection] or by an electrokinetic technique. For the latter approach, the electrode compartment at the high voltage end is replaced with the sample vial. Application of high voltage for a short amount of time allows the insertion of a small amount of sample into the capillary (combined action of electrophoresis and electroosmosis). The sample reservoir is then exchanged by the buffer compartment before further application of current. Since the first report in 1981,[34] many instruments featuring fused-silica capillaries of 10 to 100 μm have been assembled in the researchers laboratories. The first commercial instrument emerged in 1988 and there are currently about 12 companies manufacturing electrokinetic capillary instrumentation featuring capillaries of 25 to 75 μm i.d. and on-column optical sample detection (Table 17-3). Many of these apparatuses are fully automated, comprising an autosampler as well as data gathering and evaluation with a computerized data station. Some instruments are modularly built, whereas others are operated manually. Thus far, all commercial instruments are equipped with a single capillary, and capillaries can simply be exchanged by the operator. In addition, Bio-Rad offers disposable, preassembled capillary cartridges that can only be used in their instruments. Setups comprising multiple capillaries in parallel are currently in development. Along with the commercialization of fused-silica capillary-based instrumentation, capillaries with different surface properties (Table 17-4) and kits containing complete protocols, precut capillaries, conditioning reagents, and prefabricated (proprietary) running buffers became available as well (Table 17-5). Furthermore, several companies that manufacture CE instrumentation also offer ready-made buffers for CZE and CIEF. These consumables provide the basis for a widespread applicability of CE technology to the analysis of biopolymers, such as proteins, oligosaccharides, and oligonucleotides, and many classes of small molecules.

CE on miniaturized instrumentation[36-38] (i.e., CE on a glass chip on which separation channels, sample injector, and detection are combined on an area of a few square centimeters, also referred to as a miniaturized total chemical analysis system[38]) respresents the newest generation of CE setups. In that approach, fluid flow is driven electrokinetically through a network of intersecting small channels that have been fabricated on planar glass substrates by photolithographic masking and chemical etching techniques and have been formed by bonding the etched substrate to a plain glass plate. Capillaries of 30 to 70 μm width, about 10 μm height, and a few centimeters in length are typically employed. Microfabricated linear and cyclic capillary structures and

TABLE 17-4. Selected Commercial Fused-Silica Capillaries[a]

Product/Supplier	Capillary Coating	Characteristics	Comments
TSP CE capillaries/Polymicro Technologies	Untreated	Plain capillary	Without detection window ($8–12 per m)
CE capillaries/Instrument suppliers	Untreated	Precut capillary	With detection window (expensive)
BioCAP LPA-coated capillary/ Bio-Rad	Linear polyacrylamide	Low surface charge	General protein analysis, CIEF
PVA-coated capillary/Hewlett Packard	Polyvinyl alcohol	No electroosmosis	Protein analysis, CIEF
MicroSolv PEG CE/Scientific Resources	Polyethylene glycol	Low surface charge	Protein analysis
MicroSolv PVA CE/Scientific Resources	Polyvinyl alcohol	Low surface charge	Separation of closely related compounds
CElect-H1/H2/Supelco	Bonded C8/C18 phases	Reduced anionic surface charge	Protein analysis[b]
CE-100-C18/Isco	Bonded C18 phase	Reduced anionic surface charge	Protein analysis[b]
CE-200/Isco	Glycerol		Protein/peptide separations
μSIL DB-1/J&W Scientific	Dimethyl polysiloxane (DB-1)	No electroosmosis	CIEF
μSIL DB-17/J&W Scientific	(50%-phenyl)-methyl polysiloxane (DB-17)	No electroosmosis	Protein analysis, CITP

[a] Capillaries can be mounted into most instruments. Bio-Rad also offers preassembled, disposable capillary cartridges that can only be employed with their instruments.

[b] In combination with dynamically adsorbed surfactants (e.g., Brij-35) as described by Towns and Regnier.[53]

619

TABLE 17-5. Selected Commercial Capillary Electrophoresis Kits for Protein Analysis[a]

Product/Supplier	Capillary Coating	Principle	Application
CE-SDS protein kit/Bio-Rad	Untreated	Dynamic sieving hydrophylic polymer	Size separations for 14- to 200-kDa proteins
eCAP SDS 14-200 kit/ Beckman	Polysaccharide coating	Dynamic sieving hydrophylic polymer (about 50% reduced eo)	Size separations for 14- to 205-kDa proteins
eCAP amine kit/Beckman	Polyamine coating	Cationic surface (reversal of eo)	Analysis of basic (positively charged) proteins
eCAP neutral kit/Beckman	Polyvinyl alcohol coating	Neutral surface	Protein analysis, CIEF
MicroCoat kit/Applied Biosystems	Untreated, dynamic coating with cation	Cationic surface (reversal of eo)	Analysis of positively charged proteins
ProFocus IEF kit/Applied Biosystems	Coated (DB-1)	No electroosmosis	CIEF
ProSort SDS kit/Applied Biosystems	Untreated?	Dynamic sieving hydrophylic polymer?	Size separations for 14- to 205-kDa proteins

[a] Kits typically include complete protocols, buffers, standards, and capillaries.

capillary arrays have been successfully applied to very rapid separations (few seconds) of dyes, labeled amino acids, and DNA fragments, compounds that were monitored by on-column laser-induced fluorescence detection.[38] No protein separations have been reported so far. This technology, although still at the research stage, is very promising, and commercial chip-based instrumentation with high sample throughput will certainly become available in the years to come.

17.3 CAPILLARY ELECTROPHORESIS OF PROTEINS

Proteins are macromolecules composed of a large number of amino acids. They possess hundreds of ionizing groups and are amphoteric. Thus, depending on pH, proteins exhibit a positive, negative, or no charge. Isoelectric points, titration curves, and electrophoretic mobilities of proteins are most accurately determined experimentally. Reasonable approximations can, however, be obtained by calculation using the Henderson–Hasselbalch equation if the amino acid composition is known.[39] This approach assumes that any specific ionizing group has the same pK everywhere on the molecule. Calculated titration curves of four proteins, cytochrome c (CYTC), ovalbumin (OVA), β-lactoglobulin B (BLB), and ribonuclease A (RNase), are presented in Figure 17-5. According

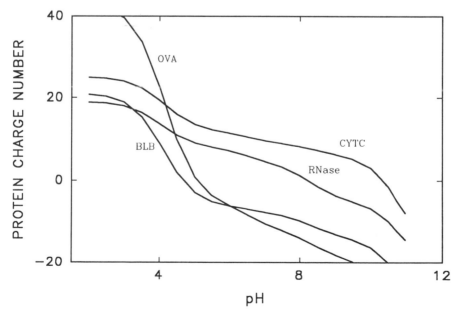

FIGURE 17-5. Calculated charge vs pH relationships of cytochrome c (CYTC), ovalbumin (OVA), ribonuclease A (RNase), and β-lactoglubulin B (BLB). Data were generated according to the procedure given in Ref. 39.

to these data, it is anticipated that these proteins should be easily separable by CIEF and, by proper selection of the pH value of the running buffer, also by CZE and CITP. Furthermore, the diffusion coefficient of a protein is small, which favors the realization of high plate numbers in CE. The highly charged proteins (particularly positively charged proteins in negatively charged fused-silica), however, can interact unfavorably with the capillary walls, a process that causes band broadening or, if strongly adsorbed, prevents their analysis by CE. Moreover, interactions based on hydrophobicity and hydrophilicity are possible as well. Thus, for a successful choice of suitable CE conditions for the sample protein(s) (i.e., the interplay of capillary size, material, inner surface, and electrophoretic buffer), this basic understanding of chemical properties of proteins has always to be kept in mind.

17.3.1 Capillary Electrophoresis in Plastic Capillaries

Compared to fused silica and glass, one advantage plastic capillaries offer is their higher hydrophobicity and somewhat lower and different origin of surface charge, both making protein–wall interactions less pronounced. Furthermore, electroosmosis can virtually be abolished by dynamic coating with a neutral surface-active agent added to the buffer.[40] Conversely, plastic capillaries suffer from several disadvantages, including low heat dissipation and the difficulty in manufacturing tubes with constant inside diameters smaller than about 200 μm. With the exception of dynamic coating with surfactants, modification of wall surface properties can be difficult. Nevertheless, narrow-bore PTFE (Teflon) and FEP tubes of 0.2 to 0.8 mm i.d. have been employed successfully for high-resolution CITP analyses of proteins.[2,3,41] For CZE[9,40,42–44] and CIEF,[9,45,46] however, they have hardly been explored. Electropherograms obtained with model proteins are presented in Figure 17-6. They were monitored with the Tachophor 2127 capillary analyzer (LKB,

---

FIGURE 17-6. CE data obtained in a 0.5-mm i.d. Teflon capillary. (A) Cationic CZE separation of CYTC (1), carbonic anhydrase (2), and RNase (3) at pH 3.50. A 0.5-μL sample with protein quantities of 1 μg (1), 2.5 μg (2), and 6.5 μg (3) was injected, and the run was executed at a constant 300 μA. (B) CIEF separation of CYTC (1, 0.18 mg/mL), RNase (2, 0.36 mg/mL) and Hb (3, 0.08 mg/mL) in pH 3.5–10 Ampholine (1%) with 8 mM arginine. Focusing (F) was performed at a constant voltage of 2,000 V, and cathodic elution (E) occurred at a constant current of 100 μA. (C) Cationic CITP data showing the steady-state distribution of RNase (1), BLB (3), and OVA (5) in the absence (left) and presence (right) of tetrapentylammonium (2), γ-aminobutyric acid (4), and Tris (7) as low molecular mass spacers. Zone 6 and Na$^+$ represent impurities. L and T refer to leader and terminator, respectively. A constant 150 μA was applied. While recording data, the current was reduced to 50 μA. Absorption measurements were made at 277 nm. Conductivity data are expressed as increasing resistance R.

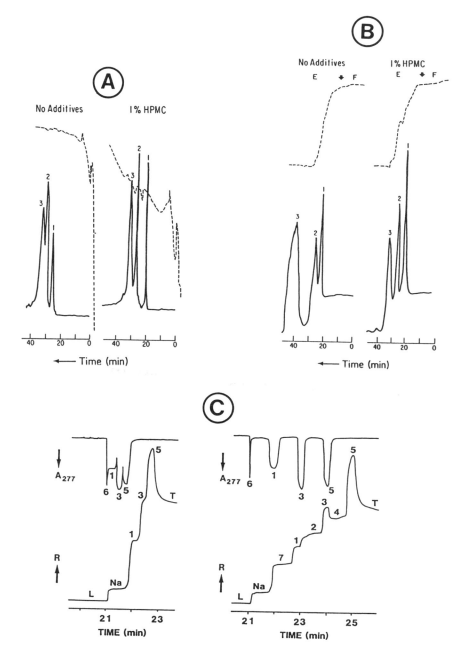

Bromma, Sweden), featuring both on-line conductivity and absorbance detection at the end of a 0.5-mm i.d. PTFE column.

Electropherograms of a cationic CZE separation of equine CYTC (1), bovine carbonic anhydrase (2), and bovine RNase (3) in 10 mM HCl and Tris buffer at pH 3.50 are shown in Figure 17-6A. Without the addition of a fluid-stabilizing agent, no complete separation of the three proteins is achieved with a column length of 25 cm, whereas the separation is markedly improved in presence of 1% HPMC. It is interesting to note that the cationic protein peaks are sharper than those obtained in the same capillary at alkaline pH (anionic separation, data presented elsewhere[9]). However, plate heights smaller than about 0.1 mm were rarely observed. Figure 17-6B depicts electropherograms obtained for the CIEF separation of equine CYTC (1, 0.18 mg/mL), bovine RNase (2, 0.36 mg/mL), and canine hemoglobin (Hb) (3, 0.08 mg/mL) in pH 3.5–10 Ampholine (1%, w/v) with 8 mM arginine. A 10-cm column was used together with conditions outlined elsewhere.[46] Finally, the cationic CITP data of bovine RNase, BLB, and OVA depicted in Figure 17-6C were obtained in a capillary of 28 cm length using 0.01 M potassium acetate and acetic acid (pH 4.75) as the leader and 0.01 M acetic acid as the terminator. The impact of low molecular mass spacers on the separation pattern is also shown. For CITP, no buffer additive was employed. For CZE and CIEF, the impact of the addition of other polymers, such as PVA, LPA, and PEG, and solubilizing agents, such as urea and CHAPSO, on the buffers was also explored. The use of 1% HPMC, however, was found to produce the best results for these proteins. Despite obvious shortcomings, the Tachophor 2127 is still employed for the investigation of buffer systems for preparative electrophoretic separations and for the validation of theoretical predictions.[39,47]

It is evident that the performance of the 0.5-mm i.d. Teflon capillary for CZE and CIEF of proteins is characterized by a rather low efficiency, long analysis times, and the need for a viscosity enhancing/electroosmosis diminishing agent. Having plastic capillaries of smaller inside diameters would certainly provide increased resolution and separation speed, as was shown for CZE of low molecular weight compounds in PTFE capillaries of 0.2 mm i.d.[40] Much improved efficiency was demonstrated using a 56-μm i.d. (350 μm O.D.) polypropylene capillary produced by meltspinning.[48] For chymotrypsinogen A, a plate height of about 5 μm was observed when the hydrophobic capillary was dynamically coated with polyethoxylated ($n = 23$) lauryl alcohol. This performance is similar to that obtained in well-conditioned fused-silica capillaries of comparable physical dimensions. It is hoped that this promising example will encourage further research and developments with plastic capillaries.

17.3.2 Capillary Zone Electrophoresis in Fused-Silica Capillaries

Because of the shortcomings associated with the use of plastic capillaries (Section 17.3.1), a very large number of studies have been (and many more still are) conducted in capillaries of fused silica that permit effective heat removal and fabrication of capillaries with very small inner diameters (10 to 100 μm). Difficulties arise, however, when proteins are analyzed, as they can interact unfavorably with the surface of the column walls. The inner wall of fused silica is typically negatively charged due to the dissociation of silanol groups. Thus, positively charged sites of a protein interact electrostatically with the capillary surface, causing peak broadening (Figure 17-7A) and possibly also a complete loss of the polypeptide due to adsorption. A number of strategies to minimize or prevent protein–wall interactions have been developed (for reviews and references, refer to Refs. 49 and 50 and various chapters in Ref. 5). First, by adjusting the pH of the buffer above the isoelectric points of the proteins (where they possess a net negative charge) results in coulombic

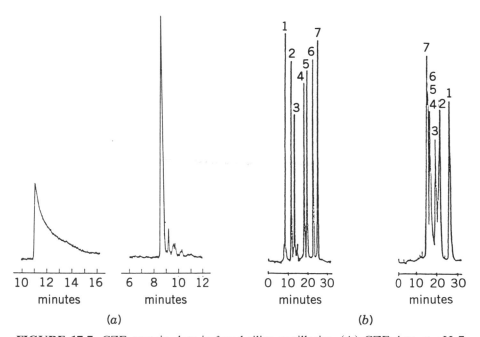

FIGURE 17-7. CZE protein data in fused-silica capillaries. (A) CZE data at pH 7 of CYTC in uncoated (left; fourth injection, for injections 1 to 3 no peak could be recorded) and C-18/Brij-coated (right; first injection) 75-μm i.d. capillary of 85 cm effective length. The applied voltage (current) was 30 kV (30 μA). (Reprinted with permission from Ref. 53. Copyright American Chemical Society.) (B) CZE data of model proteins at pH 9.5 using vinyl-bound, polyacrylamide-coated (left) and uncoated (right) 50-μm i.d. capillaries of 45 cm effective length. The applied voltages (currents) were 20 kV (15 μA) and 10 kV (7 μA), respectively. Key: 1, insulin chain A (porcine); 2, bovine serum albumin; 3, OVA; 4, porcine insulin; 5, bovine α-lactalbumin; 6, bovine casein; 7, insulin chain B (porcine). (From Ref. 50.)

repulsion. Protein–wall interactions are thereby strongly reduced and efficient separations can be achieved (Figure 17-7B). Also, Coulombic attraction between proteins and fused silica can be reduced by working at very low pH where silanol groups are undissociated, by having high concentration of salts in the buffer, and by employing various types of buffer additives that provide a dynamic coating onto the capillary wall, by interacting with charged groups of the proteins, or by inducing a conformational change in the protein structure. Examples of additives include nonionic surfactants and denaturants, as well as zwitterionic, polycationic, and polyanionic salts. The latter compounds are believed to form ion pairs with the positively charged proteins. Last but not least, surface charge alteration (including reversal) can be achieved through the application of radial electric fields.[51]

Dynamic capillary coating comprises the dynamic adsorption of neutral (e.g., HPMC, MC, Brij-35, PVA) or ionic hydroxylic polymers onto the capillary walls during wash cycles or during the run when added to the buffer. Nonionic adsorbed agents mask the surface charges, thereby reducing or eliminating electroosmosis as well as protein–wall interactions. Adsorption of charged surfactants may drastically alter the surface properties and can even reverse the surface charge (e.g., the dynamic adsorption of a cationic agent to an untreated fused-silica capillary typically reverses surface charge and electroosmosis and thus makes the analysis of cationic proteins possible). Such a procedure was commercialized in a kit format (Table 17-5; MicroCoat of Applied Biosystems). Dynamic capillary coating is a simple approach with the advantage that the capillary surface is regenerated for each run. Also, using PVA, it has been shown that dynamic coatings can be as effective as permanent capillary coatings for protein analysis.[52]

An enormous amount of work has been devoted to developing many different coatings formed by polymers that are covalently attached to the silanol groups of fused silica (for comprehensive reviews, refer to Refs. 49 and 50). Hydrophylic polymers, including those based on polyacrylamide (Figure 17-7B), polyoxyethylene, and cellulose derivatives, have been used the most often. In addition, the use of dynamic coatings onto chemically modified capillary walls (particularly on C18 hydrophobically coated capillaries, Fig. 17-7A) was reported to be very effective and versatile.[53] The advantage of all these approaches is that a large variety of surface properties can be achieved. This is an important fact, as proteins are a heterogeneous class of compounds with widely differing characteristics. Unfortunately, some of the coatings developed thus far are not sufficiently stable over a wide pH range and are difficult to prepare with a high batch-to-batch reproducibility, thus somewhat hampering the widespread use of coated capillaries. Nevertheless, a number of companies are currently offering coated capillaries and CE kits for protein analysis (Tables 17-4 and 17-5, respectively).

17.3.3 CE Using Polyacrylamide Gels and Other Sieving Media

One of the main analytical applications of slab gel electrophoresis is the determination of the molecular mass of proteins using SDS and a gel formed by cross-linked polyacrylamide (SDS–PAGE). In this process, proteins form complexes with dodecyl sulfate to yield molecules of constant surface charge density and thus equal electrophoretic mobility. Separation is based on size or molecular mass differences via sieving through the gel matrix. Capillary gel electrophoresis (CGE) has been introduced by direct adaptation of SDS–PAGE to the capillary format, using cross-linked polyacrylamide matrices that were covalently attached to the capillary inner wall. Excellent resolution was achieved.[54] However, the gel matrix prevents protein detection at low wavelengths. Furthermore, the proper production of the gel-filled capillaries turned out to be difficult and their lifetime was limited to a few experiments due to bubble formation and deterioration of the gel structure, particularly at high field strengths. Therefore, capillaries filled with cross-linked polyacrylamide are rarely used for protein analysis. They have, however, been perfected for the analysis of polynucleotides, compounds that can be detected at higher wavelengths, such as 260 nm.[49]

For protein analysis, alternate processes were studied, employing replaceable[55,56] or *in situ*-formed[57] gels of linear (noncross-linked) polyacrylamide (LPA), which is not bonded to the capillary walls. The advantage of introducing the sieving matrix after its formation is that the capillary can be emptied easily and refilled, leading to flexible and simple operation. Although this is possible using LPA, its relatively low UV transparency prevents sensitive protein detection at low wavelengths. Thus, Ganzler and colleagues[56] explored the use of dextran and polyethylene glycol (PEG) as replaceable sieving matrices and reported successful molecular mass sieving of dodecyl sulfate–protein complexes and detection at 214 nm in these media. Subsequently, several companies developed CE kits containing hydrophylic polymers to permit dynamic sieving in which dodecyl sulfate–protein complexes separate in a size-dependent fashion with a linear correlation over the entire range (Figure 17-8; Table 17-5). Molecular masses of proteins of pharmaceutical interest determined by replaceable SDS polymer-filled capillary electrophoresis were found to be close to those obtained by conventional SDS–PAGE.[58]

17.3.4 Capillary Isoelectric Focusing in Fused-Silica Capillaries

Because of the extremely high resolving power (ΔpI of 0.005 pH units), IEF is a popular method for protein characterization. It is typically conducted in slab gels, an approach that allows the processing of 10 to 20 samples in parallel lanes and that requires protein fixation and visualization procedures after focusing. Since the mid-1980s, attention has been focused on developing IEF into a more instrumental format by using gel-free capillaries as focusing col-

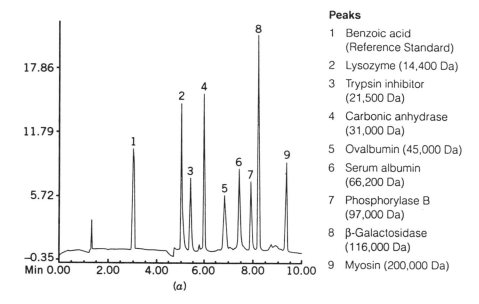

Peaks

1 Benzoic acid
   (Reference Standard)

2 Lysozyme (14,400 Da)

3 Trypsin inhibitor
   (21,500 Da)

4 Carbonic anhydrase
   (31,000 Da)

5 Ovalbumin (45,000 Da)

6 Serum albumin
   (66,200 Da)

7 Phosphorylase B
   (97,000 Da)

8 β-Galactosidase
   (116,000 Da)

9 Myosin (200,000 Da)

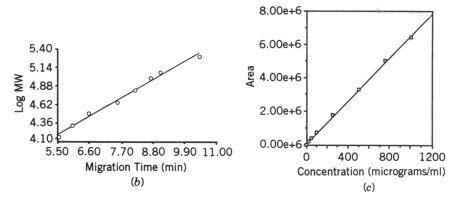

FIGURE 17-8. Dynamic sieving of SDS protein complexes in the presence of a replaceable hydrophylic polymer (CE-SDS protein kit on BioFocus 3000 with a 50-μm i.d. capillary of 24 cm length and detection effected at 220 nm) showing (A) the separation of protein standards, (B) the corresponding molecular weight calibration graph, and (C) the response linearity for carbonic anhydrase. (Courtesy of Bio-Rad, Hercules, CA.)

umns. Focusing can be conducted in stationary or flowing solutions. CIEF in a quiescent solution requires capillaries without electroosmosis. For solute detection, either an array or a scan detector has to be employed,[34,59,60] as the sample pattern must be mobilized following focusing.[61,62] Mobilization is achieved by pumping the entire IEF zone structure past a stationary monitor with the voltage applied to avoid a decay of the pattern or by electrophoretic elution. Electrophoretic mobilization toward the anode is accomplished by replacing the acid at the anode by a base or by putting salt into the acidic electrode buffer and reapplication of power. Conversely, mobilization toward the cathode is effected when the base at the cathode is replaced by an acid or a solution of the base containing salt.[62,63] In these two-step CIEF approaches, proteins are detected as they pass the on-column absorbance detector, which is mounted toward one column end. CIEF protocols comprising mobilization by application of flow were commercialized by ABI and Beckman (Table 17-5), whereas electrophoretic (or chemical) mobilization was promoted by Bio-Rad using their LPA-coated capillaries (Table 17-4).

In CIEF performed in the presence of an electroosmotic flow,[64,65] mobilization of focused zones is unnecessary because these zones are displaced toward and across the point of detection by the electroosmotic flow. Using untreated fused-silica capillaries and the CIEF procedure investigated by the author's laboratory,[65] a typical experiment proceeds as follows. First, the entire capillary is filled with the catholyte containing a small amount of a neutral polymer (e.g., HPMC) as the capillary conditioner. A sample composed of carrier ampholytes and test compounds is introduced at the anodic end, occupying about 35 to 50% of the effective capillary length. After power application, the formation of the pH gradient, the separation of the sample compounds, and the displacement of the entire ampholyte pattern toward the cathode occur simultaneously. Basic ampholytes and test substances reach the point of detection prior to neutral and acidic ones. This technique is an attractive and simple method for the analysis of normal and pathological hemoglobins.[66] Examples are presented in Figure 17-9. In the process described by Mazzeo and Krull,[64] the entire capillary is filled with the sample. To prevent the focusing of basic proteins between the detector and the cathodic capillary end, a fair amount of a strong base (i.e., TEMED) has to be added to the sample. Compared to our technique, this approach has the advantage of producing a shallower pH gradient, thereby providing better resolution.

17.3.5 Capillary Isotachophoresis in Fused-Silica Capillaries

In ITP of proteins, zones of rather high concentration (typically of the order of 10 to 20 mg/mL[67]) are formed and proteins are vulnerable to interactions with the column walls. Not surprisingly, experiments with proteins in untreated fused-silica capillaries were found to be largely unreproducible, whereas much improved data were obtained in capillaries coated with linear polyacryl-

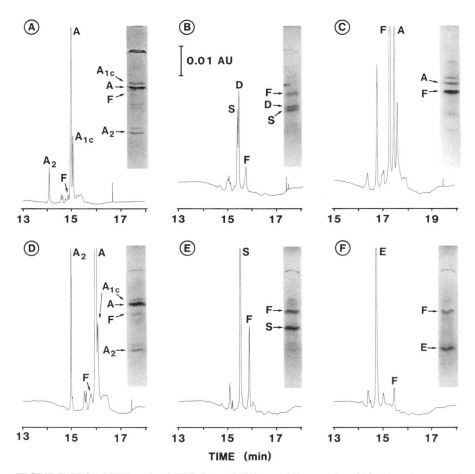

FIGURE 17-9. CIEF and gel IEF data of different Hb samples. (A) Hbs of a normal adult, (B) double heterozygote for Hb S and Hb D, (C) newborn, (D) patient with heterozygous β-thalassemia, (E) a homozygote for Hb S (patient with sickle cell disease), and (F) a homozygote for Hb E. For each sample, the gel IEF pattern is depicted as an insert. For the electropherograms and the gels the cathodes are on the left and at the bottom, respectively. The absorbance scale for A is four times larger than that of all other panels (see absorbance bar in B). Corresponding peaks and bands in the electropherograms and gel data, respectively, are denoted by the letter or letter(s) and number characterizing a specific Hb variant. For CIEF a capillary of 58 cm total length (39 cm to detector) and 75 μm i.d. was used. For other conditions, refer to Ref. 66.

amide[14,67,68] or with a bonded and cross-linked film of DB-17.[69] Dynamic coatings with additives comprising neutral macromolecules, such as HPMC, were demonstrated to largely reduce but not completely abolish protein interactions with fused silica.[70] Such an effect is clearly demonstrated with the example shown in Figure 17-10 in which the analysis of the same sample in a Teflon capillary (Tachophor, A) and an untreated fused-silica capillary (laboratory-made instrument, B) is depicted. For these experiments, the same cationic electrolyte system used for Figure 17-6C was employed. The sample was composed of two proteins, lysozyme (LYSO) and conalbumin (CAL), with creatinine (CREAT) as the spacer. In the Tachophor, analysis of a 1-μL aliquot provided plateau-shaped zones for both proteins (upper graph in Fig. 17-10A) and the spacer. In the fused-silica capillary with HPMC (0.3%) conditioning (dynamic coating) LYSO and CREAT could be nicely detected whereas CAL did not produce the expected ITP zone between CREAT and the terminator (marked with an arrow in the lower graph of Figure 17-10B), nor could it be monitored as a migrating zone within the leader or terminator. However, much higher concentrations of CAL have been shown to produce proper ITP zones.[70] Thus the small amount of CAL in the investigated sample is presumably adsorbed onto the capillary wall and lost for the analysis. LYSO appears to have a much lower or no affinity to the column material and can nicely be analyzed under the investigated conditions. Application of this technique, however, appears to be protein specific, a restriction that has to be kept in mind. Nevertheless, LPA-coated[68] and untreated capillaries dynamically coated with HPMC[71] were successfully employed to fractionate serum proteins by ITP via use of spacers. Furthermore, CITP is employed to preconcentrate (sharpen) protein zones prior to their separation by CZE.[10,11]

17.3.6 Affinity CE and Immunoassays

Affinity capillary electrophoresis (ACE) represents a combination of affinity (immuno) chemistry and CE. The chemical basis underlying ACE is a biospecific interaction between an affinity probe (ligand) and an analyte, including an antibody or a Fab' fragment of an antibody reacting with an antigen or a hapten, a lectin reacting with a glycoprotein or an oligosaccharide, and an enzyme interacting with a substrate. Traditionally, affinity electrophoresis was defined as electrophoresis of molecules interacting during the electrophoretic process. Typically, this was achieved by incorporating one of the interacting components into a medium through which the other interacting compound was electrophoresed. Here, ACE is seen in a broader sense, including assays in which an incubation of the reactants occurs prior to CE, such as in immunoassays (see later). ACE is mainly used as microscale analytical procedure or for the study of complex formation between analyte and affinity probe. In the latter case, a series of experiments is performed in which one of the reactants is applied as a small sample plug and the other is supplied in different

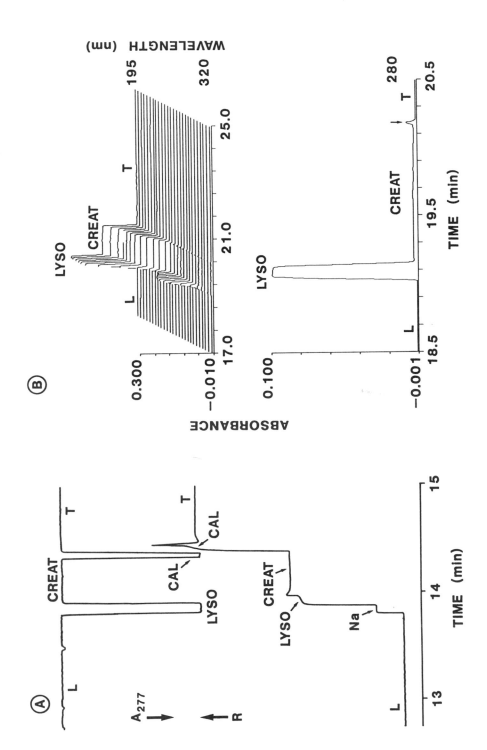

quantities as additive to the running buffer. Sample retardation as a function of the reactant present in the buffer is employed to characterize complexation of the two reactants.[72,73]

For the analysis of peptides and proteins, different strategies to perform immunoassays have been developed. They all relate to the well-known reaction schemes between antigens and antibodies or fragments of antibodies and different labeling strategies. Furthermore, the CE basis for immunoassays is the work of Nielsen and co-workers in which the CZE separation of human growth hormone (HGM), anti-HGM, and complexes of HGM and anti-HGM was demonstrated.[74] Four examples are given here. In affinity probe CE described by Shimura and Karger,[75] a fluorescently labeled affinity probe (tagged Fab′ fragment) is employed to detect the analyte as a complex after the separation of the excess of free probe by CIEF with LIF detection (Figure 17-11A). With that noncompetitive reaction approach, recombinant HGM could be successfully determined down to detection levels of about 5×10^{-12} M and separated from its mono- and dideamidated variants. Kennedy and colleagues[76] described a competitive immunoassay for human insulin using fluorescently labeled insulin as the antigen. In that approach, the free tracer and the tracer–Fab′ complexes were separated by CZE and detected by LIF. With a detection limit of 3 nM, this assay was successfully employed to determine the insulin content and insulin secretion from single rat islets of Langerhans.[77] Using the same CE conditions, this group also described a noncompetitive immunoassay for anti-insulin Fab′ fragments.[76] Finally, Reif and co-workers[78] described a noncompetitive immunoassay for IgG in human serum, employing the fluorescently labeled protein G as the affinity ligand. The FITC–protein G–h-IgG complex was quantified by CZE down to the nanomolar concentration level. Data obtained were found to correlate well with those obtained by a single radial immunodiffusion assay. Using fluorescently labeled protein A as the affinity probe, which binds selectively to the Fc region of immunoglobulins, a similar assay for the determination of IgG in cultivation media was developed[79] (Figure 17-11B). Other CE-based immunoassays, particularly for small molecules,[80] have been described as well.

FIGURE 17-10. Cationic CITP data of a model mixture composed of two proteins (LYSO and CAL, about 1 mg/mL each) and a low molecular mass spacer (CREAT) obtained (A) in a 0.5-mm i.d. Teflon capillary of 28 cm length (Tachophor; applied constant current: 150 μA) and (B) in a 75-μm i.d. fused-silica capillary (70 cm effective length) with on-column multiwavelength detection. (A) UV absorption data (277 nm, upper graph) and conductivity data (lower graph) are presented. (B) UV absorbance data from 195 to 320 nm (5-nm interval, upper graph) and for 280 nm (lower graph) at different time and absorbance scales. The sample injection was affected by gravity (height differential: 34 cm, time interval: 60 s), the applied voltage was a constant 20 kV, and the current decreased from 12 to 4 μA within the first 16 min.

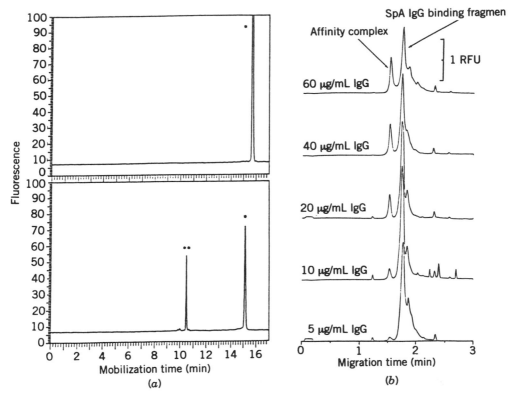

FIGURE 17-11. Immunoassay data based on CE. (A) CIEF in 75-μm i.d. LPA-coated capillary (anodic chemical mobilization and LIF detection) of purified affinity probe (top graph) and its separation from the complex with met-rhGH (lower graph). Affinity probe and complex are marked with * and **, respectively. (Reprinted with permission from Ref. 75. Copyright American Chemical Society.) (B) CZE determination (50-μm i.d. untreated capillary, pH 10.5 buffer, LIF detection) of increasing amounts of monoclonal mouse IgG dissolved in cultivation medium using 9.6×10^{-7} M of FITC-labeled protein A as the affinity probe. (From Ref. 79.)

17.3.7   The CE Capillary as Ultramicroreactor: Application to Enzyme Assays

CE has been shown to lend itself as a unique and efficient microreactor for chemical reactions followed by the on-line assay of a reaction product. In electrophoretically mediated microanalysis (EMMA[81–84]), electrophoretic mixing is utilized to merge zones containing the analyte and analytical reagents. The reaction is then allowed to proceed either in the presence or the absence of the applied electric field prior to the electrokinetic transport of the detectable product through the on-column detector. Two types of reaction schemes have been described, with one being based on a plug–plug reaction in which sample and reactant are applied separately onto the capillary, whereas the second comprises a continuous reaction with application of sample into a configuration with the reactant being present in the buffer.[85]

EMMA has been applied to the determination of enzymes. Enzymatic catalysis consists of a reaction

$$E + S = ES = E + P$$

where E, S, ES, and P refer to the enzyme, a substrate, an enzyme–substrate complex, and the product, respectively. If complex and product have different charges, then they are likely to be separable by CE. EMMA of enzymes is typically carried out in a capillary filled with substrate, buffer, and other required reagents. The enzyme is introduced as a small sample plug by electrokinetic or hydrodynamic injection. Application of power induces the mixing process and thus the start of the reaction. Product formation stops when the enzyme leaves the capillary. Two modes of operation differing in constant and periodic (stop flow) application of electric power have been described. With constant driving potential, the reactants are mixed and separated from the product continuously. Typically, enzyme catalysis occurs much faster than electrophoretic separations; thus, product formation is much larger than the amount of enzyme applied. This amplification provides the enormous sensitivity of the CE assay. For a system where the product is the only compound responding for detection, the height of the plateau recorded is directly proportional to the enzyme concentration and inversely proportional to the electric power applied (Figure 17-12A, upper graph). In the zero potential (stop flow) approach, product is accumulating because there is no electrophoretic transport of the enzyme and no separation of complex and product. When power is reapplied, a peak on top of the plateau is detected, thus augmenting the sensitivity of the assay (Figure 17-12A, lower graph). Using this approach for the determination of alkaline phosphatase (ALP) in a gel-filled capillary, a detection limit of 7.6×10^{-12} M was reported.[82] The same approach was employed for the simultaneous monitoring of two enzymes, ALP and β-galactonidase.[82] As a real world application, leucine aminopeptidase levels in human serum, human urine, and *Escherichia coli* supernatent samples were determined using the enzyme cleaving reaction of L-leucine-4-methoxy-β-naphthylamine (nonfluorescing compound) to 4-methoxy-β-naphthylamine (4-MBNA, fluorescing compound). For the exclusion of interferences during detection, time-resolved LIF detection was employed. Furthermore, 4-MBNA, which is neutral and thus separates well from the enzyme, was used as an internal standard.[83] Similarly, isoenzymes of lactate dehydogenase (LDH) were monitored in single and multiple human erythrocytes via the enzymatically catalyzed lactate to pyruvate conversion in the presence of NAD^+.[86] The resulting NADH activity (proportional to LDH present) was determined by LIF. In that approach, LDH isoenzymes were first separated electrophoreti-

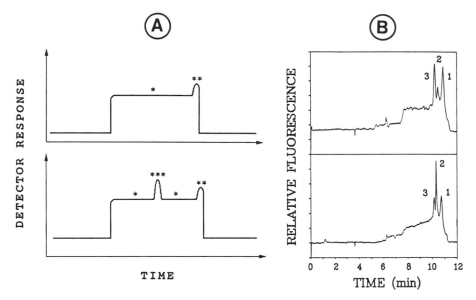

FIGURE 17-12. Electrokinetic capillary enzyme assays. (A) Schematic representation of data obtained in EMMA under continuous reaction conditions and having constant (top graph) and periodically interrupted (bottom graph) applied power. The case with $v_{ES} > v_P$ is considered only. The plateau, initial peak, and peak produced by zero potential are marked with *, **, and ***, respectively. (Adapted from Ref. 81.) (B) Electropherograms of a LDH isoenzyme standard mixture (12.3, 6.7, and 10.2 nIU of isoenzymes 1, 2, and 3, respectively; top graph) and LDH isoenzymes in lysed red blood cells (equivalent of about nine cells; bottom graph). A 20-μm i.d. capillary was employed. (Reprinted with permission from Ref. 86. Copyright American Chemical Society.)

cally prior to an incubation period of 2 min and electrokinetic transport of the product zones produced by the isoenzymes past the point of detection (Figure 17-12B). This example effectively demonstrates the EMMA capability of rapidly measuring ultratrace enzymes within an environment as small as a red blood cell (about 90 fL).

17.3.8 Hyphenated Techniques: CE-MS and LC-CE

Analytical chemists are faced with the challenge of increasing sample complexity and decreasing sample quantities. Thus, not unlike in other areas of separation science, significant developments of hyphenated techniques, including CE-MS and LC-CE, and their application to protein separation and analysis have been undertaken. The on-line coupling of CE with MS combines the extremely high resolving power of CE with structural information (molecular masses and/or fragmentation patterns) provided by MS detection (for a recent, comprehensive review, refer to Ref. 87). An example dealing with proteins is presented in Figure 17-13. These data demonstrate how accurately the molecular mass of lysozyme could be determined by CZE-MS using electrospray ionization. Although (1) the flow rates from the CE column are compatible with on-line coupling to MS, (2) numerous papers have appeared, and (3) several instrumental companies offer interfaces for MS (Table 17-3), CE-MS is still not widely accepted for routine use. The major limitation is the relatively high concentration detection limit of CE, a handicap that can be somewhat compensated by the stacking procedures referred to earlier[10,11] and elsewhere,[88,89] including on-line transient CITP precondensation of the solutes, which permits injection of sample volumes two to three orders of magnitude higher than is usual in CZE.[88,89] Nevertheless, using electrospray ionization detection limits for a full-scan CZE/MS analysis to about 10^{-7} M of proteins have been demonstrated.[88]

All the CE techniques discussed thus far, including CE-MS, are one-dimensional separation techniques with insufficient peak capacity for the separation of highly complex samples, particularly those of biological origin. Two-dimensional separations provide peak capacities corresponding to about the product of the peak capacities of the two methods involved.[90] Slab gel electrophoresis, in which separations based on charge (IEF, first dimension) and size (SDS–PAGE, second dimension) are combined, represents the most widely employed two-dimensional method. Its resolution is unmatched by any other technique, permitting the detection of hundreds or thousands of individual protein species, thus providing the best insight into the true complexity of protein systems. However, this technique is labor-intensive. In comprehensive LC-CE, as pioneered by Jorgenson's group,[91] a completely instrumental format of combining an LC separation followed by CZE analysis of the effluent has been developed. For protein analysis, microcolumn size exclusion chromatography (SEC) was coupled with CZE.[91] SEC-CZE gray scale image data obtained with protein standards are presented in Figure 17-14A. Incomplete resolution of all sample components was achieved when either technique was performed independently and under the same experimental conditions, whereas complete resolution was noted for the coupled approach. SEC-CZE data of a more challenging sample, reconstituted human serum, are presented in Figure 17-14B. Again, compared to single-dimension runs under the same conditions, higher resolution was reported for the two-dimensional approach. However, due to insufficient detection sensitivity for many proteins and the loss of some proteins adsorbed to the untreated fused-silica capillary, only a relatively small number of protein spots were detected. The use of coated capillaries in which protein–wall interactions would be diminished would have greatly enhanced the performance of that two-dimensional separation system. Nevertheless, data demonstrate that the two-dimensional separation principle can be applied to proteins. Further work in that field, including the hyphenation of LC-CE with other detectors (e.g.,

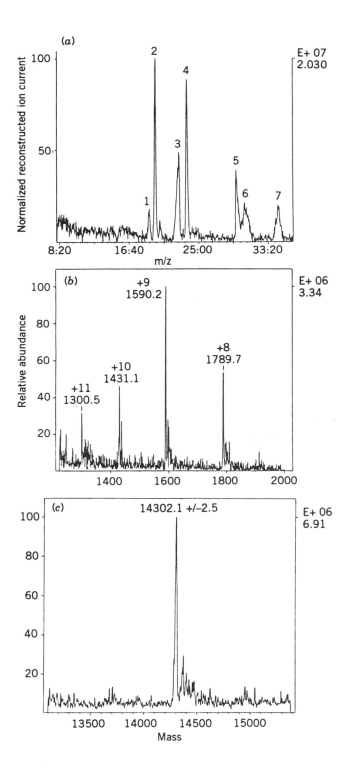

MS), will possibly pave the way toward the routine use of instrumental two-dimensional separation procedures.

17.4 OTHER APPROACHES WITH APPLICABILITY TO PROTEINS

17.4.1 Electrical Field Flow Fractionation Techniques

The overall setup of FFF techniques is schematically presented in Figure 17-15A. The relation to an HPLC system in which the chromatographic column is replaced by the FFF trough is obvious. An electrolyte pump providing the buffer flow and a flow-through detector are mounted at the two ends of the separation capillary. A computer for the control of the pump and the power supply, as well as for data acquisition, data storage, and data handling, permits versatile programming in time of the entire process. Detection of eluted solutes occurs by measuring specific physical properties such as optical and electrochemical responses using flow-through cells as in liquid chromatography and flow injection analysis.

In EFFF, developed by Giddings and co-workers,[16-18] the separation channel consists of two parallel membranes that separate the flowing stream from the electrode compartments (Figure 17-1B). Chamber thicknesses of 0.25 to 0.35 mm having channel widths of 15 to 25 mm were used with a length of about 50 cm. Reasonable protein separations could be achieved using flow rates of about 5 mL/h and electric fields between 2 and 20 V/cm. However, analysis times of several hours were typical with this kind of equipment. Alternatively, Davis and colleagues[92] designed an apparatus with an annular channel utilizing concentric porous Vycor glass tubes of various diameters. For this device, experimental retention ratios are reported to depart considerably from predictions based on theoretical EFFF. Similar problems were encountered in EPC in which the separation trough consists of an anisotropic ultrafiltration hollow fiber of circular cross section (about 50 cm long, 0.5 mm i.d., nominal molecular mass cutoff of 5,000 Da). The fiber is bathed in a buffer and subjected to a transverse electric field.[19,20] Carrier flow rates of

---

FIGURE 17-13. CZE-MS data of model proteins using electrospray ionization. (A) Full scan (m/z 600 to 2,000) reconstructed ion electropherogram of a 150-nL injection of 12 μM each of lysozyme (1), cytochrome c (2), ribonuclease A (3), myoglobin (4), β-lactoglobulin A (5), β-lactoglobulin B (6), and carbonic anhydrase (7) dissolved in water. (B) Mass spectrum of lysozyme taken from averaging the scans under the peak. (C) Deconvoluted spectrum of lysozyme, a 14.306-kDa protein. CZE conditions: LPA-coated 75-μm i.d. capillary of 50 cm length, buffer composed of 20 mM 6-aminohexanoic acid and acetic acid (pH 4.4), and applied voltage (current) of 18 kV (6 μA). (Reprinted with permission from Ref. 88. Copyright American Chemical Society.)

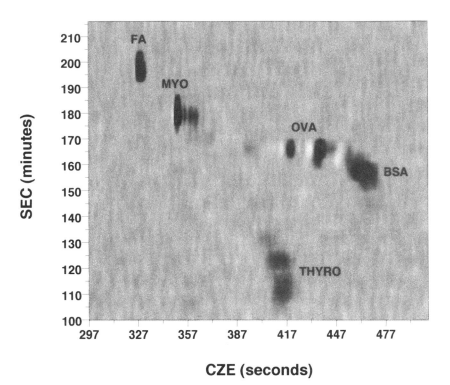

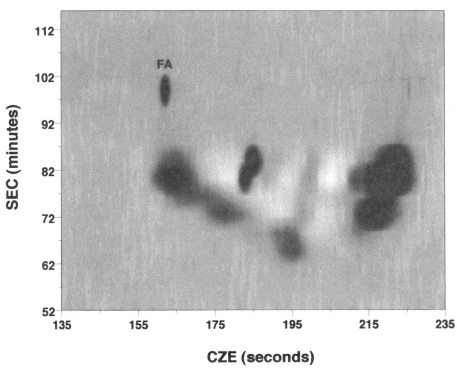

about 12 μL/min were admitted together with field strengths of 5 to 10 V/cm in order to achieve separations without complete electroretention in about 2 h. At these low voltages retardation was found to be quite consistent with the theoretical prediction. Higher field strengths, however, were shown to immobilize macromolecules in the column. The totally retained components at the hollow fiber of the wall could be eluted only by decreasing the applied voltage for the case of reversible sorption. By programming in time of the electric field, separations according to differences in electroretention values are achieved. This is an analogy to the chromatofocusing principle. For a great number of biopolymers, however, electroretention has to be avoided because of sorption nonreversibility, causing a substantial loss of material.

In EHFFF, contact between most proteins and the channel walls is avoided[22-24] by the concomitant presence of a pH and an electrical potential gradient representing the same gradient combination as in IEF. Solute layers having different depths from the walls are thereby established (Figure 17-15B). This method, also termed isoelectric focusing FFF,[23] could not be made operational in a ribbon-like (rectangular cross section) channel thus far. However, protein separations executed in a trapezoidal cross-section channel (Figure 17-15C) of 0.875 mL volume and 25 cm length and with a buffer flow rate of 40 μL/min (elution in less than 1 h) have been described.[23] A fractogram obtained with three proteins, ferritin (FER), myoglobin (MYO), and CYTC, is presented in Figure 17-16A. For the sake of comparison, CIEF data in a setup featuring electroosmotic zone displacement are also shown (Figure 17-16B). The two focusing methods in flowing streams have interesting similarities as well as differences. EHFFF and CIEF are methods that utilize hydrodynamic and electroosmotic flow, respectively, for the elution of focused isoelectric zones and their detection by conventional detectors that have been developed for HPLC and CE, respectively. In the two methods, the electric field and the flow (column axis) are oriented differently to each other, with the two vectors being perpendicular and parallel in EHFFF and CIEF, respectively. With today's instruments, the experimental procedure and setup for CIEF are simpler than that of EHFFF. For separations of proteins, CIEF

FIGURE 17-14. SCE-CZE gray scale images of (A) protein standards and (B) human serum proteins. The chromatographic separations (first dimension) were performed with flow rates of 180 and 360 nL/min, respectively, employing 250-μm i.d. fused-silica capillaries of 105 cm length that were packed with a slurry of 6-μm spherical Zorbax GF450 particles. CZE was conducted in 50-μm i.d. capillaries of 40 and 20 cm, respectively, effective length and using a pH 8.23 buffer composed of 10 mM tricine, 25 mM Na$_2$SO$_4$, and NaOH. The electrokinetic sample injection was affected every 4.5 min. Proteins were detected by UV absorption at 214 nm. FA, formamide; THYRO, thyroglobulin; BSA, bovine serum albumin. The large spot in B represents albumin. (From Ref. 91.)

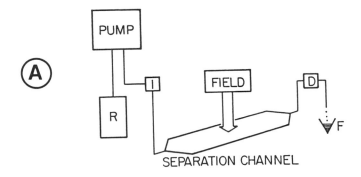

PUMP

FIELD

R

I

D

F

SEPARATION CHANNEL

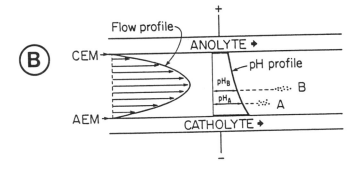

B

Flow profile

+

ANOLYTE ➤

CEM ➤

pH profile

pH$_B$ ········ B

pH$_A$ ····· A

AEM ➤

CATHOLYTE ➤

−

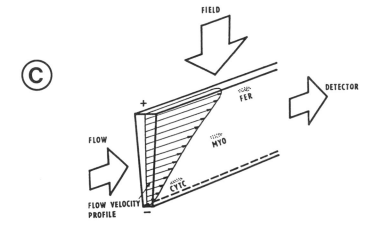

C

FIELD

DETECTOR

FER

+

MYO

FLOW

CYTC

FLOW VELOCITY
PROFILE

−

provides higher resolution and efficiency. Advantages of EHFFF comprise the requirement for lower voltages and its simpler application to micropreparative purposes. The time intervals required for separation and analysis in the two methods are comparable.

17.4.2 Analytical Continuous Flow Electrophoresis

In CFE (Figure 17-1C), a small stream of sample is admitted into the laminar flow of the buffer, except for IEF when the sample is evenly mixed with the carrier ampholytes prior to electrophoresis. For all techniques the liquid flow and, to a smaller extent, the applied electric field strength define the transit time of the sample. Resolution is given in terms of differences in mobility or pI values, respectively, sample quantity and composition, current applied, flow velocity, effectiveness of cooling, length of electrophoresis chamber, and width of the flowing fluid film. If samples are recovered, resolution is also dependent on the density of outlet tubes of the separation chamber. For analytical purposes it can suffice to monitor the solutes directly through a window at the end of the separation chamber with an optical scanning system. Good reproducibility and very short analysis times are achieved. Further assays can be performed on the fractions collected. On-line detection with a video camera or a scanning sensor also provides the means for feedback control and automation.

CFE instrumentation,[25,26] particularly for preparative applications, is currently available only under the tradename OCTOPUS from Dr. Weber GmbH, Ismaning (near Munich), FRG (formerly Elphor VaP systems of Bender & Hobein, Munich, FRG). The separation cells of these instruments are rectangular chambers in which laminar flow conditions are maintained by a small (0.3 to 0.8 mm) gap between two parallel plates. Typical chamber dimensions are 500×100 mm, and running conditions include field strengths up to 300 V/cm and sample transit times of 20 to 3600 s. Miniaturized CFE has been designed for continuous sample pretreatment and separation. CFE quartz channel devices with an internal height and width of 48 to 110 μm and 2 cm, respectively, and lengths of 2 to 7.8 cm have been constructed by Mesaros and co-workers.[94] This new approach to CFE has the potential to continuously sample and separate analytes from volume-limited microenvironments. Thus

---

FIGURE 17-15. Schematic representation of (A) the EHFFF setup, (B) the EHFFF separation of sample components A and B in a ribbon-like channel of a rectangular cross section in which the flow profile is parabolic, and (C) the EHFFF separation of three proteins (CYTC, MYO, and FER) in a trapezoidal cross-section channel with the cathode on the narrower and the anode on the wider part of the channel (B and C from Refs. 24 and 23, respectively). R, buffer reservoir; I, sample inlet system; D, flow through detector; F, fraction collector; CEM, cation-exchange membrane; AEM, anion-exchange membrane.

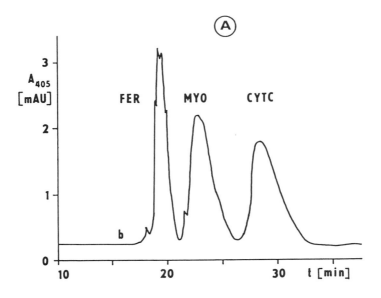

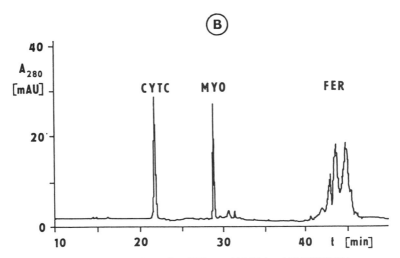

FIGURE 17-16. Separation of CYTC, MYO, and FER by (A) EHFFF in a trapezoidal cross-section channel of 25 cm length (volume 0.875 mL) and off-column solute detection and (B) CIEF with electroosmotic zone displacement using a fused-silica capillary of 90 cm total length and on-column detection at 70 cm (i.d. 75 μm; column volume 4 μL). The detection wavelengths were 405 and 280 nm, respectively. The voltages applied were 10 V and 20 kV, respectively (power levels of about 1 and 0.1 W, respectively). In EHFFF the protein load was higher (about 18-, 25- and 50 fold for CYTC, MYO, and FER, respectively) than that applied in CIEF. The protein load per unit column volume, however, was higher in the CIEF experiment. Ampholine (pH 3.5 to 10; 2.0 and 2.5%, respectively) was employed for both experiments. (From Ref. 93.)

far, the principle has been shown using dansylated amino acids for which sample introduction (but not separation) is carried out using conventional CZE, and eluting analytes are monitored at the channel exit by laser-induced fluorescence using two linearly arranged fiber optic arrays.[94] Electrochemical detection using an electrode array has also been discussed.[95] In another approach, Raymond and co-workers[96] reported the successful integration of a CFE device with comparable cell dimensions (50 μm $\times$ 10 mm $\times$ 50 mm) onto a silicon chip. The channel systems were etched into the silicon, to which a glass cover plate was anodically bonded to form the channels. Solute detection toward the chamber end was effected by laser-induced fluorescence using a scanning detector assembly. For that case, system performance was discussed employing labeled amino acids.

17.5 BRIEF DISCUSSION AND FUTURE PROSPECTS

Thirty years after the inception of conventional CE, a considerable amount of progress has been made, particularly with the emergence of fused-silica, capillary-based instrumentation since the mid-1980s. Many commercial instruments, capillaries, reagents, and entire CE kits are now available that provide the possibility of adopting CE as an attractive alternative to HPLC and slab gel electrophoresis. However, CFE and the FFF methods (i.e., EFFF, EPC, and EHFFF) are much less explored.

It is apparent that CE performed in a capillary of 10 to 100 μm i.d. can be applied successfully to a very large number of challenging analytical problems, including the analysis of biopolymers, such as proteins, oligosaccharides, and oligonucleotides. In fact, the application range of CE is considerably broader. It can also be effectively employed for separation, characterization, and determination of (1) small molecular mass substances of clinical, pharmacological, nutritional, environmental, and industrial interest, (2) synthetic polymers and particles, and (3) living cells. CE also contributes to the basic exploration of new horizons, namely the sequencing of entire genomes such as the human genome, the chemical characterization of nanoliter environments, including the content of single living cells, the on-line study of chemical reactions with the capillary being a unique ultramicroreactor, and the investigation of separations and reactions on a millisecond time scale, to name but a few.

In the field of protein separation and analysis, characterization and modification of capillary walls and the development of suitable running conditions for many different proteins were essential to understanding and preventing undesired protein–wall interactions. As a result of these efforts, CE of many proteins can now be regarded as a fast and high-resolution substitute of slab gel electrophoresis, including SDS–PAGE, gel IEF, and HPLC. Although separations and solute detection can be achieved within a few minutes, sample throughput in CE is typically lower than obtained in slab gel electrophoresis.

With the availability of instrumentation comprising multiple capillaries in parallel, however, sample throughputs comparable or even higher to those obtained in multilane gel analyses will be possible. Furthermore, CE of proteins is not at all limited to the application of different electrophoretic separation techniques, such as CZE, CITP, and CIEF. The instrumental capillary format has been shown to be well suited for (1) the investigation of biospecific affinity and other reactions involving proteins, (2) the performance of highly sensitive immunoassays and on-column reaction-based chemical analyses (e.g., enzyme assays, even within a single cell), (3) the determination of protein mass via use of a sieving environment or by CE-MS (the latter also providing structural information of the electrophoresed compound), and (4) the resolution of complex protein samples by two-dimensional LC-CE. Last but not least, (1) comprehensive and automated analyzers for proteins of clinical interest (e.g., serum proteins) are in the process of being developed,[97,9?] (2) interrupted or continuous collection systems have been designed to permit the collection of proteins emerging from capillaries for subsequent characterization by another analytical method and/or by microsequencing at the picomole level,[99,100] and (3) data obtained with bovine trypsinogen[101] suggest that CE is a powerful method for studying protein folding, unfolding, and refolding (i.e., to detect and resolve changes in protein covalent structure).

ACKNOWLEDGMENTS

Many thanks are due to those collaborators and colleagues who have contributed to the work described and who have offered advice, criticism, and discussion on the topic of capillary electrophoresis. The generous loan of electrophoretic equipment by LKB/Pharmacia (Bromma/Uppsala, Sweden) and by Bio-Rad Laboratories (Richmond/Hercules, CA, USA) is gratefully acknowledged. This work was supported by the Swiss National Science Foundation.

17.7 REFERENCES

1. B.P. Konstantinov, O.V. Oshurkova, *Sov. Phys. Tech. Phys., 11,* 693–704 (1966).

2. F.M. Everaerts, J.L. Beckers, Th.P.E.M. Verheggen, *Isotachophoresis,* Elsevier, Amsterdam, 1976.

3. P. Boček, M. Deml, P. Gebauer, V. Dolník, *Analytical Isotachophoresis,* VCH, Weinheim, 1988.

4. S.F.Y. Li, *Capillary Electrophoresis Principles, Practice and Applications,* Elsevier, Amsterdam, 1992.

5. N.A. Guzman, ed., *Capillary Electrophoresis Technology,* Marcel Dekker, New York, 1993.

6. F. Foret, L. Křivánková, P. Boček, *Capillary Zone Electrophoresis,* VCH, Weinheim, 1993.

7. P.D. Grossman, J.C. Colburn, eds., *Capillary Electrophoresis: Theory and Practice,* Academic Press, San Diego, 1992.

8. W. Thormann, R.A. Mosher, M. Bier, In *Chemical Separations,* C.J. King, J.D. Navratil, eds., Litarvan Literature, Denver, CO, 1986, Vol. 1, pp. 153–168.

9. W. Thormann, M.A. Firestone, In *Protein Purification, Principles, High Resolution Methods, and Applications,* J.-C. Janson, L. Ridén, eds., VCH, New York, 1989, pp. 469–492, and references cited therein.

10. P. Gebauer, W. Thormann, P. Boček, *J. Chromatogr., 608,* 47–57 (1992).

11. M. Albin, P.D. Grossman, S.E. Moring, *Anal. Chem., 65,* 489A–497A (1993).

12. M. Bier, O.A. Palusinski, R.A. Mosher, D.A. Saville, *Science, 219,* 1281–1287 (1983).

13. R.A. Mosher, D.A. Saville, W. Thormann, *The Dynamics of Electrophoresis,* VCH, Weinheim, 1992.

14. S. Hjertén, K. Elenbring, F. Kilár, J. Liao, A.J.C. Chen, C.J. Siebert, M. Zhu, *J. Chromatogr., 403,* 47–61 (1987).

15. K. Otsuka, S. Terabe, T. Andro, *J. Chromatogr., 348,* 39–47 (1985).

16. K.D. Caldwell, L.F. Kesner, M.N. Myers, J.C. Giddings, *Science, 176,* 296–299 (1972).

17. L.F. Kesner, K.D. Caldwell, M.N. Myers, J.C. Giddings, *Anal. Chem., 48,* 1834–1839 (1976).

18. J.C. Giddings, G. Lin, M.N. Myers, *Sep. Sci., 11,* 553–568 (1976).

19. J.F.G. Reis, E.N. Lightfoot, *AIChE J., 22,* 779–785 (1976).

20. A.S. Chiang, E.H. Kmiotek, S.M. Langan, P.T. Nobile, J.F.G. Reis, E.N. Lightfoot, *Sep. Sci. Technol., 14,* 453–474 (1979).

21. A.B. Shah, J.F.G. Reis, E.N. Lightfoot, R.E. Moore, *Sep. Sci. Technol., 14,* 475–497 (1979).

22. J. Chmelík, M. Deml, J. Janca, *Anal. Chem., 61,* 912–914 (1989).

23. J. Chmelík, W. Thormann, *J. Chromatogr., 600,* 305–311 (1992).

24. W. Thormann, M.A. Firestone, M.L. Dietz, T. Cecconie, R.A. Mosher, *J. Chromatogr., 461,* 95–101 (1989).

25. K. Hannig, *Electrophoresis, 3,* 235–243 (1982).

26. H. Wagner, R. Kessler, *GIT Lab.-Med., 7,* 30–35 (1984).

27. T. Tsuda, *LC-GC Intl., 5,* 26–36 (1992), and references cited therein.

28. W.D. Pfeffer, E.S. Yeung, *J. Chromatogr., 557,* 125–136 (1991).

29. N. Catsimpoolas, *Electrophoresis, 1,* 73–78 (1980).

30. F. Foret, M. Deml, V. Kahle, P. Boček, *Electrophoresis, 7,* 430–432 (1986).

31. W. Thormann, D. Arn, E. Schumacher, *Electrophoresis, 5,* 323–337 (1984).

32. W. Thormann, *J. Chromatogr., 334,* 83–94 (1985).

33. J. Wu, J. Pawliszyn, *Anal. Chem., 66,* 867–873 (1994).

34. J. Wu, J. Pawliszyn, *Electrophoresis, 16,* 670–673 (1995).

35. J.W. Jorgenson, K.D. Lukacs, *Anal. Chem., 53,* 1298–1302 (1981).

36. D.J. Harrison, K. Fluri, K. Seiler, Z. Fan, C.S. Effenhauser, A. Manz, *Science, 261,* 895–897 (1993).

37. S.C. Jacobson, R. Hergenröder, L.B. Koutny, J.M. Ramsey, *Anal. Chem., 66,* 1114–1118 (1994).

38. A. Manz, E. Verpoorte, D.E. Raymond, C.S. Effenhauser, N. Burggraf, H.M. Widmer, In *Micro Total Analysis Systems,* A. van den Berg, P. Bergveld, eds., Kluwer Academic Publishers, Dordrecht, 1995, pp. 5–27.

39. R.A. Mosher, P. Gebauer, W. Thormann, *J. Chromatogr., 638,* 155–164 (1993).

40. Th.P.E.M. Verheggen, A.C. Schoots, F.M. Everaerts, *J. Chromatogr., 503,* 245–255 (1990).

41. S. Hjalmarsson, A. Baldesten, *CRC Crit. Rev. Anal. Chem., 11,* 261–352 (1981).

42. F.E.P. Mikkers, F.M. Everaerts, Th.P.E.M. Verheggen, *J. Chromatogr., 169,* 1–20 (1979).

43. W. Thormann, J.P. Michaud, R.A. Mosher, In *Electrophoresis 86,* M. Dunn, ed., VCH Verlagsgesellschaft, Weinheim, 1986, pp. 267–270.

44. M.A. Firestone, J.P. Michaud, R. Carter, W. Thormann, *J. Chromatogr., 407,* 363–368 (1987).

45. R.A. Mosher, W. Thormann, M. Bier, *J. Chromatogr., 436,* 191–204 (1988).

46. M.A. Firestone, W. Thormann, *J. Chromatogr., 436,* 309–315 (1988).

47. J. Caslavska, P. Gebauer, W. Thormann, *Electrophoresis, 15,* 1167–1175 (1994).

48. M.W.F. Nielen, *J. High Res. Chromatogr., 16,* 62–64 (1993).

49. G.J.M. Bruin, A. Paulus, *Anal. Meth. Instrument., 2,* 3–26 (1995), and references cited therein.

50. M.V. Novotny, K.A. Cobb, J. Liu, *Electrophoresis, 11,* 735–749 (1990), and references cited therein.

51. P. Tsai, C.S. Lee, In *Capillary Electrophoresis Technology,* N.A. Guzman, ed., Marcel Dekker, New York, 1993, pp. 475–488, and references cited therein.

52. M. Gilges, M.H. Kleemiss, G. Schomburg, *Anal. Chem. 66,* 2038–2046 (1994).

53. J.K. Towns, F.E. Regnier, *Anal. Chem., 64,* 2473–2478 (1992).

54. A.S. Cohen, B.L. Karger, *J. Chromatogr., 397,* 409–417 (1987).

55. A. Widhalm, C. Schwer, D. Blaas, E. Kenndler, *J. Chromatogr., 549,* 446–451 (1991).

56. K. Ganzler, K.S. Greve, A.S. Cohen, B.L. Karger, A. Guttman, N.C. Cooke, *Anal. Chem., 64,* 2665–2671 (1992).

57. D. Wu, F.E. Regnier, *J. Chromatogr., 608,* 349–356 (1992).

58. K. Tsuji, *J. Chromatogr. B, 662,* 291–299 (1994).

59. W. Thormann, R.A. Mosher, M. Bier, In *Electrophoresis 84,* V. Neuhoff, ed., Verlag Chemie, Weinheim, 1983, pp. 118–121.

60. W. Thormann, A. Tsai, J. Michaud, R.A. Mosher, M. Bier, *J. Chromatogr., 389,* 75–86 (1987).

61. S. Hjertén, M. Zhu, *J. Chromatogr., 346,* 265–270 (1985).

62. S. Hjertén, J. Liao, K. Yao, *J. Chromatogr., 387,* 127–138 (1988).

63. M. Zhu, R. Rodriguez, T. Wehr, C. Siebert, *J. Chromatogr., 608,* 225–237 (1992).

64. J.R. Mazzeo, I.S. Krull, *Anal. Chem., 63,* 2852–2857 (1991).

65. W. Thormann, J. Caslavska, S. Molteni, J. Chmelík, *J. Chromatogr., 589,* 321–327 (1992).

66. S. Molteni, H. Frischknecht, W. Thormann, *Electrophoresis, 15,* 22–30 (1994).

67. W. Thormann, M.A. Firestone, J.E. Sloan, T.D. Long, R.A. Mosher, *Electrophoresis, 11,* 298–304 (1990).

68. S. Hjertén, M. Kiessling-Johansson, *J. Chromatogr., 550,* 811–822 (1991).

69. R.D. Smith, S.M. Fields, J.A. Loo, C.J. Barinaga, H.R. Udseth, C.G. Edmonds, *Electrophoresis, 11,* 709–717 (1990).

70. P. Gebauer, W. Thormann, *J. Chromatogr., 558,* 423–429 (1991).

71. J. Caslavska, P. Gebauer, W. Thormann, *Electrophoresis, 15,* 1167–1175 (1994).

72. L.Z. Avila, Y.-H. Chu, E.C. Blossey, G.M. Whitesides, *J. Med. Chem., 36,* 126–133 (1993).

73. N.H.H. Heegaard, *J. Chromatogr. A, 680,* 405–412 (1994).

74. R.G. Nielsen, E.C. Rickard, P.F. Santa, D.A. Sharknas, G.S. Sittampalam, *J. Chromatogr., 539,* 177–185 (1991).

75. K. Shimura, B.L. Karger, *Anal. Chem., 66,* 9–15 (1994).

76. N.M. Schultz, R.T. Kennedy, *Anal. Chem., 65,* 3161–3165 (1993).

77. N.M. Schultz, L. Huang, R.T. Kennedy, *Anal. Chem., 67,* 924–929 (1995).

78. O.-W. Reif, R. Lausch, T. Scheper, R. Freitag, *Anal. Chem., 66,* 4027–4033 (1994).

79. R. Lausch, O.-W. Reif, P. Riechel, T. Scheper, *Electrophoresis, 16,* 636–641 (1995).

80. L. Steinmann, J. Caslavska, W. Thormann, *Electrophoresis, 16* 1912–1916, (1995), and references cited therein.

81. J. Bao, F.E. Regnier, *J. Chromatogr., 608,* 217–224 (1992).

82. D. Wu, F.E. Regnier, *Anal. Chem., 65,* 2029–2035 (1993).

83. K.J. Miller, I. Leesong, J.Bao, F.E. Regnier, F.E. Lytle, *Anal. Chem., 65,* 3267–3270 (1993).

84. B.J. Harmon, D.H. Patterson, F.E. Regnier, *Anal. Chem., 65,* 2655–2662 (1993).

85. L.Z. Avila, G.M. Whitesides, *J. Org. Chem., 58,* 5508–5512 (1993).

86. Q. Xue, E.S. Yeung, *Anal. Chem., 66,* 1175–1178 (1994).

87. J. Cai, J. Henion, *J. Chromatogr. A, 703,* 667–692 (1995), and references cited therein.

88. T.J. Thompson, F. Foret, P. Vouros, B.L. Karger, *Anal. Chem., 65,* 900–906 (1993).

89. F. Foret, E. Szökö, B.L. Karger, *Electrophoresis, 14,* 417–428 (1993).

90. J.C. Giddings, *Anal. Chem., 56,* 1258A-1270A (1984).

91. J.P. Larmann, A.V. Lemmo, A.W. Moore, J.W. Jorgenson, *Electrophoresis, 14,* 439–447 (1993).

92. J.M. Davis, F.R.F. Fan, A.J. Bard, *Anal. Chem., 59,* 1139–1348 (1987).

93. J. Chmelík, W. Thormann, *J. Chromatogr., 632,* 229–234 (1993).

94. J.M. Mesaros, G. Luo, J. Roeraade, A.G. Ewing, *Anal. Chem., 65,* 3313–3319 (1993).

95. A.G. Ewing, J.M. Mesaros, P.F. Gavin, *Anal. Chem., 66,* 527A–536A (1994).

96. D.E. Raymond, A. Manz, H.M. Widmer, *Anal. Chem., 66,* 2858–2865 (1994).

97. F.-T. Chen, J.C. Sternberg, *Electrophoresis, 15,* 13–21 (1994).

98. G.L. Klein, C.R. Jolliff, In *Handbook of Capillary Electrophoresis,* J.P. Landers, ed., CRC Press, Boca Raton, 1994, pp. 419–457.

99. C. Fujimoto, K. Jinno, In *Capillary Electrophoresis Technology,* N.A. Guzman, ed., Marcel Dekker, New York, 1993, pp. 509–523, and references cited therein.

100. Y.-F. Cheng, M. Fuchs, D. Andrews, W. Carson, *J. Chromatogr., 608,* 109–116 (1992).

101. M.A. Strege, A.L. Lagu, *J. Chromatogr. A, 652,* 179–188 (1993).

18 Large-Scale Electrophoretic Processes

WOLFGANG THORMANN

Department of Clinical Pharmacology
University of Bern
CH-3010 Bern, Switzerland

SUMMARY

In this chapter attention is focused on preparative electrophoresis of proteins, an area of increased recent interest in conjunction with the purification and isolation problems of polypeptides from fermentation broths, proteinaceous extracts, and body fluids. Past and current technologies using solid support

Protein Purification: Principles, High-Resolution Methods, and Applications, Second Edition.
Edited by Jan-Christer Janson and Lars Rydén.
ISBN 0-471-18626-0. © 1998 Wiley-VCH, Inc.

media or free solution approaches are briefly described and compared. Major emphasis is focused on providing a broad overview on free fluid preparative electrophoretic techniques on current commercial instruments, including continuous flow and recycling large-scale processes.

18.1 INTRODUCTION

Electrophoresis is defined as the transport of electrically charged particles in liquid media under the influence of a direct current electrical field. It encompasses a wide variety of systems that are characterized by distinct, current-induced moving or stationary concentration boundaries in electrolytes. One of the most important application of this spectrum of techniques is the separation and analysis of complex mixtures of biological macromolecules, particularly proteins and polynucleotides. Electrophoresis is mainly identified with gels, which offer excellent resolution and are widely used in clinical analysis, biomedical research, and quality control in a variety of processes, including isoelectric focusing, disk, sodium dodecyl sulfate (SDS), and two-dimensional electrophoresis. Gels are used because of their ability to effectively suppress fluid convection. Electrophoresis in free fluids represents a more instrumental format and is steadily gaining popularity. Both analytical approaches (capillary electrophoresis, refer to Chapter 17) and large scale processes (continuous flow and recycling electrophoresis[1-9] are currently being explored on a broad basis.

As discussed in Chapter 17, there is a fundamental unity that underlies all electrophoretic processes, including the four basic modes of separation: moving boundary electrophoresis, zone electrophoresis, isotachophoresis (displacement electrophoresis), and isoelectric focusing. The superimposition of additional constraints, such as specific affinities, molecular sieving, cross-, co-, or counterflow, magnetic fields, and fixed charges, yield all the electrophoretic methods in use. In moving boundary electrophoresis, zone electrophoresis, and isotachophoresis, separation is based on differences in electrophoretic mobilities, whereas in isoelectric focusing, separation is based on differences in isoelectric points. These methods differ in terms of initial and boundary conditions, which in turn determine the character of the migrating sample zones. Zone electrophoresis is conducted in a uniform buffer, whereas isotachophoresis and moving boundary electrophoresis are performed in discontinuous buffer systems. In isoelectric focusing, amphoteric samples are sorted according to their isoelectric point in a pH gradient. Formation and character of the pH gradient are crucial for focusing. Immobilized and dynamically established gradients are distinguished.

Although most of the usage of protein electrophoresis is confined to analytical applications, there has been an increased interest in using preparative electrophoresis for the purification and isolation of peptides and proteins from fermentation broths, proteinaceous extracts, and body fluids.[4-9] The goals of that work are (1) gentle purification and isolation of proteins without denaturing of the macromolecules, (2) efficient and costeffective purification, and (3) high purity of the purified products. Preparative electrophoresis is mainly suitable for the concentration and final polishing of proteins, as well as the direct isolation of small amounts of a pure protein. For industry it represents an interesting approach for the purification of products on the gram scale (e.g., rDNA proteins, monoclonal antibodies, and trace proteins from body fluids). Preparative electrophoresis is an attractive alternative to chromatographic technology. Purification by electrophoresis on the kilogram to ton scale is not realistic. For that purpose, a number of other methods, such as precipitation, extraction, and adsorption, are available.

In this chapter attention is focused on large-scale preparative electrophoresis of proteins, with major emphasis being directed toward the use of free fluid approaches, such as continuous flow and recycling techniques. An overview of current commercial instrumentation is also presented.

18.2 PREPARATIVE ELECTROPHORESIS OF PROTEINS IN SOLID SUPPORT MEDIA

18.2.1 Electrophoresis in Coherent and Granulated Gels

Coherent or granular gels provide efficient fluid stabilization suitable for electrophoretic separations. The former are most commonly polyacrylamide or agarose and the latter is usually Sephadex. Regardless of the electrophoretic mode used, the preparative gel procedure is most often a batch operation and employs a cylindrical, flat bed, or annular separation chamber. Advantages of gel systems comprise (1) unsurpassed resolution (isoelectric focusing in polyacrylamide gels is capable of resolving proteins that differ in pI by 0.001 pH units[5,10] and (2) application of unique separation principles incorporating molecular sieving, chemical, and immunological affinity (electrochromatography), as well as fixed charges (e.g., immobilized pH gradient for isoelectric focusing[5]) in addition to electrophoretic transport. Gel electrophoresis is a milligram to gram preparative methodology.

In essence, there are two different approaches to batch preparative gel electrophoresis. The first involves the migration of the sample of interest to the end of the gel and from there into a buffer chamber where it can be eluted continuously or discontinuously into a buffer stream. Many arrangements have been described (see Refs. 11–13 for reviews). Commercially, a complete, automated instrument is available (Model 230A HPEC system of Applied Biosystems Inc., Foster City, CA, USA) in which tube gel electrophoresis is combined with continuous sample elution, real-time visualization, quantitation, and automatic sample collection. A somewhat simpler, less expensive system permitting a continuous collection of separated proteins has been introduced by another instrument manufacturer (Model 491 Prep Cell, Bio-Rad, Richmond, CA, USA). For separations based on native polyacrylamide gel electrophoresis (PAGE) or SDS–PAGE, up to 50 mg of total protein can be loaded onto this cell. A dilution and therefore the risk of unrecognizing the sample, especially with slowly migrating proteins, is a disadvantage of this tube gel technology.

The second approach consists of zone excision and protein elution, often followed by concentration and further purification. Several protein bands can be simultaneously excised and extracted from the same gel irrespective of their mobilities. A handicap in most applications of this technology is the requirement of sample visualization (staining) on the gel. An important consideration for large-scale preparative gel electrophoresis is the degree to which these methods can be scaled up. The efficient dissipation of joule heat restricts the maximum thickness of the gel (e.g., to 18 mm polyacrylamide for a separation within a single workday[14]) and therefore the protein load.

18.2.2 Continuous Preparative Electrophoresis in Solid Support Media: Electrochromatography

Methods for continuous preparative electrophoresis using a supporting medium have hardly been explored thus far. Two promising and recent developments are based on the usage of rotating annular columns, not unlike those introduced for continuous chromatography.[15] They also permit the combined exploitation of electrophoretic and chromatographic separations often referred to as continuous electrochromatography (i.e., continuous electrophoresis performed in a stabilizing medium that also has some specific adsorption properties). The first apparatus has an annular bed of inert or interactive media that very slowly (e.g., 0.01 rpm) rotates about its axis (Figure 18-1). The sample is continuously introduced at a single point on the top of the bed, and the buffer is admitted at multiple points[16] or as a complete reservoir[17,18] along the entire circumference of the annulus. An electric field is applied axially (i.e., along the direction of flow) and retards or accelerates charged solutes. As a result of this configuration, separated fractions appear as helical bands. The products are drawn off through ports located on the base of the annulus.

The second approach is very similar but has the electric field perpendicular (radial) to the flow direction. This arrangement requires a thicker bed compared to that with an axial field and can be regarded as a true two-dimensional separation process because electrophoresis and chromatography are conducted perpendicular to each other. This concept has been demonstrated for the separation of bovine hemoglobin (60 kDa) and Blue Dextran 2000 (2,000 kDa) using the gel filtration material Bio-Gel P 150 as column resin.[19] Tests were carried out with an 8.8-cm-diameter column that had a 30-cm active height, an annulus of approximately 3.8 cm thickness, and a continuous sample feed of 0.25 mg/min of each polymer. At an applied voltage of 10 V, the apparent electrophoretic velocity of the more rapidly moving hemoglobin fraction was 2.8 μm/s. Blue Dextran 2000 was not significantly affected by the

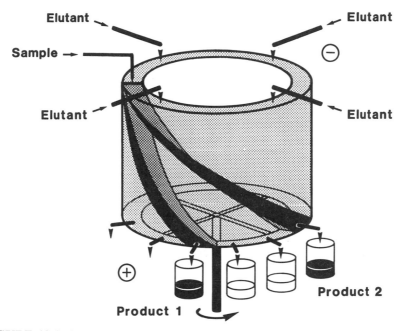

FIGURE 18-1. Schematic representation of a rotating, annular bed column system for continuous electrophoresis–electrochromatography with an axial electric field.

electrical field. There was almost complete angular separation between these two solutes (continuous chromatography) as well as radial separation of the hemoglobin components (continuous electrophoresis).

The performance of these devices has been analyzed theoretically and tested with rather simple sample mixtures. Only one reference could be found describing the high-resolution fractionation of proteins.[18] The achieved protein throughput of 600 mg per day, however, seems to be fairly modest. Further scaleup is certainly feasible, but removal of heat and complexity of the apparatus remain rate-limiting obstacles.

18.2.3 Counteracting Chromatographic Electrophoresis

Counteracting chromatographic electrophoresis (CACE), first described by O'Farrell,[20] is conceptually a composite of chromatography and electrophoresis in which opposing chromatographic and electrophoretic mass transport are counterbalanced to obtain a steady state.[21] A gradient of a chromatographic matrix (e.g., gel permeation matrices of various porosities) is the prerequisite for separation. In such an electrophoretic column, the matrix influences the movement of the sample driven by buffer flow such that electrophoretic displacement will dominate within part of the column while movement with the

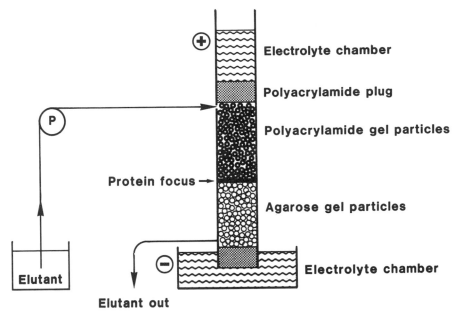

FIGURE 18-2. Principle of counteracting chromatographic electrophoresis (CACE) with a sharp discontinuity between two packing materials.

flowing buffer will dominate in another part of the separation space. Similar to isoelectric focusing, there is a restoring force. Solutes focus at characteristic equilibrium points with the matrix change (and not the pH gradient) being responsible for focusing.

The simplest implementation of CACE comprises a sharp discontinuity between two packing materials (Figure 18-2). Under suitable operating conditions of applied electric field, volumetric flow, and buffer composition (including pH), a protein of interest is accumulated at the interface from which it can be withdrawn. Other proteins that do not have the right molecular properties (for interaction with the support materials) and electrophoretic mobility will not accumulate at this location. Using a column of 50 cm in length and 0.7 cm in diameter with a two-component BioGel A-50m/BioGel P-10 filling, ferritin was brought to equilibrium with up to 1 g of protein per run.[20] CACE has the potential for further scaleup but is limited due to inefficient heat removal and the availability of ideal focusing media.[9] CACE has also been extended to continuous operation by modifying the batch column.[22]

18.3 PREPARATIVE ELECTROPHORESIS OF PROTEINS IN FREE FLUIDS

For a scaleup beyond what can be obtained in flat slabs or columns of coherent or granular media, separations have to be carried out in free fluid systems.

Electrophoresis in free fluids also offers a more instrumental format, and thereby simpler access to automation, but suffers from disturbances originating (1) from electrohydrodynamic (electroosmosis and conductivity-dielectric instability) and density driven flows, and (2) from interactions of biopolymers with column walls as well as any membranes and screens placed in the current path. Promising approaches for preparative electrophoresis of proteins have been known for many years, whereas others are currently being refined in research laboratories. Furthermore, several new commercial instruments have been introduced and there is an increasing number of purification examples reported in the literature (Tables 18-1 and 18-2, respectively). These developments indicate a renaissance of free fluid electrophoresis.

18.3.1 Purification of Proteins by Electrokinetic Membrane Processes: Forced Flow Electrophoresis

Many years ago, the effective control of electrophoretic protein separations by means of membrane or filter elements has been demonstrated in the preparative approaches of electrodecantation,[23] electrophoresis convection,[24] and forced flow electrophoresis.[25] Very large throughputs and division into two fractions, a concentrated fraction of electrophoretically mobile proteins and an isoelectric fraction at its original concentration, are characteristic for these processes (Figure 18-3). In electrodecantation and electrophoresis convection, gravity is utilized as a driving force to cause rapid decantation of the electrophoretically concentrated proteins near the membranes. Forced flow electrophoresis, first described by Bier in 1957,[25] is a continuous and versatile method that takes advantage of the combined effect of electrophoresis and filtration. It has been used for large-scale electrophoretic fractionation and concentration of proteins, *in vivo* isolation/removal of proteins, such as immunoglobulins, from blood maintained in extracorporeal circulation, cross-flow electrofiltration, electroosmotic concentration and desalting of macromolecular systems and electroadsorption.[26] Only forced flow electrophoresis is of current interest for large-scale protein purification.[27] A schematic representation of a double-stage processing unit is presented in Figure 18-4. It comprises a stack of alternate protein-impermeable membranes and protein-permeable filters, defining narrow channels inserted in parallel between the electrodes. The electrical field is perpendicular to the axial flow direction and can prevent particle deposition at the filters. The Joule heat is dissipated by the process fluid which is recycled through a cooling system. A recent study shows the purification of lentil lectins from lens culinaris using a single- and a double-stage apparatus.[27] The latter approach, in which the effluent filtrate of the first stage becomes the feed for the second stage (Figure 18-4), was found to be more effective. From 750 g of dry lentil 888 mg of dried pure lectins was harvested within a few hours of operation having a total of five parallel processing cells with dimensions of 5.9 by 16.7 cm (98.5 cm²) each. Scaleup without loss of resolution can be achieved by increasing the number of cells in the assembly and/or by expanding the area of the compartments. In the past, up to 100 cells in parallel and cells

TABLE 18-1. Typical Specifications of Selected Commercial Preparative Free Fluid Electrophoresis Instruments

Instrument Name Instrument Manufacturer	OCTOPUS (VaP 22) Dr. Weber GmbH	RF3 Protein Technol.	MinipHor Protein Technol.	IsoPrime Hoefer	Rotofor Bio-Rad
Instrument type	Continuous flow	Recycling	Recycling	Recycling	Rotating
Max. protein load (g)	n.a.	3–5	1	1	2
Protein throughput[a] (g/h)	0.05–30	0.05–5	0.05–1	0.05–0.25	0.05–0.2
Processing fluid volume (mL)	n.a.	100–500	30	30/segment	15–60
Voltage gradient[a] (V/cm)	100–300	100–300	200–400	50–300	60–120
Power dissipation[a] (W)	50–600	30–100	20–50	2–4	10–15
Chamber geometry	Rectangular	Rectangular	Rectangular	Cylindrical	Annulus
Chamber dimensions[b] (cm)	10 × 50	4 × 20	2.6 × 10.5	1 × 3.2/segment	15 × 1.9/15 × 3
Fluid layer thickness (mm)	0.5	0.75	0.75	n.a.	n.a.
Chamber volume (mL)	25	6	1.95	8/segment	18/60
Compartmentation	None	None	None	Membrane	Screen
Fraction number	Up to 96	30	20	1/segment	20
On-line monitors	Optional	c	c	c	n.a.

[a] Parameters that are strongly dependent on electrophoretic separation principle applied.[7]

[b] The first number refers to the separation axis (parallel to electric field). For the IsoPrime and Rotofor the second number represents the diameter of the cylindrical cell and the outer diameter of the annular cell, respectively. The inner diameter of the Rotofor cell is 1 cm.

[c] Flow-through monitors can be inserted into recirculating loops by the user.

TABLE 18.2. Selected Protein Purification Examples Using Preparative Free Fluid Electrophoresis

Purified Product	Source	Instrument[a]	Refs.
rh Tissue plasminogen activator	Yeast	Elphor VaP 21	47
rh Leucocyte interferons	*E. coli*	RIEF	56
Monoclonal antibodies to PCP	Mouse ascites fluid	Rotofor/RIEF	55
Monoclonal antibodies	Mouse ascites fluid	Elphor VaP 22	45
Monoclonal antibodies	Tissue culture medium	Elphor VaP 22	45
Antimicrococcal carbohydrate antibodies	Rabbit antisera	RIEF	57
Fibrinolytic enzyme	Snake venom	RIEF	58
Lectins LcH-A and LcH-B	Lentil seeds	BIOSTREAM/ VaP22/RIEF	38
Lectins LcH-A and LcH-B	Lentil seeds	FFE	27
Fibronectin	Porcine plasma	BIOSTREAM	91
Trypsin	Porcine pancreatic extract	BIOSTREAM	36
IgG, transferrin, albumin	Human plasma	BIOSTREAM	34
Monoclonal IgM	Tissue culture	BIOSTREAM	37
α-Amylase	*E. coli*[b]	Elphor VaP 22	46,92
Formate dehydrogenase, formaldehyde dehydrogenase, methanol oxidase	*Candida boidinii*	Elphor VaP 22 and VaP 220	44
α-Amylase	*Bacillus subtilis*	Autofocuser	85
Uricase	*Candida utilis*	Autofocuser	84
Peroxidase	Horseradish	Autofocuser	86
Human monoclonal antibodies to HIV gp 41	Hybridoma cell line	IsoPrime	65,66
Recombinant *N*-acetyl eglin C	*E. coli*	IsoPrime	65
rh growth hormone	*B. subtilis*	IsoPrime	67
Glucoamylase G1	*Aspergillus niger*	IsoPrime	100
Acidic form of leukoregulin	Human lymphocytes	Rotofor	59
Phospholipase D	*Corynebacterium, pseudotuberculosis*	RIEF	61
S1 nuclease	*A. oryzae*	RF3	70
Ovalbumin, lysozyme	Commercial ovalbumin	RF3	77,78
Transferrin	Human serum	RF3	73
Human serum fractionation	Human serum	RF3	73,76
Aldose reductase	Bovine and porcine lenses	Rotofor	60

[a] RIEF is the prototype instrument of the ISOPREP and ATIsolator. FFE stands for forced flow electrophoresis.

[b] α-Amylase was added to the *E. coli* protein matrix.

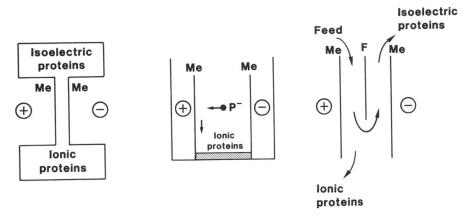

FIGURE 18-3. Schematic representation of electrophoresis convection (left), electrodecantation (center), and forced flow electrophoresis (right). ME, protein-impermeable membrane; F, protein-permeable filter.

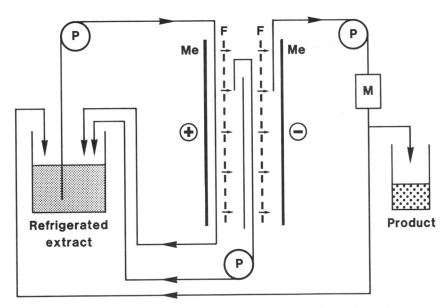

FIGURE 18-4. Schematic representation of a forced flow electrophoresis apparatus with a double-stage cell in which the effluent filtrate of the first stage becomes the feed of the second stage. The second and third subcompartments of the electrophoresis chamber are separated with a cut-down (20% at the top) protein-impermeable membrane. Me, protein-impermeable membrane; F, protein-permeable filter; P, peristaltic pump; M, on-line monitor.

of 100, 500, and 1000 cm^2 cross section were used.[26] Forced flow electrophoresis requires only low capital investment and is not expensive to operate.

Continuous protein separations have also been described in an electrophoretic setup termed Gradiflow which comprises a supported polyacrylamide gel membrane as a size- or charge-dependent separator placed between two buffer streams of equal or different composition.[28,29] In analogy to forced flow electrophoresis, protein-impermeable membranes are used to separate the cell from the electrode compartments and two fractions are obtained. Thus far, the Gradiflow apparatus has been applied to the separation of model proteins and of ovalbumin from egg white.[28] For the assessment of that approach in comparison to forced flow electrophoresis or other techniques, more work is required. In particular, the performance and stability of the supported polyacrylamide gel membranes will have to be characterized in great detail before such an approach can be adopted for protein purification in a downstream processing laboratory.

18.3.2 Purification of Proteins by Continuous Flow Electrophoresis

In 1940, Philpot[30] attempted to extend the scale of the U-tube designed by Tiselius[31] for the analysis of serum proteins by adapting it to continuous separations using a separator stabilized by a density gradient. In the following decades, inumerable approaches to continuous flow electrophoresis have been proposed, including systems to be operated in the microgravity environment of the space shuttle.[1-9] This chapter only describes systems that lead to commercial instrumentation.

In a unique approach proposed by Philpot,[32] continuous operation was accomplished under stable conditions by applying a transverse stress field across a thin annular separation chamber. This apparatus was fully developed by Mattock and colleagues[33] at the Harwell Laboratory in England and was commercially available under the tradename BIOSTREAM separator from CJB Developments Limited (Portsmouth, UK). In this approach, samples are typically separated by the principle of continuous flow zone electrophoresis. The static inner wall (cathode) and the rotating (150 rpm) outer wall (anode) are permeable for small ions and isolate electrode chambers from the separation space (Figure 18-5). In operation, a precooled, low conductivity buffer solution is continuously pumped from bottom to top and the sample is continuously infused through a narrow circumferential slit at the base of the inner wall. Under the influence of the electric field and while being transported toward the top, negatively charged sample components migrate at least partly across the 3-mm annulus. A transverse sectional view through the top of the annulus would show separated fractions as a series of concentric rings. At the top the fluid is diverted through 90° and is split, with minimal mixing, into 29 fractions by a stack of "maze plates." In this way, the film of liquid closest to the inner wall of the separator passes through the lowest disc in the stack,

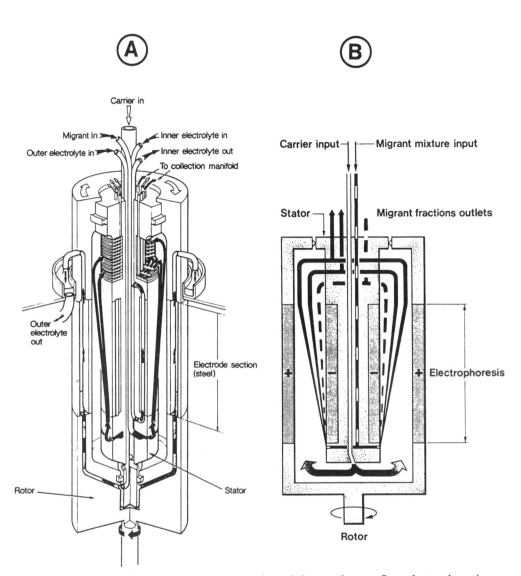

FIGURE 18-5. Diagrammatic representation of the continuous flow electrophoretic BIOSTREAM separator. (A) Cutaway drawing of the system showing the construction of the rotor, stator, and main flow channels. (B) Diagram of the system to demonstrate the separation of three components as they are carried up through the annulus. (Reprinted from Ref. 33 by courtesy of Marcel Dekker Inc.)

the next film of liquid passes through the next disc gap, and so forth. The residence time is in the order of 15 to 60 s.

The BIOSTREAM separator has the largest capacity of any electrophoretic instrument ever built, with the ability to process as much as 150 g protein per hour. A number of protein purification examples have been published[33–39] (Table 18-2), ranging from the processing of desalted human plasma for the quantitative separation of immunoglobulin G and transferrin from albumin[34] to the separation of trypsin from a porcine pancreatic extract with a capacity of 100 g/h of crude trypsin.[36] Wenger and co-workers[38] have used the BIOS-TREAM separator to purify lectins from a crude acidified extract of lentil seeds in a recycling mode. In this mode of operation no carrier buffer was utilized. The purified lectin product was drawn off from channels 1 to 3 and the products from channel 4 to 26 were recycled back through the separator. Contaminants were directed to waste from channels 27 to 29. In this way purified lectins could be collected at the rate of 0.5 g/h. The BIOSTREAM was also shown to be operational in the isotachophoretic mode.[39]

Other preparative continuous flow electrophoresis instruments included the Elphor VaP-21, VaP-22, and VaP-220 apparatuses from Bender & Hobein (Munich, FRG) in which electrophoretic separations are performed in a thin fluid film flowing between two parallel plate[40,41] (Figure 18-6). In these machines the separation chamber is formed by a narrow gap between a Plexiglas plate at the front and a mirrored, thermostated copper plate at the back. The three versions differ in the dimensions of the separation chamber (L,W), namely 25 × 20 cm, 50 × 10 cm, and 100 × 16 cm, respectively, and the thickness of a polypropylene spacer (0.3 to 1 mm), which keeps the two plates apart. Both carrier electrolyte and sample are continuously admitted at the top of the chamber and flow perpendicular to an applied electric field. The solutions exit the separation chamber through an array of up to 90 outlet tubes at the bottom and are collected in the same number of individual containers. Cellulose acetate or ion-exchange membranes separate the separation chamber from the electrode compartments. To monitor the separation process (optional), solutes and/or buffers can directly be detected through a window at the end of the separation chamber with an optical scanning system comprising a scanning light beam together with appropriate interference filters and photo sensors, allowing the monitoring of light scattering, absorbance, or fluorescence. Different continuous flow techniques can easily be implemented on the Elphor instruments, such as zone electrophoresis, isoelectric focusing, isotachophoresis, and field step electrophoresis.[41]

Instead of the Elphor instruments there is now similar but less expensive instrumentation commercially available through Dr. Weber GmbH, Ismaning (near Munich, FRG). In the two modular instruments of that company, termed OCTOPUS CHIEF and OCTOPUS PZE and designed for high-resolution continuous flow isoelectric focusing and zone electrophoresis, respectively. Buffer flow is from bottom to top and the separation cell can be run horizontally, vertically and in any desired angle in between. Instead of requiring

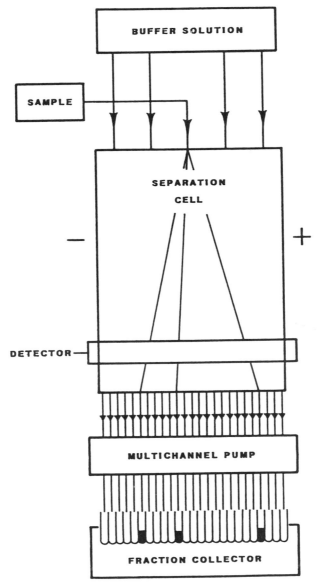

FIGURE 18-6. Schematic representation of the continuous flow parallel plate instrumentation as implemented on the Elphor instruments.

the somewhat delicate multiperistaltic fluid pumping system employed in the Elphor machines (Figure 18-6), one or several rugged syringe pumps provide buffer and sample infusion in this configuration.[42,43] Otherwise, the specs of these instruments are similar to those of the Elphor VaP 22 (Table 18-1). Continuous flow instrumentation of the Elphor/OCTOPUS type was successfully employed for the fractionation of cells and subcellular particles[40-42] and for the high-resolution purification of proteins[38,43-47] (Table 18-2), including enzymes by zone electrophoresis from a supernatant of a 50% wet weight *Candida boidinii* homogenate in which yields of up to 95% and purification factors from 3 to 7 were obtained together with a throughput of up to 500 mg/h,[44] monoclonal antibodies from mouse ascites fluid and tissue culture medium by isotachophoresis,[45] and α-amylase from an *Escherichia coli* protein matrix by field step electrophoresis.[46]

18.3.3 Purification of Proteins by Recycling Electrophoresis

In recycling electrophoresis, instrumental designs differ from those described in continuous flow electrophoresis. In principle, outlet and inlet ports of the separation cells are connected by closed circuitry loops for process fluid recirculation by means of a multichannel peristaltic pump. This allows the fabrication of a relatively small electrophoresis cell with external heat dissipation. With recycling of the fractionation solution, separation is attained gradually.[7,8,48] Recycling instruments can also be employed for electrodialysis. All recycling devices are modularly built, having heat exchange reservoirs in the recycling loops and perhaps also refrigerated separation chambers. For visualization of separation, as well as automation of the entire process, the recycling loops may further contain sensors for monitoring of UV absorption, pH, temperature, conductivity, and so on. Figure 18-7 shows a schematic representation of a complete recycling system.

Recycling isoelectric focusing was described in 1977 and 1978 in three different modes, employing (1) monofilament nylon screens for the compartmentation of the separation chamber,[49,50] (2) isoelectric membranes for compartmentation,[51] and (3) focusing in a noncompartmented, thin film of liquid.[52] In isoelectric focusing, ampholytes are separated in a pH gradient in which they migrate to the column location where the local pH equals the isoelectric point. In recycling isoelectric focusing, in each pass the proteins migrate toward the equilibrium positions until the final distribution is achieved. The attainment of steady state defines the time of fraction collection. The discontinuous, batch operating principle opens the attractive possibility of rerunning single fractions, thus providing a cascading, multistep purification procedure. An inherent limitation of isoelectric focusing is the tendency of proteins to precipitate during the process. Therefore, samples should be desalted and precipitates removed prior to processing, and if the protein of interest becomes insoluble, a variety of additives (such as glycine, glycerol, and nonionic detergents) can be used to promote solubility. Instrumentation and purification examples for the three approaches are described below.

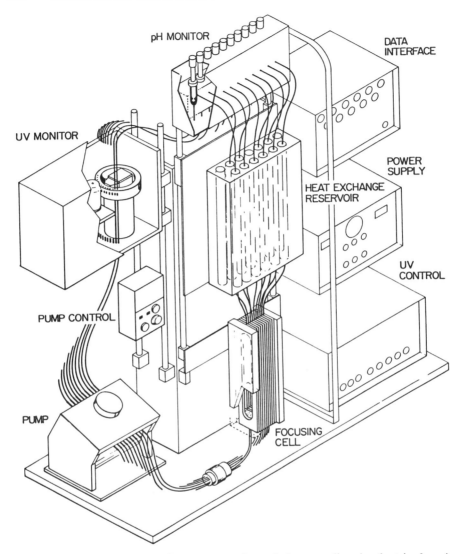

FIGURE 18-7. Diagrammatic representation of the recycling isoelectric focusing (RIEF) apparatus according to Ref. 50.

The RIEF (recycling isoelectric focusing) apparatus of Bier and co-workers[49,50,53] (Figure 18-7) has been applied to a number of fractionation problems. It comprises a compartmented focusing cell with monofilament nylon screens between each channel. The segmentation of the separation cell is essential for fluid stabilization but imposes a stepwise pH gradient. Because of the modular design of the apparatus, it can be scaled to industrially meaningful quantities.[53] It was commercialized as ISOPREP through Ionics (Watertown, MA, USA) and as ATIsolator through Surgimedics/ATI (The Wood-lands, TX, USA). The RIEF devices allow the processing of about a gram of protein per hour and require sample volumes of >150 mL. As a scaled-down version, a horizontally rotating apparatus without recycling and having screen elements for compartmentation of an annular focusing space, the Rotofor[54] (Figure 18-8), handles volumes of 35 to 60 mL and is manufactured by Bio-Rad, (Richmond, CA, USA). A Mini Rotofor cell with a sample volume of 18 mL that is interchangeable with the standard Rotofor chamber has been commercialized as well. The RIEF and Rotofor apparatuses were shown to separate protein components differing in their isoelectric points by as little as 0.05 to 0.1 pH units and have been applied to a number of relevant purification problems[7,8,48,55–61] (Table 18-2).

Recycling isoelectric focusing using isoelectric membranes is based on the pioneering work of Martin and Hampson[51] and its further development by Wenger and Javet.[62] Wenger and colleagues[63] reported the first amphoteric polyacrylamide membranes based on Immobiline technology, and Faupel and co-workers[64] were able to design and run the first device with such membranes. This new approach does not require any carrier buffer as was envisaged in the approach of Martin and Hampson. Incorporation of buffer constituents significantly decreases the efficiency of the process. The Immobiline mem-branes are the pH-defining elements which also have a characteristic buffer

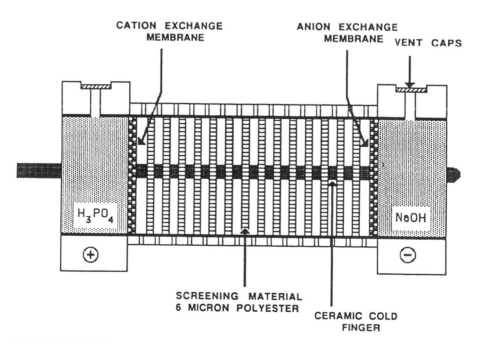

FIGURE 18-8. Diagrammatic representation of the Rotofor chamber. (Courtesy of Bio-Rad, Hercules, CA, USA.)

capacity. This recycling isoelectric focusing approach was termed preparative isoelectric membrane electrophoresis (PrIME). In its simplest operation, a three-compartment electrolyzer is used for the removal of charged contaminants from a recycled protein solution. The modular design of the focusing cell also allows the use of an array of compartments[5,65] (Figure 18-9). PrIME has been applied to a number of relevant purification problems[5,65,66,67] (Table 18-2 and references cited in Ref. 68), including the successful isolation of isoforms of a human monoclonal antibody against the transmembrane qp41 protein of the immunodeficiency virus HIV-1[65,66] and the purification of recombinant human growth hormone.[67] The first commercial PrIME apparatus, the IsoPrime with up to eight chambers, an 8-cm^2 membrane-separating area, and a protein-loading capacity of up to a few 100 mg (Table 18-1), can be purchased from Hoefer Scientific Instruments (San Francisco, CA, USA). A new generation of hydrolytically stable and hydrophylic nonpolyacrylamide-based isoelec-

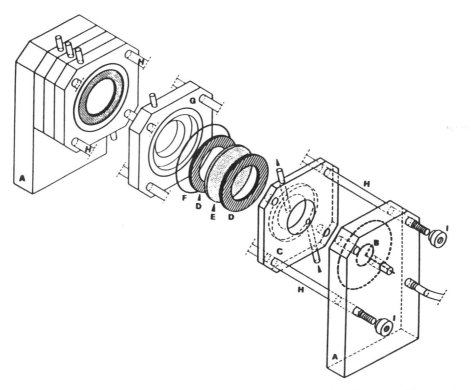

FIGURE 18-9. Exploded view of the separation chamber of IsoPrime multicompartment electrolyzer. A, supports; B, Pt electrodes; C, terminal flow chamber; D, rubber rings for supporting the membrane; E, Immobiline membrane on glass-fiber support; F, O-ring; G, one of the sample flow chambers; H, four threaded metal rods for assembling the apparatus; I, nuts for fastening. (From Ref. 66.)

tric membranes have been designed for that technology.[68] PrIME has the potential for further scaleup. However, the RIEF apparatus could also be run in the PrIME configuration through substitution of the nylon screens with the Immobiline membranes and without employment of carrier buffers.

Using the thin fluid film approach with recycling between parallel plates (Figure 18-10) permits the establishment of a linear pH gradient. In addition, the absence of screens or membranes prevents the entrapment of particles and precipitated proteins. The so-called recycling free flow focusing apparatus (RF3), which has been designed and constructed by Bier and co-workers,[7,69] is based on that principle having a fluid layer thickness of 0.25 to 0.75 mm (0.75 mm is standard). The focusing cell is oriented vertically with fluid flow from bottom to top. The fluid residence time in the cell is only a few seconds with the characteristic Reynolds number being approximately 600. This assures a remarkable stability of the laminar flow. Five different chamber designs were implemented, with various overall length, width, and number of closely spaced ports. The purification of S1 nuclease from *Aspergillus oryzae* comprises an interesting application for this apparatus.[70] RF3 with a processing fluid volume of 100 to 500 mL,[71] as well as a smaller version for about 30 mL called the MiniphHor, are commercialy available through Protein Technologies, Inc. (Tucson, AZ, USA) (Table 18-1) and are distributed by Rainin, Instrument Co. (Woburn, MA, USA). These instruments have been effectively used with pH gradients produced by commercial carrier ampholytes (such as Ampholine and Pharmalyte of Pharmacia-LKB, Uppsala, Sweden) or with simple ampholytes of high concentration.[72,73] A number of defined small molecule buffers have been designed and shown to permit separations in narrowrange pH gradients.[74] Such buffers are commercially available as OptiFocus buffers (Protein Technologies, distributed through Rainin) and as RotoLytes for the Rotofor (Bio-Rad Laboratories). An OptiFocus buffer composed of MOPS and GABA (50 mM each) has been used for the fractionation of dialyzed and filtered human serum as well as for the isolation of transferrin from a transferrin-enriched fraction obtained by recycling isotachophoresis.[73]

The RF3 has also been employed in a recycling isotachophoresis mode.[75–79] In its simplest implementation, recycling isotachophoresis constitutes a largescale batch process with a specified amount of sample solution being injected near one column end, at the location of the initial interface between leader and terminator. Immobilization of the advancing zone structures is achieved with the application of counterflow of the leading electrolyte, making the migration path essentially infinitely long. Furthermore, a continuous mode of operation in which (1) the sample is continuously infused, (2) individual components come to rest at their appropriate position in an isotachophoretic stack held stationary by counterflow, and (3) a particular product of interest is steadily withdrawn from the appropriate channels was conceived.[78] Recycling isotachophoresis has been successfully applied to the purification of egg proteins[77,78] and to the fractionation of human serum [79].

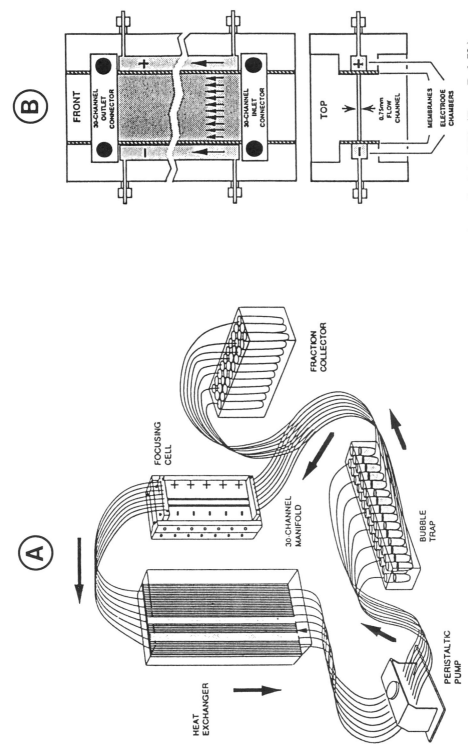

FIGURE 18-10. Schematic representation of (A) the fluid flow pattern and (B) the separation cell of the RF3. (From Ref. 72.)

The RIEF and RF3 apparatuses were also found to be useful when operated in a continuous feed zone electrophoresis mode.[7,48] While the fluid in the apparatus is still recycled, crude sample is infused into a central channel and withdrawn from the two end channels, maintaining constant volume within the apparatus. The technique has been called feed and bleed (FAB) for recycling infuse and withdraw. FAB can be applied to diverse biological samples. The principle requires the careful selection of the pH of the running buffer and was shown to be operational for the fractionation of human serum. This technique results in the sample being split into two fractions, those proteins that are more acidic than the operating pH and those that are less acidic. A similar recycling process was implemented into the BIOSTREAM separator for the purification of lentil lectins from a crude acidified extract of lentil seeds[38] (see earlier). This recycling process allowed the simultaneous purification and concentration of the desired product. It was described as a batch process but can be operated equally well in a continuous mode. Ivory and co-workers[80,81] proposed an approach where the effluent from each channel is not reinjected into the corresponding influent port but is backshifted by a specified number of input channels. This gives rise to an effective counterflow against the electromigration of the solutes and, with proper adjustment of the current, two product streams are obtained. The sample is continuously infused into a central channel. Slower moving components are carried downstream by the counterflow and the faster moving are collected upstream. In principle, any two proteins with different mobilities can be separated by proper adjustment of the electric field and the shift. Although this process has been discussed in detail, only fcw applications have been published thus far.[81]

18.3.4 Purification of Proteins in Free Fluid Columns without Externally Imposed Fluid Flow

A large number of free fluid devices without externally imposed fluid flow have been designed for the electrophoretic preparation of microgram to gram quantities of proteins. These were most often used for batch isoelectric focusing and were not commercialized. This chapter restricts itself to mentioning approaches that either lead to or are the most promising for the development of commercial instrumentation. Density gradients, produced by concentration gradients of a neutral molecule such as sucrose or sorbitol in vertical columns (LKB, Bromma, Sweden and ISCO, Lincoln, NE, USA) can be employed for fractionation of up to a gram of protein. Larger sample capacity is obtained with (1) membrane subcompartmented electrolyzers with the membranes preventing bulk flow between adjacent compartments while passing proteins and buffer constituents under current flow and (2) the saw-tooth approach with process-induced density gradient stabilization as proposed by Valmet.[82] The Autofocuser (Realizing Centre of Slovak Academy of Sciences, Kosice, Czechoslovakia) developed by Sova[83] is the most recent example of using the latter technology for large-scale protein purification. Different apparatus for various volumes from 75 to 1,000 L were examined and applied to the isolation of uricase from yeast,[84] of α-amylase from *Bacillus subtilis*,[85] and of horseradish peroxidase from horse radish.[86] Screen-segmented vertical columns for isoelectric focusing[87] and isotachophoresis[76] constitute another area under development with potential applications in the field of electrophoretic large-scale protein purification. The screens were found to confine convective disturbances to column segments defined by the screen material while permitting free electrophoretic transport.

Convective mixing in free solution can be prevented using the principle of rotational stabilization first proposed by Hjertén[88] and modified for preparative work by Bier.[89] In these approaches, separation is carried out in a glass tube slowly rotating around its horizontal axis. As the direction of the gravity vector is continuously changing with reference to the fluid column, cross convection is prevented. The Rotofor,[54] marketed by Bio-Rad (Richmond, CA, USA), can be viewed as an up-scaled version of these devices. It features a combination of screen segmentation and slow (1 to 7 rpm) horizontal rotation and is designed for isoelectric focusing, but can also be employed in the ITP mode.[76] It employs an annular chamber that is divided into 20 subcompartments (Table 18-1). A cooling tube runs through the center along the focusing axis (Figure 18-8). This instrument has proven to be very useful for the treatment of small samples (e.g., the subfractionation of fractions obtained from the RIEF or other larger scale preparative apparatuses[7]) and for the research scale purification of many different proteins (Table 18-2). The Rotofor is economically priced (approximately $2,500 without power supply). Thus, a few thousand units could already be sold worldwide, making this instrument by far the best selling and most popular free fluid preparative electrophoresis apparatus. In comparison, no more than a total of a few hundred continuous flow and recycling instruments have been distributed in the same time period.

Excellent fluid stability is obtained in tubes of very small inner diameters. Although the nature of capillary techniques is mainly analytical (see Chapter 17), various reports in the area of capillary electrophoresis promulgated fractionation procedures for the subsequent performance of a further assay or for sequencing. The fractionation methods described include (1) application of coflow, (2) employment of an elution chamber at the capillary end, (3) use of a microsyringe or a fractionation valve, (4) elution onto a cellulose acetate strip, and (5) electroosmotic elution from fused-silica capillaries into small vials or onto a membrane (for a review, see Ref. 90).

18.3.5   Miscellaneous Techniques

Continuous separation of charged proteins in an electrical split-flow thin cell approach was described by Levin and colleagues.[93] Evolved from electrical field flow fractionation (see Chapter 17), this method uses a ribbon-like channel of high aspect ratio (e.g., 28.8, 2, and 0.0635 cm as channel length, width, and height, respectively[93]) with stream splitters at one or both ends. The electric field is applied perpendicular to the flow across the small gap of about 0.6 mm; channel bottom and top are formed by semipermeable membranes. The outlet stream splitter divides the thin film of liquid flowing through the channel into two laminae. This represents a small-scale preparative approach with some similarity to forced flow electrophoresis[94] and the Gradiflow[28,29] method described in Section 18.3.1. In the electrical split-flow thin cell technique, two sample streams (fractions) are obtained as well. However, no filter is incorporated to divide the separation cell into two compartments. To provide higher resolution, splitters providing more than two sample streams could be designed (difficult endeavor) or multiple channels could be linked in sequence, each operating at a different pH.

Two-phase partitioning of proteins in the presence of an electric field across the interface between two liquid phases formed by different water-soluble but incompatible polymers represents a hardly explored approach for protein fractionation.[95,96] Marando and Clark[96] designed a cylindrical chamber of 3.8 and 5.1 cm in diameter and height, respectively, having electrode chambers (separated by ultrafiltration membranes from the separation cell) on bottom and top. Using a polyethylene glycol/dextran aqueous two-phase system at pH 6, this instrument was shown to effectively separate hemoglobin and albumin (about 10 mg each) within about 2 h. Compared to the same configuration without external field, application of the electric field could be shown to significantly improve separation. Clark and co-workers[97] have also shown the possibility of employing this two-phase partitioning technique for continuous fractionation of proteins on a large scale. In a setup similar to that employed in the Gradiflow,[28,29] the phase interface is utilized to prevent convective mixing of the two sample streams. Thus, the setup does not require a porous membrane (filter) in the center of the chamber. The two-phase electrophoresis partitioning technique has the potential to be scaled up to commercially meaningful quantities.

Last but not least, the design of a continuous flow electrophoresis instrument of the Elphor type featuring a chamber of 12.7 mm thickness (10.8 cm height and 7.62 cm width) with an internal heat exchanger made of layers of closely spaced Teflon tubes has been reported.[98,99] In this method, referred to as capillary free flow electrophoresis, the sample stream is essentially kept away from the chamber walls and electrodynamic and thermal distortions are greatly reduced. It has been characterized by the separation of model proteins with a throughput of about 500 mg/h. Further scaleup should be possible without any convection problems by increasing the number of layers of capillary cooling tubes and thereby the thickness of the separation cell. Whether such an approach will be suitable for protein purification and isolation from complex samples remains to be evaluated.

18.4 LIMITATIONS OF LARGE-SCALE PREPARATIVE ELECTROPHORESIS AND FUTURE OUTLOOK

Several problems are associated with preparative gel-based methods. Scaleup of such techniques is difficult because of the slow dissipation of the Joule heat generated by the electric current. The heat is proportional to the volume of the device whereas cooling is proportional to the surface. The manipulation and preparation of the support are time-consuming and tedious. There is often a problem with adherence of proteins to the matrix which results in a decreased recovery. In addition, the matrix must be removed from the recovered sample, cannot generally be reused, and therefore becomes another expense. Gel-based methods are adequate for most research laboratory requirements. The practical upper limit of protein throughput is at best a few grams per day which is inadequate for most industrial purposes. Also, there is no simple access to automation.

Electrophoretic protein separations in free fluids are confounded by gravitationally induced convection, gravity-independent electrohydrodynamic disturbances (electroosmosis and convection originating at conductivity and dielectric gradients), and sample adsorption onto column walls or membranes placed in the current path. Fluid stabilization as a link to resolution is the key issue in preparative free fluid electrophoresis, with recovery of proteins and the sample loading capacity (throughput) being two other important aspects of this technology. Engineering of suitable equipment, which allows processing of proteinaceous solutions on the gram scale per hour, has been successfully promoted in the past decade (Table 18-1). High resolution separations aimed at significantly larger throughputs are unlikely to be accessible through electrophoretic means. Disturbances mentioned earlier hinder a further scaleup. However, compared to electrophoresis in a solid matrix (such as a gel), free fluid electrophoresis has three major advantages: It allows the processing of larger amounts of sample per unit time, the construction of simpler fractionation devices, and the automation of the whole process. The effectiveness of free solution preparative electrophoresis for the gentle purification and isolation of proteins without denaturing of the macromolecules has been documented. Table 18-2 lists selected purification examples using continuous flow, recycling or rotating instrumentation. Current efforts are aimed at the further development and introduction of these techniques. They are believed to be capable of replacing one or several purification steps in a downstream processing protocol, a sequence of purification procedures which, to date, are mainly based on complex and expensive chromatographic principles.

Thus far preparative electrophoresis has been shown to be suitable for the concentration and final polishing of proteins, as well as the direct isolation of small amounts of a pure protein. This represents an interesting approach when purification of a product on the gram scale has to be achieved. Purification by electrophoresis on the kilogram to ton scale is not realistic. For that purpose, a number of other methods, such as precipitation, extraction, and adsorption, are available (see the other chapters in this book).

ACKNOWLEDGMENTS

The author acknowledges many enlighting discussions with Professor Milan Bier and Drs. Richard A. Mosher and Ned B. Egen. Many thanks are due to those collaborators and colleagues who have offered advice, criticism, and discussion on the topic of preparative electrophoresis. Support provided by the Swiss National Science Foundation is gratefully acknowledged.

18.5 REFERENCES

1. M. Bier, M. ed., *Electrophoresis, Theory, Methods and Applications,* Vols. I and II, Academic Press, New York, 1959 and 1967, respectively.

2. Z. Deyl, ed., *Electrophoresis: A Survey of Techniques and Applications: Part A: Techniques,* Elsevier, Amsterdam, 1979, and *Part B: Applications,* Elsevier, Amsterdam, 1983.

3. P.G. Righetti, C.J. Van Oss, J.W. Vanderhoff, eds., *Electrokinetic Separation Methods,* Elsevier, North Holland, Amsterdam, 1979.

4. W. Thormann, R.A. Mosher, In *Electrophoresis 88,* C. Schafer-Nielsen, ed., VCH Verlagsgesellschaft, Weinheim, FRG, 1988, pp. 121–140.

5. P.G. Righetti, M. Faupel, E. Wenisch, *Adv. Electroph., 5,* 159 200 (1992).

6. M. Bier, In *Frontier in Bioprocessing,* S.K. Sikdar, M. Bier, P. Todd, eds., CRC Press, Boca Raton, 1990, pp. 235–243.

7. M. Bier, N.B. Egen, G.E. Twitty, R.A. Mosher, W. Thormann, In *Chemical Separations,* C.J. King, J.D. Navratil, eds., Litarvan Literature, Denver, CO, 1986, Vol. 1, pp. 133–157.

8. R.A. Mosher, W. Thormann, N.B. Egen, P. Couasnon, D.W. Sammons, In *New Directions in Electrophoretic Methods,* M. Philips, J. Jorgenson, eds., American Chemical Society, Washington DC, ACS Symposium Series 335, 1986, pp. 247–262.

9. C.F. Ivory, *Sep. Sci. Technol., 23,* 875–912 (1988).

10. P.G. Righetti, *Isoelectric Focusing Theory, Methodology and Applications,* Elsevier Biomedical, Amsterdam, 1983.

11. A. Chrambach, *The Practice of Quantitative Gel Electrophoresis,* VCH Publishers, Weinheim, 1985.

12. R. Horuk, *Adv. Electroph., 1,* 361–379 (1987).

13. C.J. Holloway, R.V. Batterby, *Methods Enzymol., 104,* 281–301 (1984), and references cited therein.

14. A. Chrambach, T.M. Jovin, P.J. Svendson, D. Rodbard, In *Methods of Protein Separation,* N. Catsimpoolas, ed., Plenum Press, 1976, Vol. 2, pp. 27–144.

15. C.D. Scott, R.D. Spencer, W.G. Sisson, *J. Chromatogr., 126,* 381 (1976).

16. R.A. Yoshisato, L.M. Korndorf, G.R. Carmichael, R. Datta, *Sep. Sci. Technol.*, *21*, 727–753 (1986).

17. T.R.C. Boyde, M.A. Remtulla, *Anal. Biochem.*, *55*, 492–508 (1973).

18. B. Lammel, *Electrophoresis*, *2*, 39–45 (1981).

19. C.D. Scott, *Sep. Sci. Technol.*, *21*, 905–917 (1986).

20. P.H. O'Farrel, *Science 227*, 1586–1589 (1985).

21. B.R. Locke, R.G. Carbonell, *Sep. Purif. Meth.*, *18*, 1–64 (1989).

22. C.F. Ivory, W.A. Gobie, *Biotechnol. Prog.*, *6*, 21–32 (1990).

23. W. Pauli, *Biochem. J.*, *152*, 355 (1924).

24. J.G. Kirkwood, *J. Phys. Chem.*, *9*, 878–879 (1941).

25. M. Bier, *Science, 125*, 1084–1085 (1957).

26. M. Bier, In *Membrane Processes in Industry and Biomedicine*, M. Bier, ed., Plenum Press, New York, 1971, pp. 233–266.

27. M. Bier, T. Long, H. Ryu, *Sep. Sci. Technol.*, *25*, 997–1005 (1990).

28. Z.S. Horvath, G.L. Corthals, C.W. Wrigley, J. Margolis, *Electrophoresis*, *15*, 968–971 (1994).

29. J. Margolis, G. Corthals, Z.S. Horvath, *Electrophoresis*, *16*, 98–100 (1995).

30. J.S.L. Philpot, *Trans. Faraday Soc.*, *39*, 38 (1940).

31. A. Tiselius, *Trans. Faraday Soc.*, *33*, 524 (1937).

32. J.S.L. Philpot, In *Methodological Developments in Biochemistry*, E. Reid, ed., Longman Group, London, 1973, Vol. 2, pp. 81–85.

33. P. Mattock, G.F. Aitchison, A.R. Thompson, *Sep. Purif. Meth.*, *9*, 1–68 (1980).

34. BIOSTREAM Technical Note No. 2, C.J.B Developments Limited, Portsmouth, England.

35. D.E.G. Austen, T. Cartwright, C.H. Dickerson, *Vox Sang.*, *44*, 151–155 (1983).

36. C.H. Dickerson, J.R. Birch, T. Cartwright, In *3rd General Meeting of ESACT*, Oxford 1979, Develop. Biol. Standard 46, 67–74 (S. Karger, Basel, 1980).

37. BIOSTREAM Technical Note No. 4, C.J.B. Developments Limited, Portsmouth, England.

38. P. Wenger, A. Heydt, N.B. Egen, T.D. Long, M. Bier, *J. Chromatogr.*, *455*, 225–239 (1988).

39. W. Thormann, M.A. Firestone, J.E. Sloan, T.D. Long, R.A. Mosher, *Electrophoresis*, *11*, 298–304 (1990).

40. K. Hannig, *Electrophoresis*, *3*, 235–243 (1982).

41. H. Wagner, R. Kessler, *GIT Lab.-Med.*, *7*, 30–35 (1984).

42. B. Bondy, J. Bauer, I. Seuffert, G. Weber, *Electrophoresis*, *16*, 92–97 (1995).

43. G. Küllertz, S. Meyer, G. Fischer, *Electrophoresis*, *15*, 960–967 (1994).

44. S. Nath, H. Schütte, G. Weber, H. Hustedt, W.-D. Deckwer, *Electrophoresis*, *11*, 937–941 (1990).

45. G. Schmitz, A. Böttcher, H.-G. Kahl, T. Bruning, *J. Chromatogr.*, *431*, 327–342 (1988).

46. R. Kuhn, H. Wagner, *Electrophoresis*, *10*, 165–172 (1989).

47. F. Barth, M.G. Gruetter, R. Kessler, H. Manz, *Electrophoresis* 7, 372–375 (1986).

48. R.A. Mosher, N.B. Egen, M. Bier, In *Protein Purification: Micro to Macro,* R. Burgess, ed., Alan R. Liss, New York, 1987, pp. 315–328.

49. M. Bier, N.B. Egen, In *Electrofocus,* H. Haglund, J.G. Westerfield, J.T. Ball, eds., North Holland, Elsevier, Amsterdam, 1979, pp. 35–48.

50. M. Bier, N.B. Egen, T.T. Allgyer, G.E. Twitty, R.A. Mosher, In *Peptides: Structure and Biological Function,* E. Gross, J. Meienhofer, eds., Pierce Chemical Co., Rockford, IL, 1979, pp. 79–89.

51. A.J.P. Martin, F. Hampson, *J. Chromatogr., 159,* 101–110 (1978).

52. K. Hannig, H. Wirth, R.K. Schindler, K. Spiegel, *Hoppe-Seyler's Z. Physiol. Chem., 358,* 753–763 (1977).

53. M. Bier, In *Separation, Recovery and Purification in Biotechnology,* J.A. Asenjo, J. Hong, eds., American Chemical Society, Washington D.C., ACS Symposium Series 314, 1986, pp. 185–192.

54. N.B. Egen, W. Thormann, G.E. Twitty, M. Bier, In *Electrophoresis, 83,* H. Hirai, ed., Walter de Gruyter, Berlin, 1984, pp. 547–550.

55. N.B. Egen, M. Bliss, M. Mayersohn, S.M. Owens, L. Arnold, M. Bier, *Anal. Biochem., 172,* 488–494 (1988).

56. T.L. Nagabhushan, B. Sharma, P.P. Trotta, *Electrophoresis, 7,* 552–557 (1986).

57. S.B. Binion, L.S. Rodkey, N.B. Egen, M. Bier, *Electrophoresis, 3,* 284–288 (1982).

58. N.B. Egen, F.E. Russell, D.W. Sammons, R.C. Humphreys, A.L. Gnaw, F.S. Markland, *Toxicon, 25,* 1189–1198 (1987).

59. C.H. Evans, A.C. Wilson, B.A. Gelléri, *Anal. Biochem., 177,* 358–363 (1989).

60. J.M. Petrash, L.J. DeLucas, E. Bowling, N. Egen, *Electrophoresis, 12,* 84–90 (1991).

61. N.B. Egen, W.A. Cuevas, P.J. McNamara, D.W. Sammons, R. Humphreys, J.G. Songer, *Am. J. Vet. Res., 50,* 1319–1322 (1989).

62. P. Wenger, P. Javet, *J. Biochem. Biophys. Meth., 13,* 259–303 (1986).

63. P. Wenger, M. de Zuanni, P. Javet, C. Gelfi, P.G. Righetti, *J. Biochem. Biophys. Meth., 14,* 29–43 (1987).

64. M. Faupel, B. Barzaghi, C. Gelfi, P.G. Righetti, *J. Biochem. Biophys. Meth., 15,* 147–168 (1987).

65. P.G. Righetti, E. Wenisch, M. Faupel, *J. Chromatogr., 475,* 293–309 (1989).

66. P.G. Righetti, E. Wenisch, A. Jungbauer, H. Katinger, M. Faupel, *J. Chromatogr., 500,* 681–696 (1990).

67. C. Ettori, P.G. Righetti, C. Chiesa, F. Frigerio, G. Galli, G. Grandi, *J. Biotechnol., 25,* 307–318 (1992).

68. M. Chiari, M. Nesi, P. Roncada, P.G. Righetti, *Electrophoresis, 15,* 953–959 (1994).

69. M. Bier, G.E. Twitty, J.E. Sloan, *J. Chromatogr., 470,* 369–376 (1989).

70. J.A. Ostrem, T.R. van Oosbree, R. Marquez, L. Barstow, *Electrophoresis, 11,* 953–957 (1990).

71. J.A. Ostrem, G. Ward, L. Barstow, R. Marquez, T.R. van Oosbree, B.A. Rhodes, In *Frontiers in Bioprocessing II,* P. Todd, S. Sikdar, M. Bier, eds., American Chemical Society, Washington DC, 1992, pp. 399–411.

72. M. Bier, T. Long, *J. Chromatogr., 604, 73–83 (1992)*.

73. J. Caslavska, W. Thormann, *Electrophoresis, 15,* 1176–1185 (1994).

74. M. Bier, J. Ostrem, R.B. Marquez, *Electrophoresis, 14,* 1011–1018 (1993).

75. J.E. Sloan, W. Thormann, G.E. Twitty, M. Bier, *J. Chromatogr., 457,* 137–148 (1988).

76. J. Caslavska, P. Gebauer, A. Odermatt, W. Thormann, *J. Chromatogr., 545,* 315–329 (1991).

77. J. Caslavska, P. Gebauer, W. Thormann, *J. Chromatogr., 585,* 145–152 (1991).

78. J. Caslavska, W. Thormann, *J. Chromatogr., 594,* 361–369 (1992).

79. J. Caslavska, P. Gebauer, W. Thormann, *Electrophoresis, 15,* 1167–1175 (1994).

80. W.A. Gobie, J.B. Beckwith, C.F. Ivory, *Biotech. Prog., 1,* 60–68 (1985).

81. C.F. Ivory, *Electrophoresis, 11,* 919–926 (1990).

82. E. Valmet, *Sci. Tools, 16,* 8–13 (1969).

83. O. Sova, *J. Chromatogr., 320,* 213–218 (1985).

84. T. Dobransky, O. Sova, *J. Chromatogr., 358,* 274–278 (1986).

85. T. Dobransky, L. Polivka, I. Haas, O. Sova, E. Petrvalsky, *J. Chromatogr., 411,* 486–489 (1987).

86. T. Dobransky, O. Sova, M. Teleha, *J. Chromatogr., 474,* 430–434 (1989).

87. N.B. Egen, G.E. Twitty, W. Thormann, M. Bier, *Sep. Sci. Technol., 22,* 1383–1403 (1987).

88. S. Hjertén, *Chromatogr. Rev., 9,* 122–219 (1967).

89. M. Bier, T.T. Allgyer, In *Electrokinetic Separation Methods,* P.G. Righetti, C.J. Van Oss, J.W. Vanderhoff, eds., Elsevier/North-Holland, Amsterdam, 1979, pp. 443–469.

90. B.L. Karger, A.S. Cohen, A. Guttman, *J. Chromatogr., 492, 585–614 (1989)*.

91. BIOSTREAM technical note No. 3, CJB, Developments Limited, Portsmouth, England.

92. R. Kuhn, S. Hoffstetter-Kuhn, H. Wagner, *Electrophoresis, 11,* 942–947 (1990).

93. S. Levin, M.N. Myers, J.C. Giddings, *Sep. Sci. Technol., 24,* 1245–1259 (1989).

94. H.-W. Ryu, M. Bier, *Sep. Sci. Technol., 25,* 1007–1020 (1990).

95. M.L. Levine, M. Bier, *Electrophoresis, 11,* 605–611 (1990).

96. M.A. Marando, W.M. Clark, *Sep. Sci. Technol., 28,* 1561–1577 (1993).

97. W.M. Clark, M.A. Marando, C.W. Theos, Poster G2, *Recovery of Biological Products* VI, An Engineering Foundation Conference, Interlaken, Switzerland, September 20–25, 1992.

98. Y. Tarnopolsky, M.C. Roman, P.R. Brown, *Sep. Sci. Technol., 28,* 719–731 (1993).

99. P. Painuly, M.C. Roman, *Appl. Theor. Electroph., 3,* 119–127 (1993).

100. E. Wenisch, P. Schneider, S.A. Hansen, R. Rezzonico, P.G. Righetti, *J. Biochem. Biophys. Meth., 27,* 199–213 (1993).

INDEX

A

Acid-base properties
  in ion exchange chromatography, 158
Acid phosphatase, 302–303
Activation of
  agarose, 396–412
  polyacrylamide, 412–413
  silica, 413–415
Activated CH-Sepharose®, 388
Activated glass fiber membranes, 593
Acylpeptide hydrolase, 306
2′,5′-ADP, 381
Adsorption chromatography, 65–69
Adsorption isotherm, 65, 255
Aerogels, 45
Affinity adsorbents, 387–395
  ideal matrix properties, 387
  preactivated matrices, 388–390
  evaluation of, 393–395
  storage of, 395
Affinity capillary electrophoresis, 631
Affinity chromatography, 375–442
  applications
    avidin-biotin system, 436, 438
    high performance liquid affinity
      chromatography, 435–437
    purification of
      kinases and dehydrogenases,
        430–431
      lectins, 430, 432
      IgG, 432, 433
      catechol-0-methyltransferase, 433
      angiotensin converting enzyme, 434,
        435
  chromatographic techniques, 416–422
  ligands
    biomimetic, see dye ligands
    group-specific low molecular, 380–382
    group-specific macromolecular, 385
    monospecific low molecular, 380
    monospecific macromolecular, 382
Affinity elution in ion exchange
      chromatography, 150, 177, 185–186
Affinity interactions, 379–387

Affinity partitioning, 443–460
  applications
    phosphofructokinase, yeast, 457
    glucose-6-phosphate dehydrogenase,
      yeast, 458
    lactate dehydrogenase, swine muscle,
      458
  batch extraction in, 452
  countercurrent distribution in, 452
  free ligand effects in, 450–451
  influence of electrolytes, 446
  examples of ligands used in, 448
  practical procedures for, 452–457
  theory for, 449–450
  salt and pH effects in, 446–447, 450
Agar/agarose, 469
  molecular structure, 47–49
Agarose gel
  in affinity chromatography, 391
  as a chromatography matrix, 46–49
  in gel electrophoresis, 469
  and gradient drift, 511
  in IEF, 503, 506
  in immunoelectrophoresis, 544–545
  immobilization techniques for, 396–412
  for ion exchangers, 163
  SEM of, 49
  as support in IMAC, 314
AH-Sepharose® 4B, 389
ÄKTA® chromatography system, 37
Albumin
  large scale purification by IEC, 197,
    199
  by IMAC, 330, 334
Aldose reductase, 659
α_1-Antitrypsin, 327, 331, 333, 519
α-Amylase, 659
Allergen, wheat grain, characterization of,
    485
Amberlite® IRC-50, 162
Amido black, 596
Amino acid(s)
  analysis of affinity adsorbents, 394
  composition analysis, 599
  hydrophobicity of, 288

679